THE NIGHT SKY IN JUNE

The Evolving Universe

The Evolving Universe

AN INTRODUCTION TO ASTRONOMY

by Donald Goldsmith

THE BENJAMIN / CUMMINGS PUBLISHING COMPANY, INC.
Menlo Park, California · Reading, Massachusetts
London · Amsterdam · Don Mills, Ontario · Sydney

Sponsoring Editor: Mary Forkner
Production Editor: John Hamburger
Cover and Book Designer: Michael Rogondino
Artists: Kathleen Tyler, Ralph Lao, and Charles Sullivan

Front cover: Composite photograph of Jupiter and its four largest moons—Io, Europa, Ganymede, and Callisto—courtesy of NASA and the Lick Observatory (star background).

Back cover: Saturn was photographed by the Voyager 1 spacecraft in natural color (top). The images were also computer processed to bring out additional details (bottom). In this processing, ultraviolet was recorded as blue and violet was made into red. (Photographs courtesy of NASA.)

Library of Congress Cataloging in Publication Data
Goldsmith, Donald.
The evolving universe.

Previous ed. (c1976) published under title: The Universe.
Bibliography: p.
Includes index.
1. Astronomy. I. Title.
QB45.G63 1981 520 80-24277
ISBN 0-8053-3327-4

ABCDEFGHIJ-HA-89876543210

The Benjamin/Cummings Publishing Company, Inc.
2727 Sand Hill Road
Menlo Park, California 94025

Contents

Preface

The Evolving Universe is a basic astronomy textbook for one-quarter or one-semester introductory courses, requiring no mathematics beyond high-school algebra. This book has developed from my previous textbook *The Universe,* published in 1976. New discoveries in planetary astronomy, x-ray astronomy, radio astronomy, and astrophysics since that time have led to new insights about the planets and the stars around us. For this reason, I have made a thorough revision of the original work, increasing the amount of material on the solar system to include results from the Viking, Pioneer, and Voyager spacecraft missions, as well as those of the Einstein and Copernicus satellites. The book also includes new material on the history of astronomy, the development of telescopes, and the wave theory of light to give students an appreciation for our advances in knowledge.

The Evolving Universe has an accompanying *Student Study Guide* and an *Instructor's Resource Manual.* The *Student Study Guide,* by Alex Maksymowicz, summarizes the main theme of each section of the text. The *Instructor's Resource Manual* contains alternative course outline suggestions and a resource guide to astronomy teaching materials. Both the study guide and the instructor's manual include sample test questions for the material covered in each chapter.

Like its predecessor, *The Evolving Universe* examines the universe first in its general aspects, and in later chapters focuses on a more detailed study of galaxies, stars, and planets. Though I feel that this organization provides students with the best approach to understanding the cosmos, instructors can use Parts 2 through 5—The Universe at Large, Stellar Evolution, The Solar System, and Life in the Universe—as self-contained units in any order desired, once Part 1 has been covered. An introductory four-color photo essay provides a convenient way to outline the scale of distances in the universe and the types of objects it contains. I have introduced the physics relevant to each topic at appropriate points in the text to make it more accessible to the student.

As much as possible, the book avoids using algebraic formulas and instead presents the basic concepts of astronomy in descriptive form. The only exceptions to this appear in highlighted boxes at appropriate points in the text, where more detail will improve student mastery of the material. I have employed the units of the metric system and the Kelvin temperature scale. In addition to the tables found in the text, the appendices contain additional tabular information, as well as a guide to stargazing and discussions of the powers-of-ten notation and the metric system.

In preparing this book, I have received generous assistance from many astronomers and friends. I would like most of all to thank Lawrence Anderson, Roger Angel, Halton Arp, Bart Bok, Reginald Dufour, Jack Eddy, Owen Gingerich, Paul Goldsmith, George Herbig, Paul Hodge, James Liebert, Frank Miller, Ronald Moore, Tobias Owen, Arthur Page, Arno Penzias, J. William Schopf, Frank Shu, Hyron Spinrad, Sumner Starrfield, Larry Toy, Virginia Trimble, Sidney van den Bergh, and Robert Wagoner for their kindness. The book was read in manuscript by Sherwood Harrington, Robert Koch, Alex Maksymowicz, Tobias Owen, Alex G. Smith, Charles Tolbert, and Larry Toy, who attempted to set me on what they considered the proper path. Additional reviews and helpful comments were made by Dave Alexander, Hady H. Aly, Bernard Bopp, Micheal Breger, Michael Chriss, Andrew Fraknoi, Sid Freudenstein, Sheldon Kaufmann, Hubert Owen, Terry Roark, T. Sarachman, M. H. Sherman, Michael L. Stewart, and Raymond White. To all of these I must say, the help is yours, the mistakes mine. I hope the latter are as few as the former is large.

Donald Goldsmith

Photo Credits

American Science & Engineering, Inc.: Fig. 8-40

Dr. Roger Angel: Fig. 7-22

Dr. Halton Arp and Dr. Jean Lorre, Mount Wilson and Las Campanas Observatories, and Jet Propulsion Laboratory: Fig. 7-10 (center and bottom)

Dr. Halton Arp and Dr. Jean Lorre, Mount Wilson and Las Campanas Observatories: Fig. 7-16

Association of Universities for Research in Astronomy, Inc., The Cerro Tololo Inter-American Observatory: Color Plates G, H, and R. Copyright © by Association of Universities for Research in Astronomy, Inc., The Cerro Tololo Inter-American Observatory.

Mr. and Mrs. James M. Baker: Fig 11-34

Dr. Bruce Balick: Figs. 9-16 and 9-28 (top right)

Bausch & Lomb: Fig. 3-21 and Color Plate N

Bell Laboratories: Figs. 4-17 and 7-1 (left)

Bettmann Archives, Inc.: Fig. 11-29

Big Bear Solar Observatory, California Institute of Technology: Figs. 8-38, 8-43, 8-47

Dr. Bart Bok and Cerro Tololo Interamerican Observatory: Fig. 8-1

British Library: Fig. 9-25

Brookhaven National Laboratory and Dr. Raymond Davis: Fig. 8-37

Cerro Tololo Inter-American Observatory: Figs. 1-5, 5-14, 6-1, 6-7, 10-7

Dr. David DeYoung: Fig. 7-11

Dr. Reginald Dufour: Color Plates V, W, X, and Y

Charles Eames, photographed at the Royal Society, England, 1973: Fig. 3-7

Charles Eames, photographed at the Jagiellonian Library, Cracow, Poland, 1972: Fig. 2-24

Fermi National Accelerator Laboratory: Figs. 7-4 and 10-3

Dr. Owen Gingerich: Figs. 2-8, 2-23, 2-29

Griffith Observatory: Star maps on endpapers

Harvard College Observatory: Figs. 5-11, 5-29, 6-13, 8-4 (main photo), 8-6 (top)

Dr. V. P. Hessler, Geophysical Institute, University of Alaska: Fig. 8-45

Dr. Paul Hodge: Figs. 5-13 and 6-5

Kitt Peak National Observatory: Plate J, Chapter 3 opener, Chapter 4 opener, Chapter 5 opener, Chapter 6 opener, Part 1 opener, Part 3 opener (insert), Figs. 1-2, 1-6, 3-11 (center left), 4-4, 4-22, 5-3, 5-8, 5-10 (left), 5-12, 5-19, 6-2, 6-10, 6-20, 7-14 (top left), 8-23, 9-1, 9-24 (right), 10-6

Dr. Jerome Kristian, Mount Wilson and Las Campanas Observatories, Carnegie Institution of Washington (5 meter Hale telescope, Palomar Observatory, California Institute of Technology): Fig. 10-21

Lick Observatory: Figs. 2-10 (part), 6-3, 6-4 (right), 6-15, 6-22 (left), 6-30 (left), 7-10 (top), 8-18, 9-14 (left), 9-18, 9-22, 9-23 (top left), 11-21, 12-8, 13-24, 13-25, Color Plate P, Chapter 11 opener, Part 5 opener

Dr. James Liebert: Fig. 8-33

Dr. William Liller: Fig. 13-22

Lowell Observatory: Fig. 12-13

Lund Observatory, Sweden: Fig. 6-19 (bottom)

Maine Department of Marine Resources: Fig. 11-18

Dr. Carter Mehl: Fig. 4-2

Meteor Crater Enterprises, Inc.: Fig. 11-33

Metropolitan Museum of Art: Fig. 8-34

Frank Miller: Color Plates K and L

William C. Miller: Fig. 9-26

Dr. Ronald Moore: Figs. 8-38 and 8-43

Mount Wilson and Las Campanas Observatories, Carnegie Institution of Washington: Figs. 3-22, 5-5, 5-6, 5-7, 5-10 (right), 8-6 (bottom), 8-9, 8-25, 8-41, Chapter 8 opener

Drs. S. S. Murray, G. Fabbiano, A. C. Fabian, A. Epstein, and R. Giacconi: Fig. 9-28 (top left)

National Academy of Sciences: Fig. 4-7

National Aeronautics and Space Administration: Color Plates A, B, C, D, and E, Part 4 opener, Chapter 12 opener, Chapter 13 opener, Chapter 14 opener, Figs. 3-11 (top right and center right), 12-3, 12-5, 12-10, 12-11, 12-14, 12-16, 12-17, 12-18, 12-21, 12-24, 12-25, 13-1, 13-3, 13-4, 13-9, 13-10, 13-11, 13-12, 13-13, 13-15, 13-16, 13-17, 13-18, 13-19, 14-15, Front and Back Cover Photographs

National Astronomy and Ionosphere Center: Fig 14-14

National Radio Astronomy Observatory: Figs. 3-11 (top left), 6-22 (right), 6-30 (right), 7-1 (right), 7-11

Dr. Gerry Neugebauer: Fig. 6-24

Palomar Observatory, California Institute of Technology, copyright by the California Institute of Technology and the Carnegie Institution of Washington, reproduced with permission: Color Plates I, M, O, Q, S, T, and U.

Palomar Observatory, California Institute of Technology: Part 2 opener (part), Figs. 4-5, 4-9, 5-1, 5-2, 5-9, 5-24, 6-4 (left), 6-11, 7-6, 7-12, 7-13, 7-14 (top right, bottom left, and bottom right), 9-14 (right), 9-23 (bottom), 9-24, 9-28 (center), 10-1, 11-27, Chapter 7 opener, Chapter 9 opener

Dr. Arthur Page: Color Plate F

Dr. Allison Palmer: Fig. 14-9

Allen W. Parker: Fig. 11-34

Planetary Patrol photographs furnished by Lowell Observatory: Figs. 12-2 and 12-12

Astrid Preston: Fig. 2-1

Dr. P. B. Price: Fig. 9-29

Ben Rose: Fig. 12-1

Royal Observatory, Edinburgh: Figs. 6-3 and 8-4 (inset), Chapter 10 opener

Royal Society of London: Fig. 3-7

Dr. Martin Ryle and Dr. P. J. Hargrave: Fig. 7-7

Dr. Th. Schmidt-Kaler and Dr. W. Schlosser, Ruhr-Universität Bochum (super wide-angle photograph): Fig. 6-19 (top)

Dr. J. William Schopf: Fig. 14-5

Emil Schulthess from Black Star: Fig. 2-7

Scientific American: Fig. 14-13

Dr. Fredrick Seward: Fig. 9-28 (right)

Dr. John S. Shelton: Figs. 11-16 and 12-22

Dr. Alex G. Smith: Fig. 7-17

Dr. Hyron Sprinrad: Fig. 5-25

Stanford Linear Accelerator Center: Fig. 4-15

Paul Trent: Fig. 14-16

Dr. Virginia Trimble: Fig. 9-23 (bottom)

United States Geological Survey: Fig. 12-19

University of Chicago Press: Fig. 9-28 (right)

Dr. William van Altena: Fig. 9-23 (top right)

Dr. Sidney van den Bergh: Figs. 9-28 (bottom) and 10-1

Dr. M. Waldmeier: Fig. 8-46

Dr. Joseph Weber: Fig. 10-25

Yerkes Observatory: Figs. 2-10 (part), 3-1, 3-2, 3-5, 10-9, 11-32

Jacob Zeitlin: Fig. 2-25

To my daughter Rachel.

PART 1

Understanding the Universe

Our ancestors marveled at the heavens, seeking to find the connection between the motions of the sky and the seasons on Earth. Almost every culture created an elaborate mythology to explain how the heavens rule the land.

But as our knowledge increased, science replaced myth, and we lost the connection between our lives on Earth and the universe around us. Yet our knowledge shows that we consist of matter that has evolved with the universe. We derive our energy from the output of our sun, a thermonuclear reactor of astronomical size.

Astronomy offers an opportunity to see how we relate to the universe. Once we make the effort to learn about astronomy, we can understand what it means to live on a small planet circling a representative star, one star in a galaxy of 400 billion, one galaxy among billions in the cosmos.

Color Plate A The Earth was photographed by the Apollo 17 astronauts from a distance of 80,000 kilometers. *Light travel time: 1/4 second*

1

A Look at the Universe

Our planet Earth, so immense to us, floats in space like a speck of dust, one of nine planets orbiting the sun. The Earth, 12,756 kilometers in diameter, is the fifth largest of the sun's nine planets, larger only than Mercury, Venus, Mars, and Pluto. From our small world we look outward to a vast and complex universe.

Generations of human observation and deduction have given us a chance to unravel some of the mystery of the cosmos. This book will take you on a journey through space and time to show you the universe of galax-

Color Plate B
The Eagle lander returns to the Columbia orbiter on the first Apollo lunar flight; the Earth is in the background. *Light travel time to Earth: 1 1/3 seconds*

ies and quasars, and of stars, nebulae, and planets within galaxies. We shall return to our solar system and to our home, the Earth, with a knowledge of our origins and our connection with the cosmos. Astronomy offers the chance to see what the universe tells us about ourselves: our past, our present, and our future.

Let us then embark on a brief tour of the cosmos, from our closest celestial neighbor, the moon, to a distant cluster of galaxies. The majesty and mystery of the universe lie before us.

The first stop on our celestial tour is the moon, the Earth's natural satellite that accompanies the Earth on its yearly journey around the sun. Since the moon is 400,000 kilometers away, light waves and radio waves, traveling at 300,000 kilometers per second, take 1 1/3 seconds to travel from the Earth to the moon. Yet the distance from the Earth to the sun exceeds the distance from the Earth to the moon by 400 times. Light from the sun therefore takes more than 8 minutes to reach the Earth.

Color Plate C
Viking photographed Mars and landed on the planet's surface, guided by radio commands from Earth. *Light travel time: 19 minutes*

Mars, the next planet outward from the sun, is never closer to the Earth than 40 million kilometers. This red planet, named for the Roman god of war, has only half the Earth's size. Yet the possibility of life on Mars has fascinated humanity throughout history, and in 1976 two automated space-

Color Plate D The Voyager spacecraft photographed Jupiter and its large satellites in 1979. *Light travel time: 43 minutes*

craft landed on Mars. They sent back thousands of pictures of the surface, and carried out radio commands to dig into the Martian soil, analyze its contents, and send the results back to scientists on Earth. Although photographs show branching valleys on Mars seemingly carved by water, no definite signs of Martian life have appeared. At the time of the landing, Mars was so far from Earth that radio signals took 19 minutes to travel from Earth to Mars.

Mercury, Venus, Earth, and Mars form the inner planets of the sun's family, our solar system. The four giant planets—Jupiter, Saturn, Uranus, and Neptune—are giant balls of gas, much larger than the Earth and much more distant from the sun than Earth. Jupiter, the innermost and largest of the giant planets, has eleven times the Earth's diameter, and is five times farther from the sun than Earth. The two Voyager spacecraft that photographed Jupiter and its satellites in 1979 traveled more than a billion kilometers in over two years to reach the planet. Sunlight takes nearly 45 minutes

Color Plate E In 1980, the Voyager spacecraft passed by the planet Saturn. *Light travel time: 85 minutes*

to reach Jupiter, where the sun appears more than 5 times smaller and nearly 30 times fainter than on Earth.

If we travel still farther from the sun, almost twice as far as Jupiter, we reach Saturn, with its spectacular system of rings. Light takes more than an hour to reach Saturn, which is 1.4 billion kilometers from the sun. In these outer reaches of the solar system, the sun shrinks to a small, faint disk of light, nearly 10 times smaller and almost 100 times fainter than the sun as it appears from Earth.

Though Saturn may seem an immense distance from the sun, three of the sun's nine planets lie still farther away—Uranus, Neptune, and tiny

Color Plate F
A wide-angle photograph of our Milky Way galaxy shows the Eta Carinae nebula in the lower left. *Light travel time: 4000 years*

Pluto, a frozen snowball smaller than our own moon. Pluto is so far away that sunlight takes 5 1/2 hours to reach it. Vast though Pluto's distance may seem—almost 6 billion kilometers, 40 times the Earth-sun distance—it shrinks to almost nothing in comparison to the distances to the stars. Our sun, a typical star in the Milky Way galaxy, orbits around the center of the galaxy, once every 240 million years. The sun's closest neighboring stars, 7000 times farther away than Pluto, are so distant that their light takes more than 4 years to reach us. The starlight from the Big Dipper travels for a century on its way here; the light from Polaris, the north star, takes eight centuries. The center of the Milky Way, in the direction of the constellation Sagittarius, is so far away that light from the center takes 30,000 years to reach us.

Strewn among the stars of the Milky Way are great clouds of gas and dust called nebulae. Most nebulae are either groups of stars in formation, or old stars shedding their outer layers. The Eta Carinae nebula contains a group of young stars that light up their cocoon of gas with a flood of energy. There a star cluster is being born whose bright stars will last a few million years. The Milky Way contains perhaps 10,000 such star clusters, young and old.

Beyond the Milky Way, the frigid depths of intergalactic space hold a few stars and only a trace of gas. Most of the matter in the universe has condensed into galaxies like the Milky Way—"island universes" that may

Color Plates G, H
Two high-resolution photographs of the Eta Carinae nebula show a region of star formation.
Light travel time: 4000 years

Color Plate I
The Andromeda galaxy, much like our own Milky Way, has a diameter of 100,000 light years. *Light travel time: 2 million years*

contain billions of stars. The galaxies themselves form part of "galaxy clusters," and the clusters belong to "superclusters." Our Milky Way is part of a small cluster of galaxies called the Local Group. Within the Local Group are the Magellanic Clouds, satellites of our own galaxy, as well as the famous spiral galaxy in Andromeda, which also has two satellite galaxies. The Andromeda galaxy closely resembles the Milky Way, but is so far from us that its light takes 2 million years to reach us. Thus we see the Andromeda galaxy as it was before our ancestors understood the use of fire.

The Andromeda galaxy's distance from us is 20 times the diameter of the Milky Way. Despite this enormous distance, astronomers can see much more distant galaxies, thousands of times farther from us than the Andromeda galaxy. These galaxies are so distant that their light has taken billions of years to reach us—a journey that spans much of the history of the universe. By studying such galaxies, we hope to unravel not only the past but also the future of the universe.

In this book, we shall study our place in the cosmos—how we came to be, adrift on a bit of matter, circling a larger speck in a whirlpool galaxy that resembles billions of other galaxies. Let us begin, then, to know the stars and the universe that brought us here.

Plate J A cluster of galaxies in the constellation Centaurus contains various types of galaxies. *Light travel time: 200 million years*

1.1 Connecting with the Universe

If you stand outside on a clear autumn evening and look toward the west, you will see three bright stars that form a sprawling triangle (Figure 1-1). When you look at stars such as these, you are seeing the results of 15 billion years of cosmic evolution. We and the stars have emerged from the development of the universe: In a sense, the universe has learned to look at itself.

Human understanding of the heavens has proceeded at an enormously rapid pace, compared with the time scale of most events in the universe. Our Earth formed, along with the sun and its other planets, almost 5 billion years ago. Mammals appeared on Earth only 200 million years ago, and hominids (humanlike primates) only a few million years ago. Records of astronomical observation go back in time only a few thousand years—less than one-millionth of the existence of the universe. In that short time, we have learned how the heavens appear to move, hour by hour, day by day, month by month. Eventually, we drew the conclusion that our Earth does not form the center of the universe, as people once believed. Instead, we inhabit a small speck of matter, in orbit around a typical star, our life-giving sun.

The Solar System

Together with the Earth and its satellite, our moon, eight other planets (so far as we know) orbit the sun (Figure 11-3). The Earth, third planet out from the sun, has an orbit almost 300 million kilometers in diameter. Jupiter, the largest planet, has an orbit five times larger, and Pluto, the outermost planet, moves in an orbit eight times larger than Jupiter's. Mercury and Venus, the two innermost planets, have no moons, but the other seven planets possess a total of 35 known satellites, some of which exceed our own in size and mass.

At the center of the solar system, with more than 99% of the system's total mass, our sun provides the gravitational force that keeps the planets in orbit (see page 371). Planets have such small sizes compared to stars that we have yet to detect with certainty *any* planets besides those in our solar system. We cannot be sure whether our solar system forms the exception or the rule. However, most astronomers believe that since our sun is a typical star, something like our solar system must exist—with various changes, to be sure—around many of the other stars in our galaxy.

The Milky Way Galaxy

Stars in the universe congregate into galaxies, huge agglomerations that contain millions, billions, or even trillions of individual stars (Figure 1-2). Our sun belongs to the **Milky Way** galaxy, as do 400 billion other stars. Among the nearest of these stars are the three stars that form the triangle shown in Figure 1-1. These three stars, Vega, Altair, and Deneb, are each a million times or more farther from us than the sun. Deneb, the most distant of the three, has 100 million times the sun's distance! Yet Deneb, so far from us that its light takes 1600 years to reach us, occupies almost the same part of the Milky Way galaxy that we do (Figure 1-3). Our galaxy's diameter equals almost 100 times the distance from the sun to Deneb.

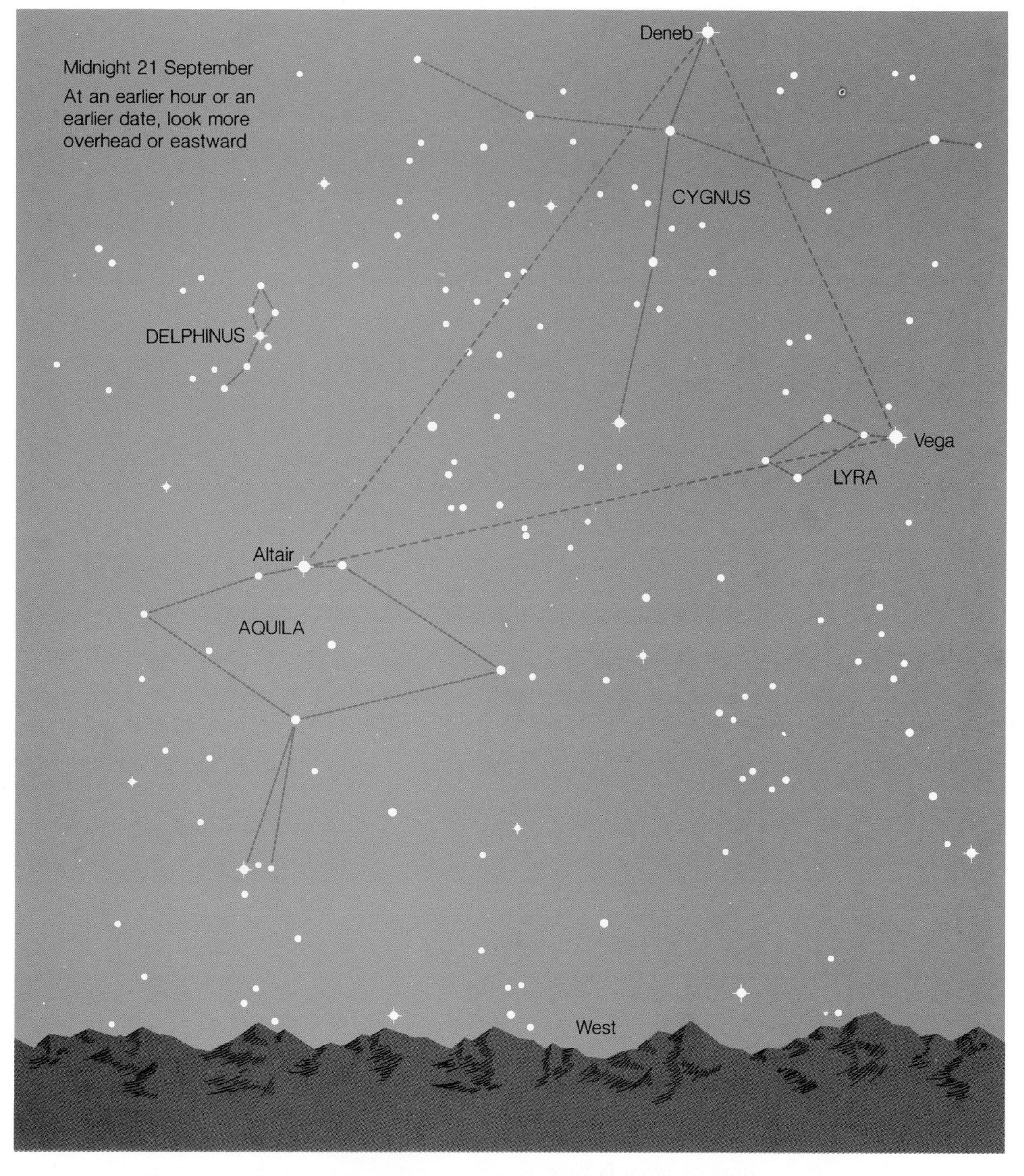

Figure 1-1 A view of the western sky at about midnight in mid-September, or about 10 P.M. in mid-October. The Summer Triangle consists of Vega (the brightest star in Lyra, the Lyre), Altair (the brightest star in Aquila, the Eagle), and Deneb (the brightest star in Cygnus, the Swan).

Figure 1-2 This spiral galaxy in the constellation Cepheus, seen almost face-on, has a general resemblance to our own Milky Way galaxy.

Figure 1-3 An artist's side view of our Milky Way galaxy shows its flattened, disklike shape, with a bulge toward the central regions. Our sun and Deneb lie far from the center, relatively close to the outer reaches of the Milky Way.

The Naming of Stars

Farther back in time than any written records can take us, human beings must have looked at the skies and wondered at the stars. Any culture with myths and legends of heroes, giants, evil spirits, and magical animals could easily have found star groups that resembled such figures. These star groups then became the heavenly symbols for a society's legends. Our modern constellations (from the Latin *constellatus*, meaning "star-studded") originated in the Euphrates Valley and date back to legends at least 6000 years old. Claudius Ptolemy's star catalog, the *Almagest*, listed 48 constellations; the parts of the sky never visible from Egypt later provided several dozen more. Today astronomers divide the sky into 88 constellations, each with an arbitrary boundary, so that every star belongs to just one constellation (with a few exceptions on the boundaries).

The brightest individual stars were named in classical antiquity. Some of these names are still in use: Castor and Pollux, Regulus, Spica, Arcturus, Sirius, Bellatrix, Polaris, Procyon, Capella, and Mira (the "marvelous one," a variable star) are the best known. Islamic scientists, who preserved the *Almagest* while Europe suffered through its Dark Ages, added more names to the list. These Arabic names, often garbled in transliteration, have become Vega, Betelgeuse, Rigel, Aldebaran, Achernar, Shaula, Eltanin, Altair, Deneb, Menkalinan, Alderamin, and Fomalhaut. Additional Arabic names—Zubenelgenubi and Zubeneschemali—appear in the Scorpion's claws, now considered part of Libra, and in the stars of the Big Dipper, Dubhe, Merak, Phecda, Megrez, Alioth, Mizar (with its companion Alcor), and Alcaid. Algol, the "ghoul" or demon star, is a famous variable in the constellation Perseus.

The first systematic naming of stars by constellation consisted of attaching the letters A, B, C, and so forth, to the brighter stars, followed by the name of the constellation. In 1603, Johannes Bayer published the first real star atlas, the *Uranometria*, meaning "astronomy measured." Bayer introduced the practice, still followed, of giving the stars designations in which the brightest star of each constellation was called Alpha, the next-brightest Beta, and so on through the Greek alphabet to Omega. The name of the constellation appears in Latin, and in the genitive case. Thus Aldebaran is Alpha Tauri, the brightest of Taurus; Bellatrix is Gamma Orionis, third-brightest of Orion; and so forth. Some errors crept in, so that Castor became Alpha Geminorum although Pollux, Beta Geminorum, has a greater apparent brightness. Likewise, Betelgeuse, Alpha Orionis, has less apparent brightness than Rigel, Beta Orionis. When the 24 Greek letters had been exhausted for a given constellation, numbers could be used. John Flamsteed, the first Astronomer Royal of England, codified this practice in his *British Catalogue* of stars, published posthumously during the 1720s.

Later star catalogs used different numbering systems, but all the brighter stars of a constellation have either a Greek letter or a Flamsteed number. Special sorts of objects, such as variable stars, have specialized naming systems that may overlap the general scheme.

Beyond the Milky Way

The Milky Way belongs to a small group of galaxies called the **Local Group.** The Local Group contains about twenty galaxies, of which the largest are our own and the famous Andromeda galaxy (Color Plate I). The stars that form the Andromeda galaxy have distances from us about 20 times greater than the diameter of the Milky Way (Figure 1-4).

When we look outside the Local Group, we find galaxies, typically grouped into clusters, as far as we can see. Some of these clusters contain thousands of member galaxies and may be hundreds of times larger than the Local Group (Figure 1-5). The distances to these galaxy clusters range from

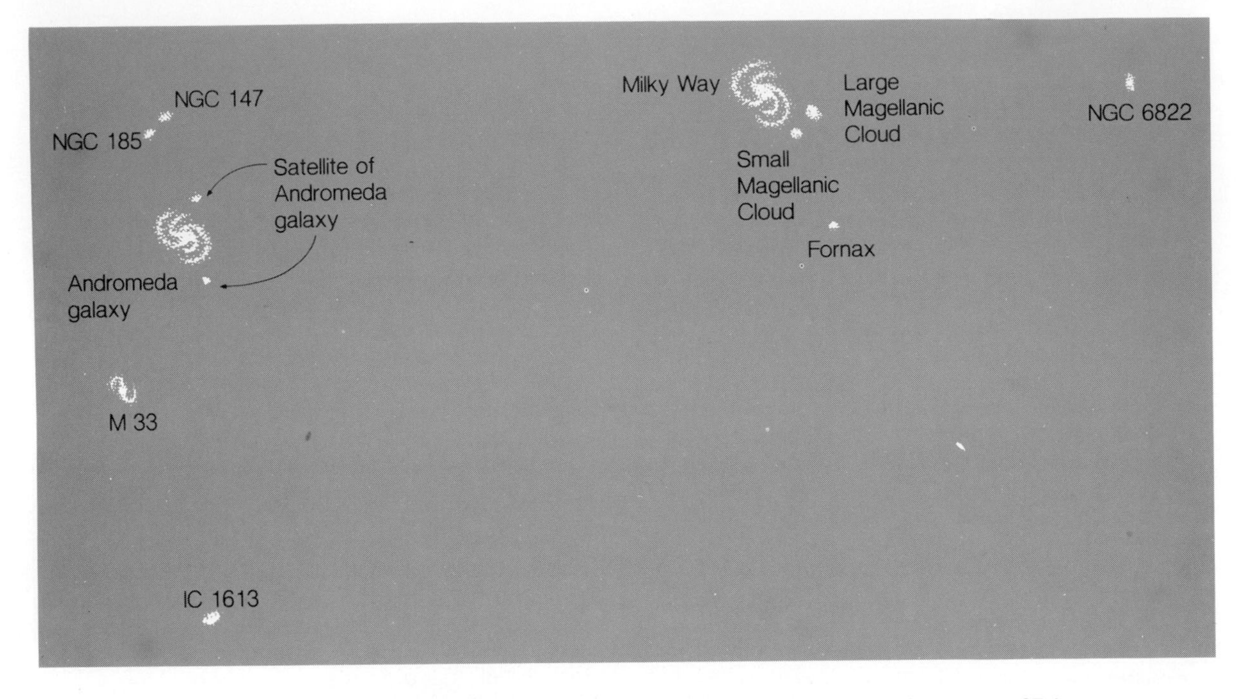

Figure 1-4 The Milky Way belongs to the Local Group of galaxies. (See page 167 for a complete list of the galaxies in the Local Group.)

Figure 1-5 This galaxy cluster in the constellation Centaurus has more than 1000 member galaxies.

20 times the distance to the Andromeda galaxy up to 4000 times this distance for the farthest clusters yet detected.

The Delayed Messages of Starlight

We learn about stars and galaxies by studying the light from them. Since light does not travel through space instantaneously, our information must inevitably be out of date. Light consists of tiny particles called **photons,** each of which travels at about 300,000 kilometers per second. At this speed, a photon can travel around the Earth seven times in a second. For a photon to travel from the sun to the Earth requires 8 minutes. We would therefore have 8 minutes of grace if the sun suddenly disappeared! Light from Polaris, the pole star, takes almost 800 years to reach us. Hence we see Polaris not as it is now, but as it was in the early part of the thirteenth century.

Astronomers have grown used to the fact that our view of the universe is a view of the past, and of different times in the past for different objects. When we study the Andromeda galaxy, we can discover only what the galaxy was like 2 million years ago, at a time when our ancestors roamed the gorges of Africa in search of insects and roots to eat. More distant galaxies appear to us as they were many millions, or even billions, of years ago. If you stop to think about it, you will realize that we can never see an entire galaxy as it was at one particular time, since we see the "near" side of the galaxy as it was many thousands of years after the light left the galaxy's "far" side (Figure 1-6).

Figure 1-6 Light from this spiral galaxy in the constellation Virgo has taken about 50 million years to reach us. The light from the near side of the galaxy left about 50,000 years later than the light from the far side.

You might think that astronomy would suffer from our not being able to see stars and galaxies as they are now, but in fact this is not true. A few million years form only a fleeting instant in the history of the universe. Our sun has lasted for 4.6 billion years and will look much the same a billion years from now as it does today. Furthermore, astronomers have learned how to use the travel time of light to their advantage. When we look at faraway clusters of galaxies, we see the universe as it was many billions of years ago. By comparing this information with our studies of the universe at more recent times (from observations of less distant galaxies), we can learn a great deal about the evolution of the universe. The changes that the universe has undergone can themselves be used to predict the future of the universe, not with certainty, but with better accuracy than would be the case if we could *not* look back into the past.

What are the oldest messages we receive—the photons that have been traveling the longest? These turn out to be the photons that were produced during the early years of the universe, long before any stars or galaxies had formed. The photons that form this cosmic background, described in Chapter 4, have been traveling for 15 billion years. They take us as far back in time as anything we can observe, and help us to reconstruct the early history of the cosmos.

The enormous span of time in the universe may seem as overwhelming as the tremendous distances, for we cannot easily think about millions, let alone billions, of years. But such thought is essential if we are to understand the cosmic evolution that produced us. We can assimilate the universe's time scales by imagining a train ride from San Francisco to New York. Suppose New York represents "now," while San Francisco stands for the time when the photons in the cosmic background first appeared. Then the Milky Way formed when the train passed through Utah, but the sun and its planets did not appear until Detroit. Life on Earth emerged near Cleveland, mammals got on in Newark, and human beings have been aboard for about half a block. The total recorded history of humans would occupy the last 10 meters, and an average human lifetime would span 2 centimeters of the journey.

In this analogy, each of us has 2 centimeters to embrace the universe. We can decipher the messages that photons bring us. We can reconstruct the history of the cosmos. We can even begin to see how part of the universe turned into ourselves. With this book, I invite you to ride that train from the edge of darkness, to see how humans once looked at the universe and how we see things today. With this knowledge, we can face the future better acquainted with the universe, and perhaps wiser in our sense of connection with it.

1.2 Distances in the Universe

The universe shows us a tremendous range in the distances to the various objects we can see, though all of them seem immensely far by human standards. If we imagine the distance between the Earth and the sun as a line 1 centimeter long, then the distance from the Earth to Altair would be 10 kilometers, and the distance to Deneb would be 1000 kilometers, equal to the distance from Detroit to Boston! On this scale, the Milky Way would span 60,000 kilometers, five times the Earth's diameter, and the Andromeda galaxy would be three times farther away than the moon!

Table 1-1 gives the distances to various celestial objects, as well as the number of times farther from us each object is than the preceding object in the list. The story of *how* these distances have been measured (or estimated, in the case of the greatest distances) takes us through the history of astronomy. Even today, astronomers work hard to refine their knowledge of how far away the stars and galaxies are. For some objects, such as the mysterious quasars (see page 223), some doubt remains as to whether astronomers know the distances to within a factor of 10!

Measuring the Distances to the Closest Objects

We know the distances within our solar system with great accuracy, thanks to our ability to bounce radar waves off the sun, the moon, and the closer planets. By timing the interval needed for the signal to return, we can determine the distances to these objects, since we know the speed with which radar waves travel through space. (This speed equals the speed of light, about 300,000 kilometers per second.) Thus in the case of at least one star—our sun—we can measure distance by the straightforward technique of radar echo. This method, however, does not work for other stars, since they are too far away to return a detectable radar echo. Instead, we must use a slightly more subtle technique to determine stellar distances: the parallax effect.

The Parallax Effect: Measuring Distances of the Stars

As the Earth moves in its orbit around the sun, its changing position in space produces a change in the stars' apparent positions on the sky. This change,

Table 1-1 Distances to Various Objects in the Universe

Object	Distance from Earth (Earth–Sun Distance = 1)	Distance in Light Years[a]	Number of Times Distance Exceeds Distance to Preceding Object
Moon	0.0026	4×10^{-8}	
Sun	1	1.6×10^{-5}	400
Pluto (most distant planet in solar system)	40	6×10^{-4}	40
Altair	10^6	16	25,000
Deneb	10^8	1,600	100
Center of Milky Way	1.8×10^9	30,000	18
Andromeda galaxy	1.4×10^{11}	2.2×10^6	78
Virgo cluster (nearest large cluster of galaxies)	3×10^{12}	5×10^7	21
Coma cluster of galaxies	2.3×10^{13}	4×10^8	8
Most distant cluster of galaxies now known	6×10^{14}	About 10^{10}	25

[a]See page 25 for explanation of light years.

Powers of 10

Astronomers and other scientists often must work with extremely large or small numbers. They therefore find the **powers of 10** notation a great convenience, since it allows them to write large or small numbers with a minimum of difficulty. By using positive **exponents** of 10, that is, positive numbers placed above and to the right of the number 10, we can designate how many zeros should follow the number 1 in a large number. Thus 1 million (1,000,000) becomes 10^6, and 1 trillion (1,000,000,000,000) is written as 10^{12}. Negative exponents, on the other hand, denote the number 1 *over* the number indicated by the positive exponent. Thus, 10^{-3} is 1/1000, and 10^{-5} is 1/100,000.

An additional advantage of using exponents appears when we multiply or divide large or small numbers. We simply add the exponents to multiply, or subtract them to divide. Hence

$$10^2 \times 10^3 = 10^5$$
$$10^{-4} \times 10^{-6} = 10^{-10}$$
$$10^2 \div 10^3 = 10^{-1}$$
$$10^{-4} \div 10^{-6} = 10^2$$

If we want to write 4 million in powers of 10, we write 4×10^6. The number 0.02, on the other hand, can be written as 2×10^{-2}. We can then easily multiply the first number by the second to obtain 8×10^4. If we divide the first by the second, we obtain 2×10^8.

Figure 1-7
If we ride on a lakeside merry-go-round and look at the boats on the lake, we see the nearer boats appear to shift their positions against the background of more distant boats. This apparent shift is called the parallax effect.

called the **parallax effect**, corresponds to the changes in position we see from a lakeside merry-go-round as we look at the boats on the lake (Figure 1-7). As we move in a circle, we see the nearer boats appear to shift their positions relative to the more distant boats. If we imagine the Earth to be on such a merry-go-round, 300 million kilometers in diameter, with the sun at its center, then the stars would be the boats. We can measure the amount by which the parallax effect changes the stars' positions, and then use our knowledge of trigonometry to determine the stars' distances from us.

The difficulty with measuring the amount of the change in the stars' positions—called the **parallax shift** of the stars—comes from the stars' immense distances. In our model, if the merry-go-round has a diameter of 10 meters, the closest star to the sun would have a distance of 1400 kilometers, several states away. Hence the parallax shift of the closest star to the sun, Alpha Centauri, amounts to *only 3/4 of 1 second of arc*, back and forth from its average position, as the Earth moves from one side of its orbit to the other (Figure 1-8).

Astronomers measure angles in **degrees of arc** (360° in a complete circle), **minutes of arc** (each 1/60 of a degree), and **seconds of arc** (each 1/60 of one minute). The sun and moon each have an apparent diameter of half a degree, or 30 minutes of arc. Thus the parallax shift of the sun's closest neighbor, Alpha Centauri, equals only 1/2400 of the apparent diameter of the sun or the moon. This angular displacement corresponds to looking from

Angular Measurement

Mathematicians long ago divided an entire circle into 360 degrees of arc (abbreviated °) and subdivided each degree into 60 minutes of arc, with each minute made of 60 seconds of arc. This use of 60 and 360 comes from Babylonian tradition and may seem a bit confusing to those of us raised on the decimal system. We are, of course, familiar with the minutes and seconds of a clock, and we must be careful not to confuse minutes and seconds of arc with the more common minutes and seconds of time.

When we make angular measurements on the sky, we compare the angular size in question with the 360° that would make a full circle. For example, when we say that the sun and the moon each cover half a degree (30 minutes of arc), we mean that it would take 720 suns or moons placed next to one another to span a full circle around the horizon. Similarly, when we say the planet Jupiter has an angular diameter of 40 seconds of arc (2/3 of a minute of arc), we mean it would take 32,400 Jupiters to form a full circle all the way around the sky.

Any object's angular size depends on its true size and on its distance from us. If we moved the moon to twice its present distance, it would have only half its present angular size. Jupiter has a diameter 41 times greater than the moon's and would have the same angular size as the moon if it were 41 times farther away than the moon. Jupiter in fact has about 1800 times the moon's distance from us. Therefore its angular diameter is about 1800/41, or 43, times less than the moon's.

The closest stars to the sun are 280,000 times farther away from us than the sun. If these stars have the same size as the sun, their angular diameters should be 1/280,000 of the sun's 1800 seconds of arc (30 minutes of arc). This would give the stars an angular diameter of 0.006 seconds of arc. Since the smallest angles we can measure on the sky are about 0.02 or 0.01 seconds of arc, we cannot hope to see these stars as anything but single points of light, with no measurable angular size at all.

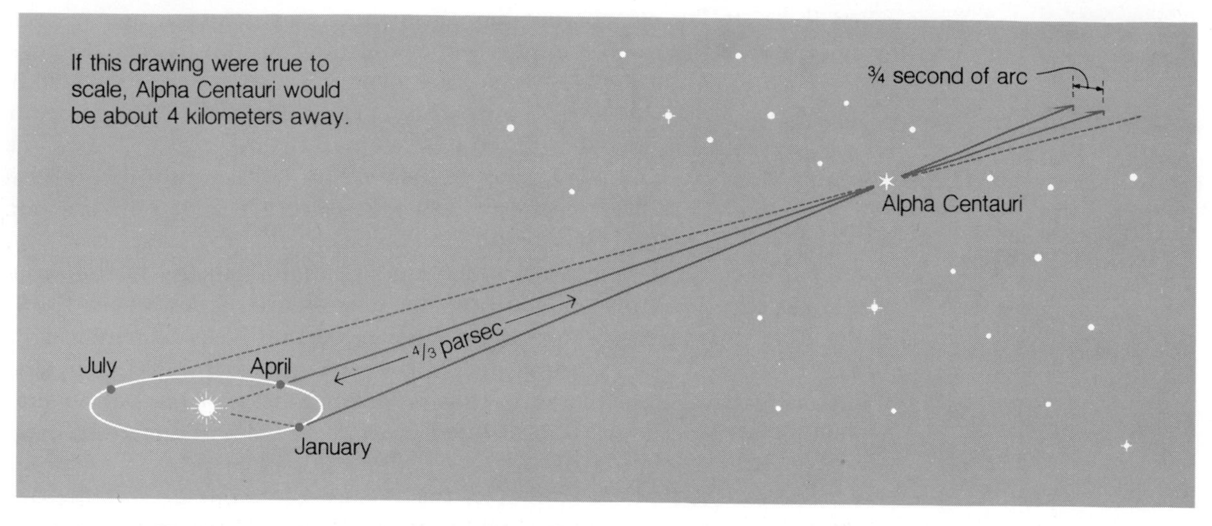

Figure 1-8 The vast distances to all stars but our sun means that these stars will show only tiny parallax shifts. The star closest to the sun, Alpha Centauri, has a parallax shift no larger than ¾ second of arc back and forth from its average position during the course of a year.

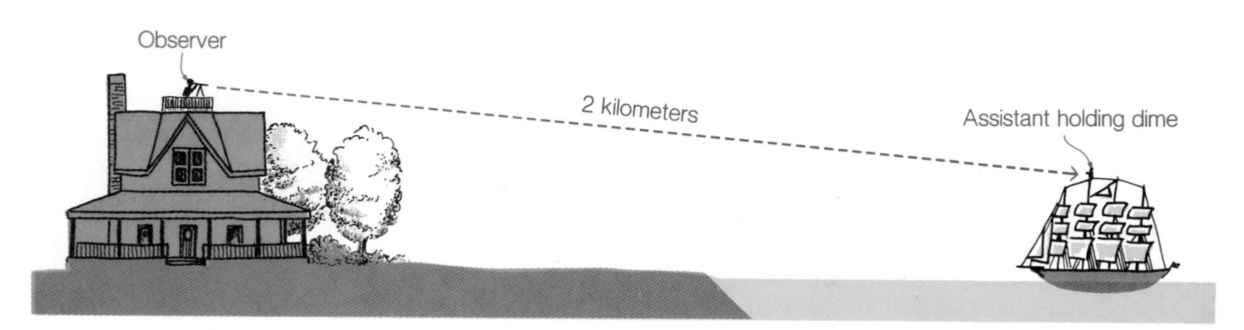

Figure 1-9 One second of arc equals the angular size of a dime held 2 kilometers away. Even this tiny angle exceeds the amount of all stars' parallax shifts.

one side to the other of a dime held 2.7 kilometers away (Figure 1-9). Astronomers could not measure such small angular displacements until the 1830s, but when they did, they could determine the distances to the nearest stars in terms of the Earth–sun distance.

Trigonometry shows us that stars *farther* from the sun will have progressively *smaller* parallax shifts (Figure 1-10). A similar change occurs when you look at objects first with one eye and then with the other. You can see the objects undergo parallax shifts that grow smaller as the objects' distances increase. Thus we can estimate distances by the parallax effect in a process that by now has become entirely automatic. As babies, though, we had to learn how to integrate the parallax effect into our perception of the world around us.

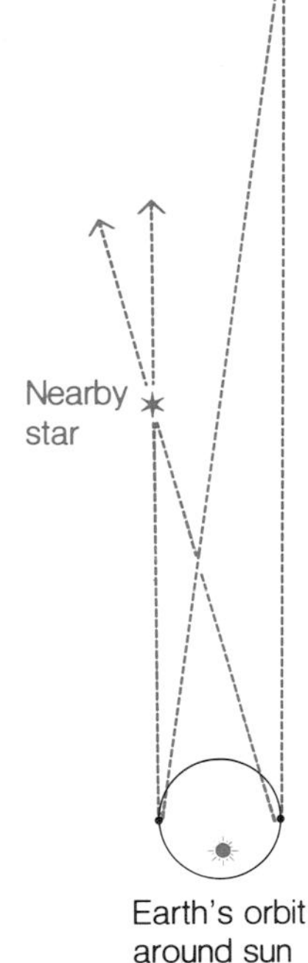

Figure 1-10
The amount of the parallax shift decreases in proportion to a star's distance: A star twice as far from us as another will have half the parallax shift of the closer star.

Parsecs and Light Years

When astronomers began to measure distances by the parallax effect, they invented a convenient measure of distance called the **parsec.** One parsec (from "*par*allax" plus "*sec*ond of arc") is the distance a star would have if the Earth's orbital motion produced an apparent shift of 1 second of arc in each direction, back and forth from the star's average position, during the course of a year (Figure 1-11). A star at 2 parsecs from us will show a parallax shift of ½ second of arc, back and forth from its average position. *The distance to any star, in parsecs, equals 1 over the amount of its parallax shift.* Note that we always measure the parallax shift in either direction away from the average position (Figure 1-11).

One parsec equals 206,265 times the distance from the Earth to the sun, or about 32 trillion kilometers. Light travels at nearly 300,000 kilometers per second and takes 3.26 years to cover a distance of 1 parsec. Astronomers also like to measure distances in **light years,** the distance light travels in one year. Since 1 parsec equals 3.26 light years, 1 light year equals 0.31 parsec, or about 10 trillion kilometers.

The Breakdown of the Parallax Effect

Of all the stars, only the sun lies within 1 parsec of the Earth. Alpha Centauri, whose parallax shift equals 3/4 second of arc, has a distance of 4/3 parsec. Light from Alpha Centauri takes 4.4 years to reach us; in other words, Alpha Centauri's distance also equals 4.4 light years. Altair, the closest star in the Summer Triangle (Figure 1-1), has a parallax shift of 1/5 second of arc and a distance from us of 5 parsecs. Vega's parallax shift equals 1/8 second of arc, so Vega's distance is 8 parsecs or 26 light years. Deneb is considerably farther from us—so far, in fact, that we cannot measure its distance through the parallax effect. Even today, astronomers cannot measure parallax shifts smaller than about 1/50 second of arc, or perhaps 1/100 second of arc in favorable cases. As a result, stars farther than 50 to 100 parsecs away must have their distances estimated, rather than accurately measured.

To estimate the distances to faraway stars, we can compare the apparent brightnesses of stars thought to have almost the same luminosity or total energy output. Any object's apparent brightness—the intensity of light that we measure on Earth—decreases in proportion to the square of the object's distance from us. Thus if two stars produce the same energy output per second but have a ratio of 100 in their apparent brightnesses, the star that seems fainter must be 10 times farther away. If we know the distance to the closer star, we can estimate the distance to the more distant star. This method works only if we are fairly sure the two stars *do* have the same energy output per second. Just how we determine such a fact rests on what we know about the ways in which stars produce energy (see page 268).

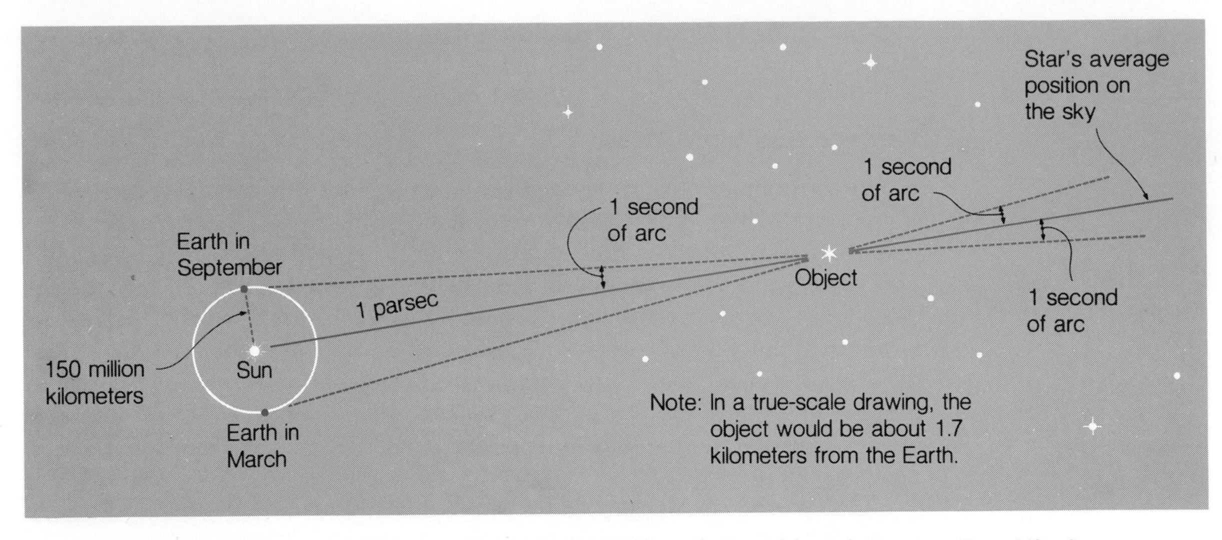

Figure 1-11 By definition, 1 parsec is the distance that would produce a parallax shift of 1 second of arc in each direction from the average position, as the Earth orbits the sun during the course of a year.

When we consider the distances to other galaxies, each of which lies hundreds of times farther from us than Deneb, we can see that it would be hopeless to try to use the parallax effect. Instead, we must rely on the method of comparing the apparent brightnesses of two stars with the same luminosity: one, in our own galaxy, whose distance is known, and the other, in the distant galaxy, whose distance we want to find. This method, discussed in some detail in Chapter 5, does not work as well as astronomers would like, but it is all we have to estimate distances to faraway galaxies.

Summary

Our Earth, one of the sun's nine planets, makes a yearly journey around our sun, a typical star. The Milky Way galaxy, which contains 400 billion stars, includes the sun in its outer reaches. Other galaxies spread throughout the universe, typically appearing in clusters. Our Milky Way belongs to a small cluster of about twenty galaxies called the Local Group.

Because light takes time to travel immense distances, our view of the universe is a view of the past. We look 16 years back in time when we observe Altair, 1600 years back when we observe Deneb, and more than 2 million years back when we observe the Andromeda galaxy. The light from the most distant galaxies we can see has taken 10 billion years to reach us, more than half the age of the universe and twice the age of our sun and Earth.

Astronomers measure the distances to stars in light years (the distance light travels in one year) and in parsecs (1 parsec equals 3.26 light years).

Parsecs are named after the parallax effect, the apparent shift in stars' positions caused by the Earth's orbital motion around the sun. This motion would produce an apparent shift of 1 second of arc, back and forth from the average position, in a star 1 parsec from the sun. A star 5 parsecs away would have a proportionately smaller parallax shift of 1/5 second of arc. We can use the parallax method to determine stellar distances up to 50 or 100 parsecs. For greater distances, the size of the parallax shift becomes too small for accurate measurement.

The apparent brightness of any object decreases in proportion to the square of its distance. Hence if two stars have the same luminosity, but one is three times more distant from us, the farther star will appear nine times fainter than the nearer star. If we know that the two stars produce the same energy output, we can determine the distance to the farther star once we know the distance to the nearer star.

Key Terms

degree of arc	Local Group	parallax effect	photon
exponent	Milky Way	parallax shift	powers of 10
light year	minute of arc	parsec	second of arc

Questions

1. Why does the amount of the parallax shift in stellar positions *decrease* as the distance to stars increases?
2. The planet Jupiter orbits the sun at five times the Earth's distance from the sun. If an astronomer on Jupiter observes Altair during the course of one orbit, will that astronomer find the same parallax shift for Altair that we do on Earth? Explain.
3. The unaided human eye can detect changes in position as small as 1 minute of arc. If an object had a parallax shift (back and forth) of 1 minute of arc, how far away would that object be, in parsecs? The star closest to the sun has a parallax shift of 3/4 second of arc. How far from us, in parsecs, is that star?
4. Light takes about 3 1/4 years to travel 1 parsec of distance. How long does the light from Arcturus (10 parsecs away) take to reach us? How many years does it take for the light from the galaxy M 87 (10 million parsecs away) to reach us?
5. One year contains about 32 million seconds. If we were to compress the entire history of the universe (about 16 billion years) into a single year, how much of the year would the history of our solar system (4.6 billion years) occupy? How many seconds would the history of the human species (about 3.2 million years) occupy?

Projects

1. *Cosmic Distances.* Take a survey among your friends in which you ask them to compare the distances to the moon, the sun, and the stars closest to the sun. How do the average distance ratios compare with those listed in Table 1-1? What does this tell you about the way we think about celestial objects?
2. *A Model for the Universe.* Using the distances given in Table 1-1, design a model for the universe in which the Earth–sun distance is represented by the size of a hydrogen atom (10^{-8} centimeter). How many times has the model reduced the scale of the true universe, in which the Earth–sun distance equals 150 million kilometers? In the model, what is the distance from the Earth to the Andromeda galaxy? What is the distance to the most distant galaxy cluster we can see? Can you calculate how large a person would be in this model?
3. *Parallax.* Observe a distant object, such as a faraway tall building, first with one eye and then with the other. Do this while holding your index finger at arm's length. Why does your finger seem to shift its position relative to the distant object? Determine the distance of your finger from your eyes by estimating the amount of the parallax shift. You can do this at night when the moon is full by seeing how many times the moon's diameter your finger appears to shift. (The moon has an apparent angular diameter of ½°.)

   Suppose we define a unit of distance called the "pardeg" as the distance to an object whose parallax shift when you blink your eyes equals 1°. How many pardegs away is the tip of your finger? If your head has an average size, one pardeg equals 1.8 meters. How far away is your finger in meters? What does the distance between your eyes (6.6 centimeters) correspond to in the astronomical use of the parallax effect? What sources of error enter the determination of the distance to your finger by use of the parallax method?

Further Reading

Bernhard, H. J.; Bennett, D. A.; and Rice, H. S. 1948. *New Handbook of the Heavens.* New York: Mentor Books.

Ley, Willy. 1969. *Watchers of the Skies.* New York: Viking Compass Books.

Rey, H. A. 1952. *The Stars: A New Way to See Them.* Boston: Houghton Mifflin.

Whitney, Charles. 1974. *Whitney's Star Finder.* New York: Alfred A. Knopf.

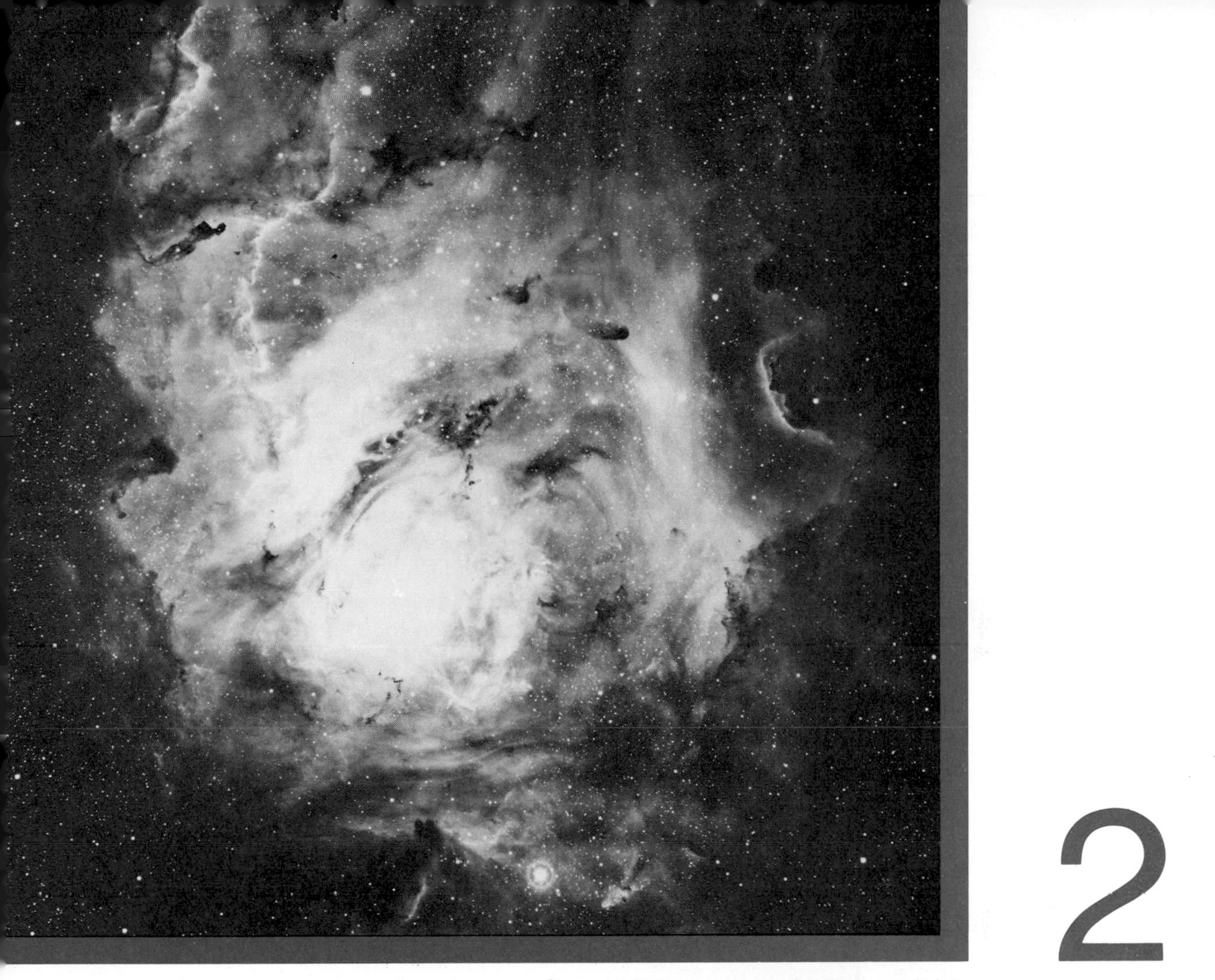

Stars are being born even today within the Lagoon Nebula.

2

Understanding the Heavens

Ever since human beings looked up at the sky with the capacity to wonder, they must have marveled at the sun, the moon, and the changing stars. Every day the sun rises and sets, producing the cycle of day and night that governs life on Earth. Every 29½ days the moon completes a cycle of its phases, rising and setting at different times of the day and night. Every year brings a new cycle of the seasons; each summer the sun rises higher in the sky, and stays above the horizon longer, than in winter. And every clear night brings a background of points of light, as thousands of stars move through the heavens, changing their orientation in the sky through the course of a year.

2.1 The Calendar

Our ancestors studied the skies out of necessity, for their lives were tied to the cycles of the day, month, and year far more firmly than our own. In early food-gathering societies, hunters and fishers depended on the life cycles and migrations of the animals they caught. Once agriculture became important for survival, a precise knowledge of the planting, growing, and harvesting seasons grew essential as well. The moon's phases provided growers and hunters with a convenient way to subdivide the yearly cycle of the seasons, and the sun's motion through the sky furnished a natural way to keep track of the passage of a day.

Early interest in astronomy thus sprang from a need to tell time and to predict the seasons. Even before the introduction of clocks, most people could tell the time within a half-hour simply by observing the sun's position in the sky. The use of months ("moonths") to subdivide the year goes back at least 50 centuries. The Jewish Midrash, first begun in the fourth century B.C., says that "the moon has been created for the naming of days."

Since agriculturally based societies ceased their labors in winter, most ancient **calendars** began with the coming of spring. Thus, our own months of September, October, November, and December refer to the numbers seven, eight, nine, and ten, reminders of the time when the year began in March. Geography has also played a role in calendar making. The Eskimos of Labrador have no names to divide the dark time of the year, when they do little outdoors, but break the remainder into fourteen periods. The Polynesians, on the other hand, had a regular calendar based on full-moon intervals, since they could farm, hunt, and fish in both winter and summer.

2.2 The Celestial Sphere

We now know that the sun rises and sets because the Earth spins once each day. But ancient observers of the sky, lacking such knowledge, sought other explanations for the mystery of day and night. In Egyptian mythology, Nut, the goddess of the sky, swallowed the sun every evening. By morning, the sun had traveled through Nut's body to be reborn (Figure 2-1). People in other cultures imagined that the sun rides in a boat or tunnels through the Earth during the night, only to reappear each morning in the eastern sky.

These ancient observers imagined the sky as a vast bowl, spinning once each day. Although we know this picture is false, it nevertheless remains useful. If we think of the night sky as a great spherical shell—what astronomers call the **celestial sphere**—we can watch the bowl of night turn majestically, mirroring the Earth's rotation in the opposite direction (Figure 2-2). While the stars appear to move, they maintain their position on the celestial sphere relative to one another over many years. The distances between stars are so vast that even though the stars are moving through space, we cannot discern changes in their relative positions unless we use a telescope or wait many centuries.

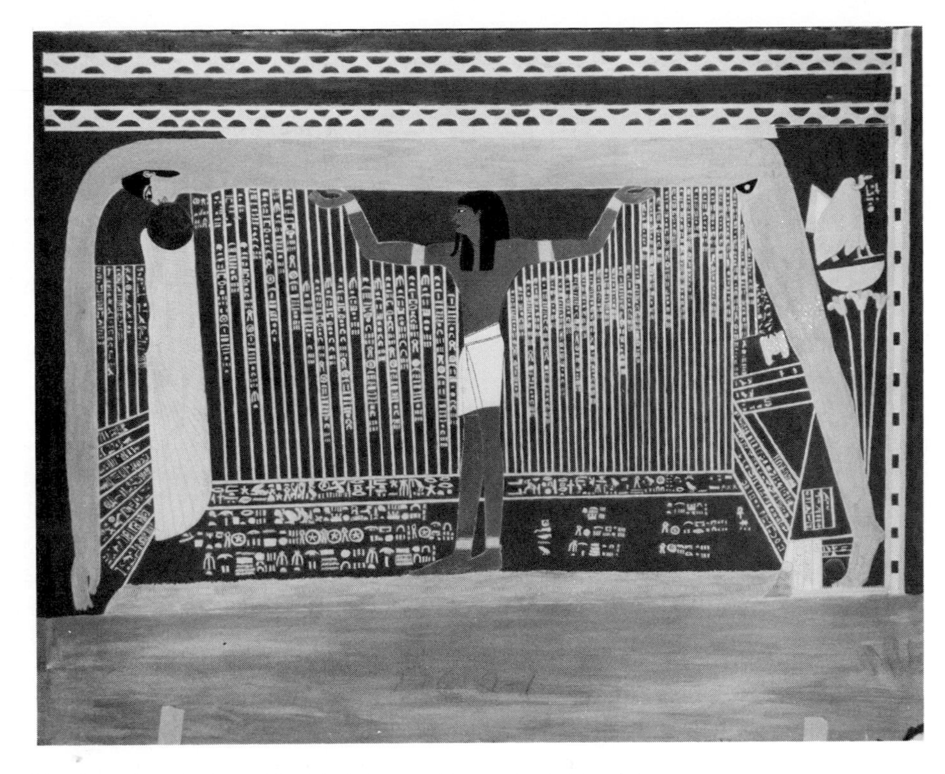

Figure 2-1 The Egyptians believed that Nut, the goddess who bridged the sky, swallowed the sun each evening, only to give birth to the sun again each morning.

The apparent rotation of the celestial sphere leaves just two points on the sky stationary: the north and south **celestial poles.** These two points, which are directly above the Earth's north and south poles, mark the places where the Earth's axis intersects the celestial sphere (Figure 2-2). An observer at a given **latitude** (degrees north or south of the Earth's equator, reaching 90° at the poles) will see the north or south celestial pole at a given angular height above the northern horizon (in northern latitudes) or southern horizon (in southern latitudes). This angular height, easily located at the center of the stars' apparent motion through the night, is the same as the observer's latitude. An observer in Chicago, for example, will see the north celestial pole at 42° altitude above the northern horizon. We are fortunate to have the bright star Polaris within 1° of the north celestial pole, providing us with a bright marker to help find the pole in the sky.

Almost all of the objects we see on the celestial sphere seem to maintain their relative positions. The most obvious exceptions are the sun and the moon; less obvious exceptions are the planets. Observers of the heavens noticed these exceptions, charting their courses among the "fixed" stars at least four thousand years ago, and probably well before that.

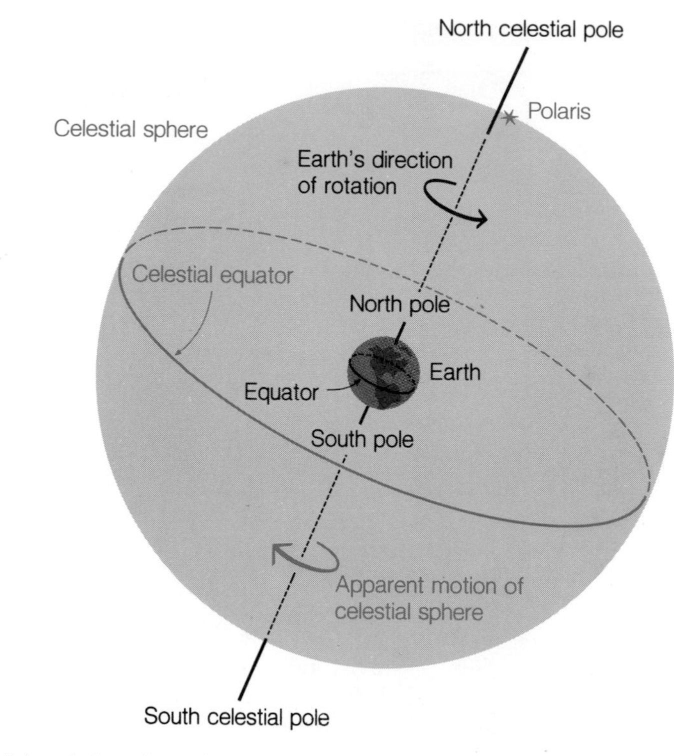

Figure 2-2 The Earth's rotation about the axis through its north and south poles makes the stars appear to move in the opposite direction. The apparent motion of the stars leaves two points stationary: the north and south celestial poles, directly above the Earth's north and south poles. Polaris, the brightest star in the Little Dipper, happens to lie within 1° of the north celestial pole.

2.3 The Sun's Path in the Sky

The yearly passage of the seasons and the corresponding changes in the sun's position in the sky must have captured the attention of those who depended on the sun for warmth and food. At various times of the year, the sun rises and sets at different points along the horizon, remains in the sky for differing lengths of time, and reaches different maximum altitudes in the sky. All of these changes occur because the Earth's rotation axis does not form a right angle with the plane of the Earth's orbit around the sun.

The Ecliptic

Ancient observers realized that at various times of the year the sun occupies different points on the sky with respect to the background of stars. They could see the different star groups at night during different seasons (Figure 2-3). These observers could reconstruct in their minds the groups of stars the

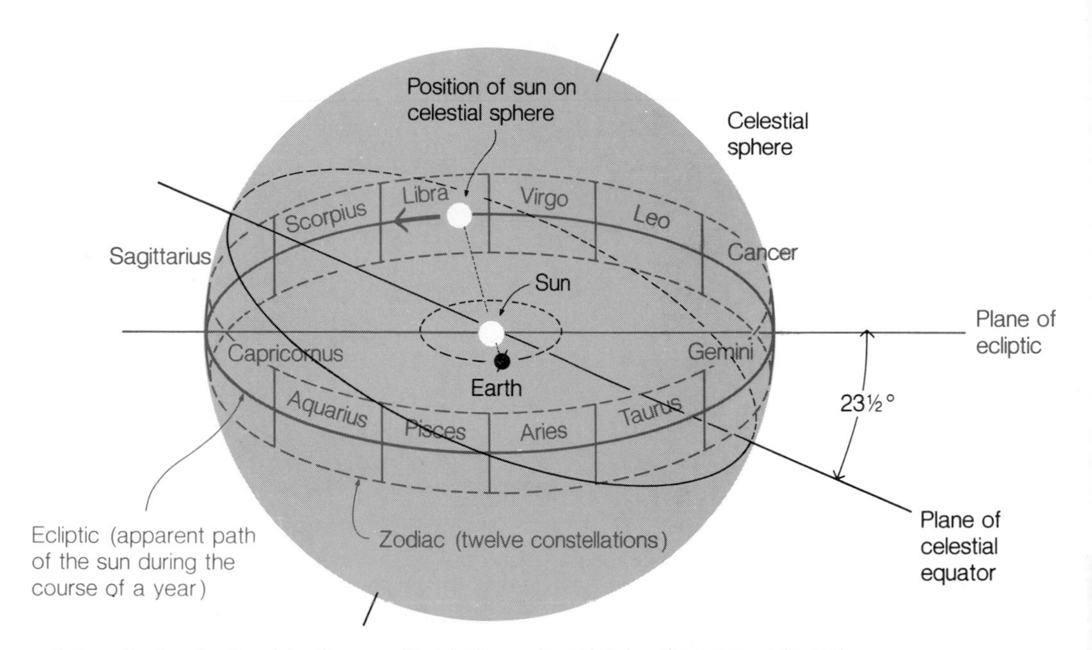

Figure 2-3 As the Earth orbits the sun, its rotation axis maintains the same orientation in space throughout the year. The rotation axis tilts by 23½° from the perpendicular to the plane of the Earth's orbit; therefore, the ecliptic, the apparent path of the sun around the sky during the course of a year, tilts by 23½° from the celestial equator. When the Earth is at different points along its orbit, we see different star groups at night, since the sun occupies a different place on the ecliptic.

sun passed through during the year using their memories of the intricate patterns formed by the stars in the sky. This impressive mental feat represents a peak achievement in many cultures and testifies to the importance ancient peoples ascribed to the motions of the heavens.

The apparent path of the sun among the stars is called the **ecliptic** (originally from the Greek). The ecliptic forms a circle around the sky that reflects the Earth's yearly motion around the sun (Figure 2-3). Every day the sun appears to move by one part in 365¼ of the way around the ecliptic. Since the Earth's rotation axis tilts by an angle of 23½° from the perpendicular to the Earth's orbit, the ecliptic tilts by an angle of 23½° from the celestial equator (Figure 2-3). If the earth's rotation axis were perpendicular to the Earth's orbital plane, the sun's position on the sky would always coincide with some point along the **celestial equator,** the points on the sky directly above the Earth's equator and halfway between the two celestial poles (Figure 2-2). As the Earth turns, different points along the celestial equator rise and set, but they always rise directly at the east point of the horizon and set directly at the west point (Figure 2-4).

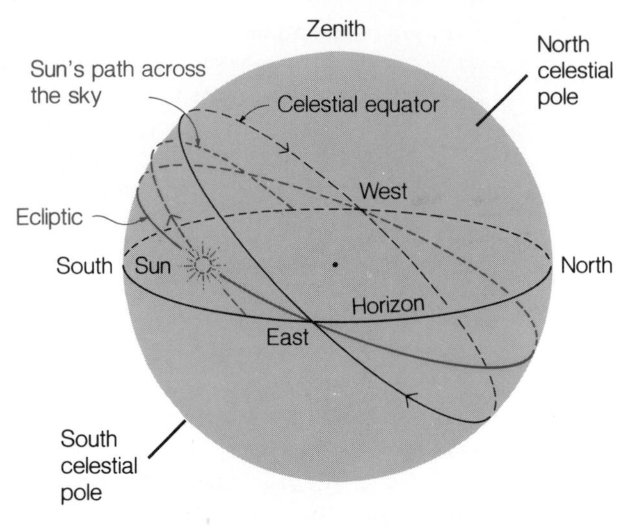

Figure 2-4 The celestial equator always rises at the east point on the horizon and sets directly at the west point. On the dates of the spring and fall equinoxes, the sun's position on the ecliptic places it on the celestial equator as well. On other days of the year, the sun's position on the ecliptic lies either north or south of the celestial equator.

The Seasons

The Spring and Autumn Equinoxes. As the sun moves along the ecliptic, it moves first to the north and then to the south of the celestial equator (Figure 2-3). North and south have useful meanings on the sky, since the north and south celestial poles are those points directly above the Earth's north and south poles (Figure 2-2). Each day the sun appears to move 1/365.2422 of the way around the ecliptic. On only two days each year, however, does the sun's position coincide with the celestial equator. On these days, called the spring and autumn **equinoxes,** there are equal amounts of day and night everywhere on Earth (hence the name *equinox,* meaning "equal night"). At the spring (or vernal) equinox (about March 21), the sun is heading north as it crosses the celestial equator; six months later, at the autumn equinox (about September 22), the sun crosses the celestial equator from north to south. The terms *spring* and *autumn* are correct for the northern hemisphere, but precisely backward for the seasons in the southern hemisphere.

The Summer and Winter Solstices. Midway between the equinoxes come the **solstices.** These are the times when the sun reaches a maximum deviation from the celestial equator and momentarily "stands still," as the name *solstice* implies, before reversing its direction of motion relative to the north and south celestial poles. The summer solstice (about June 22) brings the longest day and shortest night to the northern hemisphere, since the sun rises farthest to the north of east and sets farthest to the north of west of any day of the year. Winter solstice (about December 22) brings the shortest day to the northern hemisphere and the longest day to the southern hemisphere (Figure 2-5). The solstices also mark the sun's maximum and minimum altitudes above the horizon at the time when it crosses the **meridian,** the line from north to south that passes through the overhead point, or **zenith** (Figure 2-6). Since the vast majority of the world's population has always lived in the northern hemisphere (which contains most of the land), most people think of summer as the period of the year when it is summer (warmer) in the northern hemisphere and winter (colder) in the southern hemisphere. The naming of the solstices and equinoxes reflects this bias.

During the time between the spring and fall equinoxes, the sun always rises to the north of east and sets to the north of west, since the sun lies north of the celestial equator (Figure 2-6). From the autumn to the spring equinox, the situation reverses, and the sun always rises to the south of due east and sets to the south of due west. Observers within 23½° of the north or south pole (that is, within the Arctic and Antarctic Circles) will experience a period of anywhere from one day (just within these limits) to six months (at the poles) during which the sun never sets (Figure 2-7). These same observers will also experience an equal period of time, half a year later, during

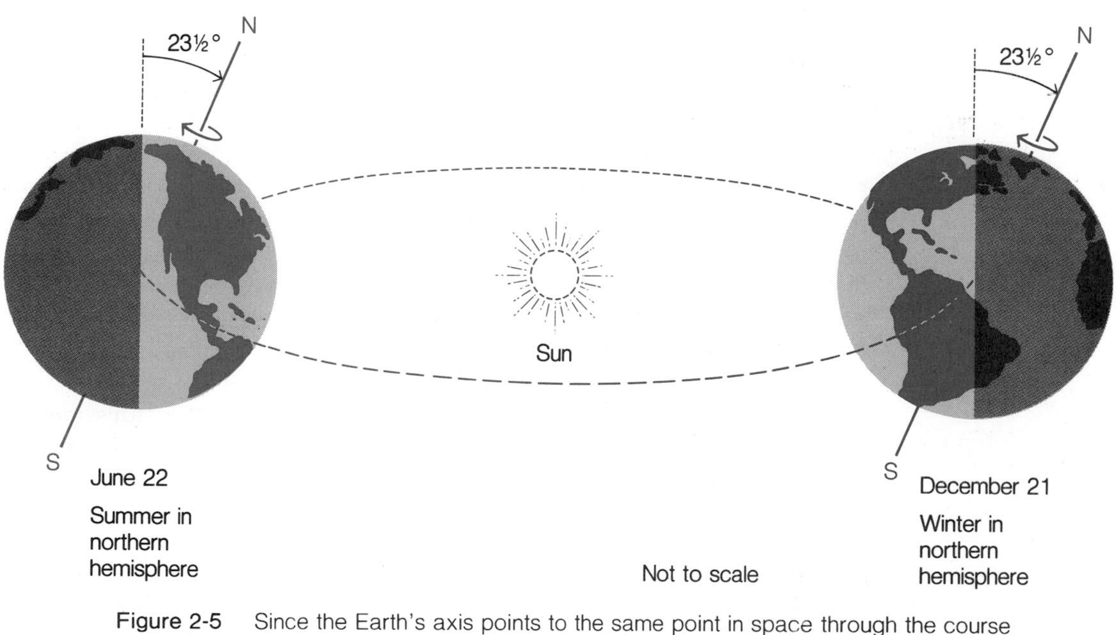

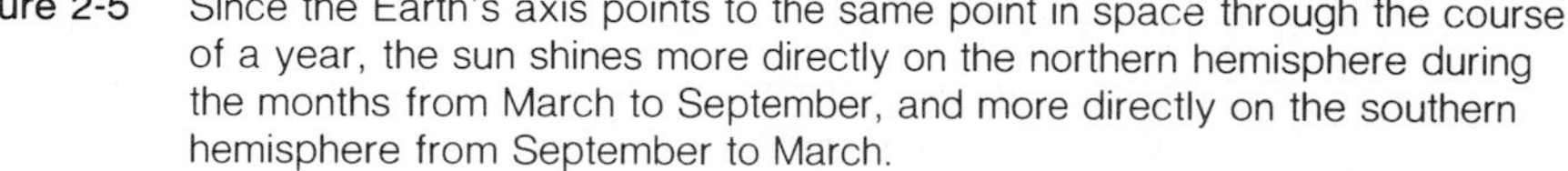
Figure 2-5 Since the Earth's axis points to the same point in space through the course of a year, the sun shines more directly on the northern hemisphere during the months from March to September, and more directly on the southern hemisphere from September to March.

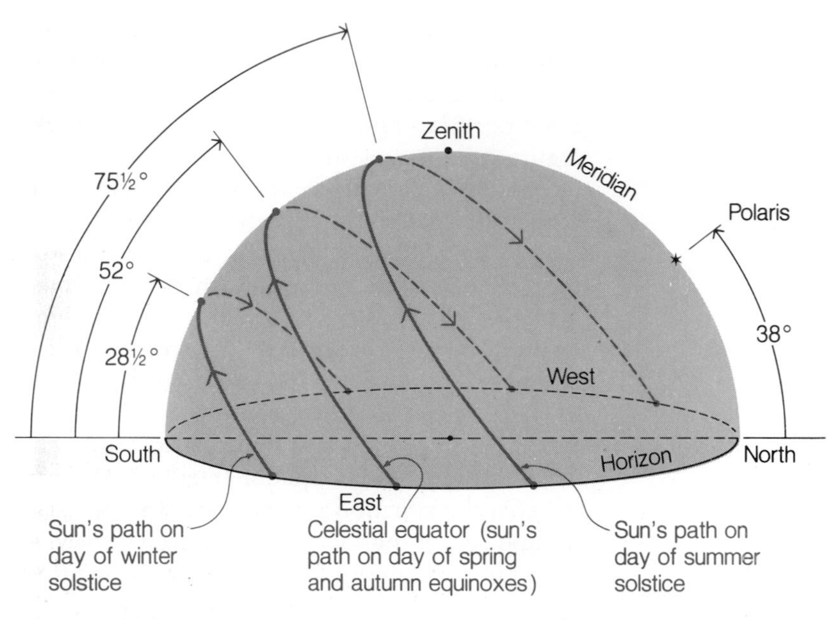

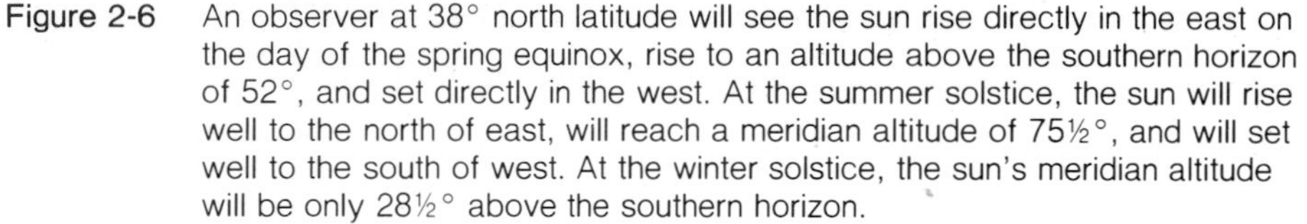

Figure 2-6 An observer at 38° north latitude will see the sun rise directly in the east on the day of the spring equinox, rise to an altitude above the southern horizon of 52°, and set directly in the west. At the summer solstice, the sun will rise well to the north of east, will reach a meridian altitude of 75½°, and will set well to the south of west. At the winter solstice, the sun's meridian altitude will be only 28½° above the southern horizon.

which the sun never rises. These long periods of daylight and darkness center around the summer and winter solstices.

Early Attempts to Chart the Seasons. Consider how ancient cultures matched the sun's position in the sky to the changing of the seasons. By using natural landmarks near the horizon—or by building artificial ones—they could note the points of sunrise and sunset on each day. The sunrise and sunset points farthest to the north marked the summer solstice, those farthest to the south marked the winter solstice, and those halfway in between marked the beginnings of spring and fall.

Associated with these changes are the changes in the sun's meridian altitude at noon, the time halfway between sunrise and sunset on any day. Ancient observers could determine the maximum altitude of the sun on a given day. Soon they realized that a correlation existed between the sun's excursions north and south as it rose and set and its altitude as it crossed the meridian.

This may seem a bit complex, but many ancient cultures determined not only the cycles of the month and year, but how to know when the equinoxes and solstices were occurring. The next step was to predict from past experience when the next equinox and the next solstice would occur. This ability was achieved by the Babylonians, the Mayas in Mexico, the Chinese, the Indians, and the Egyptians. The people who built Stonehenge in southern England around 1400 B.C. could predict the date of the summer solstice, and perhaps also eclipses and the phases of the moon (Figure 2-8). In A.D. 1000 the Anasazi (Pueblo) culture left a solstice and equinox marker carved near Chaco Canyon in what is now New Mexico (Figure 2-9).

Figure 2-7 This series of photographs was taken at one-hour intervals near the day of the summer solstice, at 70° north latitude in Norway. At its lowest point around the sky (at local midnight), the sun dipped to within 3½° of the horizon. If such a series of photographs were taken from the north pole on the summer solstice, the sun would maintain a constant altitude of 23½° above the horizon.

Figure 2-8 More than 3000 years ago, the great stone trilithons at Stonehenge were arranged in such a way that only at the summer solstice would the sun rise over the heelstone, as seen from the center of the stone circles. The arrangement of the stones also provided important sight lines for sunrises and sunsets, as well as for moonrises and moonsets, throughout the year.

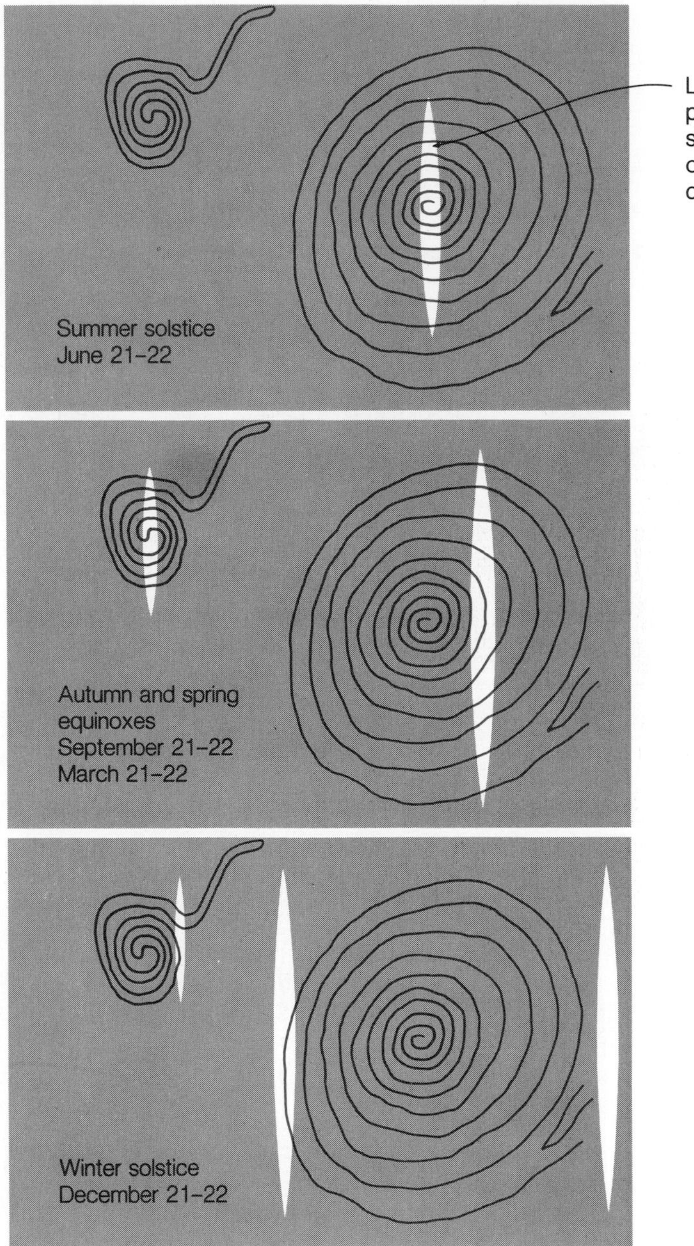

Figure 2-9 In 1979, three researchers found a solar observatory on a butte in Chaco Canyon, New Mexico. Three stone slabs were apparently chosen and arranged so that only around noon will any light pass through the slit between two slabs onto the third. The third slab of stone has two carved spirals, one of which is bisected by the slit of sunlight near noon on the day of the summer solstice, the other of which is bisected at the spring and autumn equinoxes. At winter solstice, two bands of light exactly frame the larger spiral, while a third band of light cuts the loop that extends from the smaller spiral just where the loop begins.

The Zodiac

Around 1000 B.C., Babylonian astronomer-priests divided the star groups along the ecliptic into the twelve constellations of the **zodiac** (Figure 2-3). In this way, the sun's yearly path along the ecliptic was divided into twelve equal intervals, each of which was believed to have special significance. The twelve constellations—the signs, or "houses," of the zodiac—formed the basis of Babylonian **astrology**, which we have inherited.

If the sun's 365.2422-day motion around the ecliptic turned out to equal a simple number of 29.53-day lunar cycles, our calendar would be simple. In fact, however, each year contains not 12 but 12.37 lunar cycles. Much of the history of the calendar consists of attempts to reconcile the twelve constellations of the zodiac with a lunar month that does not quite equal one-twelfth of a year. The Jewish calendar follows the phases of the moon, but extra months must be added in seven out of every nineteen years to keep the calendar in step with the seasons. The Christian calendar makes no attempt to follow the lunar phases and contains twelve months of varying durations. The Mohammedan calendar, on the other hand, obeys the lunar cycles exactly, with the result that a given holiday will occur at a different time each year.

2.4 Phases of the Moon

Perhaps the most obvious motion among the stars is that of the moon, which appears in a different place on the sky from one night to the next. As the moon moves among the stars, it also changes its **phase**, passing from a slim crescent through first quarter, waxing to full moon, and then waning through last quarter back to an invisible new moon (Figure 2-10). We now know that the lunar cycle of phases occurs because the moon orbits the Earth, or to be more precise, orbits the center of mass of the Earth-moon system. The sun always lights half of the moon, leaving the other half in darkness except for the small amount of light reflected by the Earth. As the moon orbits the Earth, always with the same side facing the Earth, we see more and more of the illuminated half until full moon, when we see the entire sunlit half. The moon takes 27.3 days to orbit once around the Earth, but since the Earth moves in its orbit during this time, it takes 29.53 days for the moon to regain the same phase, such as full moon.

Ancient cultures followed the lunar phases with keen interest, attaching great importance to the moment when the first sliver after new moon could be seen low in the western sky at dusk (Figure 2-11). Some societies believed that the sun and moon were conjugal enemies, as in the Indian myth that the moon was the sun's faithless bride, cut in two by her husband but sometimes allowed to shine in full glory.

The Moon's Path in the Sky

The plane of the moon's orbit around the Earth tilts by only 5° from the plane of the Earth's orbit around the sun (Figure 2-12). Therefore, the moon's

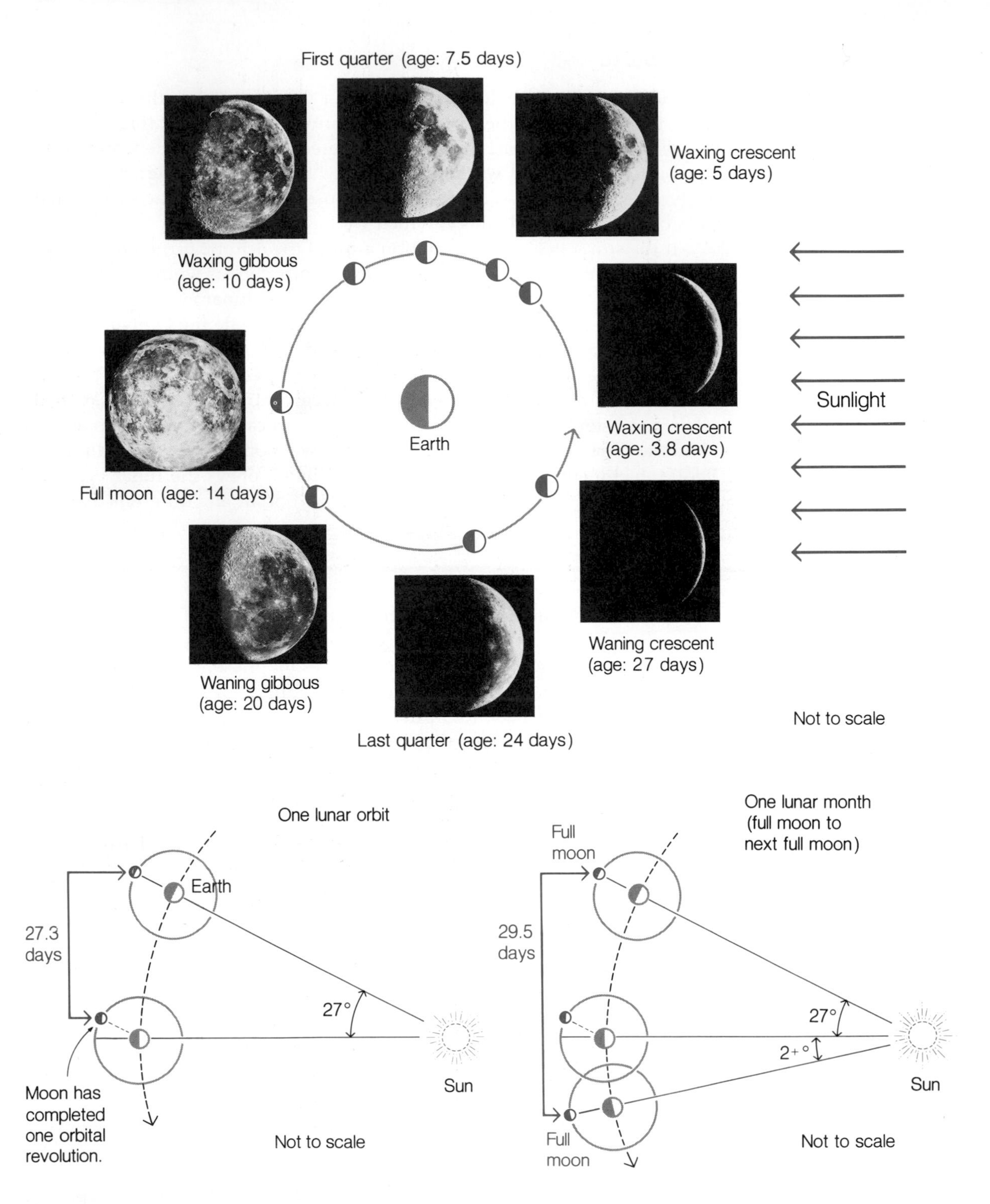
First quarter (age: 7.5 days)
Waxing crescent
(age: 5 days)
Waxing gibbous
(age: 10 days)
Sunlight
Full moon (age: 14 days)
Earth
Waxing crescent
(age: 3.8 days)
Waning crescent
(age: 27 days)
Waning gibbous
(age: 20 days)
Last quarter (age: 24 days)
Not to scale
One lunar orbit
Earth
27.3
days
27°
Sun
Moon has
completed
one orbital
revolution.
Not to scale
One lunar month
(full moon to
next full moon)
Full
moon
29.5
days
27°
2+°
Sun
Full
moon
Not to scale

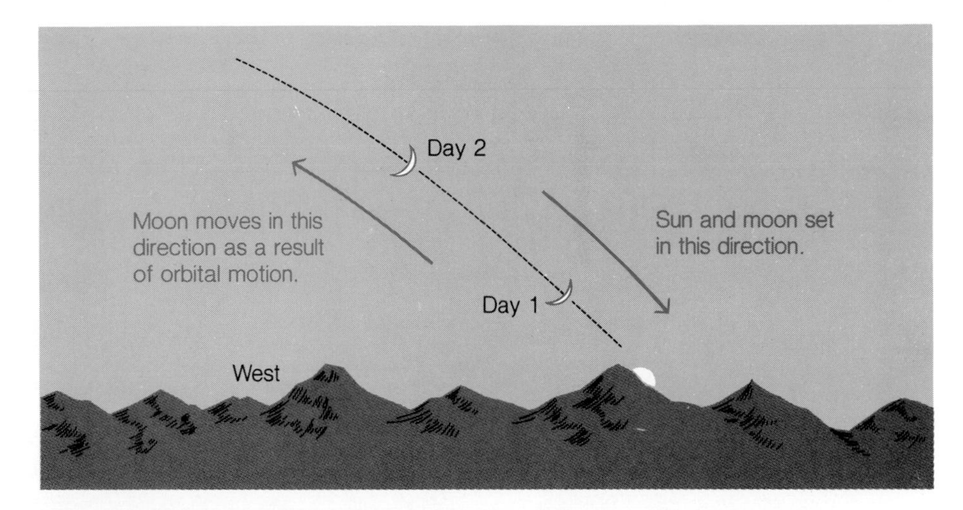

Figure 2-11 Most ancient cultures began their months with the day that the moon could first be seen, low in the west just after sunset. The moon's motion around the Earth carries the moon eastward in the skies (that is, in the direction opposite to the apparent motion caused by the Earth's rotation). The moon rises and sets 49 minutes later (on the average) on each day of the month than on the previous day.

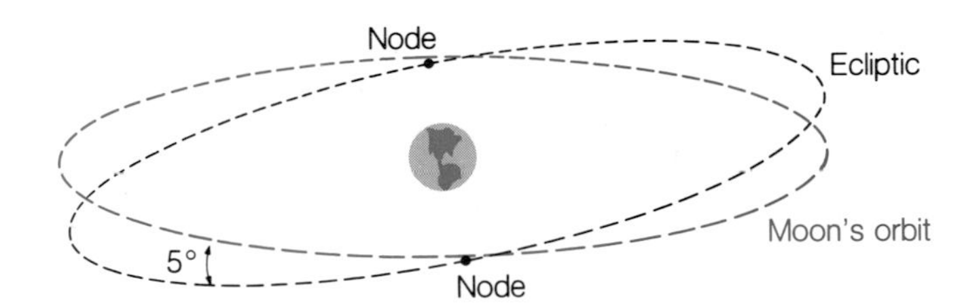

Figure 2-12 The moon's orbit around the Earth deviates by about 5° from the plane of the Earth's orbit around the sun. Therefore, the moon's orbit tilts by 5° from the ecliptic on the celestial sphere and crosses the ecliptic at just two points, called the *nodes*.

←
Figure 2-10
The moon's motion around the Earth causes us to see the *phases* of the moon, ranging from new moon, when the moon is nearly between the sun and the Earth, to full moon, when the moon is almost directly behind the Earth from the sun (top). Since the Earth and moon both orbit the sun, one complete orbit of the moon around the Earth (27.32 days) does not bring the moon back to the same point in its cycle of phases (bottom left). Instead, 29.53 days must elapse for the moon to recover the same point, relative to the sun and Earth, and thus to complete a full cycle of its phases (bottom right).

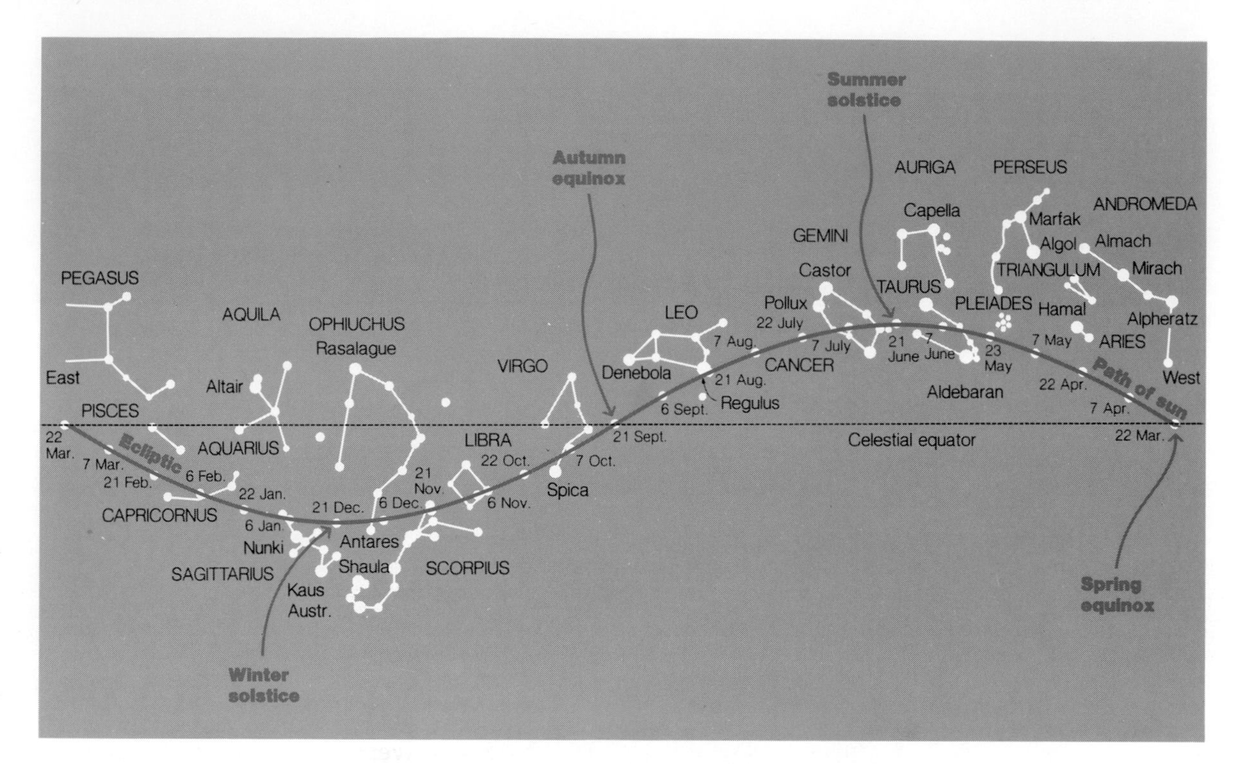

Figure 2-13 A Mercator-projection map of the celestial sphere centers on the celestial equator. The oldest division of the stars close to the ecliptic created 27 or 28 "houses." A later set of groups combined these houses into the 12 constellations of the zodiac, in which the sun was supposed to spend one month per constellation.

apparent path around the sky falls close to the ecliptic. In other words, the moon can never appear more than 5° from the ecliptic and usually it lies closer to it than this. The moon's path crosses the ecliptic at just two points. These points are called **nodes.**

The star groups close to the ecliptic gained attention as the home of the moon as well as of the sun. In fact, the oldest systems of star groups, which come from China, India, and Mesopotamia, divide the stars along the ecliptic into 27 or 28 groups. The moon would spend one day in each of these "houses" during its 27.3-day orbit around the Earth. These groups include such familiar patterns as the Pleiades, the horns of the Bull (Taurus), the feet of the Twins (Gemini), and the heads of the Twins (Figure 2-13). This division of the stars along the ecliptic, which dates back almost 4000 years, was eventually replaced by the twelve constellations of the zodiac.

Eclipses

As we discuss at some length in Chapter 10, the moon's orbit around the Earth occasionally brings it directly between the sun and Earth, or directly

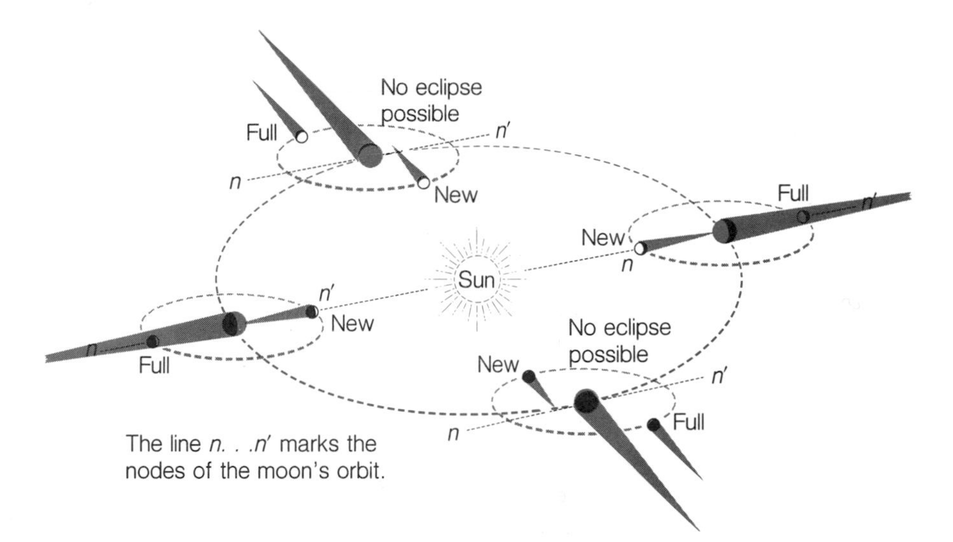

Figure 2-14
If full moon or new moon occur when the moon is close to one of the nodes of its orbit (see Figure 2-12), the moon can eclipse the sun (at new moon) or be eclipsed by the Earth (at full moon).

behind the Earth away from the sun (Figure 2-14). These positions produce **eclipses**, the temporary disappearance of the sun or moon. The word *ecliptic* is derived from the word *eclipse*, because eclipses can occur only when the moon happens to cross the ecliptic at the time of full moon (for an eclipse of the moon) or new moon (for an eclipse of the sun).

Clearly, these strange events of the heavens impressed ancient peoples. The oldest record we have of astronomy consists of a Chinese chronicle of two officials named Hsi and Ho who were executed for failing to perform the proper rites during an eclipse of the sun, apparently in 2136 B.C. We do not know, however, whether astronomers of that time were expected to predict eclipses (which demands years of observation, carefully recorded and skillfully extrapolated), or whether the execution of Hsi and Ho was simply meant to encourage their successors to determine how to make such predictions. The Chinese were the first culture to centralize recordkeeping, and most of our knowledge of astronomical events before 500 B.C. goes back to Chinese chronicles. On the other hand, we have some records from Babylonian times as far back as the reign of Sargon of Akkad (3800 B.C.), showing that even then the heavens were systematically studied.

2.5 The Rising of the Stars

All of the ancient cultures recognized the importance of the solar cycle to their way of life. To keep track of the sun's changing position on the sky, many early societies looked to the stars for their time and seasons. Since the Earth's motion around the sun makes the sun appear to move among the stars, different stars will appear at nightfall in a particular part of the sky as the year progresses. As Figure 2-15 shows, most stars will rise 4 minutes (1/365 of a day) earlier on each successive evening, because the Earth's motion in its orbit will have changed the angle between the star and the sun by 1/365 of a full circle in 24 hours.

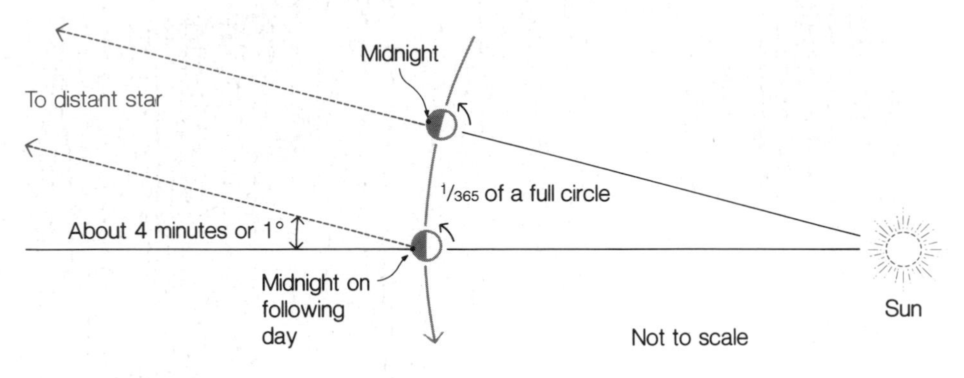

Figure 2-15
Every day the Earth moves about $^1/_{365}$ of the way around the sun. Therefore, a given point on Earth takes slightly longer to regain the same orientation with respect to the sun (24 hours) than the time it takes to regain the same orientation with respect to the stars (23 hours 56 minutes). The difference of four minutes represents the amount by which the stars gain on the sun in the time of rising and setting, until the passage of a full year brings the sun and stars back into their original orientation.

Star-Based Calendars

The Egyptians, who depended on the annual rising of the Nile for their agriculture, used a star-based calendar. The brightest star of the night, Sirius, could be seen in the east, just before sunrise, at the time when the Nile began its annual flooding. This valuable discovery led to another. The Egyptians used a 365-day year, while the actual interval between successive spring equinoxes is 365.2422 days, much closer to 365¼ days. Thus, the rising of Sirius just before dawn advanced by one day every four years on the 365-day calendar, which slowly fell out of step with the sun and the stars. After 1460 years (4 × 365) the calendar was back in step, and Sirius again was first seen at the proper date by the calendar. This led to the legend of the holy bird Phoenix, which consumed itself with fire and rose up out of the ashes every 1460 years.

Many other cultures besides the Egyptians regulated their lives by the rising and setting of the stars. The people of Java watched for Orion's belt. When it could no longer be seen at night, all work in the fields ceased. When it again appeared before dawn, the agricultural year began once more. The original inhabitants of Australia also derived their calendar from the stars. They knew that their spring (our fall) begins when the Pleiades can first be seen in the evening twilight.

Circumpolar Stars

Not all stars rise 4 minutes earlier each night; **circumpolar** stars never set at all. If a star's position on the sky places it close enough to the north celestial pole, then an observer in the northern hemisphere will see the star above the horizon at all times during the night (Figure 2-16). In order for a star to appear circumpolar ("around the pole") to a given observer, the star must have an angular distance from the north celestial pole no greater than the observer's latitude. This holds true because any observer sees the north celestial pole at an altitude above the northern horizon equal to the observer's latitude (Figure 2-16).

Circumpolar stars—and in particular the stars closest to the north celestial pole—played an important role in the navigation of the seas. To the ancient sailor these stars were a kind of compass, pointing the way north.

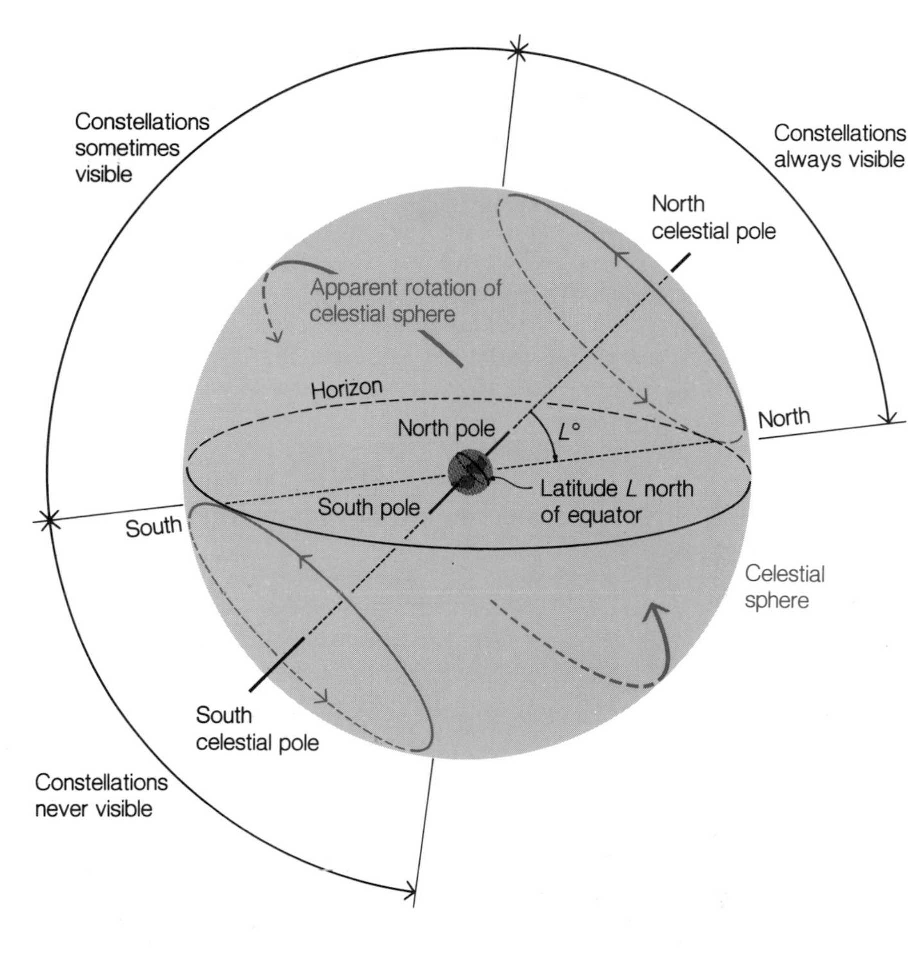

Figure 2-16
At a given latitude, L, north of the equator, the north celestial pole will maintain an altitude of $L°$ above the northern horizon. Stars within $L°$ of the north celestial pole will neither rise nor set, since the Earth's rotation never makes them dip below the horizon. Stars farther than $L°$ from the north celestial pole but also farther than $L°$ from the south celestial pole will appear above the horizon for part of the day before setting. Stars within $L°$ of the south celestial pole will never rise above the horizon.

2.6 Motions of the Planets

If you keep a careful watch on the heavens, you will soon notice a few unusually bright objects moving through the background of stars. Of course, all stars appear to move as the Earth turns, so that a single night carries them from east to west, and they all rise and set at different times during the year. But once you have worked out the rhythm of these motions, you will still find exceptional objects, the **planets,** which change their position in a matter of days, or in some cases weeks, with respect to the stars (Figure 2-17). The planets ("wanderers" in Greek) move against the backdrop of stars because they move in orbit around the sun, just as the Earth does. Much of the history of astronomy involves the effort to understand the motions of the sun, moon, and planets by observing these wanderers in the night sky.

Early Observations

The ancient astronomers could observe five planets wandering among the stars: Mercury, Venus, Mars, Jupiter, and Saturn (to use their Latin names).

They always appeared close to the ecliptic, though only occasionally exactly on it, and they moved in the same general direction along the ecliptic as the sun and the moon. But it must have seemed strange that the three planets that took the longest to move around the ecliptic—Mars, Jupiter, Saturn—would occasionally reverse direction along the ecliptic for a while before resuming their normal, eastward movement among the stars (Figure 2-18). The planets Mercury and Venus also undergo such reversals, but only at times when they are close to the sun on the sky and thus difficult to observe.

Imagine how the motions of the planets must have mystified these observers, who typically had come to worship the sun and moon as gods. The planets stood out as objects that moved among the stars. Though these motions had a rather simple basis—along the ecliptic from west to east—the patterns were actually complex. The motions did not occur precisely along the ecliptic and even showed a **retrograde,** or backward, tendency for three of the five planets, reversing direction for anywhere from a few weeks to a few months during the course of a year (Figure 2-17).

Astrology

Is it any wonder that most ancient cultures came to identify the planets with their own gods and to marvel at the mystery of the skies? To the regular

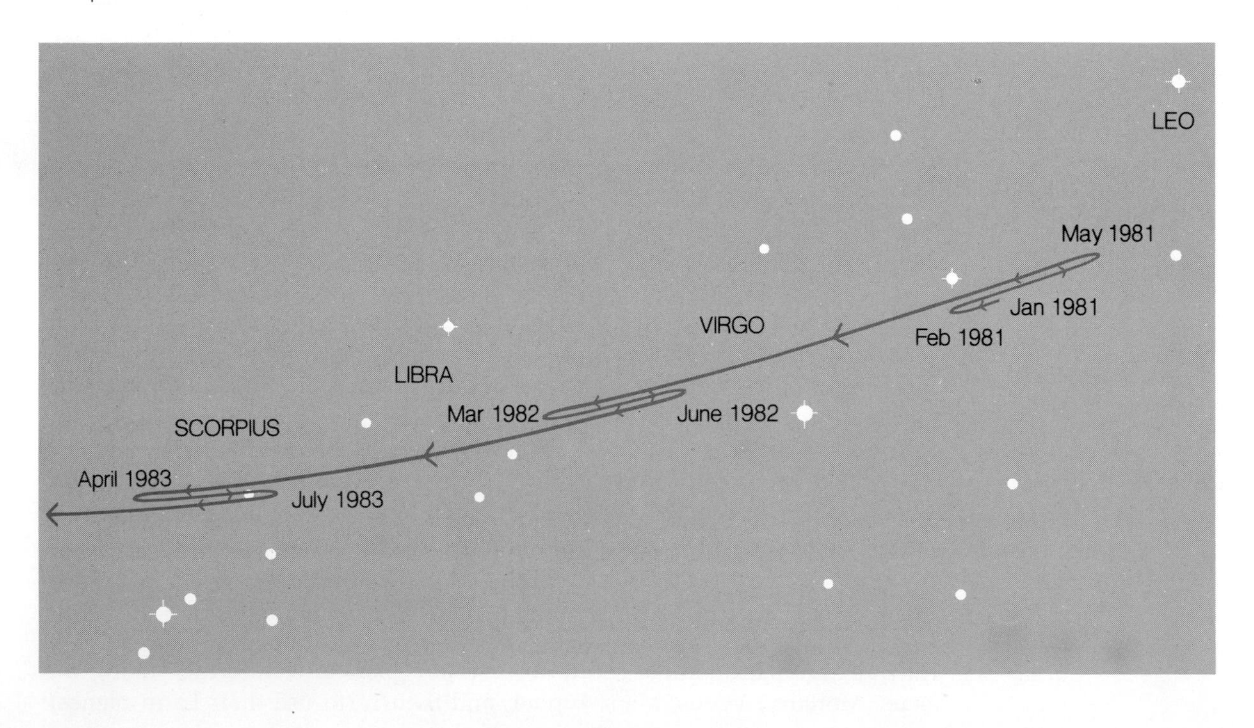

Figure 2-17
The apparent path of the planet Jupiter among the stars in 1982 and 1983 will have Jupiter "wandering" among the stars in the constellations Virgo, Libra, and Scorpius.

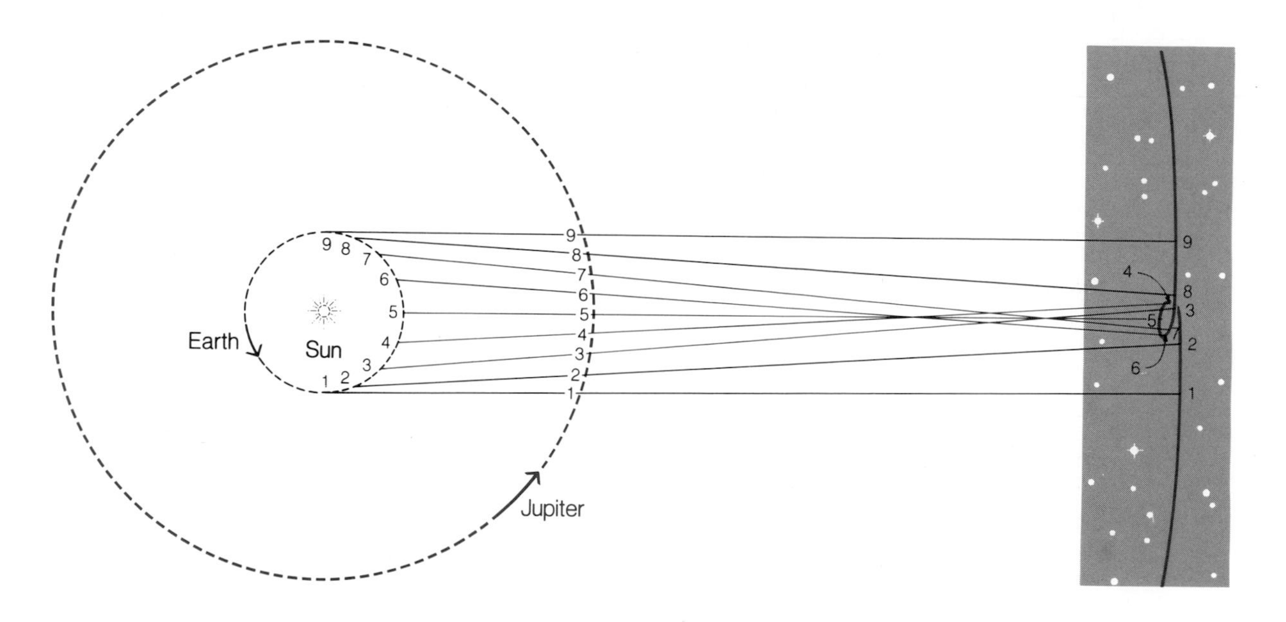

Figure 2-18
The retrograde, or backward, motion of the planets farther from the sun than the Earth occurs as the Earth, moving faster in its orbit, overtakes the outer planets. The diagram shows the position of Jupiter against the background of stars at nine points of its orbit, with the corresponding nine points of the Earth's orbit also marked. The period of retrograde motion extends from points 4 through 6. During this period, Jupiter appears to move backward (westward) among the stars, in contrast to its usual (eastward) motion during the rest of the year.

cycles of day and night, of full and new moons, of alternating seasons, they now had to add wandering planets. In China, in Egypt, and in Babylonia, people began to see their fortunes as tied to the strange motions of the planets. In these societies, however, the ruler was everything and the individual nothing; thus most of astrology related only to the state. The Chinese saw the emperor as the Son of Heaven, who could easily be offended by events on Earth. The Assyrians merged their interest in timekeeping with their belief that the spirits who ruled the Earth dwelt in the heavens, where they provided omens of good or evil for humanity.

The notion that everyone, not just the prince or the state, has a right to a horoscope had to wait for a different, more humanistic, society: the ancient Greeks. To the Greek culture of the sixth century B.C. through second century A.D. we owe not only the beginnings of modern astronomy but also the daily astrological columns of our newspapers. The Romans in particular loved the Greek idea of individual horoscopes—deductions of a person's life story from the position of the planets and moon at the moment of birth. The Emperor Augustus even issued a coin that showed his birth sign (Capricorn) to imply that he owed his fortune to the sun and stars.

The astrology that evolved from the Babylonians and Greeks puts major emphasis on the position of the sun in the zodiac at the time of a person's birth (the "birth sign") and on the constellation of the zodiac that is rising at this moment (the "rising sign"). To these basic determinants a host of omens that derive from the positions of the planets and the moon around the ecliptic must be added and interpreted, an activity that has made astrologers rich for thousands of years.

The Greek words *astronomy* and *astrology* distinguish between the "nomos," or human law, and the "logos," or divine law, that we ascribe to the stars. In other words, astronomy refers to a human understanding of the heavens, astrology to the attempts to find the divine pattern behind what we see. As astronomy, the human law, separated from astrology, the divine law, people naturally found the latter more appealing, since it dealt with their own lives. Some astronomers have resented this fact; others have merely looked on with amusement.

2.7 Early Models of the Cosmos

Possessing an intense interest in the heavens, early astronomers and astrologers (often one and the same person) made the motions of the planets a subject of deep inquiry. The ancient Greeks were apparently the first people to question what the heavens consist of, and how the motions we see can be explained by a **model**, a mental construct that duplicates what goes on in the universe. Today it seems logical to seek a physical explanation for the motions of celestial objects, but this was a bold step for the ancient world.

A Round Earth

The first great astronomical discovery of Greek culture was that our Earth is *round*, a fact known to Pythagoras and his followers in the sixth century B.C. The word *geometry* means "the measurement of the Earth," and beginning with Pythagoras, through six centuries of Greek thought, a succession of great geometers attempted to measure the Earth and even the cosmos around it.

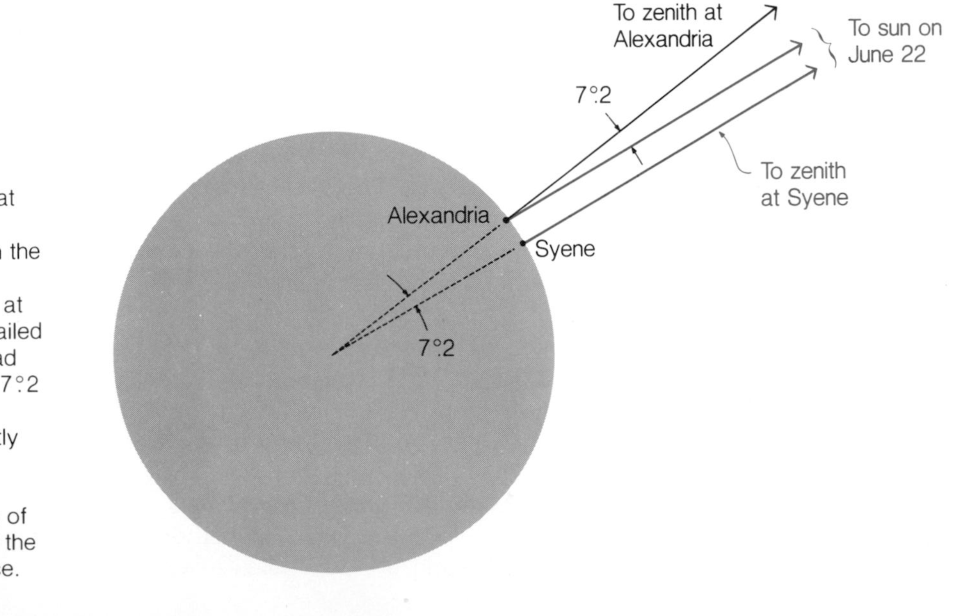

Figure 2-19
Eratosthenes saw that the sun was directly overhead at noon on the summer solstice at Syene, in Egypt; but at Alexandria the sun failed to reach the overhead point on this day by 7°.2 (1/50 of a circle). Eratosthenes correctly concluded that the amount by which Alexandria lies north of Syene equals 1/50 of the Earth's circumference.

You can tell that the Earth is not flat by looking at the sun and the stars as you travel from one place to another. A change in your latitude (distance north or south of the equator) will bring a change in the height of the north celestial pole above the horizon (Figure 2-4). This, in turn, will produce a change in the maximum altitude the sun reaches above the horizon on a given day of the year. In the third century B.C., the geometer Eratosthenes used this fact to measure the size of the Earth. Eratosthenes observed that the sun's altitude above the horizon at noon on the day of the summer solstice was different at Alexandria than at Syene. This difference came to 7°.2, or 1/50 of a circle (Figure 2-19). Eratosthenes then calculated that the distance from Alexandria to Syene must equal 1/50 of the Earth's circumference—assuming, of course, that the world was round.

Sun-Centered Versus Earth-Centered System

Once the idea of a spherical Earth took hold, it was a relatively straightforward step to imagine the Earth in daily rotation. If the Earth turns on its axis once each day, the heavens will appear to turn in the opposite direction. This fact was clearly explained by Heraclides, a pupil of the famous philosopher Plato, about the year 350 B.C. Heraclides went on to consider the motions of the planets, and in particular the fact that Mercury and Venus never appear far from the sun on the sky: no more than 28° for Mercury, no more than 47° for Venus (Figure 2-20). Heraclides suggested that Mercury and Venus move around the sun, not the Earth, as indeed they do.

A century after Heraclides, Aristarchus of Samos went on to claim that all the planets, including the Earth, orbit the sun. Although Aristarchus did not realize it, his **heliocentric**, or sun-centered, model of the solar system provides a logical explanation of the planets' retrograde motion (Figure

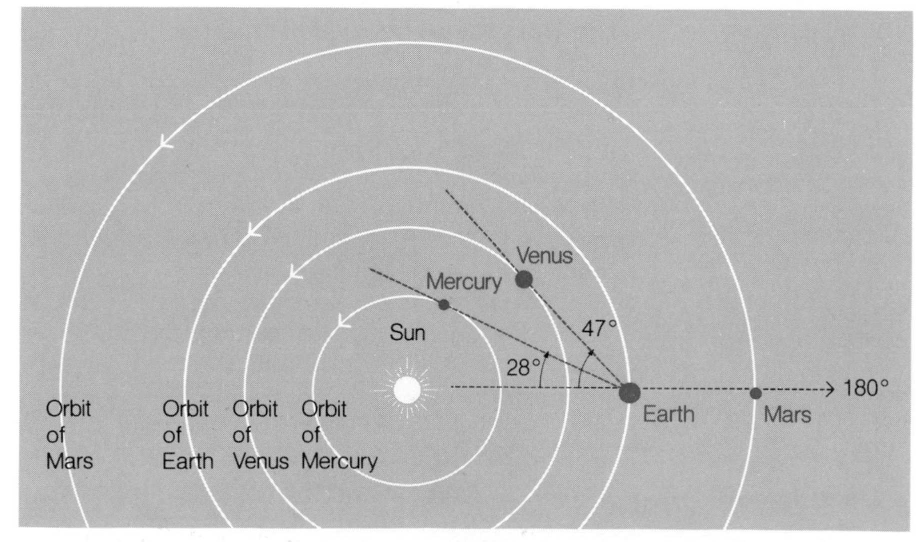

Figure 2-20
The planets Mercury and Venus, which orbit the sun inside the Earth's orbit, can never appear in the direction opposite to the sun on the sky. The greatest angular distance between Mercury and the sun equals 28°; between Venus and the sun, 47°. In contrast, Mars, Jupiter, and the other outer planets can appear as much as 180° from the sun.

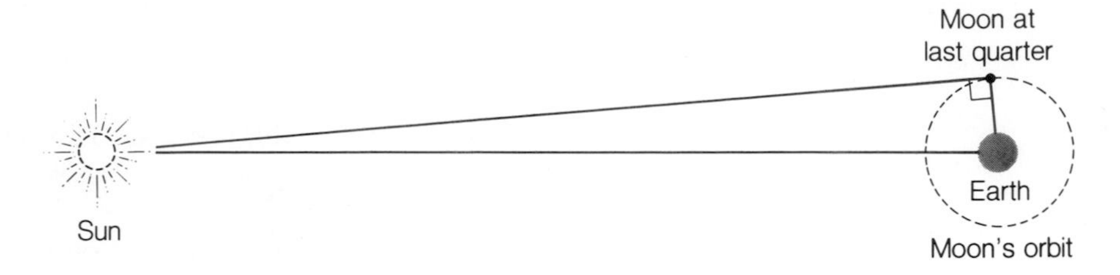

Figure 2-21 Aristarchus realized that the moment of first- or last-quarter moon, when we see half of the moon's illuminated side, represents the time when the Earth and the sun are at right angles as seen from the moon. By determining the angle between the sun and moon, as seen from the Earth at this moment, Aristarchus hoped to use trigonometry to determine the relative distances of the sun and the moon to Earth. Aristarchus estimated that the sun is 20 times farther than the moon; the correct ratio is close to 400.

2-18). But Aristarchus was more interested in finding the distances to the sun and the moon, and he devised a brilliant method to do so. By letting the Earth, sun, and moon represent the three vertices of a triangle, he was able to show, by trigonometry, that the sun's distance from us far exceeds that of the moon (Figure 2-21).

Aristarchus's heliocentric model did not gain general acceptance, and later astronomers continued to place the Earth at the center of the cosmos. Hipparchus, the greatest astronomer of the second century B.C., was able to use this **geocentric,** or Earth-centered, model to make accurate predictions of eclipses and to discover the luni-solar precession (page 389). Hipparchus also compiled the first accurate catalog of star positions, which, with some slight improvements by the astronomer Ptolemy, remained the best reference of its kind for the next 1700 years.

The Epicyclic Theory of Ptolemy

The Greek astronomer with the longest-lasting influence was Claudius Ptolemy of Alexandria, who lived three centuries after Hipparchus. Ptolemy absorbed the knowledge of his predecessors and published it, with his own improvements, in a great work entitled *Syntaxis* ("arrangement"). We know this book by its Arabic title *Almagest* ("the greatest"), a tribute to the Arabic scientists who kept learning alive in the Mediterranean area once Greek and Roman civilization had disappeared.

Ptolemy's *Almagest* provided the first comprehensive model of the universe. Although wrong in many respects, this model deserves admiration for the sweep of its coverage and its attempt to explain the cosmos in geometrical terms. Ptolemy outlined a universe in perpetual circular motion, all in harmony, the ideals of geometry made real in the heavens. Since the time of Hipparchus, astronomers had known that the motions of the planets did not form exact circles around the Earth, for the length of time the planets spent at various points near the ecliptic did not correspond to the predictions of a

The Brightness of Stars

When Claudius Ptolemy summarized all astronomical knowledge during the second century A.D., he continued the tradition of Hipparchus in assigning each visible star to one of six **magnitudes.** Stars of the first magnitude had the greatest brightness; those of the second magnitude were somewhat fainter; and so on down to the stars of the sixth magnitude, faintest of all. Astronomers still use this system. With the help of telescopes, they can now speak of tenth-magnitude stars, eighteenth-magnitude stars, and even twenty-third-magnitude stars.

Two facts about the system of stellar magnitudes give immediate difficulty. First, the *lower* the number of the magnitude, the *greater* the brightness of the star. This may contradict our desire to link higher numbers with greater brightnesses. Second, the magnitude system is logarithmic; that is, it works by *ratios* of brightness. The human eye responds to brightness differences by noting the ratio of one object's brightness to another's. Each step in magnitude turned out to correspond to a ratio of about 2½ in brightness. Thus a first-magnitude star, on the average, had 6¼ times the brightness of a third-magnitude star. Astronomers later defined the magnitude scale by deciding that a difference of 5 magnitudes in brightness would correspond to a brightness ratio of 100. Then each magnitude step would correspond to a brightness ratio of 2.512, the fifth root of 100.

Hipparchus, Ptolemy, and the astronomers who followed them dealt with the *apparent* brightness of a given star, that is, with how bright the star appears to us. For any star, the apparent brightness is a readily measurable quantity, and with the magnitude system we can specify it with an **apparent magnitude.** Deneb has an apparent magnitude of 1.23; Polaris, a second-magnitude star, has an apparent magnitude of 2.3. The brightest objects have *negative* apparent magnitudes; that is, they are brighter than first-magnitude stars. Thus Sirius has an apparent magnitude of −1.45, and Canopus has an apparent magnitude of −0.7. Canopus is therefore just 3 magnitudes, or 16 (2.512 × 2.512 × 2.512) times brighter than Polaris. The sun has an apparent magnitude of −26.7 and the full moon has an apparent magnitude of −12.5.

A star's **absolute magnitude** gives the apparent magnitude that the star would have if it were 10 parsecs from us. The sun's absolute magnitude is 4.8, greater than those of an overwhelming majority of the stars we can see without a telescope. "Greater" means fainter; so we may conclude that the stars of our night skies represent stars that are brighter than average, especially in the case of Rigel (absolute magnitude −7.0), Deneb (absolute magnitude −7.3), and Betelgeuse (absolute magnitude −6.0). Sirius has an absolute magnitude of 1.4, while Polaris has an absolute magnitude of −4.6.

The advantage of having a logarithmic scale of brightness for both absolute and apparent magnitudes becomes clear when we consider how the two kinds of magnitude are related. The algebraic answer is that for any star,

$$\text{Absolute magnitude} = \text{Apparent magnitude} + 5 - 5 \log D$$

where D is the star's distance in parsecs. Notice that if $D = 10$, the absolute and apparent magnitudes are equal, as we expect from our definition of absolute magnitude. The formula makes it easy to find the distance of a star once we know its absolute and apparent magnitudes. Likewise, if we know the distance and the apparent magnitude, we can immediately calculate the absolute magnitude.

In this book, we shall use the term **apparent brightness** rather than the more formal *apparent magnitude.* Similarly, we shall talk of a star's **luminosity,** or *true brightness,* which refer to what astronomers call the absolute magnitude. It might be easier always to use the term **flux** in place of *apparent magnitude,* to remind us that we are dealing with an incoming stream, or flux, of photons at the point of measurement. This flux clearly varies in amount as we vary our distance from the source of photons. But since the word *flux* seems a strange way to refer to the brightness of an object, we shall stick to the term *apparent brightness.*

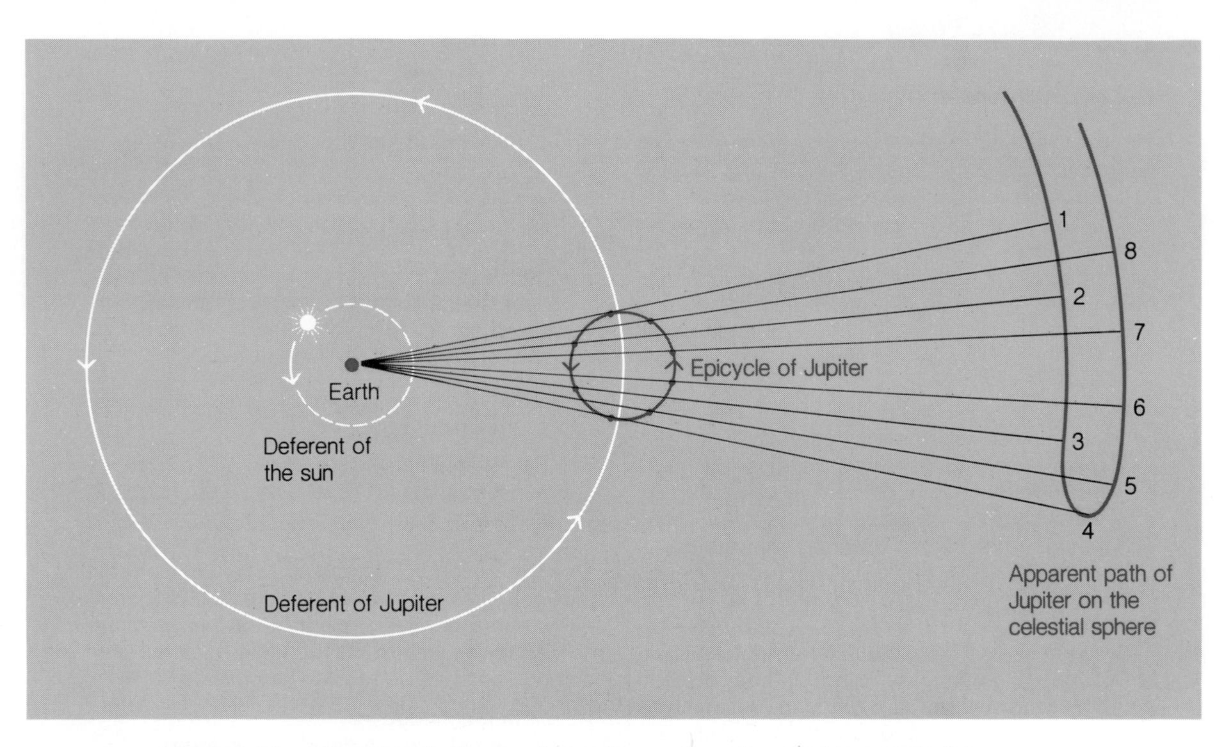

Figure 2-22 The epicyclic theory of planetary motion attempted to explain the retrograde motion of the outer planets by assuming that each of these planets moved around a small circle, or epicycle, whose center moved around the sun in a large circle, called the deferent. This assumption, if true, would indeed explain the retrograde motions observed, even though the Earth did not move at all.

theory based on circular orbits. Hence, Ptolemy advanced his theory of **epicycles,** small circles superimposed upon the large orbital circles (Figure 2-22). Planets would deviate from their basic orbits to make excursions on these epicycles, thus explaining their retrograde motions on the sky.

With the introduction of epicycles, Ptolemy could make a fairly good match between his theory and the observed motions of the planets. The epicyclic theory, based on a geocentric model of the cosmos, therefore remained in vogue for thirteen centuries. In Ptolemy's model of the universe, the stars lay beyond the planets, on the outermost sphere of the heavens. This sphere would rotate once each day, for Ptolemy rejected the idea that the Earth would rotate, imagining that such rotation would cause objects to fly off the surface.

With Ptolemy, Greek thought reached its greatest expression of the idea that mathematics, and geometry in particular, could explain the cosmos and its motions. This idea is perhaps the greatest gift of Greek culture to modern science.

2.8 Islamic Astronomy

During the Roman Empire, scholars such as Ptolemy had been able to work and study free from everyday strife. With the decline and fall of the Empire, the Mediterranean world passed into chaos, and it emerged by meeting the new religion of Islam. The scientists of the Islamic world, which included Spain, Persia, North Africa, and northern India during the seventh and eighth centuries A.D., enthusiastically adopted the Greek passion for mathematics and astronomy. They translated classic works on these subjects into Arabic, studied them, and improved upon them.

Today we pay tribute to Islamic science mainly through our use of "Arabic" numerals, which replaced the far more cumbersome Roman system. These numerals were in fact taken from the Hindu cultures of India, although Arabic scientists took the important additional step of introducing the zero ("zifr" in Arabic). Using these numerals, the greatest Islamic astronomer, Muhammad al-Khwarizmi, produced astronomical tables during the eighth century A.D. Somewhat revised, these tables were used from Spain to China. Another of al-Khwarizmi's works introduced the Arabic term *algebra* for an important branch of mathematics.

Despite their skills at building observatories and measuring the positions of objects in the sky, Islamic scientists did not produce any new astronomical theories. From our vantage point, the greatest contribution of Islamic science was the preservation and nourishment of the tradition of scientific inquiry begun by the Greeks. During the Italian Renaissance of the twelfth and thirteenth centuries, when the Dark Ages of Europe slowly evolved into the modern era, the Arabic translations of Greek books provided a priceless heritage to European scholars. Not only Ptolemy's works, but also those of Aristotle, Euclid, Archimedes, Hippocrates, and a host of other Greek scholars, first came to Europe through translations from Arabic. The preservation of these works testifies to the Islamic love of books and knowledge, as illustrated by the "house of wisdom," or library, which the Caliph al-Mamun established in Bagdad in A.D. 830. The caliph enlarged his library by paying translators the weight in gold of the books they rendered into Arabic.

2.9 Modern Models of the Cosmos

The rediscovery of ancient learning began in Europe with the Italians and spread northward and westward to France, Germany, England, and Scandinavia. From the beginning of the Renaissance until well into the eighteenth century, a typical "scientist" was likely to be a clergyman or priest with a good education in classical texts, an interest in nature, and the time to think about the world and perhaps to perform experiments. In this traditional role we find the man who led astronomy from its ancient confines into the modern epoch: Nicolas Copernicus (Figure 2-23).

Figure 2-23 Nicolas Copernicus, from a sixteenth-century woodcut.

Copernicus

Copernicus was born in Poland in 1473, twenty years after the Turks conquered Constantinople and nineteen years before Columbus rediscovered America. He studied in Italy and returned to Poland in 1503 as a minor official of the Cathedral of Frauenburg in East Prussia. There Copernicus continued to develop his interest in astronomy, an interest that was further motivated by the calendrical needs of his church.

The most important Christian holiday, Easter, falls on the first Sunday after the first full moon after the spring equinox. But there was an error of 11 days between the true spring equinox and the equinox according to the calendar. The calendar in use at that time had been introduced by Julius Caesar fifteen centuries earlier. This calendar assumed a seasonal year of 365¼ days, allotting 365 days to the calendar year with an entire day added to February every four years. The error in the Julian calendar arose from the fact that the true seasonal year equals 365.2422 days, and the calendar was therefore providing too many leap years. When Copernicus was an infant, Pope Sixtus IV had asked the greatest astronomer of the age, Regiomontanus, to resolve this error, but Regiomontanus died in the plague at Rome before he could do so. The actual reform of the calendar did not occur until 1582. Meanwhile, the continuing use of a calendar known to be wrong may have led Copernicus to wonder about the heavenly motions on which the calendar was based.

For 30 years, Copernicus attempted to derive from observations an accurate picture of the planets' orbits and to use his model to predict the future motions of the planets. The result of his labor was the rediscovery of Aristarchus's model: The planets all orbit around the sun. In Copernicus's model, the planets move in circular orbits (Figure 2-24), and like Ptolemy, Copernicus had to invoke epicycles to reconcile his theory with the observed positions of the planets over the course of many years. Today we know that although Copernicus was entirely right in supposing the planets to orbit the sun, the actual orbits are not circles but **ellipses** (see page 58).

The Gradual Acceptance of the Copernican Model

Knowing that his theory would meet opposition from theologians, Copernicus delayed publication of his great book on the solar system until the year of his death, 1543. Indeed, theologians and many others opposed the notion

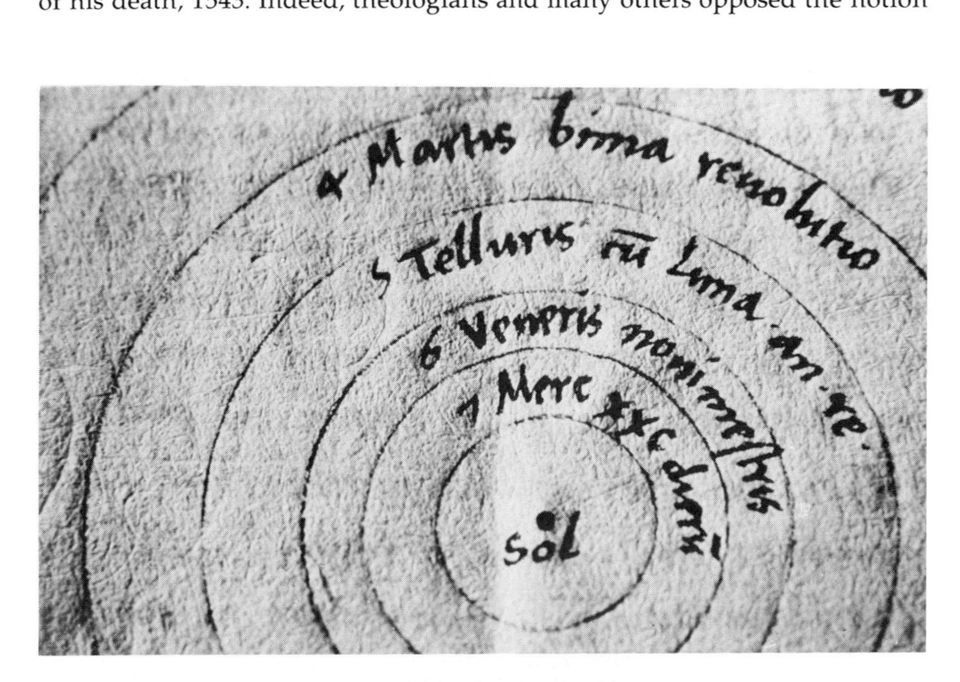

Figure 2-24 In his great work *De Revolutionibus Orbium Coelestium (On the Revolutions of the Celestial Orbs)*, Copernicus sketched his basic, correct model of the solar system. All the planets, including the Earth, orbit around the sun.

that our Earth does not form the center of the universe. If the Earth constantly changed its position in space, they argued, why don't we see a parallax effect (page 21)? Copernicus had correctly explained that the stars must all have such great distances from us that we cannot observe this effect, but not everyone accepted this argument. Worst of all, said many critics, if Copernicus were right, then the Bible must be wrong, for Joshua ordered the sun, not the Earth, to stand still over Gibeon. Martin Luther, who had begun the Protestant opposition to the Catholic church in 1517, said of Copernicus: "This fool wants to overturn the whole art of astronomy." The Catholic church tolerated Copernicus's model as a theory only. When Giordano Bruno insisted on its actual truth and even drew the conclusion that the cosmos must contain many inhabited planets, he was burned at the stake for heresy.

Though it may be hard to believe, Copernicus had dealt an awful blow to human self-esteem by dethroning the Earth as the center of the universe. If our planet was not the center of the cosmos, perhaps humanity was not the center of God's attention. The result was to delay acceptance of the Copernican model of the solar system by most scholars for several generations. Copernicus in fact had offered no convincing proof of his model, merely the argument that it provided a simpler way to understand the motions of the planets. More direct evidence had to wait for two generations. It came from the observations of the Danish astronomer Tycho Brahe, as interpreted by his one-time assistant, Johannes Kepler.

Tycho Brahe

Tycho Brahe, born three years after Copernicus died, had a rich uncle who sent him to study at various German universities. This uncle arranged for the king of Denmark to give Tycho a small island, on which Tycho ruled in medieval fashion, with a castle well equipped for astronomical observations (Figure 2-25) and a dungeon for unruly peasants. As a young man, Tycho had observed a new star in the constellation Cassiopeia in 1572. We now know that this supernova, or exploding star, must have been tremendously distant from Earth; Tycho showed that it must be at least much farther than the moon. This supernova had to represent a new feature in the cosmos; therefore, the cosmos could not be as permanent as some had supposed.

The observation of the supernova of 1572 led Tycho to his single great discovery, that repeated, careful observations represent an essential part of scientific research. Previous astronomers, convinced of the geometric harmony of the heavens, had been content to make relatively few observations. Tycho saw that observations made over and over again were likely to be of use. Besides, he liked to measure the skies and record the positions of the planets among the stars.

Tycho did not accept the Copernican model of the universe and proposed his own model, halfway between the Ptolemaic and Copernican concepts. In this model, the planets Mercury and Venus move around the sun, but the sun itself, and the outer planets, all move around the Earth. Such a model could in fact explain the planets' motions just as well as Copernicus's model, and it embodies Tycho's basically conservative attitude toward astronomical theorizing.

In 1597, Tycho moved from Denmark to Prague, where the Emperor Rudolf II made him imperial mathematician. There, in the year 1600, Tycho

Figure 2-25 A contemporary engraving shows Tycho Brahe governing the activities of his observatory. A large quadrant, marked in degrees from zero to 90, was used to measure the altitudes of stars and planets as they crossed the meridian.

hired a young assistant who had particular skill in mathematical calculations: Johannes Kepler. Kepler, then 28 years old, had become fascinated with the problem of constructing a model that would correctly explain the motions of the planets. Tycho's observations, by far the best in the world, could provide the key to the mysteries of the universe. But Tycho, perhaps envious of his assistant's abilities, was reluctant to show Kepler his observations. Then fate intervened. In 1601, after overindulgence at a banquet, Tycho died, leaving Kepler to inherit the post of imperial mathematician (though at a far lower salary than Tycho had received) and, far more important, the observations made during almost 30 years.

Johannes Kepler

Kepler went to work with vigor to discover the plan on which the heavens turned. By 1609, after countless trials and errors, he published his most important result: all the planets *do* orbit the sun, as Copernicus had said, but they move in elliptical orbits, not circular ones (Figure 2-26). This was the first of **Kepler's three "laws" of planetary motion.** Kepler's second law noted that a given planet will move more slowly when it is farther from the sun along its elliptical orbit, more rapidly when it is closer (Figure 2-27). The change in orbital velocity will occur in just such a way that the imaginary line connecting the planet with the sun will sweep over equal areas in equal intervals of time.

Some Great Astronomers

Astronomer	*Chief Accomplishments*
Hipparchus (180–125 B.C.)	First star catalog; discovery of precession
Claudius Ptolemy (A.D. 120?–190)	Epicyclic theory of planetary motion; codification of predecessors' knowledge
Nicolas Copernicus (1473–1543)	Heliocentric theory of planetary motion
Tycho Brahe (1546–1601)	Repeated, accurate measurements of positions of stars and planets
Johannes Kepler (1571–1630)	Basic laws of planetary orbits
Galileo Galilei (1564–1642)	First great telescopic discoveries; research into theory of motion
Christiaan Huygens (1629–1695)	Discovery of Saturn's rings and largest satellite; formulation of wave theory of light
Isaac Newton (1642–1727)	Basic understanding of laws of motion; universal law of gravitation; research into nature of light
William Herschel (1738–1822)	Observations of nebulae; attempts to determine size and shape of Milky Way galaxy; discovery of planet Uranus
Harlow Shapley (1885–1969)	Determination of size and shape of Milky Way galaxy
Edwin Hubble (1889–1953)	Discovery of universal expansion; first accurate determination of distances to other galaxies

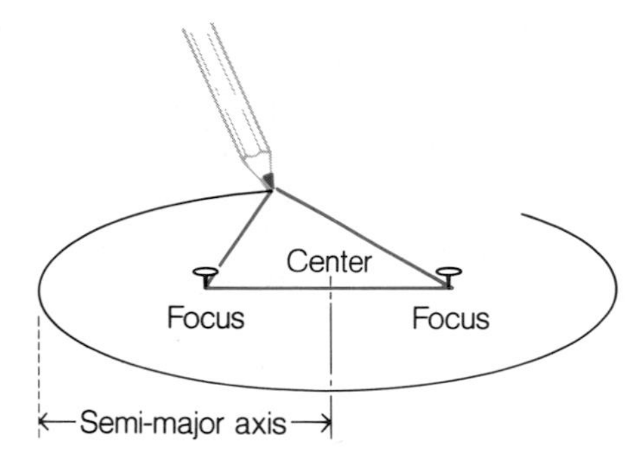

Figure 2-26 An ellipse can be drawn by marking two points, or foci, and stretching a string with a pencil so that the sum of the distances from each focus to the tip of the pencil remains the same all the way around the ellipse. If the two foci coincide, the ellipse becomes a circle. The elongation, or *eccentricity*, of the ellipse equals the distance of either focus from the center divided by the semi-major axis of the ellipse. For each planet's elliptical orbit, the sun occupies one of the two foci.

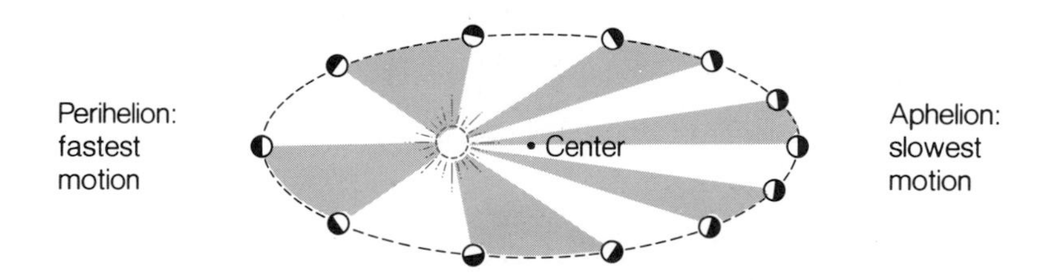

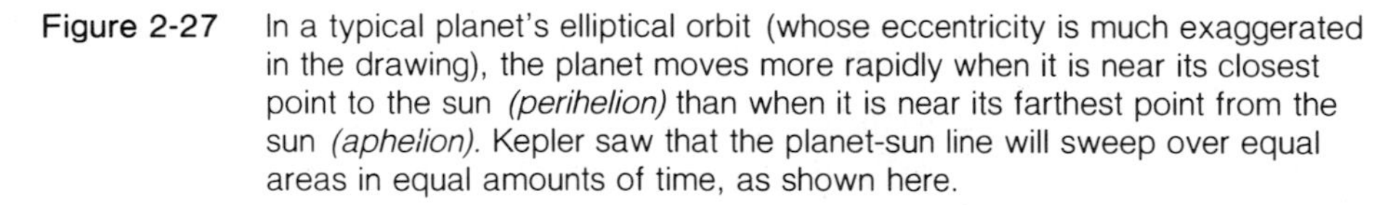

Figure 2-27 In a typical planet's elliptical orbit (whose eccentricity is much exaggerated in the drawing), the planet moves more rapidly when it is near its closest point to the sun *(perihelion)* than when it is near its farthest point from the sun *(aphelion)*. Kepler saw that the planet-sun line will sweep over equal areas in equal amounts of time, as shown here.

In 1619, ten years after his first two laws appeared in print, Kepler published his third law of planetary motion: For each planet's orbit, the *square* of the orbital period varies in proportion to the *cube* of the orbit's size. Here the size of the orbit refers to its **semi-major axis** (half of the long axis of the elliptical orbit). Jupiter, for example, takes 11.86 years to orbit the sun once, and its orbital semi-major axis equals 5.2 times the Earth's. The square of 11.86 equals the cube of 5.2. (Notice that the orbital period and orbital semi-major axis are measured with respect to the Earth's period and semi-major axis.) Similarly, Saturn takes 29.46 years for each orbit and has a semi-major axis 9.54 times the Earth's. The square of 29.46 and the cube of 9.54 are approximately equal to 868. Figure 2-28 shows a graphical representation of Kepler's third law.

Kepler's three laws of planetary motion represent the high point of astronomy before the invention of the telescope (Figure 2-29). Although the telescope appeared during Kepler's lifetime, he derived the first two of his

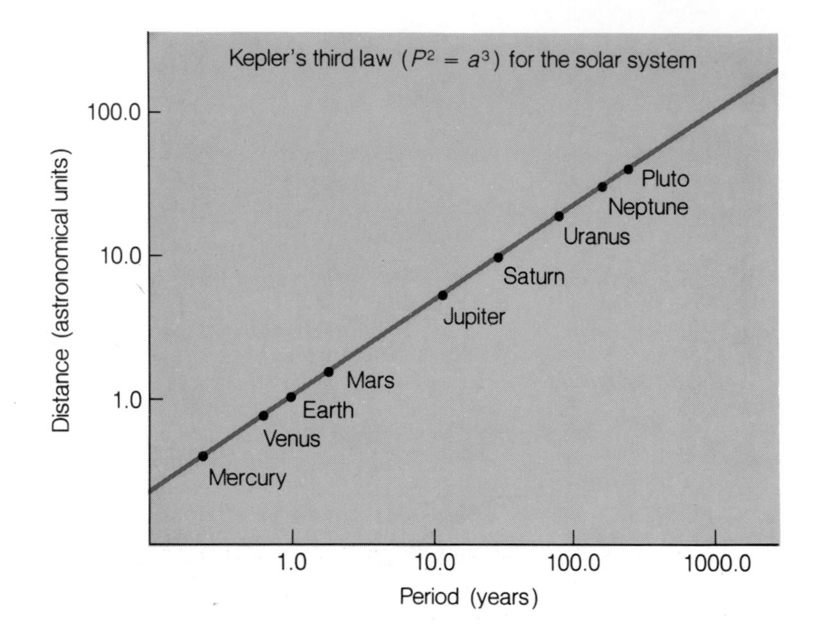

Figure 2-28 Kepler's third law of planetary motion states that the cube of the orbital semi-major axis varies in direct proportion to the square of the orbital period of the planet. This logarithmic graph expresses this relationship as $P^2 = a^3$, where P equals the period in years and a equals the orbital semi-major axis in terms of the Earth's orbital semi-major axis.

laws without ever looking through such an instrument. Later he would admire the moons of Jupiter with a telescope constructed by his great contemporary, Galileo Galilei (page 66).

We shall continue the story of our understanding of the solar system when we examine the planets in detail (page 370). Kepler's great contribution to this story was his demonstration that Copernicus was right, that the planets do orbit the sun, and that the sun, not the Earth, forms the center of the solar system. From this firm knowledge, astronomers could proceed—with centuries of effort!—to the discovery that the universe has no center at all. Kepler himself clung to a mystical belief in the harmony of the universe, while suffering some of the worst of the social and religious upheavals of his

Kepler's Three Laws of Planetary Motion

1. The orbit of each planet is an ellipse, with the sun at one of the two focal points.
2. The line joining the sun to a planet sweeps over equal areas in equal amounts of time.
3. The square of each planet's orbital period, divided by the cube of its orbital semi-major axis, equals the same number for all planets.

Figure 2-29
Johannes Kepler (1571–1630) was the first to understand the details of planets' orbits around the sun.

time. He died on a journey undertaken to resolve a dispute with his publisher, and his grave was obliterated during the wars that had begun while Kepler was still alive. Kepler had hoped that his gravestone would carry a Latin epitaph whose translation would be:

> Once I measured the skies, now the shadows of the Earth;
> My soul was from Heaven, but my body's shadow lies here.

Summary

Observations of the sun, moon, stars, and planets go back many millenia in recorded history, and no doubt well before that. The cycles of heavenly events drew rapt attention from ancient cultures. Of these cycles, the most obvious and most important for life on Earth are the daily turning of the heavens, the monthly cycle of lunar phases, and the yearly cycle of the seasons.

The daily motion of the heavens, which comes from the Earth's rotation, led to the concept of the celestial sphere, upon which the stars were placed, and which turned once a day. This still-useful picture allows us to trace the motions of the sun and moon among the stars and to locate the celestial equator and celestial poles as those points directly above the Earth's equator and poles. The sun, however, seems to move along the ecliptic, quite distinct from the celestial equator, because of the tilt of the Earth's rotation axis. This tilt produces the seasons on Earth, and the sun's motion along the ecliptic, together with the moon's motion close to the ecliptic, drew great attention to the star groups of the zodiac, the regions on the sky near the ecliptic.

Also close to the ecliptic are the planets, whose wanderings among the stars gave them their name and made them the object of centuries of observation. Attempts to make models of the motions of the sun, moon, and planets began with the assumption that the Earth was at the center of the cosmos. This assumption led to difficulty in explaining the backward, or retrograde, motion on the sky of the planets Mars, Jupiter, and Saturn. Although the epicyclic theory of Ptolemy solved this problem, the simpler explanation of Copernicus—that the planets in fact move around the sun and not around the Earth—eventually gained general acceptance, although accompanied by the pain of surrendering our imagined position at the center of the universe.

Copernicus's great successors, Tycho Brahe and Johannes Kepler, provided the observations and theory that supported his assertions. Tycho recorded the planets' positions with great accuracy in the pretelescopic era. His assistant and successor, Kepler, used these observations to determine the true orbits of the planets around the sun. These orbits consist not of circles, as Copernicus and Tycho had believed, but of ellipses, which deviate from being perfect circles by different amounts for the various planets. Kepler also discovered two more important laws about planetary motion: the planet-sun line sweeps over equal areas in equal times, and the squares of the planets' orbital periods are proportional to the cubes of their orbital semi-major axes.

Key Terms

absolute magnitude
apparent brightness
apparent magnitude
astrology
calendar
celestial equator
celestial poles
celestial sphere
circumpolar
eclipse
ecliptic
ellipse
epicycle
equinox
geocentric
heliocentric
Kepler's three laws of planetary motion
latitude
luminosity
magnitude
meridian
model
node
phases
planet
retrograde
semi-major axis
solstice
zenith
zodiac

Questions

1. Why was it important to ancient cultures to keep track of the seasons of the year? Why did they keep track of the months?
2. Why does the height of the north celestial pole above the north point of the horizon correspond to an observer's latitude north of the equator? Where would the north celestial pole appear to an observer located on the equator?
3. The moon always shows the same side to the Earth during its 27⅓-day orbit around the Earth. Does this mean that the moon is rotating? If so, how long does the moon take to rotate once?
4. Why does the moon always appear close to the ecliptic?
5. Why does the sun rise to the north of east and set to the north of west in summer? Why does the sun rise to the south of east and set to the south of west in winter?
6. What is the zodiac? Why does the zodiac play a prominent role in astrology?
7. Why does a given star rise four minutes earlier on each successive night? How does this relate to the fact that the Earth rotates once every 23 hours 56 minutes, while the day as reckoned by the sun is 24 hours long?
8. What are circumpolar stars? At a given latitude, how close (in degrees) must a star be to the celestial pole, if it is to be a circumpolar star?
9. Why are the planets easily distinguishable from stars by careful observation?
10. How does the Copernican theory of planetary motion explain the occasional retrograde, or backward, motion of the planets Mars, Jupiter, and Saturn? How did the Ptolemaic theory explain these retrograde motions?

Projects

1. *Daily Motion of the Heavens.* Observe the sun shortly after rising, or shortly before setting, or both. Sketch the path of the sun on the sky with regard to landmarks near the horizon. At night, observe a bright star that rises close to the point of sunrise, or that sets close to the point of sunset, and sketch the path of this star with respect to the same set of landmarks. How does the star's path compare to the sun's? Why?
2. *Yearly Motion of the Heavens.* Find a place where you can observe the eastern horizon without obstruction, and find a bright star just above the horizon. (You may have to wait for ten or twenty minutes to do so.) On the next evening, find the time when the star has assumed the same position in which you first observed it, relative to landmarks near the horizon. (If clouds intervene, repeat the observation on the next possible night.) How do the times compare? How can these observations be used to show that the Earth moves around the sun? Could these observations also be used to demonstrate that the sun moves around the Earth?

3. *Phases of the Moon.* Wait until the moon has recently passed through the new-moon phase, and try to establish the night when the moon has reached first quarter, showing us half its lit side and half its dark side. Why is this moment important in Aristarchus's method for finding the relative distances of the sun and moon (page 50)? Can you estimate the angle between the sun and moon, as seen from the Earth, at the time of the first-quarter moon, and so solve Aristarchus's problem?
4. *Determination of Latitude.* Try to determine your latitude by four different, though related, methods. First, measure the angle of the north celestial pole above the horizon, either by finding Polaris and using it as a rough marker of the pole, or by watching the circumpolar stars as they seem to turn and estimating the center of their motion. Second, measure your colatitude (90° minus latitude) by finding the altitude of the celestial equator at the meridian. This can be done by observing one of the bright stars that lies almost upon the celestial equator, such as the westernmost star of Orion's belt, the brightest star in the constellation Leo (Regulus), or the brightest star in Aquila (Altair), one of the three stars of the Summer Triangle. What errors enter this determination? Third, if the date is March 21 or September 22, or close to them, you can use the sun to mark the celestial equator and find the colatitude by measuring its meridian altitude. Fourth, observe a bright star as it sets (see Project 1), and try to determine the angle that its path on the sky makes with the horizon. This angle equals the observer's colatitude. Why?
5. *The Harvest Moon.* Suppose, as an approximation, that the moon's path around the sky keeps it on the ecliptic. (In fact, the moon may appear as much as 5° away from the ecliptic.) Draw the position of the celestial equator and the ecliptic as we look toward the eastern horizon in the spring (about March 21) and fall (about September 22).

   The moon moves from one day to the next by 12° along the ecliptic in an easterly direction, that is, opposite to the direction that the Earth's rotation seems to turn the heavens. Draw the moon's location at full moon, or just after, on two successive nights in early spring and early fall. Show that the moon's orbit will carry it along different tracks, relative to the horizon, on successive days in spring and fall. Can you see why although on the *average* the moon rises 50 minutes later on each successive day, the delay is less than average for the full moon of autumn (the harvest moon) and greater than average for the full moon of spring (the Easter moon)?

Further Reading

Brecher, Kenneth, and Feiertag, Michael, editors. 1979. *Astronomy of the Ancients.* Cambridge: MIT Press.

Krupp, Edwin, editor. 1979. *In Search of Ancient Astronomies.* New York: Doubleday.

Kyselka, Will, and Lanterman, Ray. 1976. *North Star to Southern Cross.* Honolulu: University of Hawaii Press.

Pannekoek, A. 1961. *A History of Astronomy.* New York: Interscience Publishers.

This cluster of galaxies in the constellation Centaurus shines with light that has taken 200 million years to reach us.

3

Unraveling the Mystery of Light

Nicolas Copernicus, Tycho Brahe, and Johannes Kepler showed that the Earth orbits the sun and does not form the center of the cosmos. Their observations and calculations laid the foundations for the science of **dynamics,** the study of objects in motion. This science was crucial to the development of astronomy. So too was the science of **optics,** the study of light.

What is light? How does it propagate through space? Why do we see various colors from different objects? These questions occupied some of the greatest scientists from the sixteenth century to the twentieth. If we hope to

Figure 3-1
Galileo Galilei (1564–1642) was the first to use the newly invented telescope to make astronomical observations.

understand the ways in which we can see the universe—from the smallest dust speck to the largest galaxy—then we must follow the story of light and attempt to perceive its true nature.

3.1 Telescopes

Three scientists who contributed greatly to the study of optics, and of dynamics as well, were Galileo Galilei (1564–1642), Christiaan Huygens (1629–1695), and Isaac Newton (1642–1727). These men—an Italian, a Dutchman, and an Englishman—opened our eyes to new ways of improving our knowledge of the universe. Although all three were deeply concerned about finding the true nature of light, they first made important advances in practical optics: They built new and better telescopes.

Refracting Telescopes

The first astronomical observations with a telescope were apparently made by Galileo (Figure 3-1). Galileo had heard about the principle on which a

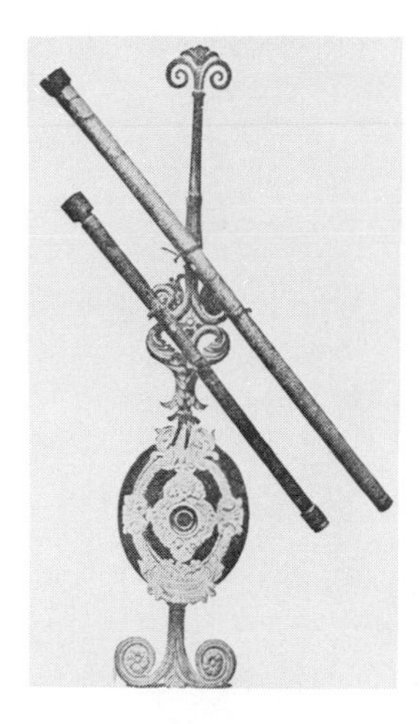

Figure 3-2
Galileo's telescopes all were refractors. Each of them consisted of a simple tube with a lens at each end.

telescope is based, and he quickly built his own (Figure 3-2). Galileo realized that if one lens bent light to a focus, a second lens could magnify the image formed at this focal point (Figure 3-3). He quickly improved on his original design, until he had telescopes capable of magnifying objects by more than 30 times. With these instruments he discovered Jupiter's four large moons, the spots on the sun, the craters of the moon, and the fact that the milky way in the sky consists of a host of stars. Galileo's telescopes were **refracting telescopes;** that is, they contained lenses to focus light and to magnify the image thus produced.

Christiaan Huygens, born when Galileo was already an old man, produced refracting telescopes superior to any of Galileo's, because he and his brother had found a better way to grind the telescope lenses. With his improved telescopes, Huygens discovered the ring around Saturn, which had been only poorly sighted in previous telescopes, and Titan, Saturn's largest satellite. (Three more satellites were found soon after by Giovanni Cassini.)

Huygens originated the **wave theory of light,** the concept that light waves consist of disturbances that spread outward from a source of light like ripples on a pond (Figure 3–4). With this theory, Huygens had great success in explaining why lenses can focus light from a source to produce an image (Figure 3-3).

He had less success in understanding why light of different colors comes to a focus at different points, a fact that made the construction of large telescopes difficult. The explanation for this **chromatic aberration** had to await the genius of Huygens's great contemporary, Isaac Newton (Figure 3-5).

Modern refracting telescopes partially overcome the problem of chromatic aberration by having their main lenses made of two different kinds of glass, each of which affects different colors in different ways. The two halves of the lens thus almost cancel out their individual effects of chromatic aberration. (Figure 3-6). The largest refracting telescopes, with lenses of 1.02-meter and 91-centimeter diameters, are in Wisconsin and California, respectively.

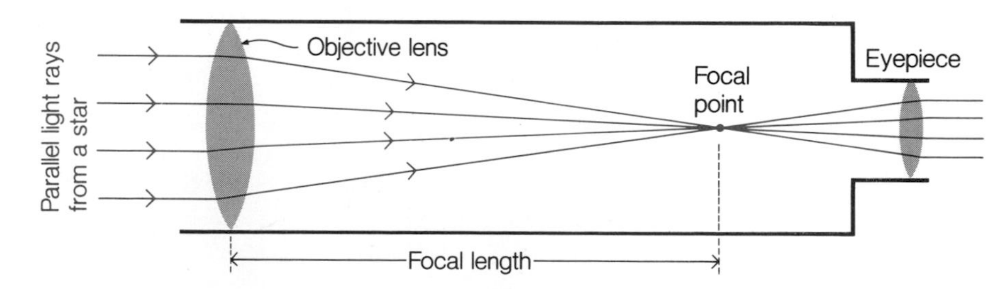

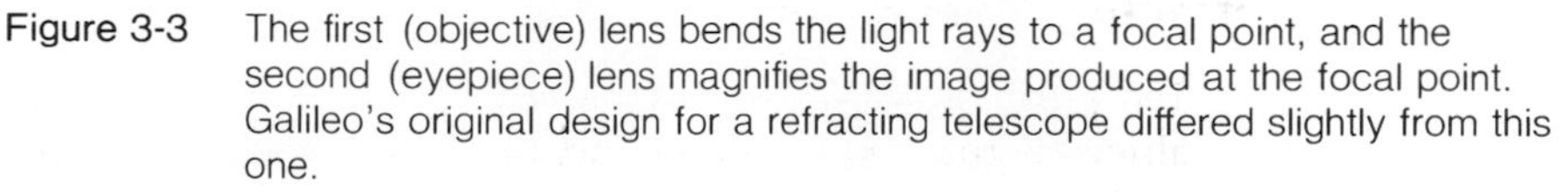

Figure 3-3 The first (objective) lens bends the light rays to a focal point, and the second (eyepiece) lens magnifies the image produced at the focal point. Galileo's original design for a refracting telescope differed slightly from this one.

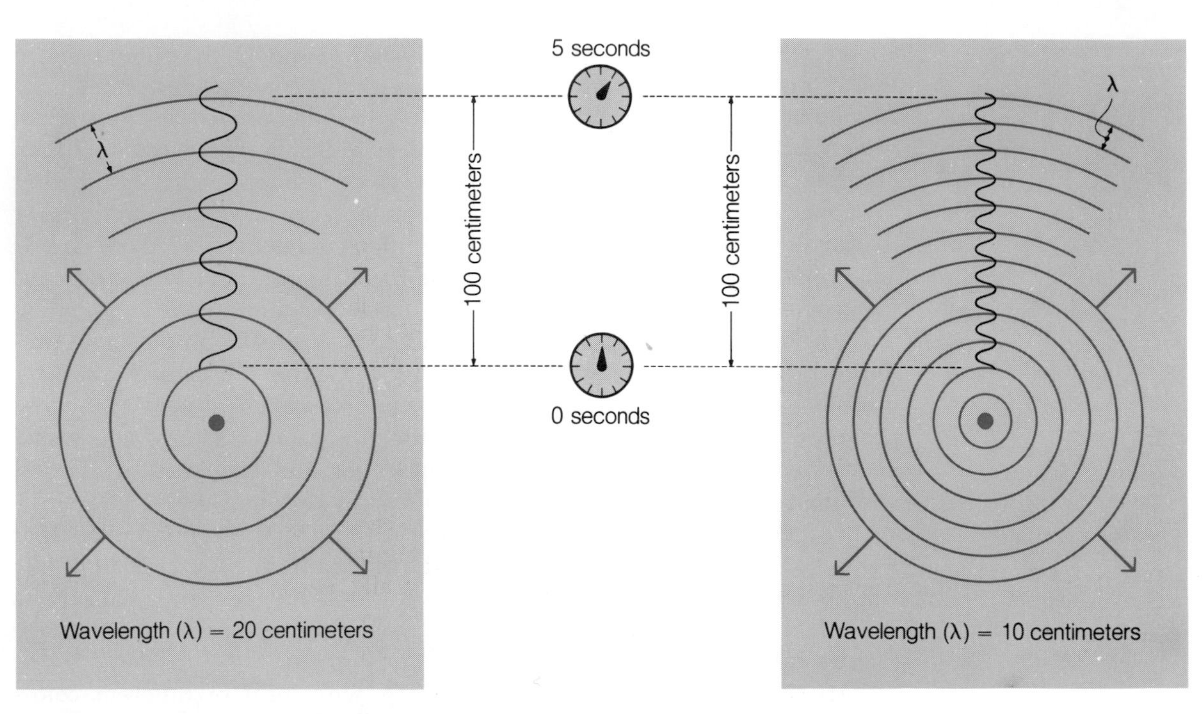

Figure 3-4 Christiaan Huygens proposed that light consists of waves that spread outward from a source like ripples on a pond. The ripples can have different wavelengths (distances between successive wave crests), but they all travel outward at the same speed.

Figure 3-5
Isaac Newton (1642–1727), born three months after his father's death, was so small at birth that his mother said he could have fit into a quart pot. From this man came the foundations of modern physics.

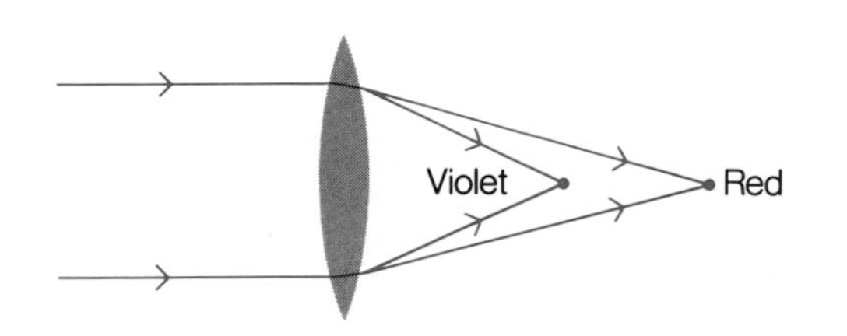

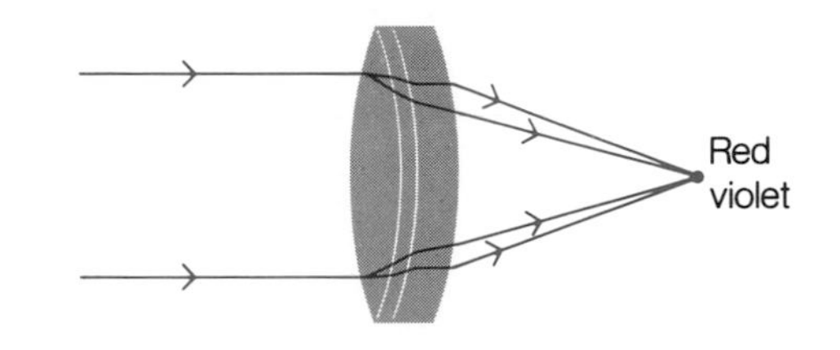

Figure 3-6 Because glass bends light of different colors by different amounts, a lens made of a single piece of glass will focus different colors at different places, producing chromatic aberration, which blurs the image. Seventeenth-century telescope makers discovered that a lens made of two different kinds of glass could correct for most of this chromatic aberration, and modern refracting telescopes embody this discovery.

Reflecting Telescopes

In 1668, Newton built the first **reflecting telescope,** in which a *mirror* rather than a lens is used to focus the light (Figure 3-7). Unlike refracting telescopes, reflecting telescopes focus light of all colors to the same point, thus avoiding the problem of chromatic aberration. As Figure 3-7 shows, Newton's basic design used two mirrors: one to focus the light, and a smaller one to reflect the image out of the main beam of light to the side of the telescope tube, where the image could be magnified. Astronomers soon realized that reflecting telescopes can be made much larger than refractors, because their mirrors can be supported from behind, whereas a lens must be supported only around its edge. Hence, the bending of both mirrors and lenses under the force of gravity can be better overcome in a reflecting telescope than in a refracting telescope.

Astronomers value large telescopes for their great light-gathering power: the capacity to provide bright images. Most of the celestial objects that astronomers study are so far from us that they seem extremely faint. Wider lenses or mirrors will form brighter images because they gather light from a larger area. Therefore a large telescope can collect more light from a given object than a small telescope can and thus can better overcome the faintness problem. Despite the difficulty and expense of building larger and larger telescopes, astronomers almost always prefer to use the largest telescope possible. The largest reflecting-telescope mirrors have diameters of 6 meters, 5 meters, and (in four cases) 4 meters. They are in Russia, California, Arizona, Australia, and (two) in Chile, respectively. A variant on the basic reflector design, the multimirror telescope in Arizona, uses six mirrors to achieve the light-gathering power of a single 4.5-meter reflecting mirror.

Any telescope must perform several functions. It must gather the radiation that reaches it, focus it, and analyze it. The job of analyzing the light or the other kinds of radiation belongs to instruments placed near the focal point of the telescope. Since the instruments that analyze the light may be too large to be attached to the telescope as it moves, many large reflecting telescopes can be used in the coudé mode, in which several mirrors reflect the focused beam of light to a separate room (Figure 3-7).

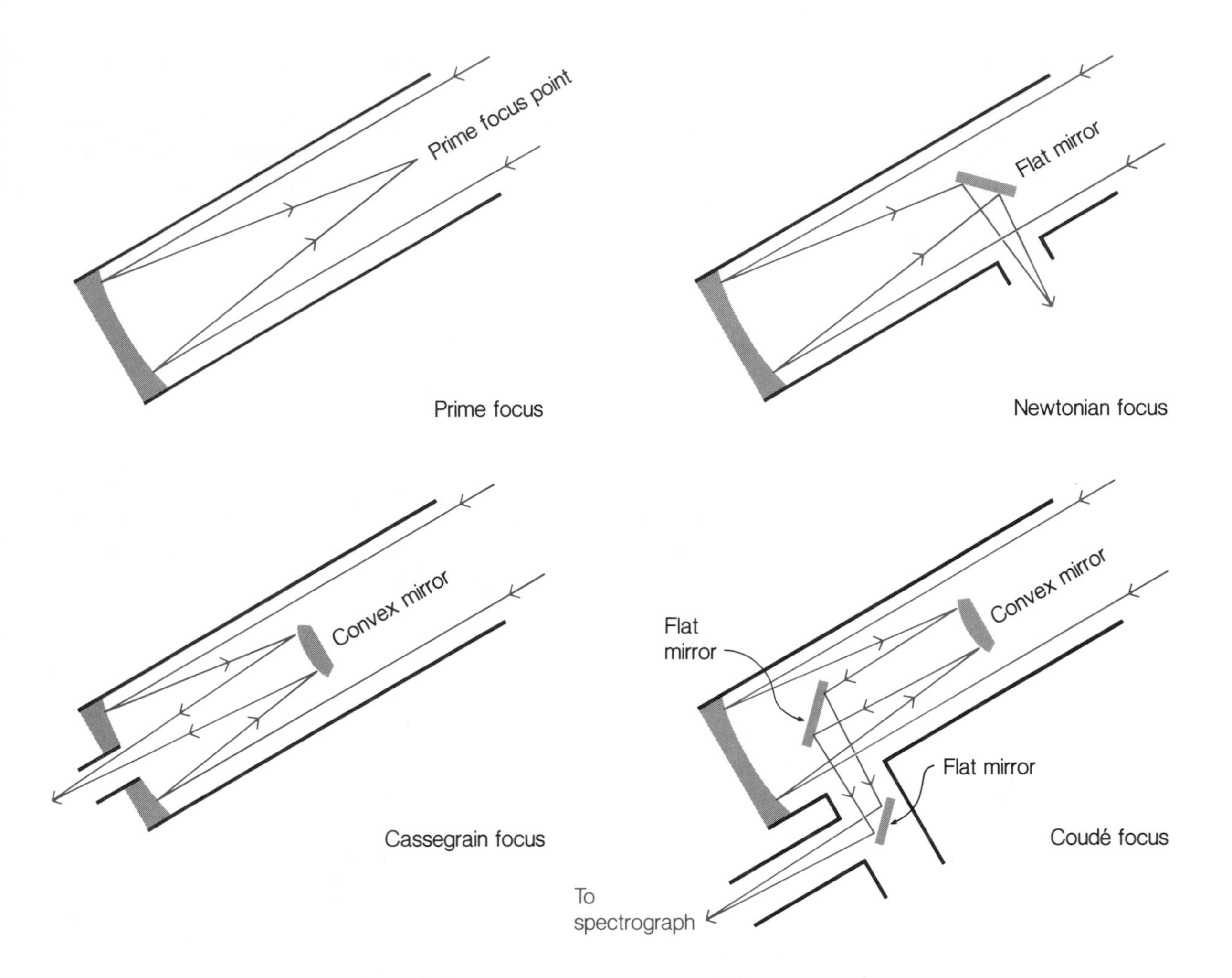

Figure 3-7
Newton's design for a reflecting telescope (top right) used a parabolic mirror to focus the incoming light, and a second, flat mirror to reflect the light out of the telescope tube to the eyepiece for magnification. Other arrangements for reflecting telescopes (above) reflect the light back through the main mirror (Cassegrain focus) or use several flat mirrors to reflect the light to another room for detailed examination.

3.2 The Nature of Light

When Newton built his first small reflecting telescope, he could not have imagined that one day there would be immense reflectors weighing hundreds of tons and costing millions of dollars. What puzzled Newton was an unsolved riddle: What is light, and what are the different colors of light?

Visible Light

In 1666, Newton had bought a glass prism at the famous fair at Stourbridge, England. Such a prism bends light of different colors by different amounts (Figure 3-8). Thus Newton found that each color of light can be characterized by the amount of bending it undergoes in a prism. He also saw that what we call white light in fact consists of light of different colors. In Huygens's wave theory of light (which Newton did not accept), different colors are characterized by different **wavelengths,** the distance between successive wave crests (Figure 3-4). Each kind of ripple—light of a particular color—has a particular wavelength: Red light has the greatest wavelength, violet light the shortest wavelength, of the light our eyes can see.

In the wave theory of light, we can define the **frequency** of light as the number of full cycles (from one wave crest to the next) that reach us each second. All light travels through space with the same speed, almost 300,000 kilometers per second. Thus we must receive *more* cycles each second of the short-wavelength light (lesser distance between wave crests) and *fewer* cycles per second of the long-wavelength light (greater distance between wave crests). For all light, the frequency times the wavelength always equals the same number, the speed of light (Figure 3-4). Thus violet light has a higher frequency (more cycles per second) than red light.

Other Kinds of "Light"

What about light with still longer wavelengths than red light, or still shorter wavelengths than violet light? Indeed such light exists, but since the word

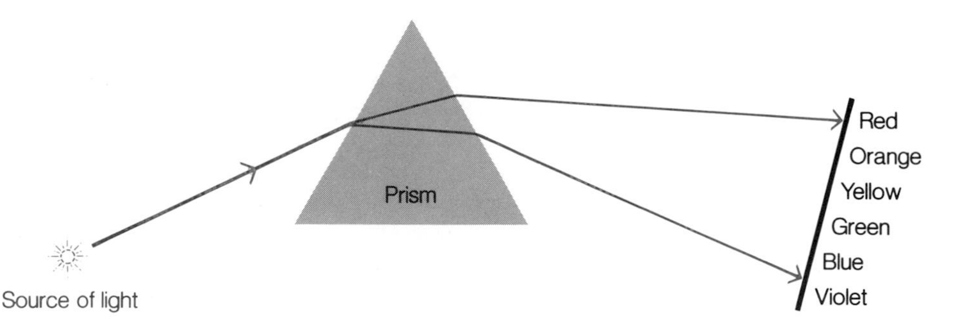

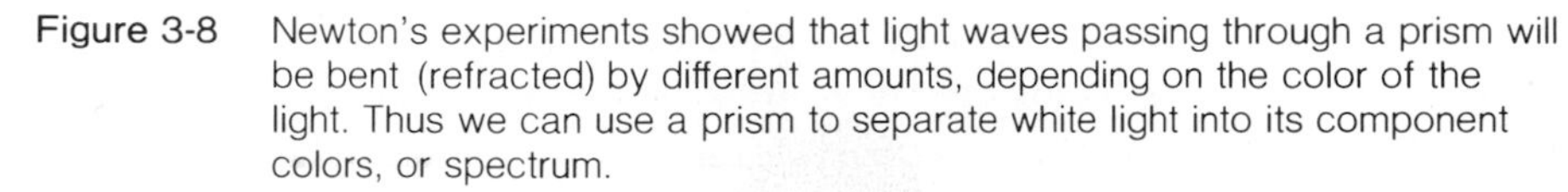
Figure 3-8 Newton's experiments showed that light waves passing through a prism will be bent (refracted) by different amounts, depending on the color of the light. Thus we can use a prism to separate white light into its component colors, or spectrum.

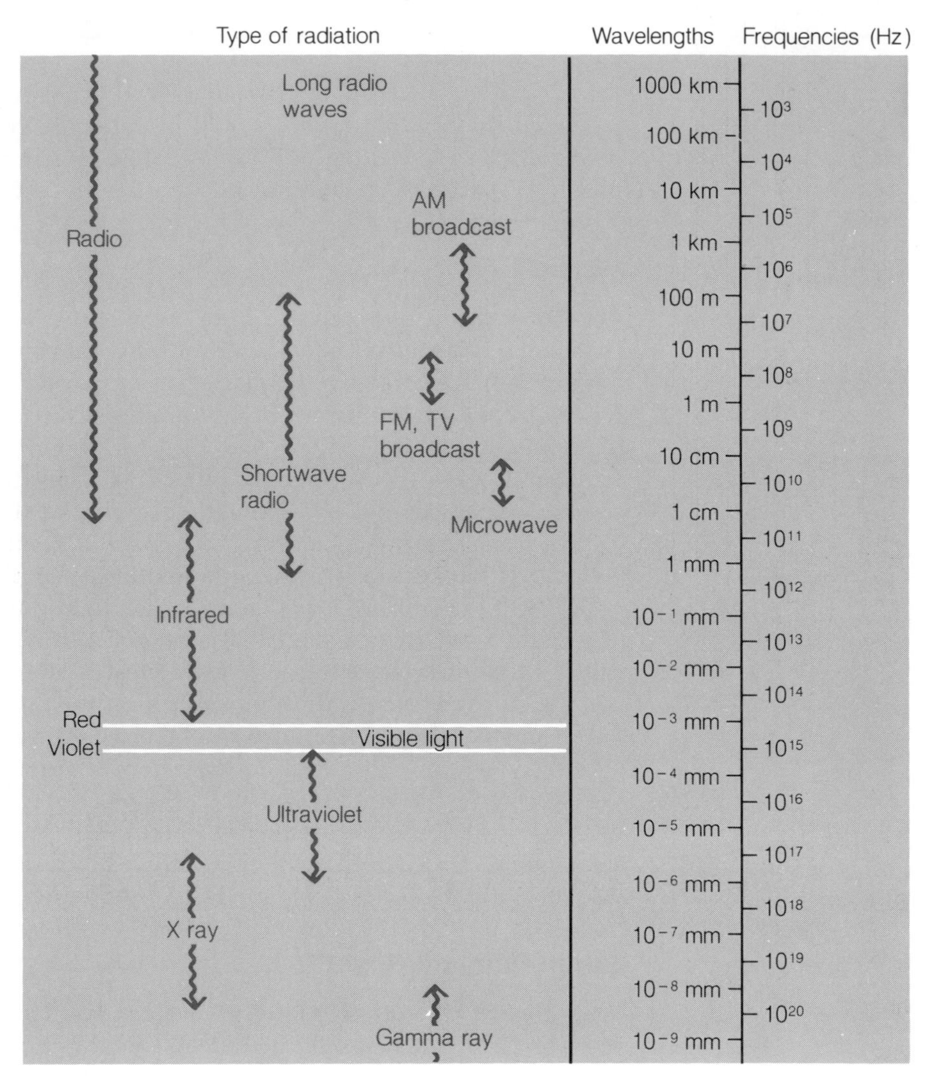

Figure 3-9
The spectrum of electromagnetic radiation extends over all possible wavelengths and frequencies. The shortest-wavelength (highest-frequency) radiation has the name gamma rays, while the longest-wavelength (smallest-frequency) radiation consists of radio waves. Visible light has intermediate wavelengths and frequencies.

light usually refers to what we can see, we may use the more general term *radiation* to refer to all of these waves. More precisely, light waves form just one kind of **electromagnetic radiation,** all of which consists of radiation with different frequencies and wavelengths. Light, or **visible light,** consists of electromagnetic radiation with wavelengths between about 3.5×10^{-5} centimeter and 8×10^{-5} centimeter. Electromagnetic radiation with wavelengths somewhat shorter than those of violet light is called **ultraviolet;** radiation with wavelengths somewhat longer than those of red light is called **infrared.**

Figure 3-9 gives an overview of the range, or **spectrum,** of electromagnetic radiation. We now know that this spectrum is a continuum, in which

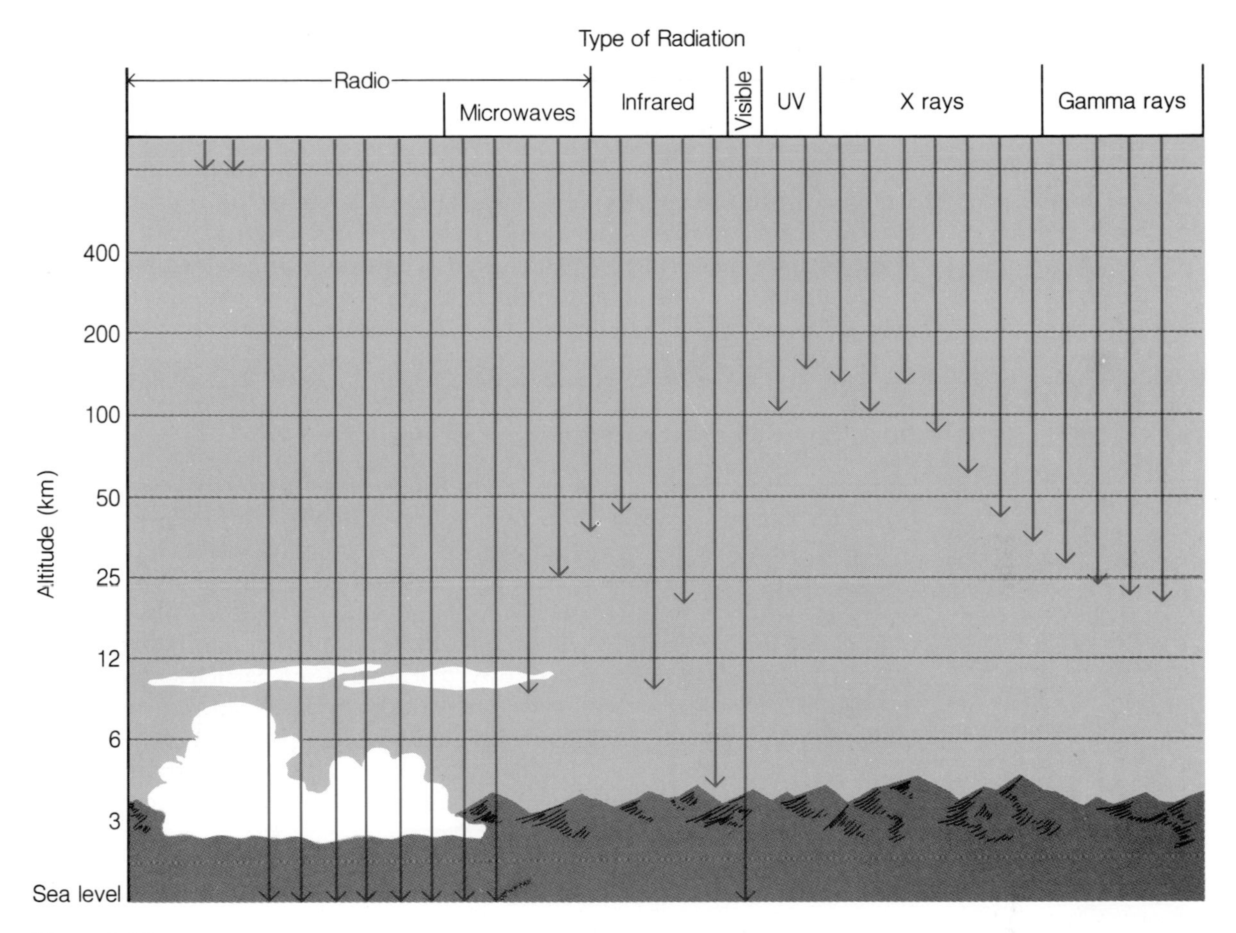

Figure 3-10
The Earth's atmosphere absorbs radiation of most wavelengths at varying heights above the ground. Only visible-light and radio waves can penetrate the atmosphere. Thus to obtain observations of the heavens at other wavelengths, we must send telescopes above most or all of the Earth's atmosphere.

one kind of radiation blends into another. Indeed, certain individuals can even see some ultraviolet light although the wavelengths are a bit too short for most human beings. Even shorter than the ultraviolet wavelengths are those of **x rays** (short for x radiation, a name that reminds us of the initial mystery surrounding its discovery), and shorter still are the wavelengths of **gamma rays.** At longer wavelengths, we find first **infrared** and then the **radio** part of the electromagnetic spectrum. (*Radio* is simply short for radiation, but it stands for a particular part of the total spectrum.) Radio waves have wavelengths ranging from millimeters to kilometers. The shortest-wavelength part of the radio spectrum sometimes bears the name of **microwaves.** But these wavelengths are still longer than those of infrared and visible-light radiation, and belong to the radio domain.

Our eyes evolved to see visible light because much of the sun's electromagnetic radiation has visible-light wavelengths and frequencies. Furthermore, the Earth's atmosphere filters out most infrared and almost all ultraviolet radiation from the sun, permitting only visible light and radio waves to pass largely unhindered (Figure 3-10).

Today, astronomers specialize, to some degree, according to the type of electromagnetic radiation they study. In Newton's day, only visible light could be detected, and all telescopes were designed to collect, focus, and analyze visible light. Now we have radio telescopes, microwave telescopes, infrared telescopes, ultraviolet telescopes, and, above the Earth's atmosphere, x-ray and gamma-ray telescopes. Each kind of telescope aims at collecting, focusing, and analyzing radiation of certain wavelengths. Most of them work on the same principle as the Newtonian reflecting telescope, though the details of their operation vary in accordance with the methods that must be used to study radiation of different wavelengths (Figure 3-11).

Does Light Consist of Waves, or of Particles?

Huygens's theory that light is a series of waves seems perfectly reasonable in many respects. It explains refraction, reflection, and a host of other phenomena. The wave theory cannot explain, however, why visible light of certain wavelengths can pass through a particular gas—hydrogen, for example—whereas light of other wavelengths cannot. Newton himself did not believe in Huygens's wave theory and developed an alternative theory of light, in which light consists of tiny "corpuscles" or particles.

Two centuries after Newton's death, modern scientists, led by Albert Einstein and Louis de Broglie, developed a theory of light that *combines* Huygens's wave theory with Newton's particle theory. Einstein and de Broglie suggested that light and other forms of electromagnetic radiation have some properties of *both* waves and particles.

How can light be both a particle and a wave? In our ordinary experience, such duality is impossible. But in the domain of **quantum mechanics**, a branch of physics that deals with minute distances and sizes, this wave-particle duality provides the only consistent theory of the nature of electromagnetic radiation.

We can best picture light as a stream of energy bundles, or **photons.** As an analogy (but only an analogy!), imagine each of these bundles as a tiny tadpole, moving at the speed of light and wiggling as it moves (Figure 3-12). Each photon has a characteristic rate of wiggling: The number of wiggles per second gives its frequency, and the distance it travels while it wiggles once gives the photon's wavelength. Since a rapidly wiggling tadpole has more energy than a slowly wiggling one, we can understand why *each photon's energy varies in direct proportion to its frequency.* Because any photon's wavelength varies in inverse proportion to its frequency, low-energy photons have long wavelengths and high-energy photons have short wavelengths. Thus violet light has more energy per photon than red light does.

If we can accept the idea that light and other kinds of electromagnetic radiation consist of these wiggling bundles of energy called photons, we can proceed to examine how light interacts with matter. But first we must consider something that was unknown to Huygens and Newton—what matter looks like at the atomic level.

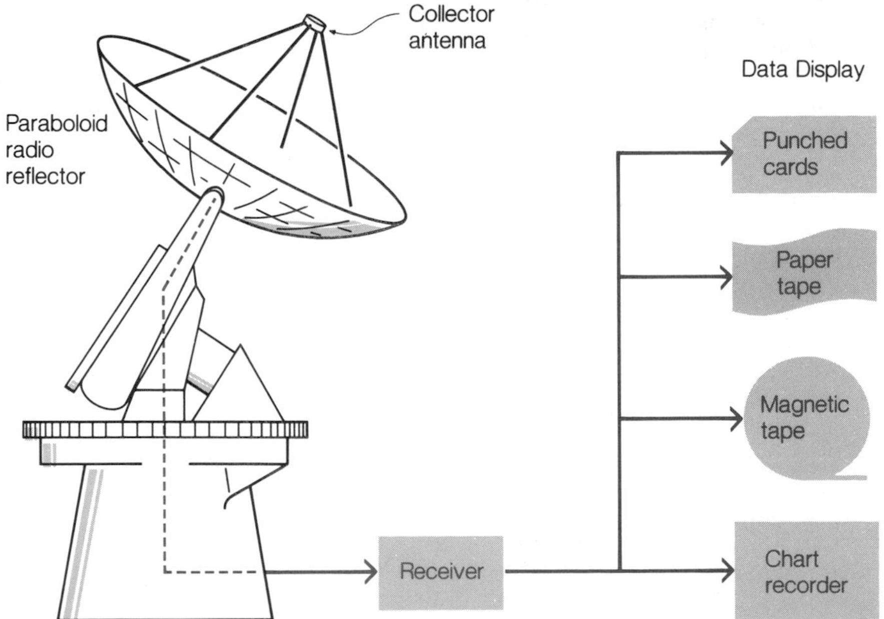

Figure 3-11
Radio telescopes usually employ a parabolic mirror to reflect radio waves to a focal point (top left and below). The Voyager spacecraft (right center) has a similar parabolic antenna, used to detect radio waves from the Earth. The Kitt Peak solar telescope uses a flat mirror to reflect sunlight down a long, slanted shaft, where a parabolic mirror focuses the light (left center). The Copernicus satellite (top right) used a parabolic mirror to focus ultraviolet light.

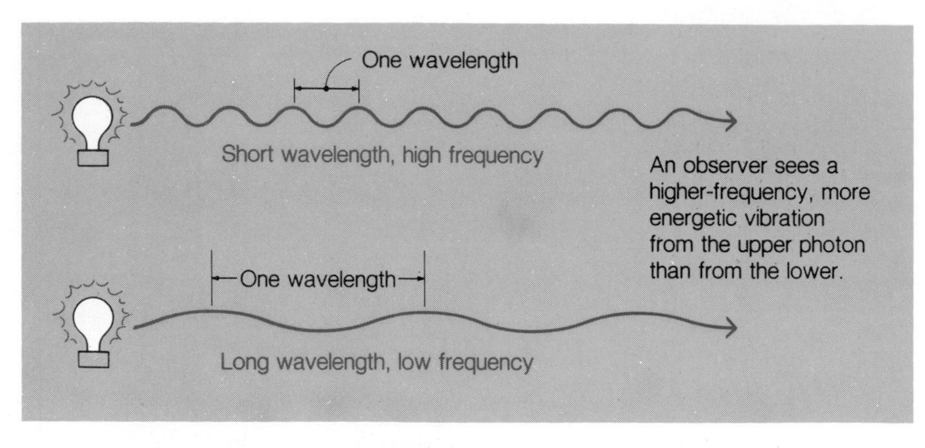

Figure 3-12 We may picture a photon as a tiny bundle of energy, traveling at the speed of light and oscillating as it moves. The photon's frequency is its number of oscillations per second, and the wavelength of the photon gives the distance that it moves during one oscillation.

The Frequency, Wavelength, and Energy of a Photon

Every kind of photon—that is, every kind of electromagnetic radiation—can be specified by its wavelength, by its frequency, or by its energy per photon. These three quantities are interrelated, so that to specify any one of them is to specify the other two.

We have seen that the frequency, f, times the wavelength, w, always equals the speed of light, c = 299,793 kilometers per second. In symbolic language, we have

$$f \times w = c$$

The relationship between energy and frequency is one of simple proportionality. If we denote the photon's energy by E, we have

$$E = h \times f$$

where h is a universal constant, called Planck's constant after the man who first proposed it. In the metric system of units, we can express h as

$$h = 6.626 \times 10^{-27} \text{ erg-seconds.}$$

One **erg** is the amount of energy needed to accelerate a 2-gram mass from zero velocity to a speed of 1 centimeter per second. (One erg is a tiny amount of energy in everyday terms. A single heartbeat typically expends 10 *million* ergs!)

Using the first of our relations, we can see that a photon's energy and wavelength are related by

$$E = \frac{h \times c}{w}$$

so that longer wavelengths imply lower energies, and vice versa.

3.3 The Structure of Atoms

Most of the matter with which we are familiar consists of **atoms,** miniature "solar systems" in which tiny particles called **electrons** orbit around a **nucleus** made of **protons** and **neutrons** (Figure 3-13). Each type of atom that makes up a particular **element** has a certain number of protons in its nucleus. The protons are tiny elementary particles, each with one unit of *posi-*

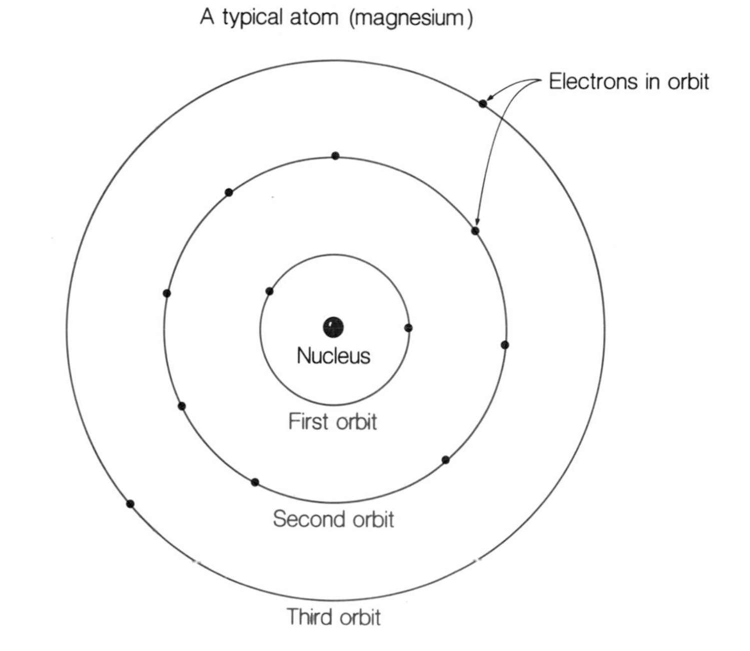

Figure 3-13
All atoms have a massive nucleus, made of protons and neutrons, surrounded by less massive electrons. The electrons can move only in certain orbits, with one particular orbit the smallest possible.

tive electric charge, and they attract an equal number of electrons, each with one unit of *negative* electric charge. Typically, the atomic nucleus also contains neutrons, which have no electric charge.

Atoms of a particular element may vary in the number of neutrons in the atomic nucleus. For example, the simplest atom, hydrogen, has one proton and one electron. Unlike all other atoms, most hydrogen atoms have *no* neutrons in their atomic nuclei. However, some hydrogen atoms do contain one or two neutrons along with the proton (Figure 3-14). These are the **isotopes** of hydrogen, variations of the basic hydrogen atom (always with the same number of protons and electrons) through the differing number of neutrons.

Since atoms are extremely small, typically 10^{-7} centimeter or so in diameter, they are hard to study, and physicists did not understand their structure until the first quarter of the twentieth century. By that time, chemists had already sorted out the basic properties of atomic elements by studying

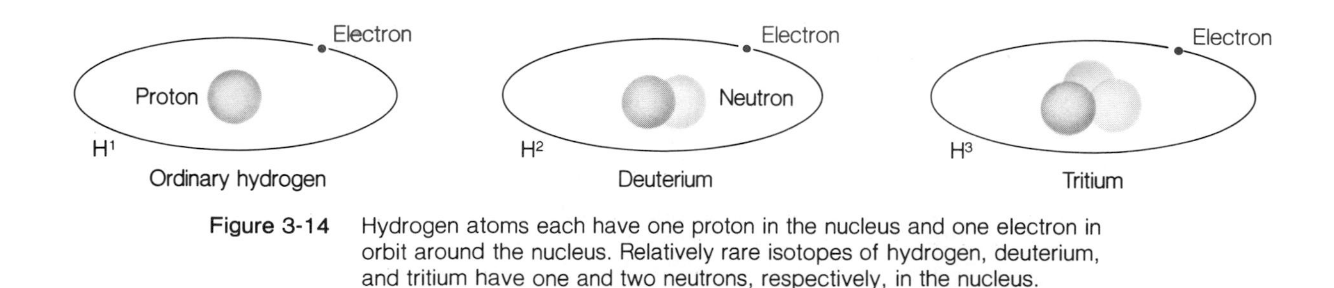

Figure 3-14 Hydrogen atoms each have one proton in the nucleus and one electron in orbit around the nucleus. Relatively rare isotopes of hydrogen, deuterium, and tritium have one and two neutrons, respectively, in the nucleus.

how the atoms interact with one another. They had determined that atoms differ in their masses, which derive primarily from the combined masses of the protons and neutrons they contain. (The mass of the electrons is negligible, since each electron has a mass almost 2000 times *less* than the mass of a proton or neutron.) Atoms also differ, as we have said, in the number of positive charges (protons) and negative charges (electrons) they contain (see Table 3-1), although the number of protons is always equal to the number of electrons. An atom from which one or more electrons have been removed is called an **ion.** Such ions have a net positive charge. They therefore exert an electromagnetic force that attempts to recover the missing negative charges by capturing electrons until the original atom reappears with its proper number of electrons.

Electron Orbits in Atoms

A great advance in our understanding of atoms came in 1913, when the Danish physicist Niels Bohr (1885–1962) suggested that *the electrons in an atom can move only in certain orbits,* and not in just any orbit around the nu-

Table 3-1 Properties of Some Common Atoms and Isotopes

Name of Element (with Isotopes)	Number of Protons[a]	Number of Neutrons	Isotopic Abundance (%)
Hydrogen:			
Hydrogen	1	0	99.998
Deuterium	1	1	0.002
Tritium	1	2	[b]
Helium:			
Helium-3	2	1	0.02
Helium-4	2	2	99.98
Carbon:			
Carbon-12	6	6	98.9
Carbon-13	6	7	1.1
Carbon-14	6	8	[b]
Nitrogen:			
Nitrogen-14	7	7	99.6
Nitrogen-15	7	8	0.4
Oxygen:			
Oxygen-16	8	8	99.76
Oxygen-17	8	9	0.04
Oxygen-18	8	10	0.20
Neon:			
Neon-20	10	10	88.9
Neon-21	10	11	0.3
Neon-22	10	12	10.8
Magnesium:			
Magnesium-24	12	12	78.7
Magnesium-25	12	13	10.1
Magnesium-26	12	14	11.2

[a] The number of electrons always equals the number of protons in any atom.
[b] These isotopes make up less than 0.001% of the total of all atoms of this element.

cleus. Bohr concentrated on hydrogen, the simplest of all atoms, with only one electron. In Bohr's model for the hydrogen atom, the smallest possible electron orbit around the proton has a radius of 5.3×10^{-9} centimeter (Figure 3-15). The second-smallest orbit has a radius four times greater, or 2.1×10^{-8} centimeter. No orbit for the electron is possible between these two. Similarly, for the electron to move in an orbit larger than the second-smallest orbit, it must move at least to the third-smallest orbit, which is nine times larger than the smallest orbit. These rules may seem to be without rhyme or reason, but they appear valid for all hydrogen atoms. Other types of atoms obey similar rules, with electrons moving only in certain definite orbits and not in any orbit whatsoever.

The Electron Shuttle: Transitions from Orbit to Orbit

The Bohr model for atoms can account for the basic features of how atoms absorb light or produce it. If we accept the notion that only certain orbits are allowed for the electrons in an atom, we can proceed to examine how an electron can move from one orbit to another. This process has tremendous importance for astronomy, because it provides much of our information about the temperature, density, and even the chemical composition of matter in stars.

Consider hydrogen, the simplest of atoms. Left to itself, any hydrogen atom will soon get its electron into the smallest possible orbit, because the attractive electromagnetic forces between the negatively charged electron and the positively charged proton will move the electron to the minimum

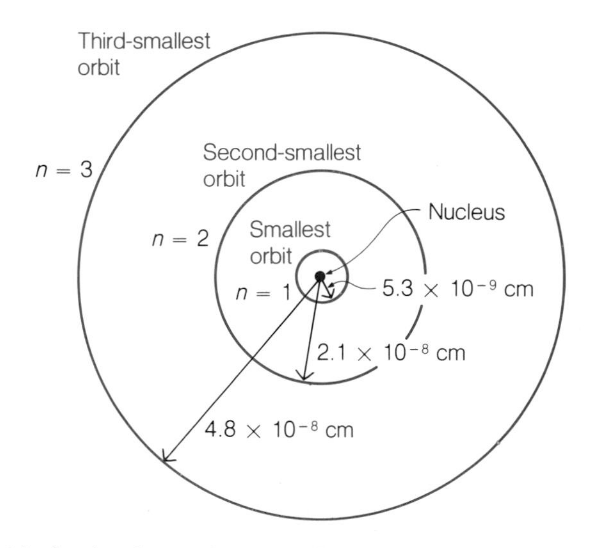

Figure 3-15 The possible orbits for the electron in a hydrogen atom have different sizes. The second-smallest orbit is four times larger than the smallest orbit, and the third-smallest orbit is nine times larger than the smallest orbit.

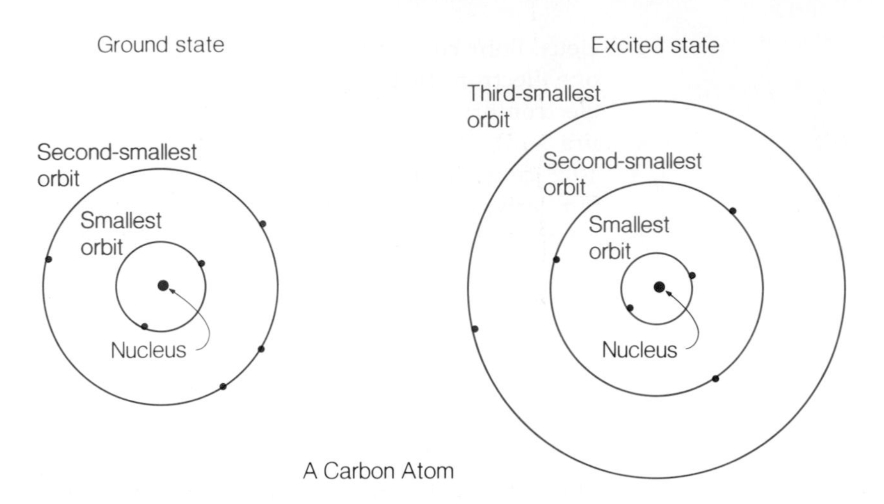

Figure 3-16 A carbon atom has six electrons. In the atom's ground state, two of the electrons are in the smallest orbit and the other four are in the second-smallest orbit. Collisions among atoms may knock one or more electrons into a larger orbit, thus producing an excited state of the atom.

possible distance from the proton. Likewise, in a helium atom, which has two protons and two electrons, the two electrons will soon enter the smallest orbit. We might expect that this same rule would apply to all atoms, but another rule of physics intervenes. Physicists have found that *each orbit can hold only a certain maximum number of electrons.* This rule, from quantum mechanics, again violates our intuition, but nonetheless, it has been repeatedly checked by experiment. The smallest orbit can hold no more than two electrons, and the second-smallest orbit can hold no more than eight electrons. Similar rules apply to the larger orbits, and the result is that atoms more complex than hydrogen and helium cannot have all their electrons in the smallest orbit. For example, a carbon atom has six electrons. The closest it can come to having all its electrons nestled by the nucleus is to have two electrons in the smallest orbit and four electrons in the second-smallest orbit (Figure 3-16). An atom with its electrons as close to the nucleus as the laws of quantum mechanics will allow is said to be in its **ground state.**

All atoms, left to themselves, will gather their electrons into the ground-state configuration. To move the electrons outward from this state requires that energy be added to the atom. Where can such energy come from? In astronomical situations, two sources play the most important roles: collisions and light rays.

Collisions among atoms can knock one or more electrons in each atom into a larger orbit (Figure 3-16). In this process, some of the energy of motion of the colliding atoms provides the energy needed to move the electrons to larger orbits, producing what is called an **excited state** for the atoms. Similarly, when light strikes an atom, the light itself can provide the energy required to move the atom's electron farther from the nucleus. This process merits our detailed attention, for it holds the key to understanding many of the mysteries of the universe.

3.4 The Interaction of Light with Atoms

When light waves pass through a gas, they have a chance to interact with the atoms that make up the gas. But to do so, the photons must have just the right amount of energy to knock an electron into a larger orbit. If a photon has just a little more energy than is needed to make the electron jump from the smallest into the second-smallest orbit, it will pass by the atom with no interaction at all. This is an all or nothing process. If the photon *does* have just the right amount of energy to make the electron jump, the photon gives up *all* its energy and disappears in the process. The photon simply cannot give *part* of its energy to the atom in producing an electron jump. In a hydrogen atom, each particular jump from a smaller to a larger orbit requires a particular amount of energy, and hence a particular frequency and wavelength, for the photon that can make such a jump occur (Figure 3-17). Photons with the wrong frequency and wavelength will pass among the atoms with no interaction, whereas those with the right frequency and wavelength will be **absorbed,** their energy used up in making the electron move to a larger orbit.

Once the hydrogen atom's electron moves to a larger orbit, it can jump back into its original orbit or into any smaller orbit. As it does so, the atom will emit a photon of precisely the same energy, frequency, and wavelength as the photon that would produce the jump outward. This photon can appear headed in any direction.

A photon may hit the atom while it is still in an excited state, that is, with its electron in an orbit larger than the smallest possible one. If this photon has the proper energy, it can make the electron jump into a still larger orbit. The eventual jump back into the smallest possible orbit may then occur in one, two, or more stages. Each jump inward produces a photon of a certain definite energy, the exact energy needed to make the electron undergo the same jump outward (Figure 3-18).

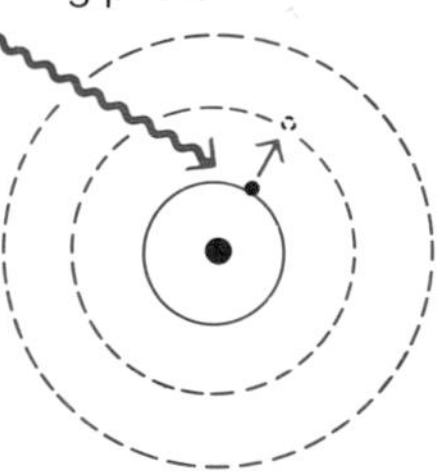

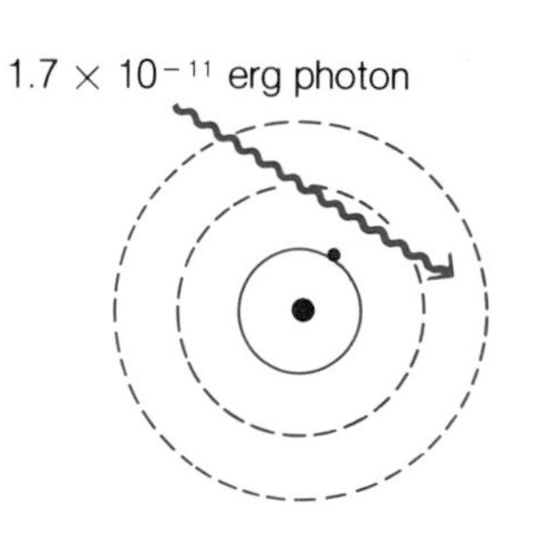

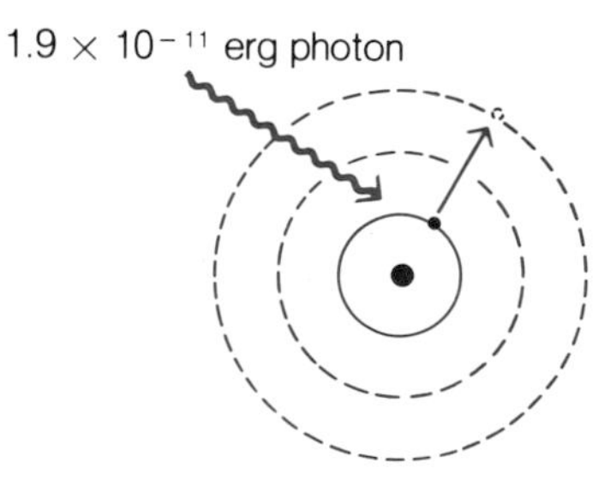

Figure 3-17 If a photon of just the right energy (1.6×10^{-11} erg) strikes a hydrogen atom in the ground state, the photon will give up all its energy to the atom and disappear. As it does so, the photon will excite the electron into the second-smallest orbit. A photon with an energy of 1.9×10^{-11} erg will excite a hydrogen atom from the ground state into the third-smallest orbit. A photon with an energy between 1.6×10^{-11} and 1.9×10^{-11} erg cannot excite a hydrogen atom out of the ground state. Such a photon will pass by the atom with no interaction at all.

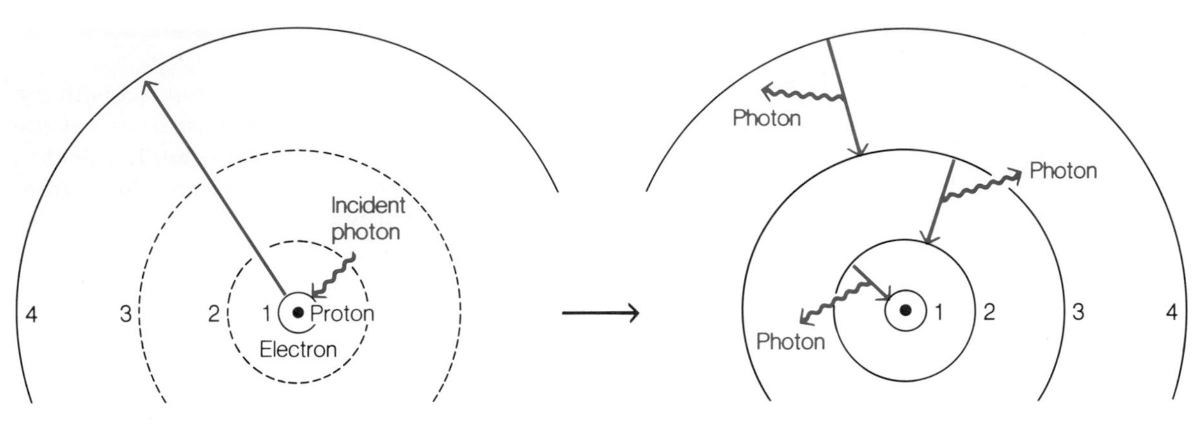

Figure 3-18 If a photon has the right amount of energy to make a hydrogen atom's electron jump from the smallest into the fourth-smallest orbit, the excited atom can jump its electron back into the smallest orbit after a fraction of a second. This de-excitation can occur in one, two, or three steps, of which the latter possibility is shown here. Each jump to a smaller orbit produces a photon with an energy, frequency, and wavelength characteristic of the transition between electron orbits.

What holds true for hydrogen holds for other atoms, and for ions and molecules as well. Helium, nitrogen, oxygen, neon—all of these consist of atoms in which the electrons move around the nucleus only in certain orbits. The exact orbits vary for different atoms, ions, and molecules, but in each case an electron can orbit only with certain limited options.

3.5 The Spectra of Starlight

Consider the light that emerges from a faraway star. Light from the stellar interior consists of photons of *all* visible-light frequencies and wavelengths, thanks to the immense number of interactions the photons have had before emerging (see page 258). The outer layers of a star have a temperature low enough for various types of ions, atoms, and sometimes molecules to exist, with each type able to absorb photons of certain definite wavelengths (Figure 3-19). Most of the atoms and ions will have their electrons in the smallest possible orbits. At any particular moment, however, a fraction of the atoms

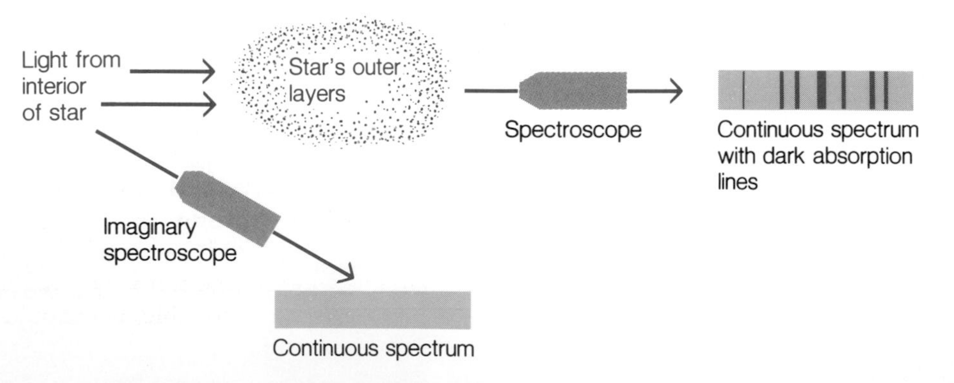

Figure 3-19
Stars produce light of all frequencies and wavelengths—a continuous spectrum—in their interiors. As this light passes through the star's outer layers, the atoms, ions, and molecules there remove photons of certain energies, wavelengths, and frequencies. The result is the characteristic spectrum of a star, with dark lines at those frequencies at which photons have been absorbed.

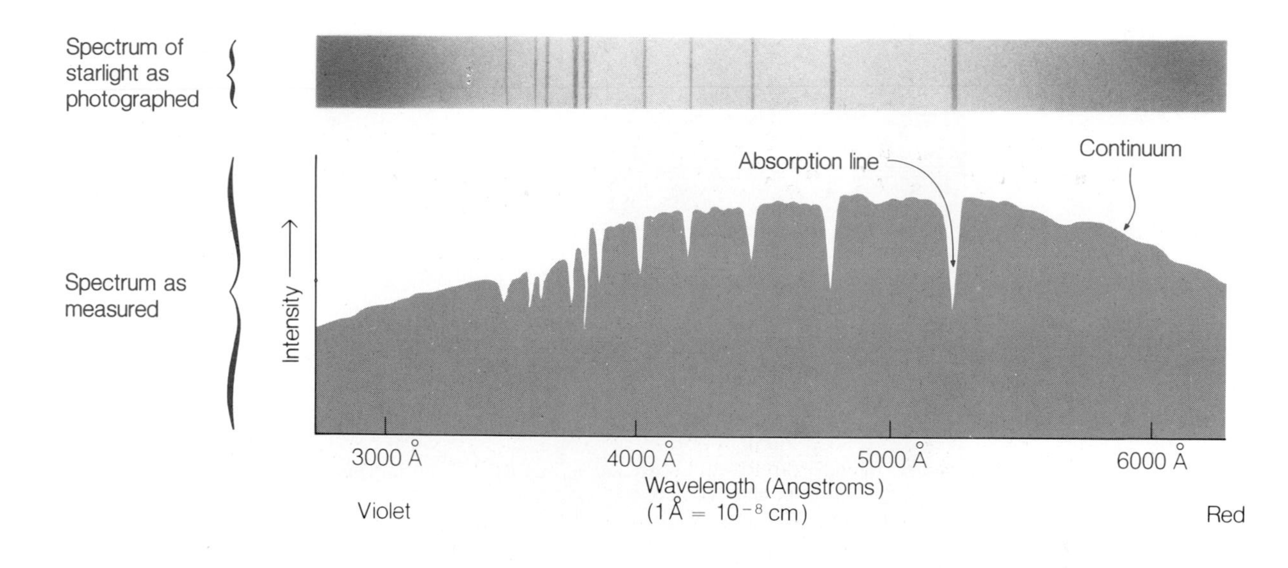

Figure 3-20 When we analyze the spectrum of light from a star, we can measure how much light of a particular wavelength has been removed. These measurements will indicate how much of a particular kind of atom, molecule, or ion is present in the star's outer layers to absorb the light that the star produces in its interior.

and ions will have had some of their electrons knocked into the second-smallest, third-smallest, or even larger orbits by collisions and will not have had time to jump their electrons back toward the smallest possible orbits, emitting photons as they do so.

As light of all energies attempts to pass through such a combination of atoms, ions, and molecules—each presenting a mixture of electron orbits—some of the photons will be absorbed. Just how much light of each wavelength will be absorbed depends on the number of atoms, ions, and molecules present to absorb that particular wavelength. At certain wavelengths—for example, those absorbed by hydrogen atoms with their electrons in the smallest possible orbits—*all* the light will most likely be absorbed. At other wavelengths—for example, those absorbed only by cesium atoms—only a small fraction of the light will be removed (Figure 3-20).

3.6 Stellar Spectroscopy

The analysis of the spectrum of light, called **spectroscopy**, can therefore tell us not only *which kinds* of atoms, ions, and molecules exist in the star's outer layers, but also *how much* of each kind. Spectroscopy can even reveal how fast the various particles are colliding, thus giving us the *temperature* in a star's outer layers, and the *density* of matter there. But that is not all. A star's spectrum can also tell us whether the star is moving toward us or away from us. Before we can see how this is possible, we should examine the spectrum of starlight in a bit more detail.

Figure 3-21 Light that reflects from a finely ruled grating will produce a pattern in which each color of light reaches a maximum intensity in a different position. (See Color Plate N.)

We have seen how a prism spreads light into its various colors, that is, into its different frequencies and wavelengths. An alternative way to display the spectrum consists of reflecting light from a grating, a mirror on which many parallel lines have been ruled at finely spaced, equal intervals (Figure 3-21 and Color Plate N). Light waves of different wavelengths will produce a maximum intensity at different points in space after reflecting from such a grating.

The Solar Spectrum

If we photograph the spectrum of light with such an instrument, we have a record far better than any our eyes can provide. A spectrum of this kind made for sunlight shows a tremendous wealth of detail (Figure 3-22). At each of thousands of different wavelengths, the partial or total absence of light occurs if photons are absorbed by one particular kind of atom, ion, or molecule as one of its electrons moves from one orbit to a larger one.

The study of the solar spectrum began during the nineteenth century, when the physicists William Wollaston and Joseph Fraunhofer recognized the existence of these **absorption lines.** We still call the chief absorption lines in the solar spectrum the Fraunhofer lines. Wollaston and Fraunhofer could not, however, explain why the sun's spectrum showed absorption lines only at certain definite frequencies and wavelengths. Such explanation awaited the genius of Niels Bohr and his collaborators a century later. By now, scientists have identified more than 60 different atomic elements in the

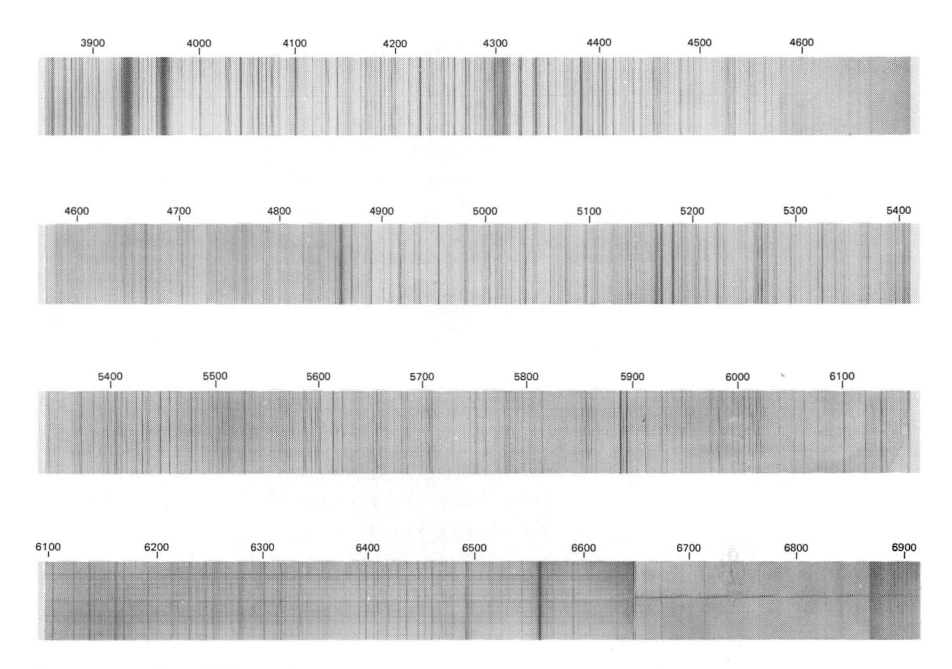

Figure 3-22 If we spread sunlight out into its component wavelengths, we find that at many different wavelengths, far less light, or even none, is present than at most other wavelengths. In this photograph of the sun's spectrum, the wavelengths are marked in units of angstroms (10^{-8} centimeter).

sun through the absorption lines they produce (Figure 3-22). It seems reasonable to assume that the remainder of the 92 naturally occurring elements also exist in the sun, but cannot produce detectable absorption lines, either because their abundance is too small or because their absorption lines do not appear in the visible-light portion of the solar spectrum.

Resemblances Among Stellar Spectra

As astronomers examined the spectrum of sunlight and compared it with the spectra of light from other stars, they found an overall similarity. For example, the absorption lines produced by hydrogen atoms appear in many spectra, always at almost the same wavelengths. By studying how much light was absorbed at each wavelength, astronomers could determine the density, temperature, and chemical composition of the stars' outer layers. A key to the accurate interpretation of stellar spectra has a tremendous importance for astronomy. This key is the Doppler effect.

3.7 The Doppler Effect

The **Doppler effect,** first studied by Christian Johann Doppler during the mid-nineteenth century, is an apparent change in the frequency and wavelength of waves emitted by a moving source. When you hear a train or ambulance that first approaches you and then recedes, you can hear the Doppler effect in the pitch of the sound: The Doppler effect produces the drawn-out, lonesome wail of an engine whistle, falling to lower pitch as the train passes by. As Doppler first correctly explained, we hear a higher pitch when a source of sound moves toward us (or we move toward the source), and a lower pitch when the source of sound moves away from us (or we move away from the source).

Sound waves, which radiate outward from the source in all directions, may be characterized by a certain frequency of vibration for each particular tone we hear. If we move toward the source of sound, or if the source moves towards us, then each successive sound wave arrives with less time delay than is the case for no relative motion. As a result, we receive more wave crests per second, and we hear a higher frequency and a higher pitch (Figure 3-23). Just the reverse occurs when the source moves away from us, or we move away from the source: We lengthen the interval between successive waves, thus lowering the frequency (number of waves per second) and the pitch that we hear.

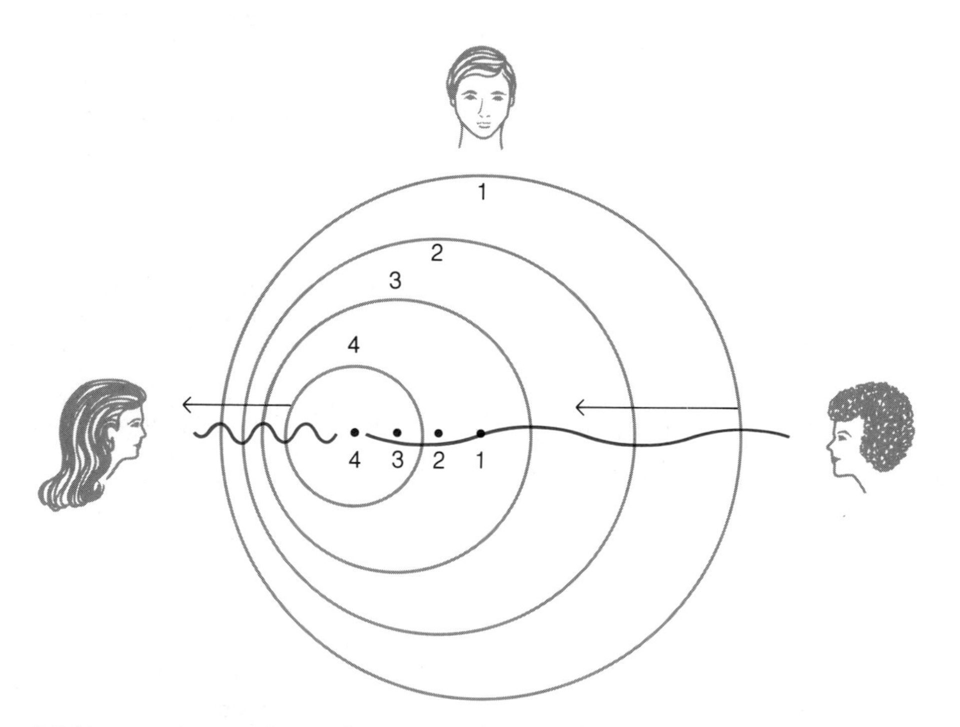

Figure 3-23 If light waves (or sound waves) come to us from a source in relative motion toward us or away from us, we perceive a different frequency and wavelength for the waves than we would in the absence of motion. Relative motion toward us crowds the waves, thus raising the frequency and decreasing the wavelength. Motion away from us increases the separation of wave crests, thus lengthening the wavelength and decreasing the wave frequency.

The Doppler Shift

The Doppler effect also applies to light waves. But because light waves travel at such high speeds (299,793 kilometers per second), we require fairly high relative velocities before the Doppler effect produces a noticeable change in the frequency of light that we observe. As shown in the box on the Doppler effect, the change in photon frequency produced by the Doppler effect will increase for larger velocities. The amount of the change in frequency, called the **Doppler shift**, will be the same in fractional terms for *all* the photons that arrive from a given source. Thus, if the Doppler effect makes the observed frequency of one particular photon 2% less than it would be if no motion occurred, *all* the photons from that source will have their frequencies reduced by 2%. We also know, from the way a photon's frequency, wavelength, and energy are linked, that in this case all the photons will have their wavelengths increased by 2%, while their energies will all be decreased by 2%, just as the frequencies are.

Measurement of Velocity Using the Doppler Effect

Because the amount of the Doppler shift changes as the relative velocity changes, we can use the Doppler effect to measure the velocities of light sources with respect to ourselves. To do this, we must find the difference between the frequency (or wavelength) of the light that reaches us and the

The Doppler Effect for Light Waves

The Doppler effect depends only on the *relative* motion of the source of waves toward or away from an observer. It does not matter whether we move toward a source of light or the source moves toward us: The relative velocity along the line of sight between the source and observer is all that counts. Larger relative-velocities will produce greater Doppler shifts, that is, greater changes in the frequencies and wavelengths that we observe compared to those we would observe if no relative motion along the line of sight existed.

The mathematical expression that links the relative velocity, v, of the source and observer to the speed of light, c, and to the frequencies that are observed in comparison to those we would observe if no motion existed is

$$\frac{\text{Frequency actually measured}}{\text{Frequency if no motion occurred}} = \sqrt{\frac{1 - (v/c)}{1 + (v/c)}}$$

Here the relative velocity, v, is taken to be positive if the motion is one of recession, and negative if the motion is one of approach. Only the velocity along the line of sight (toward us or away from us) counts in v.

The equation above shows that

$$\frac{\text{Change in frequency}}{\text{Frequency if no motion occurred}} = \sqrt{\frac{1 - (v/c)}{1 + (v/c)}} - 1$$

If v is much less than c, this gives the approximate formula

$$\frac{\text{Change in frequency}}{\text{Frequency if no motion occurred}} = -\frac{v}{c} \quad \textit{if} \quad \frac{v}{c} << 1$$

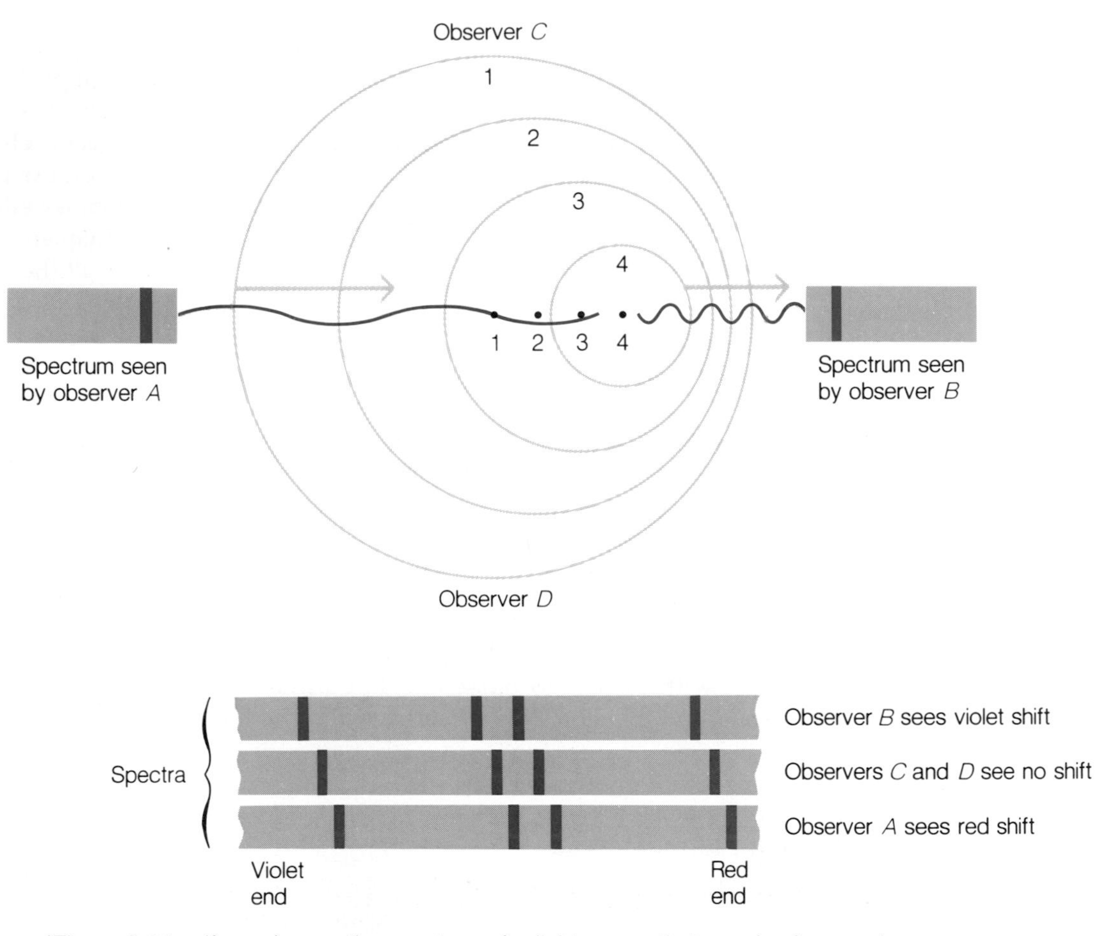

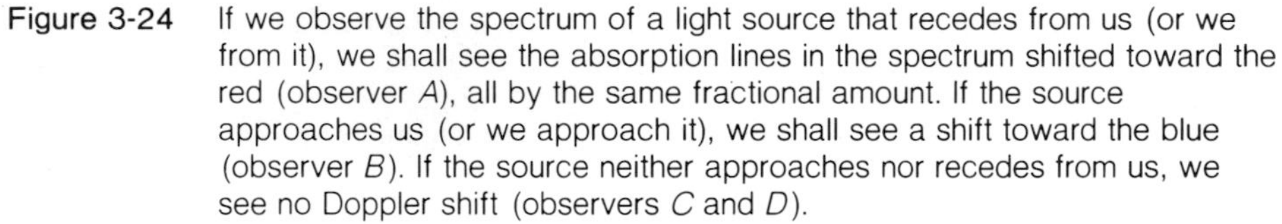

Figure 3-24 If we observe the spectrum of a light source that recedes from us (or we from it), we shall see the absorption lines in the spectrum shifted toward the red (observer *A*), all by the same fractional amount. If the source approaches us (or we approach it), we shall see a shift toward the blue (observer *B*). If the source neither approaches nor recedes from us, we see no Doppler shift (observers *C* and *D*).

frequency (or wavelength) that the light would have if no relative motion occurred. The fractional amount of this change—the size of the Doppler shift—will tell us the relative velocity of the source of light. If the frequency is lowered, we know that the source must be in relative motion away from us, like an ambulance whose siren falls in pitch as it recedes. Conversely, relative motion of the source toward us will raise the frequency of each photon we observe, and once again, the amount of the Doppler shift will allow us to determine the size of the relative velocity of approach.

To use the Doppler effect, we must compare the photon frequency that we *do* observe with the frequency that we *would* observe if no relative motion occurred. How can we do this? The answer lies in our familiarity with the photon spectrum. Suppose we photograph the spectrum of light from a star and find a similar pattern of absorption lines, reminiscent of that in Figure 3-22. If all the absorption lines are slightly displaced toward higher frequencies and shorter wavelengths, and all by the same fractional amount, we may reasonably conclude that the Doppler effect is at work. In this case, the star has a relative motion of approach: Either we are moving toward the star, or the star is moving toward us (Figure 3-24).

3.8 The Mystery of the Nebulae

During the later years of the nineteenth century, astronomers used their measurements of the Doppler effect to determine the motions of thousands of stars with respect to the Earth. They did this before they understood why the now-familiar absorption lines had certain definite frequencies; it was enough that they recognized the patterns of absorption lines that appeared in spectrum after spectrum. But in addition to the stars, astronomers discovered another class of objects, not small points of light, but diffuse, cloudy objects called *nebulae* after the Latin word for cloud (Figure 3-25).

Figure 3-25
This drawing of the Whirlpool Nebula in the constellation Canes Venatici (the Hunting Dogs) was made by Lord Rosse during the first half of the nineteenth century. Compare this drawing with the photograph of what we now know to be a galaxy (Figure 4-22).

The Messier Catalog of Nebulae

Charles Messier had made the first catalog of nebulae during the 1770s, not because he had a particular interest in them but because he did not want to mistake them for comets, his true astronomical love. Today Messier's comet searches are forgotten, and his fame lives in his list of 109 "Messier objects," the cloudy nebulae he saw with a small telescope and a great deal of patience. Astronomers designate the objects in Messier's catalog with the letter M followed by a number, such as M 1 (the Crab Nebula) or M 20 (the Trifid Nebula) or M 42 (the Orion Nebula). Appendix 4 lists the objects in Messier's catalog.

Emission-Line Spectra

When spectroscopic observations were made of the various nebulae, they presented a puzzle. Some of the nebulae showed spectra reminiscent of stars, with relatively familiar patterns of absorption lines. We now know these to be galaxies, faraway groups of billions of stars, such as the famous Andromeda galaxy (M 31). But many nebulae showed spectra dominated not by absorption lines but by **emission lines,** frequencies at which far more than the normal number of photons appeared.

Today we understand such emission-line spectra as the result of special situations in which more atoms and ions are emitting photons (by having their electrons jump into smaller orbits) than are absorbing photons (by having their electrons jump into larger orbits). These special situations usually arise in the vicinity of a star near either the beginning or the end of its life (see Chapters 9 and 10).

In the history of astronomy, the nebulae played an important, and at times confusing, role. The nebulae, we now know, include various types of objects, some within our own galaxy, others far beyond the Milky Way. The nebulae within the Milky Way are not all alike. Some are clouds of gas and dust in the process of becoming stars, others are recently formed star clusters, and still others include matter ejected from old stars. The nebulae outside the Milky Way, now called galaxies, also have different sizes, shapes, and distances. But through the telescopes of the late eighteenth and nineteenth centuries, all the nebulae seemed vaguely similar. Hence astronomers often tried to construct theories that explained the nebulae as a single class of objects.

Eventually, astronomers resolved the mystery of the nebulae by using better telescopes, spectroscopy, and the Doppler effect. The solution to this puzzle takes us into the next chapter, and into the realm of **cosmology,** the study of the universe as a whole. This fascinating tale hinges on the steps of understanding that astronomers made in studying the details of the light from stars. Like much else in science, the great leaps forward could never have been made without years of patient effort on the part of dedicated men and women. Not only the Niels Bohrs and Albert Einsteins, but also the host of scientists who catalogued the stars and nebulae, studied their spectra and their motions, and photographed the changes in stellar brightnesses, were needed for humanity to arrive at a startling conclusion: The universe is expanding.

Summary

The science of optics began with the construction of telescopes, first the refracting type and later the reflecting type. Each kind of telescope focuses light to produce an image, which can be magnified by an eyepiece. Each kind will also produce a brighter image if the telescope's light-gathering power is increased by use of a larger lens or mirror.

From the practical study of light came the question, What *is* light? Modern physics answers this question by stating that light has properties of both waves and particles. We can think of light waves as streams of tiny particles called photons. Each photon travels through space at nearly 300,000 kilometers per second. We can imagine a photon to be a tiny, wiggling tadpole, characterized by a particular kinetic energy, a particular frequency (number of cycles per second), and a particular wavelength (distance between successive wave crests). For any photon, the frequency times the wavelength always equals the speed of light.

The photons in light waves can interact with atoms, the basic units of matter. Each atom contains a nucleus, made of protons and neutrons, surrounded by electrons, whose number equals the number of protons in the nucleus. These electrons can move only in certain definite orbits, and not in any orbit whatsoever. When a photon strikes an atom, it can make one of the atom's electrons jump into a larger orbit *if* the photon has just the right amount of energy to produce this jump. If the photon does not have the proper amount of energy to produce an electron jump into a larger orbit, the photon will pass by the atom with no interaction. Once the electron has jumped into a larger orbit, it can jump into a smaller one again. In such a process, the reverse of the outward jump, the atom will *emit* a photon, whose energy exactly equals that needed for a photon to produce the outward electron jump.

Photons that pass outward from the center of a star through a mixture of atoms in the star's outer layers will, for certain photon frequencies, suffer absorption: The photons will have the right energies to make electrons jump into larger orbits and will disappear in the process. Hence an examination of the spectrum of light from a star—the number of photons at various frequencies—will show a relative lack of photons at the frequencies and energies that produce electron jumps in the most abundant atoms. Through careful study of such spectra, astronomers can deduce the density, temperature, and chemical composition in the outer layers of stars.

The absorption lines in many stars' spectra appear at frequencies slightly different from usual because of the Doppler effect. The motion of a light source relative to an observer will change the frequencies, wavelengths, and energies of the photons observed. Relative motion of approach will raise the frequencies and decrease the wavelengths, all by the same fractional amount. In contrast, relative motion of recession will lower the frequencies and increase the wavelengths, again by the same fractional amount for all photons. Astronomers can use the amount of the Doppler shift in frequency and wavelength to determine the direction and magnitude of the motion of a light source relative to ourselves.

Key Terms

absorbed
absorption lines
atom
chromatic aberration
cosmology
Doppler effect
Doppler shift
dynamics
electromagnetic radiation
electron
element
emission lines
erg
excited state
frequency
gamma rays
ground state
infrared
ion
isotope
microwaves
neutron
nucleus
optics
photon
proton
quantum mechanics
radio
reflecting telescope
refracting telescope
spectroscopy
spectrum
ultraviolet
visible light
wave theory of light
wavelength
x rays

Questions

1. In Huygens's wave theory of light, how do light waves resemble the ripples on a pond that appear when we drop a stone in the water? What do the frequencies and wavelengths of light waves correspond to in this model?
2. What did Isaac Newton discover about the way light waves bend when they pass through a glass prism? Why does the effect Newton discovered pose problems for the manufacture of refracting telescopes?
3. Why have human eyes evolved to be sensitive to the frequencies of electromagnetic radiation we call visible light?
4. What are infrared and ultraviolet radiation? Does it make sense to call these infrared light and ultraviolet light?
5. If two photons have wavelengths of 1 centimeter and 5 centimeters, respectively, which photon has more energy? By how much?
6. How do various isotopes of an element (such as hydrogen) differ? How are they similar?
7. An atom of carbon-13 has six protons and seven neutrons in its nucleus. How many electrons will orbit the nucleus in this atom?

8. Suppose a photon hits a hydrogen atom that has its electron in the smallest possible orbit and knocks the electron into the fifth-smallest orbit. Will the photon gain or lose energy as a result of this interaction? Explain.
9. What will happen to the hydrogen atom described in Question 8 once its electron has been knocked into the fifth-smallest orbit, if the atom is allowed to sit quietly by itself?
10. Why does a particular kind of atom absorb only certain definite wavelengths and frequencies of light?
11. Why does motion of a light source relative to an observer change the frequency of the light waves that the observer detects? What directions of motion can be detected with this method?
12. What does a Doppler shift to *longer* wavelengths imply about the relative motion of a light source and observer? What does the amount of such a shift tell about the relative velocity?
13. How do the atoms and molecules in the Earth's atmosphere affect the photons that try to penetrate it?

Projects

1. *Galileo and Newton.* Read the books about Galileo and Newton listed in the Further Reading section of this chapter, or visit your library and consult other books about these scientists. Compare the relationship of Galileo to the Catholic Church in Italy in the early seventeenth century with that of Newton to the Church of England in the late seventeenth and early eighteenth centuries. What changes had occurred in the relationship between science and religion during the seventeenth century? How did Newton regard the role of religion in science? How did Galileo reconcile his recantation of the Copernican doctrine with his true beliefs?
2. *The Spectrum of Light.* The light that we see consists of photons with wavelengths between about 3500 and 8000 angstroms (1 angstrom = 10^{-8} centimeter). The symbol for angstrom is Å. Determine the approximate wavelength of the photons emitted from hot sodium atoms by tossing a pinch of salt into a flame and observing the color of light produced. The *color* a person sees varies somewhat from individual to individual, so that what one person sees as orange may appear orange-yellow to another. Can you think of a better way to measure the wavelengths of photons than observing their colors?
3. *Excitation and De-excitation of Atoms.* Photons are produced by hot sodium atoms when one of the electrons jumps into a smaller orbit in the atom. You can see this by throwing a pinch of salt into a fire. (The sodium in salt is actually part of a sodium chloride molecule until it is tossed into the flame). So-called neon lights work on the same principle. An electrical discharge *excites* atoms, moving their electrons into larger orbits; when the atoms undergo *de-excitation,* the electrons' jumps into smaller orbits produce photons of a definite frequency and wave-

length. Consult your library (see encyclopedia articles on lighting) to determine which types of atoms produce the vivid red, gold, and green colors in neon lights. What are the wavelengths of these colors?

4. *The Doppler Effect.* Use the Doppler effect for sound waves to determine the speed of a car that approaches you, then recedes from you, with its horn blowing. The change in pitch (frequency) of the sound wave divided by the original frequency equals the speed of the car divided by the speed of sound (approximately 1000 kilometers per hour). Try to estimate the change in pitch by comparing the pitch you hear with the notes on a piano (middle C = 264 cycles per second; low C = 132 cycles per second; high C = 518 cycles per second).

Further Reading

Alfven, Hannes. 1966. *Worlds-Antiworlds.* San Francisco: W. H. Freeman.

Asimov, Isaac. 1966. *Inside the Atom.* New York: Abelard-Schuman.

Asimov, Isaac. 1975. *Eyes on the Universe: A History of the Telescope.* Boston: Houghton Mifflin.

De Santillana, Giorgio. 1955. *The Crime of Galileo.* Chicago: University of Chicago Press.

Frisch, Otto. 1972. *The Nature of Matter.* London: Thames and Hudson.

Goldsmith, Donald, and Levy, Donald. 1974. *From the Black Hole to the Infinite Universe.* San Francisco: Holden-Day.

Koestler, Arthur. 1959. *The Watershed.* New York: Macmillian.

Ley, Willy. 1963. *Watchers of the Skies.* New York: The Viking Press.

Manuel, F. E. 1968. *A Portrait of Newton.* Cambridge: Harvard University Press.

Weisskopf, Victor. 1979. *Knowledge and Wonder.* Cambridge, Mass.: MIT Press.

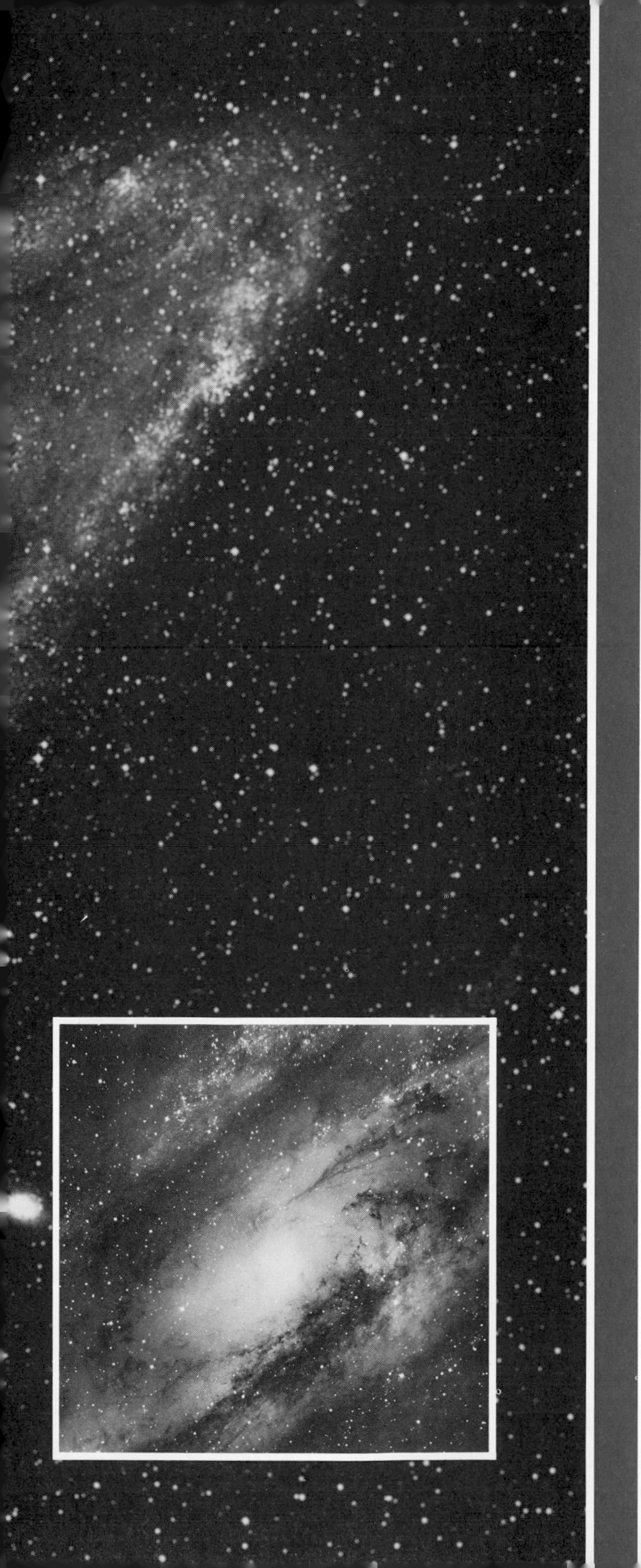

PART 2

The Universe at Large

We live within the Milky Way galaxy, a web of 400 billion stars, each of them in orbit around the center of our galaxy. Just 60 years ago, an American astronomer named Edwin Hubble was able to show that our Milky Way is only one galaxy among many, and that the Andromeda Nebula is in fact a spiral galaxy much like our own. The Andromeda galaxy, 2 million light years away from us, forms part of our Local Group of galaxies, our closest neighbors among the universe of galaxies.

Hubble's work proved that the Milky Way is not the center of the universe—an extension of the Copernican notion that the Earth is not the center of the solar system. Indeed, the universe turns out to have no center at all. Still more remarkable is the evidence that all parts of the universe are moving away from all other parts. In other words, the entire universe is expanding.

This spiral galaxy, NGC 5364, is receding from us at a velocity of 1400 kilometers per second.

4

The Discovery of the Universal Expansion

By the end of the nineteenth century and the beginning of the twentieth, astronomers had achieved a well-deserved feeling of pride in their understanding of the heavens. In the three and a half centuries since Copernicus, they had constructed an apparently correct model of the motions of the objects in the solar system. Spectroscopic studies of the sun and other stars had measured their high surface temperatures and shown that these objects consist of chemical elements identical to those found on Earth;[1] meanwhile, Doppler-effect measurements of stellar spectra told whether stars were approaching us or receding from us.

The modern science of **astrophysics** (physics applied to astronomical situations) could point to a long string of successes by the First World War. But at least two important questions remained unanswered. How do stars produce their energy? And what is the overall structure of the universe? Intriguingly enough, a fundamental part of the answer to the second question was achieved by 1929, ten years before the first received a satisfactory solution.

4.1 The Discovery of Our Galaxy

Copernicus had dethroned the Earth from its position at the center of the cosmos. Later studies showed our sun to be just one star among millions. Was there a center to the vast array of stars? Did they have a pattern in space, or did they simply appear at random? Generations of astronomers attempted to answer these questions by counting the number of stars in various directions and attempting to estimate their distances. During the late eighteenth century, William Herschel was able to show that the sun lies near the plane of the **Milky Way,** the part of the sky most densely populated by stars, first viewed through a telescope by Galileo (Figure 4-1). But was the sun close to the center of the disk of the Milky Way? And did it have a center?

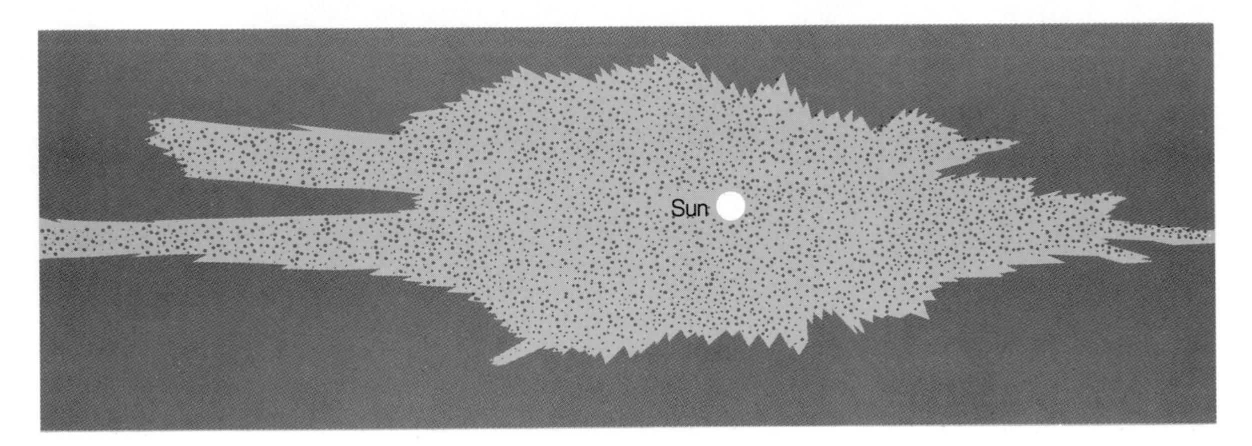

Figure 4-1
When William Herschel measured the numbers of stars he could see in various directions, he correctly realized that the Milky Way has a rather flat shape, though he incorrectly gauged the sun's location to be close to the center of the entire Milky Way.

The answers to these questions came slowly, because the absorption of starlight by interstellar dust prevents us from seeing much of the structure of the Milky Way in which we live (Figure 4-2). Only gradually did it become clear that the bulk of the material in the Milky Way lies in the direc-

[1]Helium, the second most abundant element in the universe, was first discovered from studies of the sun's spectrum (hence its name, from *helios,* the Greek word for sun). A few years later, helium was found on Earth, where it appears as a rather rare gas.

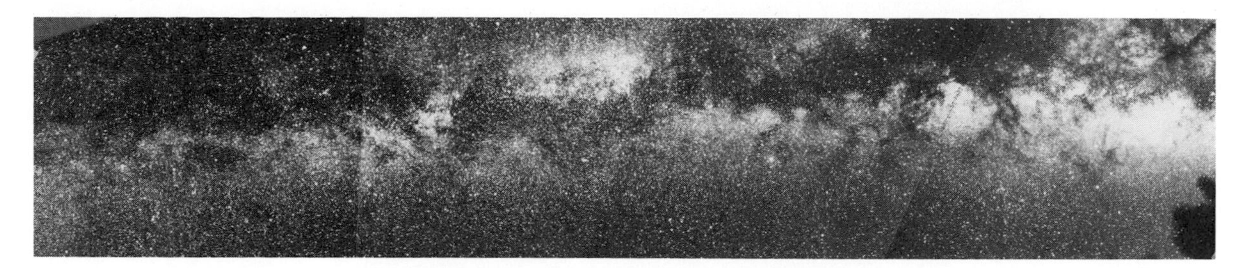

Figure 4-2 Great clouds of dust that lie within the plane of the Milky Way absorb starlight, thus preventing us from seeing the central regions of our galaxy and beyond. We therefore have difficulty in perceiving the bulk of the Milky Way and in determining our location within it.

tion of the constellation Sagittarius and that the sun must occupy a position far from the center (Figure 4-3). By the beginning of this century, astronomers had begun to call this distribution of stars the Milky Way **galaxy** (from the Greek word for "milk").

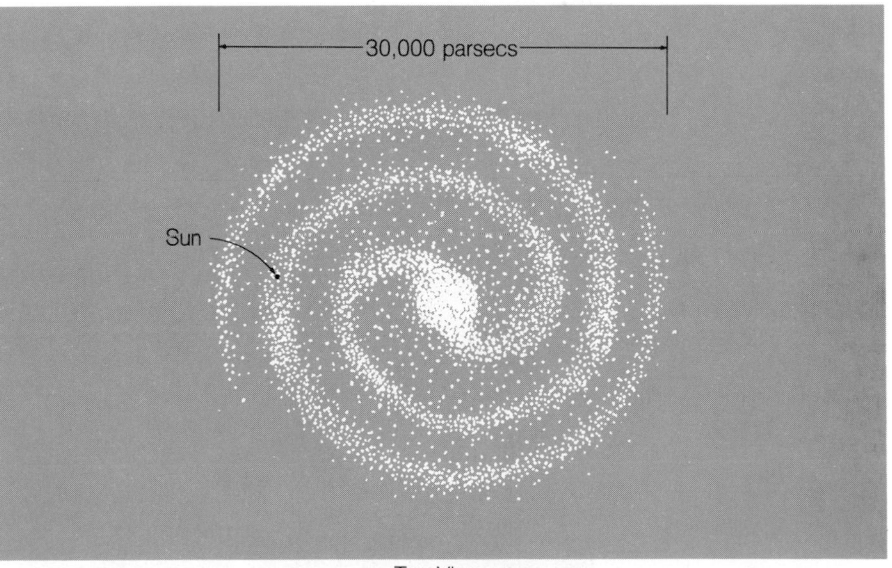

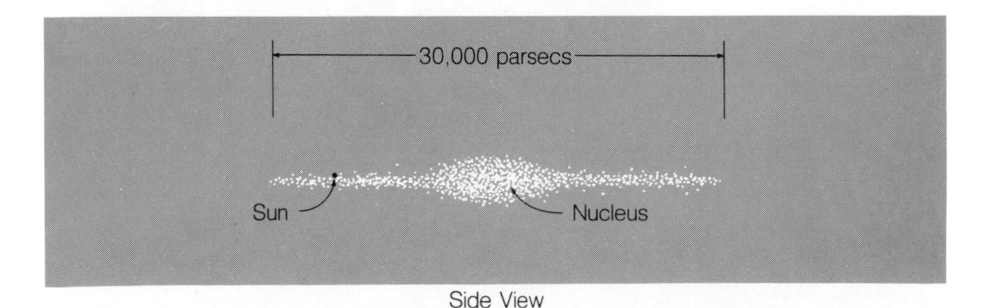

Figure 4-3
Modern observations have shown the Milky Way to be a giant spiral galaxy, basically a flattened disk with a central bulge or nucleus. Within the disk, bright young stars appear predominantly within the galaxy's "spiral arms," shown in the (imaginary) top view. The sun's distance from the galactic center is about 9000 parsecs. This complex spiral structure has been greatly simplified in this schematic diagram.

4.2 The Discovery of Other Galaxies

Does our galaxy contain everything in the universe? Or does the Milky Way form just one galaxy among many other "island universes," to use a bit of old terminology? The controversy over this issue focused on the nature of the spiral and elliptical **nebulae** ("clouds" in Latin) that astronomers had observed for 200 years and photographed for 30 (Figure 4-4).

The Great "Debate"

In 1920, the National Academy of Sciences sponsored a "debate"—actually two consecutive lectures—by Heber Curtis and Harlow Shapley. Curtis insisted that the spiral nebulae must lie far outside the Milky Way; Shapley claimed that these nebulae lie within the reaches of the Milky Way galaxy.

Shapley, the eventual loser in the debate, had previously shown that the center of the Milky Way lies far from the sun (about 9000 parsecs away, we now know). However, Shapley had been misled concerning the distances to spiral nebulae because of the incorrect interpretations of another astronomer, Adriaan van Maanen. Based on his observations, van Maanen believed that the spiral nebulae were rotating with angular speeds that would produce a complete rotation in several hundred thousand years. If van Maanen

Figure 4-4 The spiral galaxy M 33 has a distance from us of about 2 million light years (680,000 parsecs).

were correct, the spiral nebulae simply could not be at distances great enough to place them outside the Milky Way, since the rapidity of rotation he claimed would give them velocities greater than the speed of light.

Van Maanen's analysis, however, was simply incorrect—not the first or last time that careful study of the heavens has turned out to be not careful enough. The spiral nebulae *do* lie outside the Milky Way—so far away that they must be "island universes" in their own right, comparable in size to the Milky Way galaxy. Today we no longer call them spiral nebulae but rather spiral galaxies, and we recognize our own galaxy as a giant spiral quite similar in structure and size to the Andromeda galaxy (Color Plate I). Ironically, spiral galaxies *do* rotate, but in durations of hundreds of *millions* of years, not hundreds of thousands. This much slower rotation was discovered only during the 1940s, long after the Shapley-Curtis debate.

The Distance to the Andromeda Galaxy

The astronomer who proved once and for all that spiral "nebulae" form no part of the Milky Way and are indeed separate "island universes" was a Missouri-born, Oxford-educated contemporary of Shapley's named Edwin Hubble (Figure 4-5). Hubble used the 2.5-meter reflecting telescope of the Mount Wilson Observatory, then the largest telescope in the world, to study galax-

Figure 4-5 Edwin Hubble (1889–1953), shown here at the controls of the 1.22-meter reflecting telescope at Palomar Observatory, was the world's expert at estimating the distances to other galaxies.

ies such as Andromeda through night after night of long-exposure photography. In 1923, Hubble found that the Andromeda galaxy, and other spirals as well, contain **periodic variable stars**—stars with brightnesses that change on a regular schedule—whose period of light fluctuation matched those of well-studied stars in the Milky Way. Because the stars in our galaxy and in Andromeda seemed almost identical, a comparison of their apparent brightnesses would reveal the stars' relative distances. This comparison put the Andromeda galaxy well beyond the outermost reaches of our own galaxy and eliminated for good the notion that the Milky Way might contain the entire universe.

In Chapter 5 we shall examine the method used by Hubble to estimate the distances to galaxies. Hubble became the expert in this field, and working together with Milton Humason, made estimates for several dozen galaxies. When he combined these results with the Doppler-shift measurements of the velocities of galaxies, he came to a remarkable conclusion: The universe is expanding!

4.3 The Expanding Universe

To understand how Hubble could draw such a far-reaching conclusion from observations of the distances and velocities of galaxies, we must examine the observations themselves and the models of the universe we can construct in our minds.

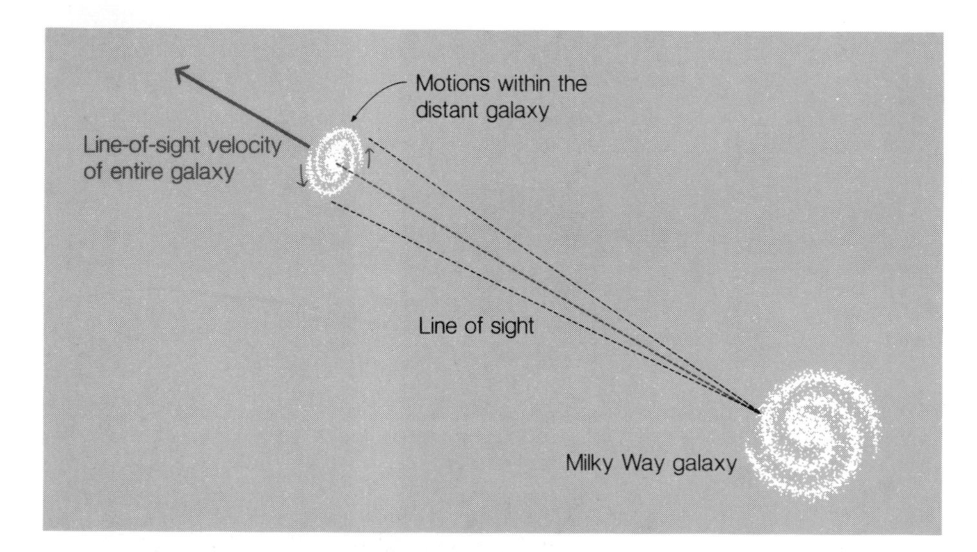

Figure 4-6 When we observe the light from an entire galaxy, we measure the total output of billions of stars, each of which has some motion within the galaxy. Nonetheless, since these motions are each much less than that of the galaxy as a whole, we can measure the relative velocity of the entire galaxy with respect to ourselves. The Doppler effect displaces the spectrum of the light from *all* the stars in the galaxy by approximately the same amount.

The Velocities of Galaxies

During the first two decades of the twentieth century, at the Lowell Observatory in Arizona, Vesto Slipher had made accurate spectroscopic measurements of the light from many galaxies. He saw in the spectra of these galaxies the same pattern of lines he had seen in his study of stars in our own galaxy. The natural conclusion seemed to be that galaxies contained masses of stars similar to those in the Milky Way. Since the motions of stars within a given galaxy amounted to only a small fraction of the speed of light, the combination of millions of stellar spectra into a single galactic spectrum preserved the basic pattern of a single spectrum. However, there is some spreading because of the Doppler effect of the stars' individual motions within the galaxy (Figure 4-6).

Slipher proceeded to measure the velocity of entire galaxies by means of the Doppler effect, measuring the amount by which the total spectrum of a galaxy was shifted toward the red (long wavelengths) or toward the blue (short wavelengths). If the familiar pattern appeared in the spectrum of a galaxy, but shifted by, say, 0.25% in all wavelengths toward the red end of the spectrum, Slipher could reasonably conclude that the entire galaxy had an overall recession velocity (away from us, or we from it) of 0.25% of the speed of light (see page 87).

The Velocity-Distance Relationship

By use of the Doppler-shift measurements, Slipher found that only a few galaxies (most noticeably, the Andromeda galaxy) have relative velocities of approach, while most have velocities of recession. To make sense of this result, Hubble's distance determinations were needed. In 1929, Hubble put together the distance and velocity measurements that produced the diagram shown in Figure 4-7. Despite a fair amount of experimental uncertainty (or

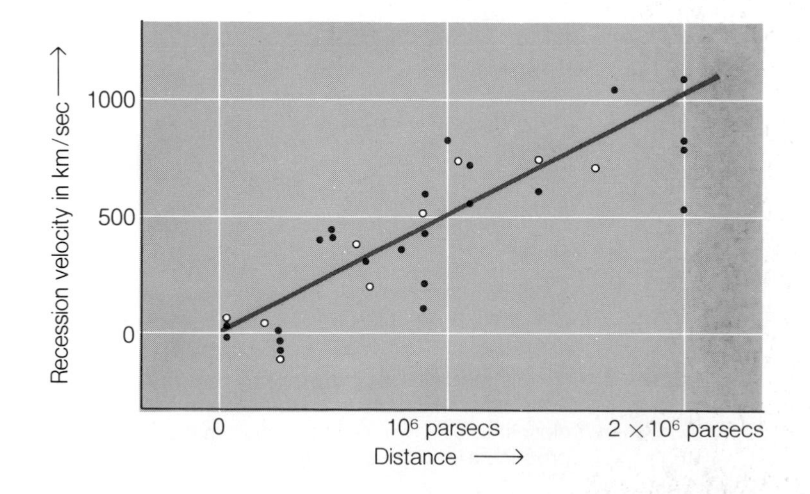

Figure 4-7
Hubble's first plot of the distances to galaxies (in parsecs) against their velocities along our line of sight (in kilometers per second) was published in 1929. This graph showed a fair amount of scatter around the general trend that larger distances implied larger recession velocities.

"scatter") in the diagram, Hubble recognized a general trend, as shown by the solid line. Clearly, more distant galaxies are receding from us more rapidly.

Hubble concluded that each galaxy has a certain random velocity of its own, relative to the other galaxies, and that this random velocity can add or subtract a relatively small amount to or from the overall trend (Figure 4-8). This explains, in part, the scatter he found, but the trend itself is the most important factor (Figure 4-9). If a galaxy's recession velocity increases in proportion to its distance from us, as Hubble found, then the entire universe must be expanding. Galaxies are rushing away from one another everywhere!

The Expanding Balloon: A Model of the Universe

To see why a theory of universal expansion follows from the diagram Hubble drew, we must consider a model of the universe. No model can be completely satisfactory, for the universe—everything that exists—can hardly be encompassed in a simple model. On the other hand, such models offer us a chance to unravel some of the mystery of the universe.

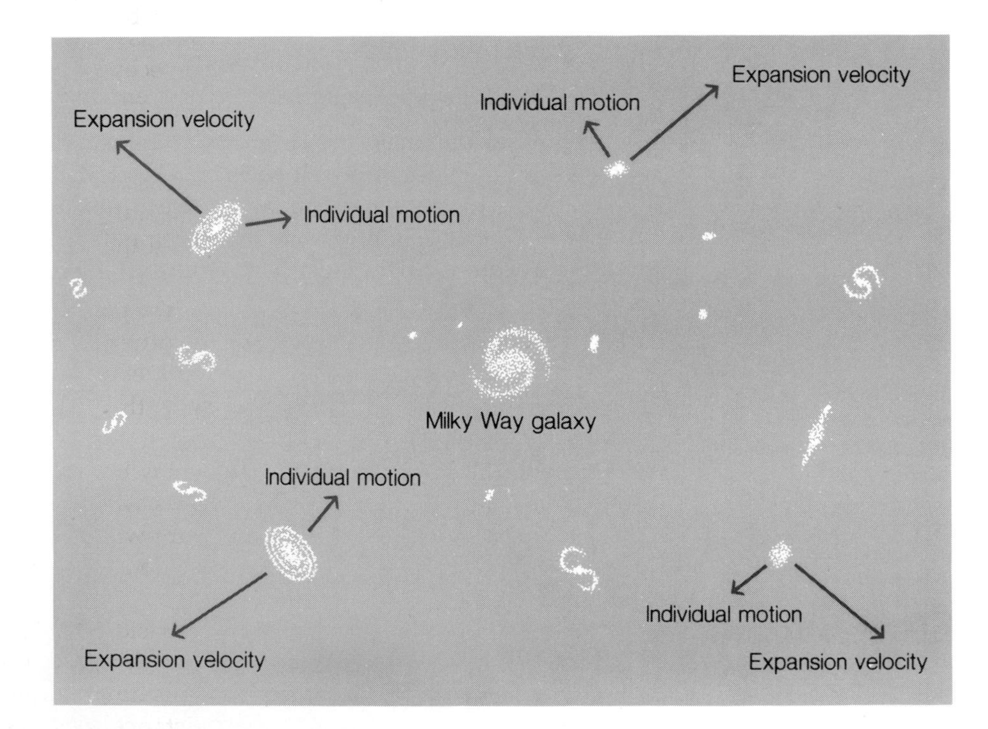

Figure 4-8 Hubble realized that each galaxy may have its own random motion that adds to, or subtracts from, the general trend. Only for those galaxies closest to our own might the random motion dominate the overall trend to produce a velocity of approach.

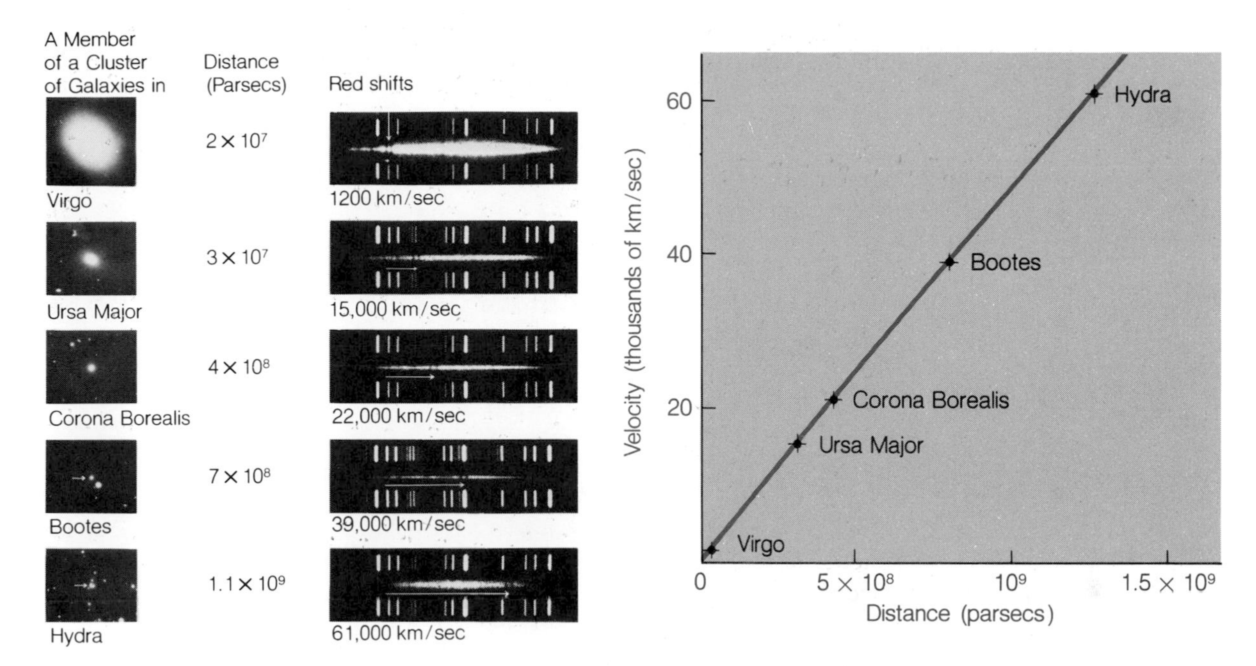

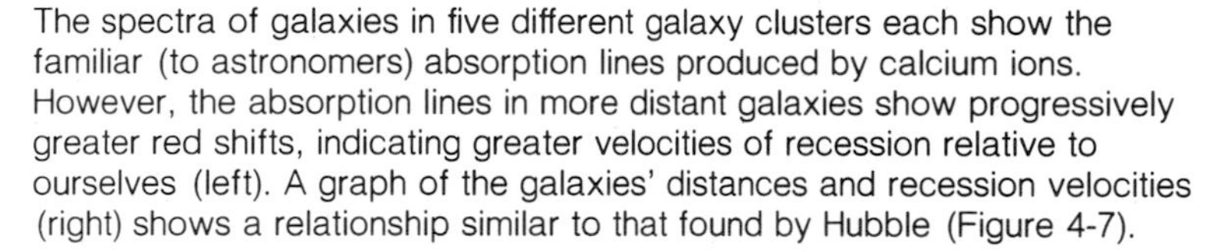
Figure 4-9 The spectra of galaxies in five different galaxy clusters each show the familiar (to astronomers) absorption lines produced by calcium ions. However, the absorption lines in more distant galaxies show progressively greater red shifts, indicating greater velocities of recession relative to ourselves (left). A graph of the galaxies' distances and recession velocities (right) shows a relationship similar to that found by Hubble (Figure 4-7).

How Does the Balloon Model Work? Picture a model of the universe in which galaxies are represented by points on the surface of a balloon (Figure 4-10). If we blow up the balloon, each point will move away from all the other points, and the speed at which any two points separate will be proportional to the distance between them. In other words, an observer on any point on the balloon (any galaxy in the universe) will see what Hubble saw: Galaxies recede at speeds that increase with the distance from that point.

How Valid Is the Model? Of course, it may be argued that Hubble's observations apply only to the motion of galaxies away from the Milky Way. Why can't we then draw the conclusion that our own galaxy is at the center of the expansion, from which other galaxies recede at speeds proportional to their distances from us?

The answer to this question lies in a basic attitude among astronomers, nurtured since the time of Copernicus, that we do not occupy the center of the universe. To be sure, **Hubble's law**—his observation that galaxies recede from us at speeds proportional to their distances from us—refers only to the motions of galaxies relative to our own Milky Way. But if our view of the universe is indeed a representative one, then every galaxy should see other galaxies receding from *that* galaxy with the same constant of proportionality

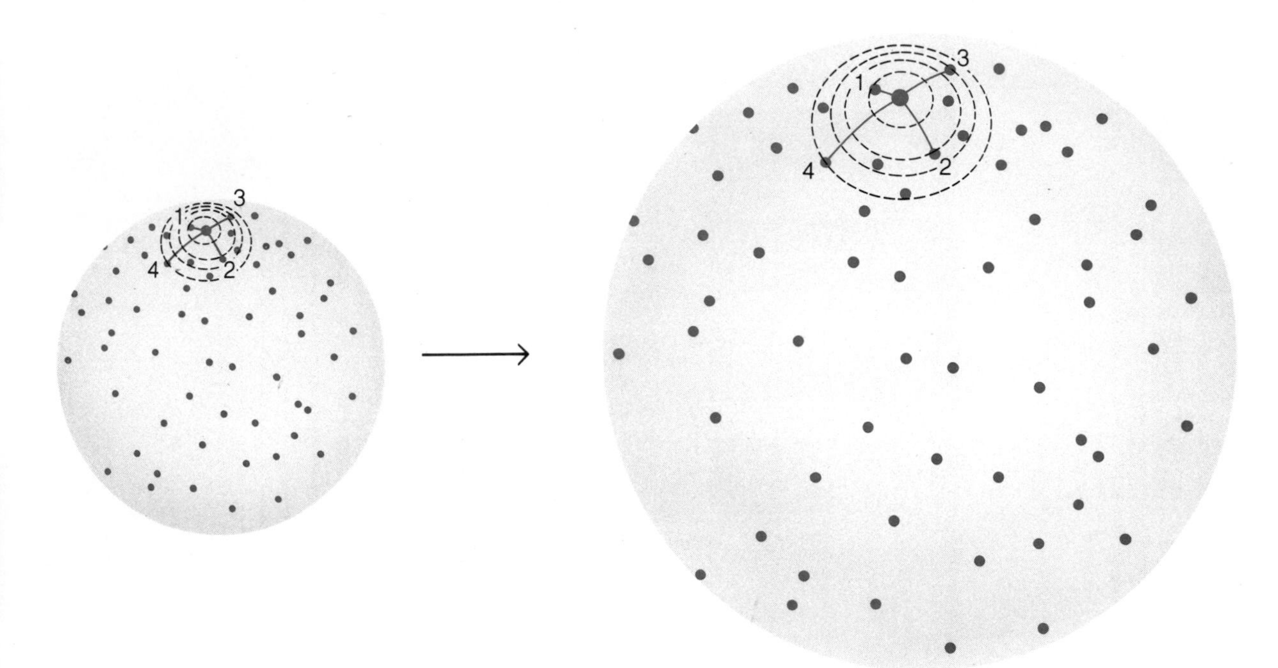

Figure 4-10 If we picture the galaxies as dots on the surface of a balloon, we can model the expanding universe by the *surface* alone. As we blow up the balloon, each dot moves away from all the other dots with a speed that is proportional to the distance between the two dots.

Figure 4-11
If we hope to imagine the curvature of the entire universe, we can compare space to the *surface* of an expanding balloon. We could imagine that we were flat creatures sliding around the balloon's surface and that nothing existed besides this surface. We could discover that the surface is finite in area, even though we could not "imagine" how this was possible.

Hubble's Law for the Expansion of the Universe

Hubble's law takes the algebraic form

$$v = H \times d$$

where v is the velocity of recession, d is the distance, and H is a universal constant, called *Hubble's constant.* Each galaxy may have some peculiar, random velocity, which adds to or subtracts from the velocity given by Hubble's law.

The best current estimate for the value of Hubble's constant gives

$$H = 55 \text{ kilometers per second per megaparsec}$$

That is, the recession velocity increases by 55 kilometers per second for every additional megaparsec (1 million parsecs) of distance.

Since Hubble's constant turns a distance into a velocity upon multiplication, H must be in units of 1 over time. Another way to see this is to remember that distance equals velocity multiplied by time, so that velocity must equal distance multiplied by 1 over time. If we use the fact that 1 megaparsec of distance equals 3.1×10^{19} kilometers, we can divide this into 55 kilometers to find that

$$H = 1.8 \times 10^{-18} \text{ per second}$$

which implies that

$$\frac{1}{H} = 5.6 \times 10^{17} \text{ seconds} = 17.9 \text{ billion years}$$

This amount of time has the name *Hubble time.* The Hubble time would give the age of the universe since the expansion began *if* the expansion had always proceeded at the same rate. Since we know that the expansion has slowed somewhat, the Hubble time gives the *maximum* age of the universe. However, our value for the Hubble constant, and thus for the Hubble time, has an uncertainty of at least 20%. So the Hubble time could be 20% greater than 17.9 billion years (21.5 billion years) or 20% less (14.3 billion years).

between distance and recession velocity that we detect. Of course, we can never prove that our view of the universe represents the universe as a whole; we might indeed be the center of the cosmos. However, judging from what we have learned since Ptolemy's day, we would do well to assume that we do *not* form the center of the cosmos.

Why the Universe Is Not a Balloon. Returning to our model of the universe, the expanding balloon, we should mention a key difference between the model and the real universe. The surface of the expanding balloon has only two dimensions, like the surface of the Earth, but the real universe has three dimensions in space.[2] The representation of space by the two-dimensional surface of the balloon is necessary if we hope to picture the entire universe (Figure 4-11). This may cause some difficulty in seeing the correspondence between the model and the real universe, but it provides a good reminder that our model cannot depict all the features of the universe, only certain important ones. The best aspect of the balloon model is that it reproduces Hubble's law. When we blow up the balloon, greater distances imply greater expansion velocities.

[2]In some respects, time may be considered a "dimension," but space itself has only three dimensions.

4.4 How Old Is the Universe?

Once we accept Hubble's law as a valid description of the universal expansion, we can run time backward in our mind and come to the beginning of the expansion, the **initial singularity,** or **big bang,** that started things off (see page 112). Simply by determining the value of the constant of proportionality in Hubble's law—**Hubble's constant**—we can estimate the length of time the expansion has taken.

Early Estimates

When Hubble first made his observations, he derived an age of the universe equal to about 1.8 billion years. This posed a problem, since many stars, as well as the Earth, already seemed older than this. Later revisions in the distances to other galaxies changed the value of the proportionality constant in the velocity-distance relationship and raised the estimated age of the universe to 15 to 20 billion years. This fits rather well with an Earth age of 4.6 billion years, our galaxy's 12 billion years, and a series of estimated stellar ages. But we should remember the large changes in past estimates and be prepared to learn of further revisions in the future.

Revised Estimates

Recently, a new technique for estimating the distances to galaxies has yielded distances somewhat less than those used to make our estimate of the age of the universe (see box, page 109). This new technique relies on correlations between the total luminosity of a galaxy and the velocity with which it rotates, as determined from the Doppler effect. For technical reasons, the rotation velocity can be determined better from observations of radio waves emitted by the galaxy (see page 195) than from observations of visible light. The new distance estimates imply that only 10 or 11 billion years have elapsed since the big bang. If this proves correct, our galaxy can hardly have an age of 12 billion years; 8 billion years would be a much better estimate. In many ways, the fact that we know any of this at all is more amazing than the fact that so much uncertainty remains about the exact length of time the universe has been expanding.

4.5 Is the Universe Finite or Infinite?

Sooner or later, any discussion of the universe leads to the question of its size. Is the universe infinitely large? Most people would say that it is, since if it does not extend indefinitely in all directions, we would encounter a barrier that must itself be part of the universe along with the space beyond the barrier. Hard-working mathematicians, however, remind us that the answer is more complex than this.

The universe may in fact be finite, containing only a finite amount of space and a finite amount of matter. If this is so, then space in the universe must be curved, bending back on itself like the surface of a balloon. We can see the balloon's two-dimensional curved surface within a three-dimensional space. If, however, all of three-dimensional space has an analogous curvature, we cannot hope to picture the curvature without a fourth dimension. Since this fourth dimension exists only in the world of science fiction, we must resign ourselves to the use of incomplete models to help us imagine the curvature imperfectly (Figure 4-11).

No one knows whether the universe is finite or infinite. The amazing thing about this problem, however, is that we can hope to make observations soon that will answer the question (see page 123). If the universe turns out to be infinite, our "intuition" about the universe may turn out to be correct. (But to imagine an infinite universe is more difficult than you might think when you first attempt it.) If, on the other hand, the universe turns out to be finite, we can be sure that the total volume of the universe extends far beyond what we can see now. In a finite universe, the fraction of the expanding space that we can see increases as the age of the universe increases (Figure 4-12). If we wait long enough, we might be able to see all of a finite universe. In an infinite universe, we would have to wait an infinite amount of time to do so.

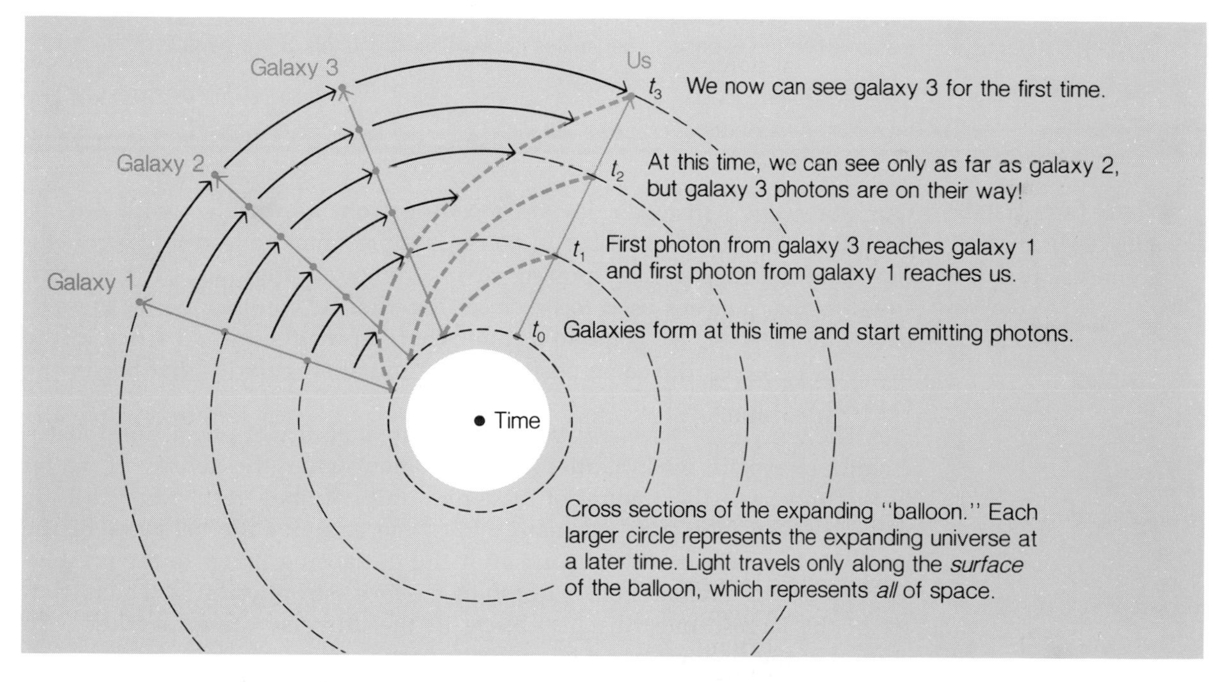

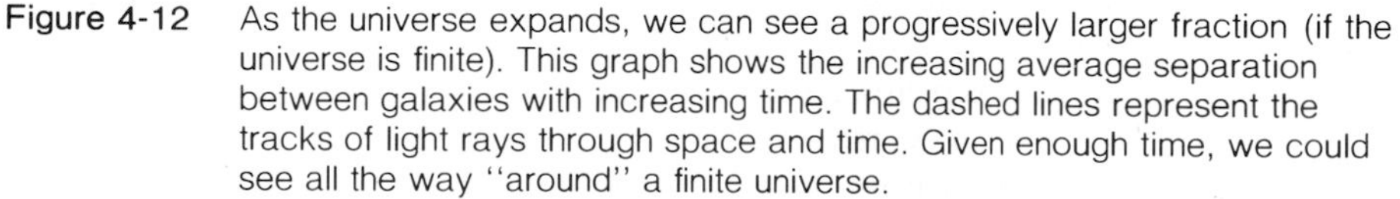
Figure 4-12 As the universe expands, we can see a progressively larger fraction (if the universe is finite). This graph shows the increasing average separation between galaxies with increasing time. The dashed lines represent the tracks of light rays through space and time. Given enough time, we could see all the way "around" a finite universe.

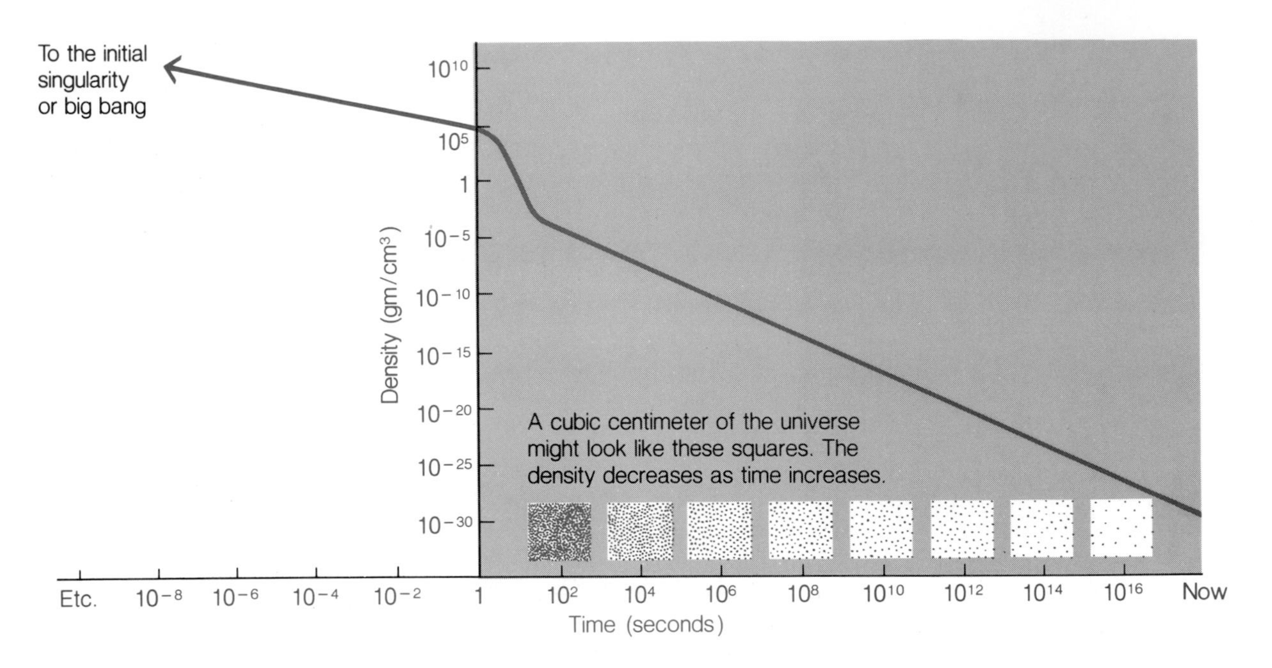

Figure 4-13 A graph of the density of matter in the universe as a function of time shows a steadily decreasing density with time. Notice that as we look back in time, the density increases without limit. The graph has been drawn so that the instant when the expansion began lies infinitely far to the left of the page.

4.6 The Early History of the Universe

Our observation that galaxies are receding from us together with our assumption that our view of the universe is representative lead to the conclusion that all galaxies are moving apart from one another.[3] This in turn suggests that galaxies used to be closer together. If we imagine far enough back into the past, we come to a time when all matter and all space in the universe occupied the same position—the initial singularity that began the expansion (Figure 4-13).

The term *initial singularity* comes from the inability of physics and mathematics to explain what happened at a moment when the density of matter in the universe—the amount of mass in a unit volume—reached an infinite value. We can, however, calculate with some accuracy the behavior of the universe from about 10^{-43} second after the initial singularity to the present. This in itself is a remarkable triumph of modern astrophysics, in which studies of the largest, most distant objects in the universe come together with studies of the smallest, most elementary particles.

[3]We discuss an alternative conclusion, the steady-state model of the universe, on page 122.

Where Was the Center of the Universe?

We must first get rid of the notion that the beginning of the universe—the big bang—happened at some particular point in space, from which the universe has been expanding ever since. In other words, we must abolish the idea that the universal expansion has a particular center; after all, we derived the fact of universal expansion precisely from the assumption that the universe has no center. If space just "sat there," then we could look for the point in space where it all began. But space participates in the universal expansion too. As the various parts of the universe recede from one another, new space appears! This concept has an analogy in the balloon model, since new surface appears as we blow up the balloon. On the other hand, our intuition tells us that space cannot grow larger or smaller, but instead simply *is.*

Another way to explain universal expansion is to say that *every* point in the universe can consider itself the center. Similarly, each point on the expanding balloon can be thought of as the center of the expansion of the balloon's surface. The actual center of the balloon lies in the third dimension, not on the surface that models space in the universe; this corresponds to a mathematical, but not physically real, center of the expanding universe in a fourth dimension of space.

Effects of the Universal Expansion

Let us picture ourselves, then, within the universe (where else could we be?) as it began to expand. All around us, matter would have been crowded together at immensely high density. Because density measures the mass per unit volume, the universe's expansion has caused a steady decline in the average density within the universe (Figure 4-13). During the early years after the big bang, no stars, galaxies, or planets existed; instead, a featureless broth of matter and radiation formed a high-temperature mixture. Only after the universe had expanded and cooled for millions of years did clumps of matter—regions with densities much higher than average—begin to form. And only after billions of years did these clumps turn into stars and galaxies. To understand how this clumping could occur, we must consider the composition and temperature of the early universe.

4.7 The First Half Hour After the Big Bang

The big bang brought forth light and matter; from where we do not know. We can, however, calculate what the basic mixture of particles and radiation must have been after the first half hour had passed. Before then, especially during the first seconds after the big bang, the entire universe had such an enormous density of matter (greater than 10^5 grams per cubic centimeter—or 50 tons per pint!) and such a tremendously high temperature (Figure 4-14) that no particle could maintain its existence for more than a tiny fraction of a

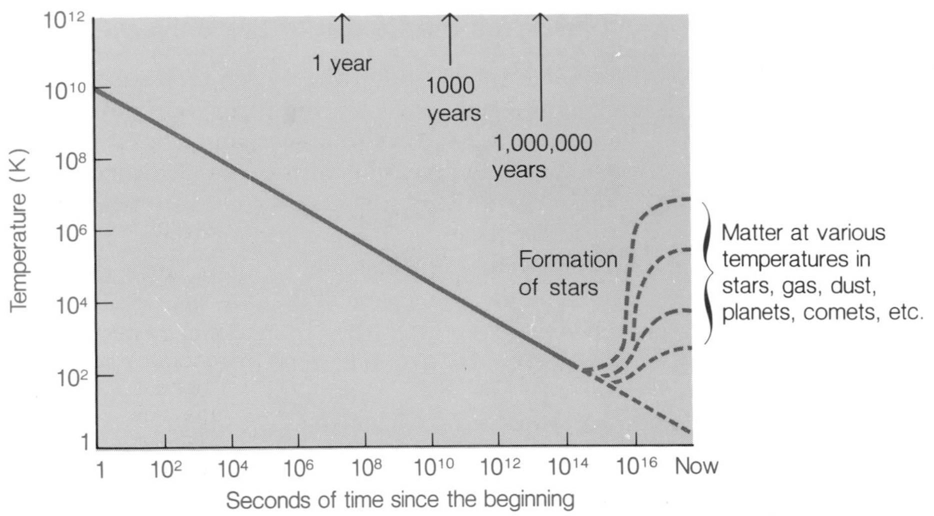

Figure 4-14
The temperature of the universe has been falling ever since the big bang. Until the time of galaxy formation, some billions of years after the big bang, all the matter in the universe had the same temperature. Once galaxies began to form, however, different clumps of matter acquired quite different temperatures.

second. Instead, rocked by high-energy collisions with all the other surrounding particles, each particle soon turned into other kinds of particles, which in turn collided and yielded still other types of particles, much as happens today inside a high-energy particle accelerator (Figure 4-15). All kinds of particles—protons, neutrons, electrons, and a host of others only recently discovered on Earth, or still to be discovered here—must have coexisted during the first brief moments of the universe.

Figure 4-15
The Stanford Linear Accelerator in Menlo Park, California, passes under Interstate Highway 280 and over the San Andreas fault. Physicists use this 3-kilometer-long machine to accelerate elementary particles to nearly the energies they had during the early minutes of the universe. Since the accelerator costs many millions of dollars per year to operate, theoreticians sometimes refer to the early universe as "the poor man's accelerator."

Heat and Temperature

The particles in a gas can move about freely; they are not bound to one another. We can specify the general state of the gas by giving the number of particles per cubic centimeter, and giving its temperature. *The* **temperature** *measures the average kinetic energy per particle.* A particle's **kinetic energy,** sometimes called **energy of motion,** equals half its mass, m, multiplied by the square of its velocity, v, or

$$\text{Kinetic energy} = \frac{1}{2}mv^2$$

(See box on Energy and Power, page 223.)
In the absolute, or Kelvin, scale of temperature, the correspondence is direct: Twice as high a temperature corresponds to twice as large an average kinetic energy per particle.

Heat *measures the total kinetic energy in a given volume.* Thus the amount of heat depends on both the temperature and the number of particles. The same amount of gas will have twice as much heat if we double its temperature (in Kelvins). But we can also double the heat by doubling the number of particles while maintaining the *same* temperature.

The **Kelvin,** or **absolute,** scale of temperature comes from the more familiar Celsius, or centigrade, scale, in which water freezes at 0° and boils at 100°. The Kelvin scale uses the same size of a degree, but it sets the zero point at **absolute zero,** at which all motion would cease (except for certain quantum-mechanical effects that do not concern us here). In the Celsius scale, absolute zero falls at −273.16°. Thus, in the Kelvin scale, water freezes at 273.16 K and boils at 373.16 K.

Astronomers prefer the Kelvin scale of temperature because it gives a direct measure of the average kinetic energy per particle. We can tell immediately that a gas at 300 K has twice the energy per particle that a gas at 150 K has. If we gave these temperatures as 27° and −123° in the Celsius scale, or as 81° and −189° in the Fahrenheit scale, we would have much more difficulty in comparing the average kinetic energy per particle. Therefore, in this book *we shall use the absolute, or Kelvin, scale throughout.* This may present some difficulty in relating temperatures in Kelvins to our familiar Celsius and Fahrenheit scales, but this difficulty is more than overcome by the ease with which we can compare energies. Actually, it is quite easy to remember that 300 K represents the temperature on a typical summer day.

Matter and Antimatter

Most physicists believe that the early universe must have contained equal amounts of matter and antimatter, equal numbers of particles and antiparticles. So far as we know, each kind of elementary particle has a corresponding kind of **antiparticle** with the same mass as the particle but the opposite electric charge. Thus, a proton has one unit of positive charge, and an antiproton has one unit of negative charge. An electron has one unit of negative charge and an antielectron (or positron) has one unit of positive charge. Neither a neutron nor an antineutron has any electric charge.

If a particle collides with its corresponding antiparticle, *both* particles disappear and are replaced by particles with no mass: photons and less familiar massless particles called **neutrinos** and **antineutrinos.** (Since photons are indistinguishable from antiphotons, we call them all photons.) Neutrinos and antineutrinos, like photons, have no mass and no electric charge, yet neutrinos and antineutrinos are different from photons and different from one another. The difference lies in the way these massless particles interact with other particles, such as protons or electrons. But like photons, neutrinos and antineutrinos travel at the speed of light. Particles *with* mass can never move quite as fast as light, though their speeds can approach it.

Because particles and antiparticles, which collectively form **matter** and **antimatter**, annihilate one another, they cannot coexist for long without turning into photons, neutrinos, and antineutrinos. A beam of photons does not have this problem, because photons and antiphotons are already the same particle type. We can tell from observations of the Milky Way that our galaxy does not contain nearly as much antimatter as matter, since otherwise we would see great bursts of photons as clumps of matter and antimatter annihilated each other. On the other hand, in other galaxies, what we call antimatter might be abundant, with our matter a rare phenomenon. Observations of photons from such a galaxy could never reveal this fact, however, since photons cannot be distinguished from antiphotons.

Cooling in the Early Universe

As the universe expanded, the matter (and antimatter) within it cooled. This cooling process occurred simply because the matter had more space in which to move. The same phenomenon occurs on a small scale when we let foam out of a spray can. The foam becomes less confined in space, under less pressure. The reduction in pressure leads to a reduction in temperature, that is, in the average speed with which each particle moves.

After the first half hour following the big bang, the temperature had fallen enough so that most collisions among particles had too little energy to turn one kind of particle into another. Only if a particle happened to meet its antiparticle did such changes occur. Otherwise, since that time, the protons in the universe have generally remained protons, and the electrons have remained electrons, even as they collided (more gently) with other particles. The exceptions to this rule have occurred deep inside stars, where temperatures rise high enough to duplicate those of the early universe (see Chapter 8).

Particles in the Universe

What were the particles that emerged from the first half hour after the big bang? The dominant particle types, which still form the bulk of the universe, turned out to be protons, electrons, and helium nuclei, along with a huge

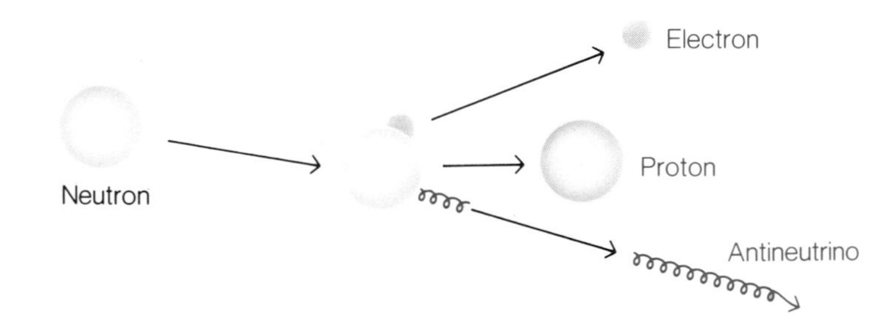

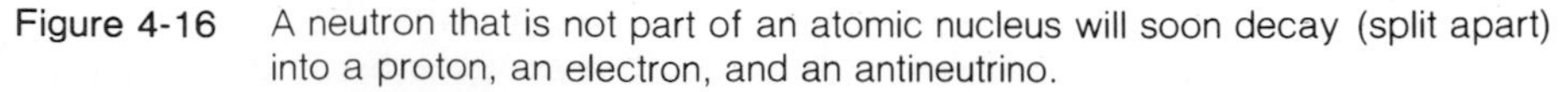
Figure 4-16 A neutron that is not part of an atomic nucleus will soon decay (split apart) into a proton, an electron, and an antineutrino.

Table 4-1 Calculated Abundance of Particles with Mass After the First Half Hour

Type of Particle	Number of Particles (Per 10^{12} Protons) Produced During the First Half Hour After the Big Bang[a]	Present Number of Particles (Per 10^{12} Protons) Found in the Solar System and in Stars Like the Sun
Proton	1,000,000,000,000	1,000,000,000,000
Electron	1,160,000,000,000	1,160,000,000,000
Helium nucleus	80,000,000,000	80,000,000,000
Carbon nucleus	1,600,000	370,000,000
Nitrogen nucleus	400,000	115,000,000
Oxygen nucleus	40,000	670,000,000
Neon nucleus	180	110,000,000
Nuclei of sodium and all heavier elements	2,500	220,000,000
Photon	1,000,000,000,000,000,000,000	1,000,000,000,000,000,000,000

[a]Different models of the early universe produce slightly different numbers of the nuclei heavier than hydrogen that emerge from the first half hour after the big bang. This column lists the *maximum* number of nuclei that could have been produced under the most favorable conditions thought to be reasonable by cosmologists.

number of massless particles: neutrinos, antineutrinos, and the photons that form electromagnetic radiation.[4]

We have already discussed protons, electrons, and neutrons in our study of the atom (see page 76). Neutrons, like protons, appear in the atomic nucleus. But the neutron has an interesting property. Outside of the nucleus, a neutron will decay into a proton, an electron, and an antineutrino (Figure 4-16). This transformation takes only a few minutes, so any neutrons left after the first half hour following the big bang soon disappeared. It should be remembered, however, that this sort of decay does not occur within the atomic nucleus. Neutrons that are part of a nucleus can last indefinitely.

Photons formed—and still form—the most abundant type of particle in the universe. Almost a billion photons exist for every particle with mass! We shall discuss the primordial sea of photons on page 119; for now, we should recognize that these photons came from the big bang, which lit the universe with an everlasting cosmic radiation. The neutrinos and antineutrinos, which also came from the first half hour, are almost as abundant as photons, but they react so little with particles that have mass that we can do little more than note their continued existence. We shall meet neutrinos again, however, when we study the processes that occur inside stars (see page 249).

Table 4-1 shows the calculated abundance of various types of particles with mass that emerged from the first half hour after the big bang. In addition to the dominant particle types—the protons, electrons, and helium nuclei—a small fraction of heavier nuclei, such as carbon, nitrogen, and oxygen nuclei, emerged from the fury of the early universe. We on Earth have a great respect for these relatively rare types of nuclei, since our planet and ourselves largely consist of them. The early universe, though, turned no

[4]If other parts of the universe consist of antimatter, there would be antiprotons, antielectrons, and antihelium nuclei in place of protons, electrons, and helium nuclei.

more than one-millionth of its particles into such nuclei. It seems clear that if we had had to rely on the first half hour after the big bang to produce the heavier nuclei so vital to us, we would not be here to talk about it. Instead, these heavier nuclei have emerged from stellar furnaces, stars that exploded to seed the cosmos with elements made there.

4.8 The Microwave Background

We have followed the development of the universe through its all-important first half hour, a period that left space permeated with a homogeneous mixture of electrons, nuclei, and massless particles. The universe was still so hot—and remained so for several hundred thousand years—that no atoms could exist for long. If an electron happened to get into orbit around a proton, the new atom soon suffered destruction from collisions with photons and other particles, which knocked the proton and electron apart. Only when the temperature of the universe had fallen to 3000 K could atoms persist, for only then did the collisions fail to occur with enough energy to break the atoms apart.

The Doppler Effect in the Expanding Universe

The decline in temperature to 3000 K took the better part of a million years (Figure 4-14). During this time, atoms simply could not exist, and all the matter in the universe was spread evenly through space, with little chance to form clumps. But when the temperature fell to 3000 K, something happened to the photons, the most abundant particles in the universe and the ones most responsible for destroying any would-be clumps. The increasing Doppler shift to longer wavelengths and thus lower energies (see page 87), the natural result of the universal expansion, robbed the photons of their energy and destroyed their ability to knock the atoms apart.

Consider a typical photon from the early universe, produced some time between one half hour and one million years after the big bang. Such photons emerged from the clash of particles during the first half hour and filled the entire universe, traveling at the speed of light. Now, as the universe continues to expand, the photons arriving at any point will undergo a Doppler shift to lower energy and lower frequency (longer wavelength), because the universe is expanding away from them in all directions. Hence from the point of view of *any* observer in the universe, the photons that permeate all of space undergo a continuous decrease in energy and a continuous increase in wavelength.

The Era of ''Decoupling''

For almost a million years after the big bang, photons still arrived with enough energy to break apart hydrogen atoms if they formed. But after that time, so few photons had the energy needed to break apart **(ionize)** an atom, or even to excite the electron into a larger orbit, that the photons basically ceased to interact with the atoms. Thus, one million years after the big bang, there began an era of **decoupling,** when the photons in the universe stopped interacting with ("decoupled" from) those particles with mass. Only then could atoms form and persist, paving the way for clumps of matter to form (much later) into stars and galaxies.

What of the photons that decoupled from the matter? They still occupy the universe, with almost a billion photons for every particle with mass. Photons left over from the early years of the universe pass through this page as you read it. They have even been detected with radio antennas and receivers, an achievement in 1964–65 that won a Nobel Prize for Arno Penzias and Robert Wilson (Figure 4-17).

The Discovery of the Microwave Background

How did Penzias and Wilson discover the sea of universal photons with a *radio* antenna? The continuing expansion of the universe increased the amount of the Doppler effect for the photons that any observer could detect. During the time interval from decoupling (less than 1 million years after the big bang) to the present (15 to 20 *billion* years after the big bang), the Doppler effect has reduced the energy of each photon by a thousand times, placing the energy of most of these photons in the radio domain of the electromagnetic spectrum (Figure 3-9). In particular, the majority of these photons have energies that correspond to the short-wavelength, or *microwave,* part of the radio spectrum. Hence, the photons that fill the universe are referred to as the **cosmic microwave background:** *cosmic* because they fill the universe, *microwave* because most of them are now microwave photons, and *background* because these photons form a background of "noise" to any other electromagnetic radiation we try to detect. In fact, Penzias and Wilson made their discovery of the cosmic microwave background through their attempts to eliminate stray "noise" in their sensitive equipment!

Great Moments in the History of the Universe

Time After the Big Bang	*Event*
Less than 1/1000 second	All types of particles in existence
4 seconds	Electrons and antielectrons begin mutual annihilation
3 minutes	Neutrons and protons fuse to form helium nuclei
30 minutes	Significant interactions among particle types end
1 million years	Background radiation decouples from matter
1 billion years (?)	Galaxies begin to form
2 billion years (?)	Stars begin to form
15–20 billion years	Present time
80 billion years (?)	Universal expansion might reverse into universal contraction

Figure 4-17 Bell Laboratories scientists Arno Penzias (right) and Robert Wilson (left) used the horn-shaped antenna behind them to detect the cosmic microwave background radiation.

The Spectrum of the Cosmic Microwave Background

Fifteen years after the discovery of the cosmic microwave background, astronomers have succeeded in measuring the number of photons in the background at various frequencies and thus have determined the spectrum of this radiation. To do this, they had to send detectors high above the Earth's atmosphere, which absorbs short-wavelength microwaves (Figure 3-10).

The spectrum found by astronomers corresponds rather well to the spectrum expected for an ideal radiator at a temperature of 2.7 K (Figure 4-18). **Ideal radiator** refers to a group of particles in energy equilibrium with the photons among them. Particles (for example, hydrogen atoms) can absorb photons and can emit photons, as we discussed on page 81. In an ideal radiator, the total photon energy emitted each second equals the total photon energy absorbed. The early universe provides a good example of an ideal radiator because the photons were constantly exchanging energy with the other particles. Once decoupling began, the universe no longer had the properties of an ideal radiator, but the photons have preserved the spectrum they obtained at that time, except for the effects of the Doppler shift.

Figure 4-18 shows the spectra that typify ideal radiators. They all have a characteristic shape, rising to a peak and then falling sharply. The peak energy—the energy at which the maximum number of photons appear—varies directly with the temperature of the ideal radiator. If we measure energy in ergs and measure temperature in Kelvins, then the peak energy always equals the temperature times 4×10^{-16}. As Figure 4-18 shows, the effect of the universal expansion on the spectrum of the cosmic microwave back-

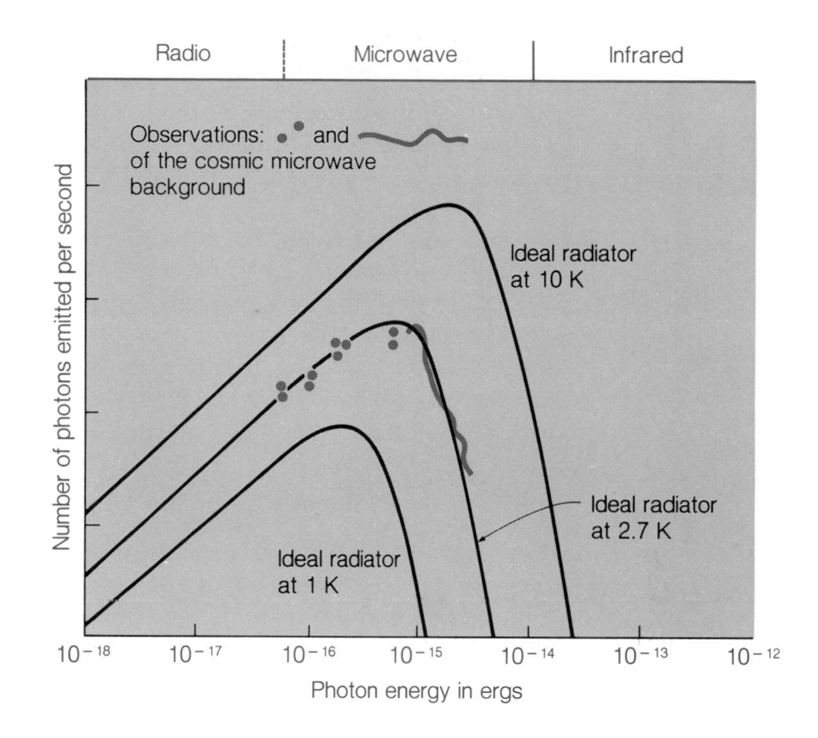

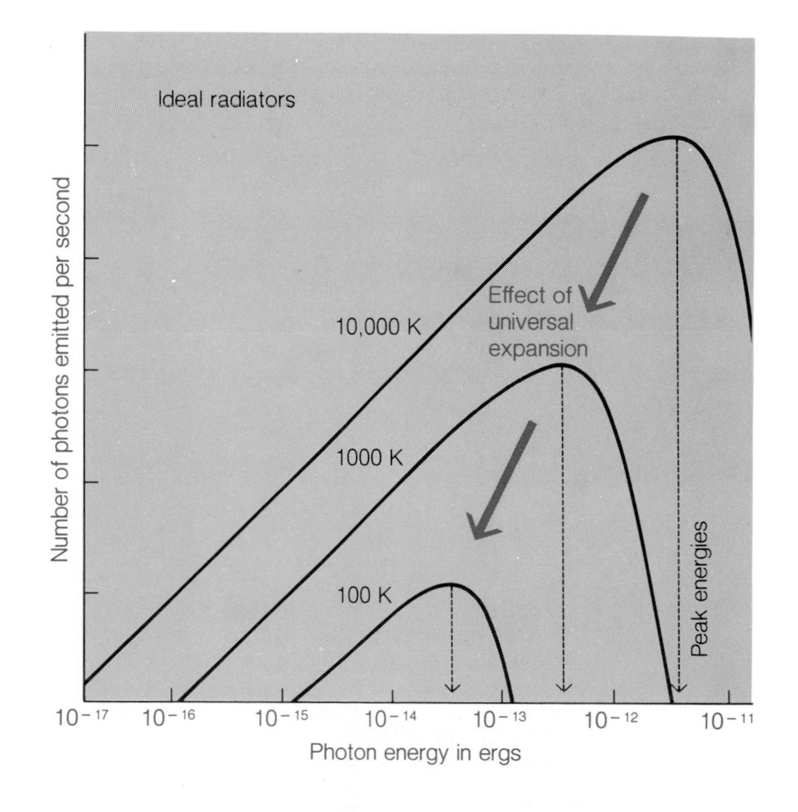

Figure 4-18
Measurements of the energy spectrum of the photons in the microwave background (top) show that they have approximately the spectrum expected for an ideal radiator at a temperature of 2.7 K. As the bottom panel shows, the spectra of all ideal radiators have a similar shape, with a peak energy that varies in proportion to the temperature of the ideal radiator. As the universe has expanded, the Doppler effect has shifted the spectrum of the cosmic microwave background, so that it has the same shape but a steadily decreasing peak energy.

ground has been a shift downward and to the left in our diagram; the *shape* of the spectrum has remained the same, but the peak energy has changed. The thousand-fold lowering of the temperature since decoupling began has produced a thousand-fold decrease in the peak energy. This corresponds to a decrease in the temperature that characterizes the ideal radiator (the early universe) from 3000 to 2.7 K.

This does not mean that the universe has a temperature of 2°.7 above absolute zero. Matter in the universe has gone its own way, not interacting with the photons to any significant degree since 1 million years after the big bang. Some matter has temperatures of millions of degrees, some of thousands, some of only a few degrees above absolute zero. The photons, however, *do* have a temperature of 2°.7 above absolute zero, since their spectrum corresponds to that of an ideal radiator at this temperature.

The Steady-State Model of the Universe

The cosmic microwave background provides the best argument against the once-popular steady-state model of the universe. In this model, originally suggested by a group of British astrophysicists, the universe produces new matter as it expands, so that the average density of matter remains unchanged (Figure 4-19). Thus, the universe looks the same at all times ("steady-state"). We may think it unreasonable that new matter could somehow just "appear," but after all, the big-bang model that we have been considering makes new matter appear too—all at once, at the initial singularity. Notice, however, that the steady-state model does not lead to a cosmic microwave background of the kind we have just described, since the universe *never* has a high-density, high-temperature state in this model. Thus the dis-

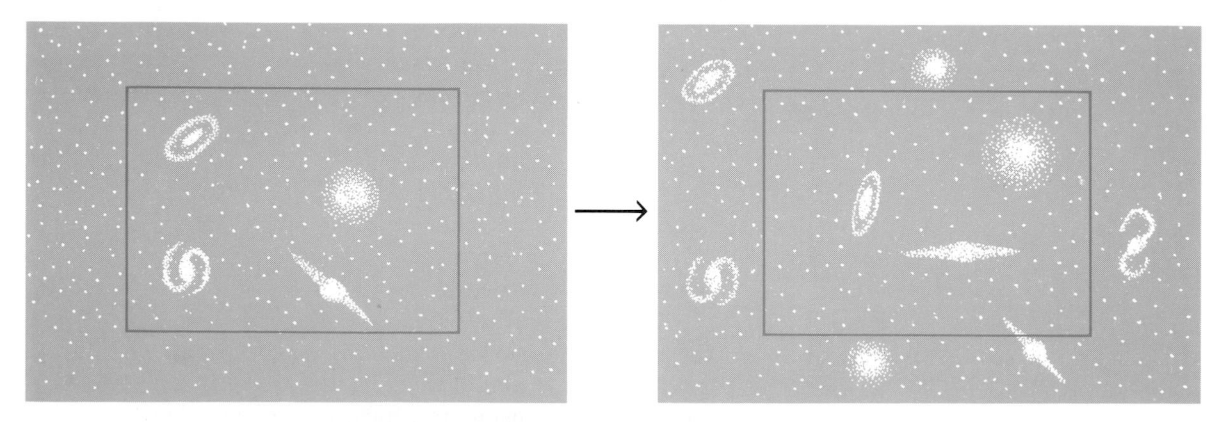

Figure 4-19 The steady-state model of the universe suggested that as the universe expands, new matter, and eventually new galaxies, appear at just the rate needed to maintain a constant average density of matter in the universe. The rate of new matter's appearance would be so small as to be undetectable, since it would amount to one new atom per cubic meter every 10 billion years.

covery of the microwave background appears to rule out the steady-state model, unless an acceptable, alternative explanation of this radiation can be found.

Where is the "Center" of the Microwave Background?

The microwave background radiation has been used to demonstrate that the universe has no center. From where do these photons arrive? From the long-vanished, early universe of 15 or 20 billion years ago. And from what direction? From all directions come these photons, and in equal numbers, each having traveled 15 to 20 billion light years to get here. Where are the photons that were produced *here* before the time of decoupling? They are 15 or 20 billion light years away, in all directions. Every observer in the universe would presumably see the "center" all around the sky, 15 or 20 billion light years away. In other words, the center must be everywhere, which corresponds to the universe having no one center of expansion.

4.9 The Future of the Universal Expansion

We have seen how the observations made by Hubble and Slipher, together with the assumption that we have a representative view of the universe, point to the conclusion—reinforced by observations of the microwave background—that the universe has been expanding for the past 15 or 20 billion years. It may give some comfort to know that our own galaxy does not expand, nor does the solar system, nor the Earth, nor our own bodies. All objects with a density greater than the average density of matter in the universe have enough self-gravitation to resist the universal expansion and will keep their present sizes despite the increasing distances between clusters of galaxies.

The important question for the future remains: Will the universe expand forever, or will the expansion someday reverse itself to produce a universal contraction, perhaps leading to another big bang (Figure 4-20)? We can begin to find the answer by imagining the entire universe to have received an enormous kick that began its expansion at the moment of the big bang. Since that time, the momentum from that kick has continued to make all parts of the universe recede from one another. On the other hand, each piece of matter in the universe exerts gravitational attraction on the other pieces, so gravity acts to overcome the expansion and to produce an eventual contraction. We can calculate that the mutual attraction among the various parts of the universe has already slowed the expansion, so that the distance between galaxies takes longer to double than it used to (Figure 4-21). But will gravity ever reverse the expansion and turn it into a contraction?

The Importance of the Density of Matter

The key factor resides in the average density of matter in the universe. If matter exceeds a certain **critical density**, which we can calculate to be close

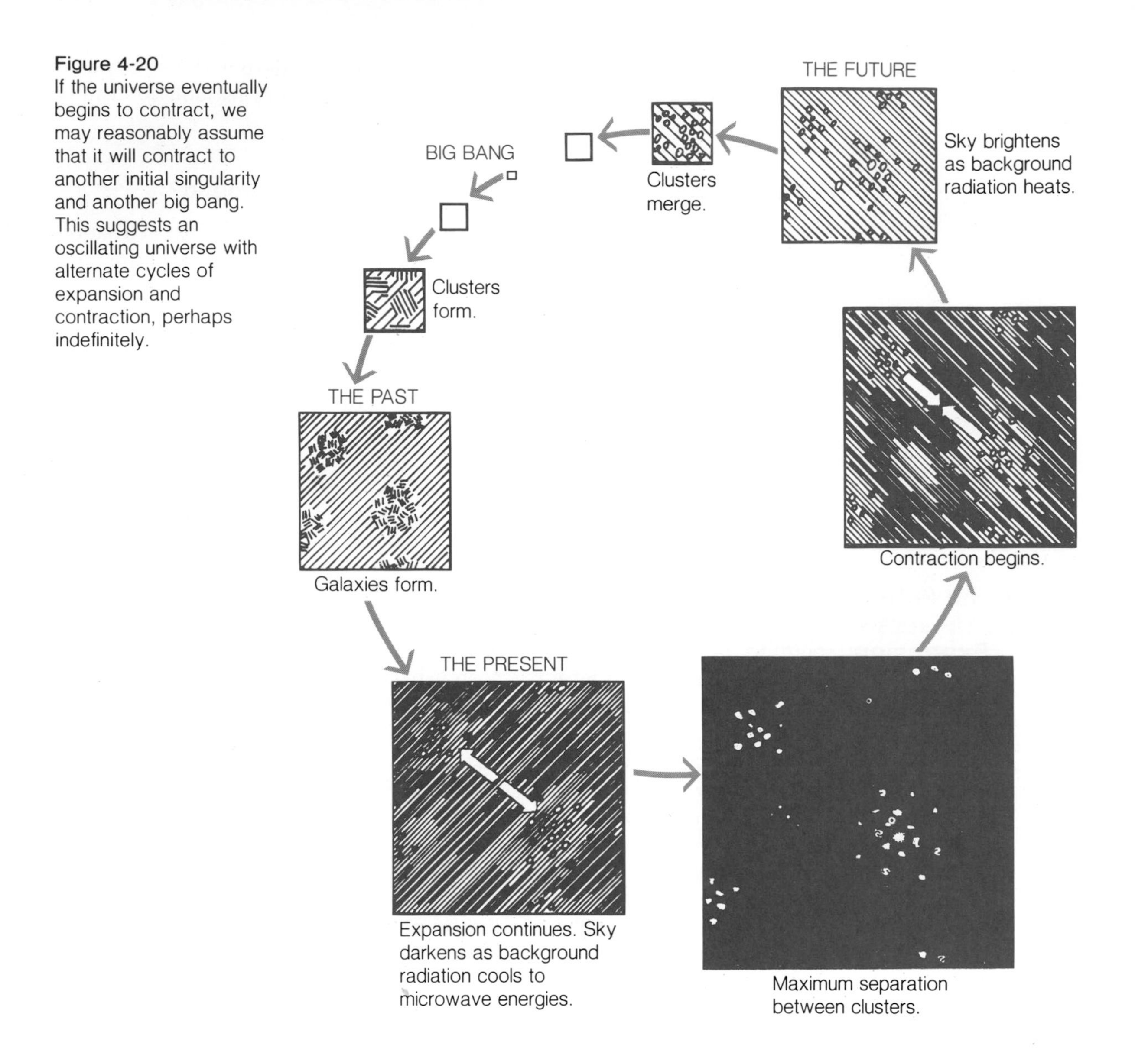

Figure 4-20
If the universe eventually begins to contract, we may reasonably assume that it will contract to another initial singularity and another big bang. This suggests an oscillating universe with alternate cycles of expansion and contraction, perhaps indefinitely.

to 10^{-29} gram per cubic centimeter, then the universe will eventually contract. This critical value decreases as the universe expands in just the same way that the real density does, so if the true density exceeds the critical value now, it exceeds it for all time. Thus our goal must be to determine the average value of the density of matter in the universe now. A value greater than 10^{-29} gram per cubic centimeter means that the expansion will eventually turn into a contraction, while a value less than the critical number implies eternal expansion, because the universe does not have enough density of matter to "pull itself together."

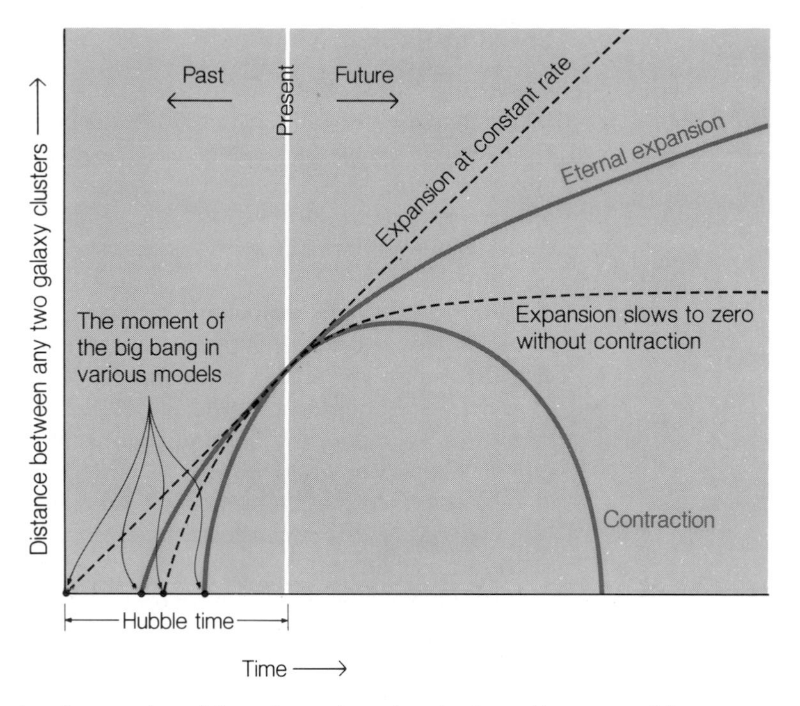

Figure 4-21 The rate of expansion of the universe has already slowed because of the mutual gravitational attraction of all parts of the universe. Hence, the time since the big bang must be less than the Hubble time, the interval that characterizes expansion at a steady rate. Whether the expansion will slow to zero—and reverse into a contraction—remains part of the undetermined future.

What do observations reveal about the average density of matter in the universe? Because astronomers specialize in observing matter that shines (by release of photon energy), their results pertain primarily to the luminous matter in the universe, such as the matter in stars and galaxies. By counting the number of galaxies in space, astronomers have found that all the matter that emits light amounts in density to no more than one-tenth of the critical value. If this matter represents all that exists—or even one-half or one-quarter of all that exists—then the universe heads for everlasting expansion.

But what about the matter that does not shine? Could the universe contain so many rocks, or clouds of dark gas, or burnt-out stars, or black holes (see Chapter 10) that the density of nonluminous matter provides the "missing mass" needed to ensure an eventual contraction? Do neutrinos and antineutrinos, as some have suggested, each have a tiny amount of mass? If so, they might provide a density sufficient to make the universe contract. To these questions there are no adequate answers, because methods of estimating the average density of nonluminous matter are simply not accurate. Yet we do have another way to determine the overall density of matter in the universe, luminous or not. To make this determination, we can look back into the past to determine the future of the universe.

Looking Back in Time

According to Hubble's law for the universal expansion, the recession velocity increases in proportion to the distance. This law clearly cannot hold for ever-increasing distances, since it would imply velocities in excess of the speed of light. Albert Einstein's theory of relativity showed how the effects of motion close to the speed of light would modify Hubble's law as we deal with galaxies so far from us that their recession velocities approach the speed of light.

But there is another factor to consider as well. When we look at distant galaxies, we are really looking far back in time. Light from a nearby galaxy, such as the great spiral in Andromeda, reaches us in 2.2 million years, just over one ten-thousandth of the age of the universe since the big bang. Light from a much more distant spiral such as M 51 (Figure 4-22) arrives after 12 million years. But when we look at faraway clusters of galaxies, we have the chance to observe light that has taken a billion years or more to reach us. Now we deal with travel times that approach a noticeable fraction of the age of the universe.

When we look a billion years or more into the past, we observe the universe as it appeared at a much earlier time. In principle, we can make

Figure 4-22 The beautiful Whirlpool galaxy (M 51) in the constellation Canes Venatici has a distance of about 3.8 million parsecs (12 million light years), so we see it as it was 12 million years ago.

accurate observations of the distances and velocities of galaxies billions of light years away and then use these observations to see how the expansion rate has changed. Calculations then show how a change in the past relates to a change in the future. As Figure 4-23 shows, we can turn these calculations into a single decisive line on the graph of galaxy distances and velocities. If the observations show a trend above the critical line, the universe will eventually contract; results below the line imply never-ending expansion.

The only hitch comes from our lack of accuracy in determining the distances to faraway clusters of galaxies. (The galaxies' velocities, found by the Doppler-effect method, have remarkable accuracy for immense distances; see Figure 4-9). Finding the distance to a cluster of galaxies so far away that we can barely see it cannot be an easy job, and only remarkably skilled work by Hubble's successors, such as Allan Sandage, has brought us as close as we are to resolving the problem. But as we see from Figure 4-23, the scatter in the

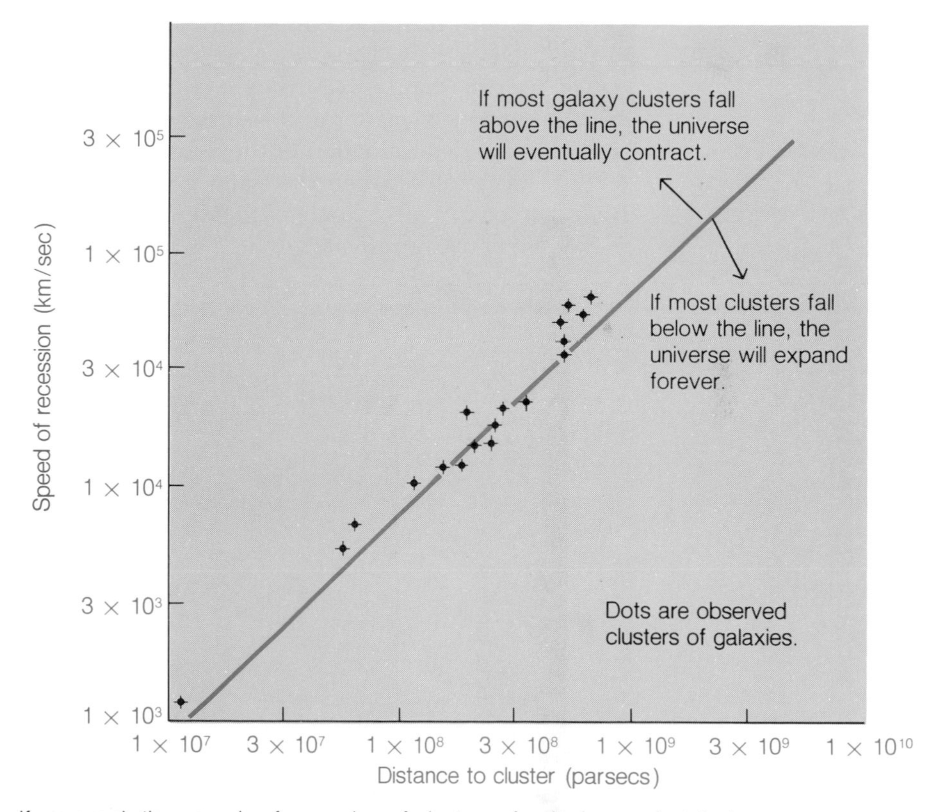

Figure 4-23 If we graph the speeds of recession of clusters of galaxies against their distances from us, different models of the universe predict slightly different trends for this relationship. Accurate observations could establish the validity of a particular model and thus distinguish between the models that predict eternal expansion and those that predict eventual contraction. Unfortunately, our present data do not allow us to select one of these models with certainty.

observational results does not yet allow a definite answer to the question of expansion versus contraction. Astronomers hope that the Space Telescope, operating above the Earth's atmosphere during the 1980s, will produce more accurate distance determinations outside the blurring blanket of air that surrounds us. Then at last we may know whether the universe will ever start contracting to "recycle" us into another state of singularity.

Deuterium: The Key to the Puzzle?

One more way to attack the problem of finding the average density of matter in the universe consists of an attempt to determine the average abundance of **deuterium.** Deuterium, an isotope of hydrogen, has an atomic nucleus containing one proton and one neutron (see page 77). Most of the deuterium in the universe outside of the stars was made, so far as we can calculate, during the first few moments after the big bang. The exact amount of deuterium nuclei formed at that time depends on the density of matter at that time. This density in turn has a direct correspondence with the density of matter in the universe now. Thus, if we can determine the abundance of deuterium nuclei, we can (thanks to astronomers' calculations) relate this directly to the average density of matter in the universe (Figure 4-24).

The best determination of the deuterium abundance in interstellar matter (the material between the stars) gives a ratio of deuterium to ordinary hydrogen of 2×10^{-5}. Some of the deuterium once in the interstellar matter may have ended up in stars, where it is easily destroyed; so the abundance

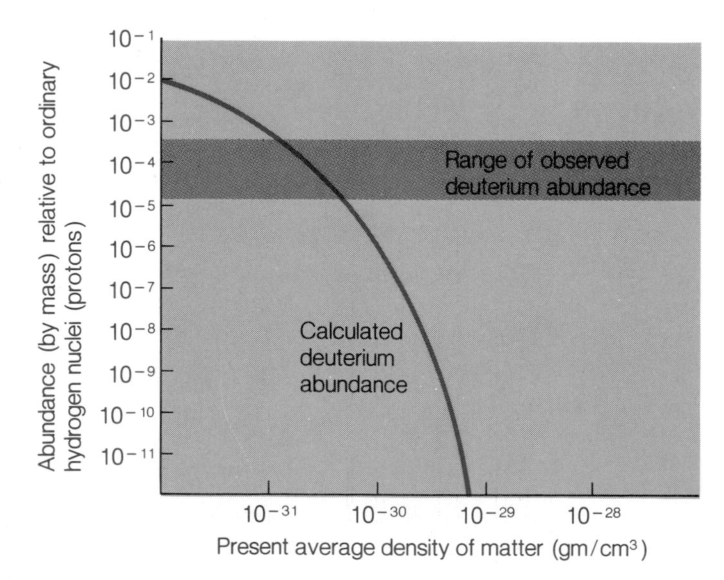

Figure 4-24 Calculations of the ways deuterium nuclei were made during the first minutes after the big bang can be used to relate the present abundance of deuterium to the present average density of matter in the universe.

without such destruction might be 50% or even 100% greater than this. Such considerations may be irrelevant, however, since even an abundance of 2×10^{-5}, relative to hydrogen, rules out an average density of matter greater than 10^{-30} gram per cubic centimeter (Figure 4-24). This density falls well below the critical value of 10^{-29} gram per cubic centimeter needed to reverse the expansion (see page 124). Thus, if we can trust our calculations and our observations of the deuterium abundance, the universe will never cease its expansion.

Implications for the Size of the Universe

If we do manage to determine the future of the universal expansion, will this tell us whether the universe has a finite size? The answer appears to be yes. In the most widely accepted cosmological models, only a finite universe can eventually contract. Thus, the discovery that the universe will not expand forever would imply that the universe must be finite, while the conclusion that the universe will expand forever would imply that the universe is, and will remain, infinite in size.

Do We Have a Representative View of the Universe?

All of the measurements that we have described—past, present, and future—can tell us about the universe as a whole only if our look at the cosmos provides a representative view. But what if we have a special perspective and the universe is already contracting in some distant region? Arguing against this is the belief that we should not assign a special nature to our view of space. Moreover, our observations show a remarkably uniform cosmos, looking at the largest distance scales. We see about the same number of galaxy clusters in all directions, and more important, we detect almost the same amount of radiation in the cosmic microwave background in all directions. If this radiation really does come from the first million years after the big bang, we have proof that a portion of the universe—the regions that produced the microwave photons we now detect—had about the same density of particles throughout. This still does not prove that the much larger (perhaps infinite) entire universe must be uniform, but it furnishes at least some support for that conclusion.

Meanwhile, the universe continues to expand. We have discovered this much by observing galaxies within a few billion light years of our own. As we push our observations farther outward into space, and thus farther back in time, we can hope to resolve the great problem of whether the expansion will continue forever. Adrift on our small planet, in orbit around a typical star in an average giant galaxy, we have come a long way in our knowledge of the cosmos. We must now turn our attention to the details of the universe—the types of galaxies, the stars and planets they contain, and how they came to be the way they are.

Is the Cosmic Background Radiation the Same from All Directions?

Fifteen years after the discovery of the cosmic microwave radiation, astronomers have greatly improved their techniques for observing the photons that arrive from all directions. By now we can detect differences of as little as one part in 3000 of the observed background radiation. With this level of sensitivity, astronomers have found an intriguing result: The amount of radiation is not exactly the same when we look at all parts of the sky.

The differences that appear when we point sensitive antennas at various parts of the sky suggest a pattern. This pattern represents exactly what we would expect if we are moving with respect to the background radiation. The Doppler effect causes us to receive a bit more energy each second from the directions along our motion, and a bit less energy each second from the directions opposite to our motion. If we interpret the observed differences in the amounts of background radiation from various directions, we find that we are moving with a velocity of several hundred kilometers per second in the direction of the constellation Leo.

We already know of several motions in which the Earth participates. Our orbit around the sun gives us our own special motion, but we move at only 30 kilometers per second, relative to the sun. More significantly, the sun and the Earth are moving around the center of the Milky Way galaxy at a speed of about 250 kilometers per second. If we take this into account in our interpretation of the differences in the background radiation, we find that the entire Milky Way galaxy must be moving at about 600 kilometers per second in the direction of the constellation Serpens Caput (the Serpent's Head), close to the constellation Hercules.

Astronomers disagree over the volume of space that shares this overall motion; certainly it must be larger than the Milky Way and its immediate neighbors. The motion of the Milky Way revealed by observations of the *anisotropy* (nonevenness) of the microwave background radiation does not contradict the observation that the background radiation should be *isotropic*, the same from all directions, *if* our galaxy has no motion with respect to the sea of radiation that surrounds it and permeates it. Once we allow for the Doppler effect produced by our galaxy's motion, the observations of the background radiation indicate that it *is* isotropic, at least to one part in 3000.

Summary

The early years of the twentieth century brought great advances in our knowledge of the overall patterns of the universe. First, astronomers showed that the disklike shape of the Milky Way galaxy, which includes our sun in its outer reaches, does not embrace the entire universe. Instead, other spiral and elliptical galaxies, once called nebulae, qualify as groups of stars equal to our own Milky Way. Edwin Hubble, the expert in estimating the distances to galaxies, combined his results with Vesto Slipher and Milton Humason's determinations of the galaxies' motions relative to ourselves. The result was Hubble's law, which states that galaxies are receding from us with speeds that increase in proportion to their distances from us.

If our view of the universe is a representative one, then any observer should see what Hubble saw: an expansion centered around the observer. If this is so, then the entire universe must be expanding, as if the galaxy clusters were dots on the surface of an expanding balloon. This expansion has gone on for the past 15 or 20 billion years.

Soon after the initial singularity, or "big bang," that began the expansion, matter in the universe was crowded together at enormous densities and temperatures. During the first half hour after the big bang, all sorts of particles were created; but afterward, the mix of particles remained almost constant: mostly protons, electrons, and helium nuclei, plus a flood of photons, neutrinos, and antineutrinos. The latter three types of particles have no mass and so draw less attention; but the photons deserve notice for their great number (almost a billion times the number of protons and electrons) and for their formation of the cosmic microwave background. This background, or universal sea of photons, had the power to destroy any atom that might form in the universe during the first million years after the big bang; but as the universe continued its expansion, the Doppler shift reduced the energy of each photon. The time of decoupling at almost a million years after the big bang marks the end of the photons' ability to interact with most atoms and prevent them from persisting. Hence, the background has remained unchanged, except for the continuing Doppler shift, until the present time, and we detect it with the spectrum characteristic of an ideal radiator. The characteristic temperature, however, has fallen to 2.°7 above absolute zero (2.7 K) because of the Doppler effect.

To tell whether the universe will ever stop expanding, we should determine the average density of matter. A value greater than about 10^{-29} gram per cubic centimeter implies eventual contraction, while lesser values are linked to eternal expansion. Luminous matter has nowhere near this critical value, so if the density to reverse the expansion does in fact exist, most of the matter must be in some nonluminous form.

Another way to attack the question of the expansion's future is to look at the past. By observing the most distant galaxy clusters, we can see the past trend in the expansion and extrapolate into the future. This method, although sound in principle, has not yet given definitive results, since it is extremely difficult to make accurate observations of faraway galaxies. The Space Telescope may resolve this problem within the next few years.

Key Terms

absolute zero
antimatter
antineutrino
antiparticle
astrophysics
big bang
cosmic microwave background
critical density
decoupling
deuterium
energy of motion
galaxy
heat
Hubble's constant
Hubble's law
ideal radiator
initial singularity
ionize
Kelvin (absolute) temperature scale
kinetic energy
matter
Milky Way
nebula
neutrino
periodic variable star
temperature

Questions

1. Why is it so difficult to see all parts of our own Milky Way galaxy?
2. What position does the sun occupy in the Milky Way?
3. How did Edwin Hubble demonstrate that the Andromeda nebula lies far outside the Milky Way?
4. How did Vesto Slipher measure the motions of galaxies toward us or away from us?
5. What did Hubble find when he looked for a possible correlation between galaxies' distances and their velocities relative to ourselves?
6. Why do Hubble's observations imply that the entire universe is expanding? What assumption is fundamental to drawing this conclusion?
7. How can we determine the approximate age of the universe since expansion began simply by observing the motions and distances of other galaxies?
8. What is Hubble's law? What is Hubble's constant? What is the Hubble time?
9. What is the big bang? How has the density of matter in the universe changed since the moment of the big bang?
10. What were the most numerous particle types to emerge from the first half hour after the big bang? What are the most numerous types today?
11. Why did the universe cool as it grew larger?
12. Where did the photons in the cosmic microwave background come from? What has happened to them during the time since the first half hour after the big bang?
13. What was the time of decoupling? When was it? Why did it occur?
14. What is an ideal radiator? Why did the early universe resemble an ideal radiator?
15. Why does the average density of matter in the universe determine whether or not the universe will expand forever?
16. Why do we have difficulty measuring the average density of matter?
17. Why can we determine the future of the universe by looking into the past? How can we look into the past? What have these observations shown so far?
18. Is the universe finite or infinite in extent? How does the resolution of this question pertain to the question of whether the universe will expand forever?

Projects

1. *The Milky Way.* On a clear night, find the "milky way" in the sky by means of star charts (see book end papers) or simply by looking at the heavens. Does the Milky Way appear to be made of individual stars? Find a lane of absorbing dust, which makes the light of the Milky Way seem to divide. (To do this, you must have a clear night outside the city, where stray light cannot dim your view.) Which do you find more con-

vincing: That dust absorbs light from the Milky Way, or that fewer stars exist where the Milky Way appears to divide? What does this tell you about the ways in which our brain and our experiences influence what we see?

2. *Hubble's Law for the Expanding Universe.* Gather together a group of friends or classmates, and have them stand in a line facing you, with a separation between adjacent persons of about 1 meter. At your signal, have each person raise his or her arms, so that adjacent persons move away from each other until the tips of their outstretched arms just touch. The separation between adjacent persons will now be about 2 meters. Perform this "intergalactic ballet" in several different ways, first holding the person at one end stationary, then the person at the other end, then the person in the middle. You can call one person the Milky Way, another Andromeda, another M 51, and so on; we are ignoring the individual random motions of galaxies. Show that each galaxy (that is, each person) will observe Hubble's law along the row of galaxies. All that you need to complete the model would be columns of people in front of, and behind, the row you have constructed, as well as more columns of people above and below each person on the floor. What about the people at the ends of the rows? How do they correspond to the "end of the universe"?

3. *Microwaves.* The short-wavelength part of the radio spectrum has the name *microwaves,* meaning "tiny waves"; the wavelengths of the photons that form this type of electromagnetic radiation range from a few millimeters to a few centimeters. Where have you encountered the term *microwave* before? Why do the household users of microwaves employ radiation of these wavelengths? (Hint: They do so for the same reason television antennas use metal rods some meters long to detect radiation with wavelengths of a few meters.) What is the wavelength at which the maximum number of photons in the cosmic microwave background are detected (Figure 4-18)? (Use the fact that a photon's wavelength, measured in centimeters, equals 1.99×10^{-16} divided by the photon's energy, measured in ergs.)

Further Reading

Berendzen, Richard; Hart, Richard; and Seeley, Daniel. 1976. *Man Discovers the Galaxies.* New York: Neale Watson Academic Publications.

Ferris, Timothy. 1977. *The Red Limit.* New York: William Morrow.

Gamow, George. 1961. *The Creation of the Universe.* New York: Bantam Books.

———. 1961. *One, Two, Three . . . Infinity.* New York: Bantam Books.

Gott, J. Richard, III; Gunn, James; Schramm, David; and Tinsley, Beatrice. 1977. "Will the Universe Expand Forever?" In *Cosmology + 1.* Edited by Owen Gingerich. San Francisco: W. H. Freeman.

Kaufmann, William. 1977. *Relativity and Cosmology.* 2d ed. New York: Harper & Row.

Muller, Richard. 1978. "The Cosmic Background Radiation and the New Aether Drift." *Scientific American* 238:5, 64.

Schatzman, Evry. 1966. *The Origin and Evolution of the Universe.* New York: Basic Books.

Struve, Otto, and Zebergs, Velta. 1962. *Astronomy of the 20th Century.* New York: Macmillan.

Webster, Adrian. 1977. "The Cosmic Background Radiation." In *Cosmology + 1.* Edited by Owen Gingerich. San Francisco: W. H. Freeman.

Weinberg, Steven. 1977. *The First Three Minutes.* New York: Basic Books.

An S0 galaxy, NGC 3718, shows absorption of light by large amounts of interstellar dust within the galaxy.

5

Galaxies

Within the expanding universe, whose density has continuously decreased with time, matter has formed clumps of much higher density than average: stars and galaxies. The stars are the basic luminous units of the universe, with masses that range from one thousand to 20 million times the Earth's mass. Galaxies consist of millions, billions, or (in extreme cases), trillions of stars, all in mutual gravitational interaction. The galaxies themselves cluster by the dozens, hundreds, or even thousands (Figure 5-1). When we examine the thousands of galaxies within, say, 100 **megaparsecs** (100 million parsecs) of our own Milky Way, we find a basic similarity among galaxies and a useful differentiation into apparently fundamental galaxy types.

Figure 5-1
The Hercules cluster of galaxies contains several thousand galaxies, some of them larger than the Milky Way galaxy in which we live.

5.1 Types of Galaxies

About 90% of all galaxies can be classified as spiral galaxies or elliptical galaxies. The remaining 10% are appropriately called irregular galaxies, since they have a far less ordered structure than spirals and ellipticals. Our own Milky Way forms a giant spiral, but since we are inside this galaxy, we have difficulty seeing its shape and must go to considerable lengths to determine its structure (Figure 5-2). On the other hand, we can photograph other spiral galaxies with ease, acquiring an intimate knowledge of their shapes and patterns (Figure 5-3).

Figure 5-2
Because the sun lies inside the Milky Way galaxy, our view of the galaxy shows a host of crowded stars and interstellar gas.

Figure 5-3
A spiral galaxy, seen almost edge-on, shows a characteristic lane of absorption caused by dust grains. These grains are most abundant close to the galaxy's median plane.

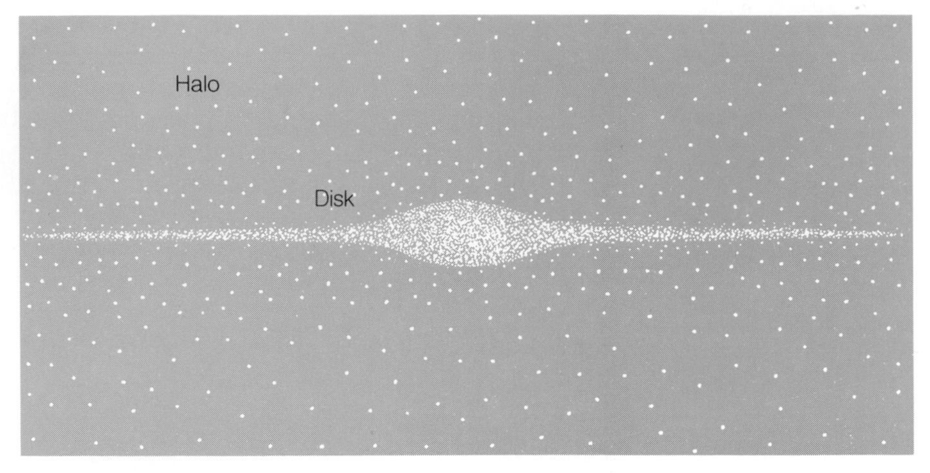

Figure 5-4
Giant spiral galaxies have enormous halos of matter surrounding their spiral disks. These halos extend for several times the diameter of the disk and may contain most of the galaxy's mass.

Spiral Galaxies

Spiral galaxies have flattened, disklike shapes, within which the pattern of **spiral arms** stands out clearly (Figure 4-22). Closer study reveals that giant spiral galaxies, such as our own Milky Way, have enormous **halos** that surround their disks. These halos apparently contain several times more mass than is in the disk (Figure 5-4). We do not yet know what form this mass is in; it may consist of trillions of extremely low-mass stars, or perhaps still smaller objects similar to the planet Jupiter.

NGC 2811 Type Sa

NGC 488 Type Sab

NGC 3031 (M81) Type Sb

NGC 2841 Type Sb

NGC 628 (M74) Type Sc

Figure 5-5
Astronomers classify spiral galaxies by the tightness of the winding of the spiral arms. Type Sa denotes the most tightly wound arms and type Sc the most loosely wound.

Seen from "above" or "below," rather than edge-on, spiral galaxies show differences in the tightness with which their spiral arms are wound (Figure 5-5). Type Sa has the most tightly wound arms; type Sc has the most loosely wound, and type Sb is in the middle. The Milky Way's galaxy type is probably Sbc, meaning between Sb and Sc.

A special kind of spiral galaxy is the **barred spiral** (Figure 5-6). Barred spiral galaxies have a flat shape like ordinary spirals, but they have an extended "bar," or spindle-shaped concentration of stars, in their central regions. Barred spirals, known as SB galaxies to astronomers, can also be classified by the tightness of the winding of their spiral arms; thus we have SBa, SBb, and SBc galaxies (Figure 5-7).

Elliptical Galaxies

In contrast to spiral galaxies, **elliptical galaxies** are all halo and no disk (Figure 5-8). Ellipticals can be either spherical or somewhat flattened (ellipsoidal) in shape, with no disklike structure and no spiral arms. Furthermore, elliptical galaxies have little or no free-floating gas and dust among their stars. This contrasts with spirals, which reveal interstellar gas and dust through

Figure 5-6 A barred spiral galaxy possesses a "bar" of stars at its center. The spiral arms begin at the ends of this bar.

NGC 175 Type SBab(s)

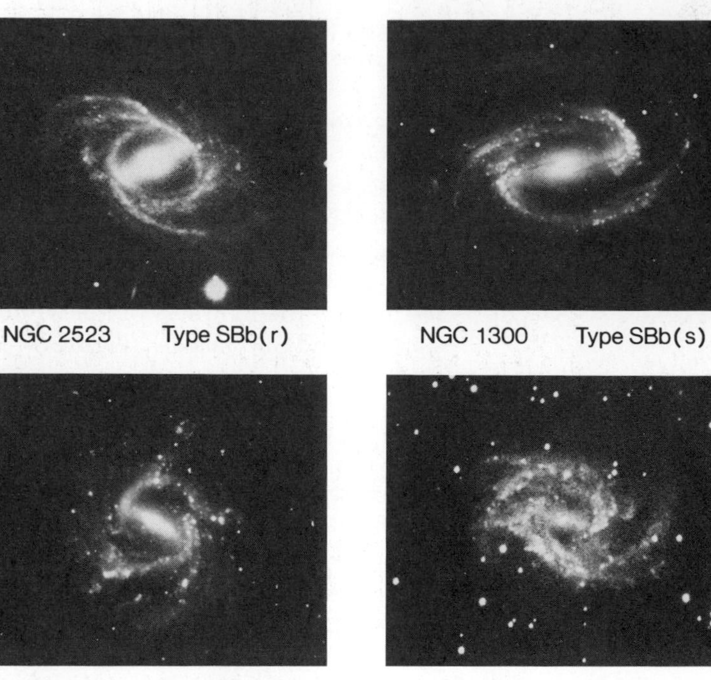

NGC 2523 Type SBb(r)

NGC 1300 Type SBb(s)

NGC 1073 Type SBc(sr)

NGC 2525 Type SBc(s)

Figure 5-7 Barred spiral galaxies, like ordinary spirals, can be classified by the tightness with which their spiral arms are wound. SBa galaxies have the most tightly wound arms; SBc the least tightly wound.

the absorption of starlight (Figure 5-3) and the emission of radio waves (page 193).

Astronomers classify elliptical galaxies by their degree of flattening, ranging from E0 (no flattening) to E7 (highly flattened) (Figure 5-9). Even the most flattened elliptical, however, falls far short of having the thin disk that characterizes spirals. In elliptical galaxies, as in spirals, the number density of stars increases toward the center of the galaxy (Figure 5-8).

S0 Galaxies

An important transitional type of galaxy, called **S0 galaxies**, has some characteristics of both spirals and ellipticals (Figure 5-10). Like spiral galaxies, S0 galaxies have a disklike shape; but they have little or no spiral-arm pattern.

Irregular Galaxies

In addition to the spiral, elliptical, and S0 galaxies, we find **irregular galaxies**, with neither the disklike shape of spirals nor the ellipsoidal shape of ellipticals. Two good examples of irregulars are the **Magellanic Clouds** that

Figure 5-8
The giant elliptical galaxy NGC 4472 contains many hundreds of billions of stars, spread in an ellipsoidal (E1) distribution. The bright star just above NGC 4472 belongs to our own galaxy.

Figure 5-9
Elliptical galaxies are classified on the basis of their flattening: E0 galaxies are spherical in shape, while E6 and E7 galaxies are highly elongated ellipsoids. This elliptical galaxy, NGC 205, is type E5.

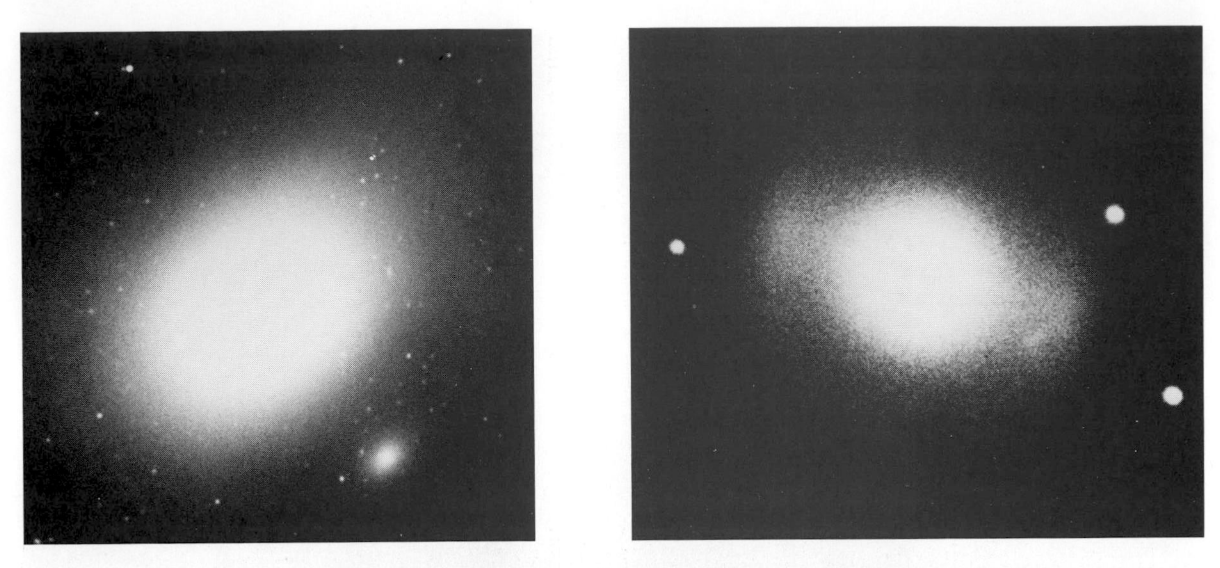

Figure 5-10 S0 and SB0 galaxies have shapes reminiscent of spiral and barred spiral galaxies, since they are highly flattened, but they show no spiral arms.

Figure 5-11 The Milky Way's two irregular satellite galaxies are the Large Magellanic Cloud (left) and the Small Magellanic Cloud (right). Each of these galaxies has a distance of about 50,000 parsecs from us.

Figure 5-12
The giant galaxy M 87 contains about a trillion stars.

orbit our own Milky Way.[1] Each consists of a loose grouping of a few billion stars (Figure 5-11).

5.2 The Masses of Galaxies

The Magellanic Clouds are as massive as any known irregular galaxies, with as much as 25 billion times the mass of the sun. In contrast, spiral and elliptical galaxies may have masses in excess of 1 trillion times the sun's mass and may contain a trillion stars or more. The Milky Way, with an estimated 400 billion stars, ranks among the most massive spiral galaxies, while the giant elliptical M 87 (Figure 5-12), with a mass that is a trillion times greater than the sun's, ranks among the largest known ellipticals.

At the low end of the mass range we find **dwarf elliptical galaxies** (Figure 5-13). These loosely bound aggregates of stars may each contain only a few million, or a few tens of millions, of stars. Thus, they have no more mass than the subunits of a large galaxy, called **globular star clusters** (Figure 5-14). Dwarf ellipticals cannot be easily seen or photographed, so we know of only a few, relatively close to the Milky Way. Nevertheless, they may represent the most common of all galaxy types. Some spiral galaxies have far less mass than the Milky Way, but they never have so little mass as the dwarf ellipticals. The least massive spirals still have the better part of a billion times the sun's mass.

[1]The Magellanic Clouds owe their name to Ferdinand Magellan, whose sailors noticed them during Magellan's voyage around the world in 1520. These galaxies cannot be seen from northern latitudes because of their position on the sky (Figure 2–16). Magellan's crew first thought the galaxies were clouds, but later realized that the "clouds" maintained a fixed position relative to the stars. They are in fact the closest galaxies to our own.

Figure 5-13
This dwarf elliptical galaxy in Sculptor contains only a few million stars, millions of times fewer than the number of stars in the largest galaxies. The photograph is printed as a negative to make the stars more visible.

Figure 5-14
The globular star cluster in the constellation Tucana (and part of the halo population of our Milky Way galaxy) contains about 1 million stars.

5.3 The Formation of Galaxies

Astronomers, who seek a coherent picture of how the universe came to its present configuration, have had a hard time explaining why galaxies exist. The root of their difficulties lies in the early universe, during the million years or so before the time of decoupling. As we have described (page 113), the early universe was basically smooth; that is, density differences between one part of the universe and another, if they existed momentarily, tended to be washed away by the constant impact from photons. So long as photons had enough energy to interact with matter, they exerted a strong *homogenizing* impulse, like a mixer that keeps milk and cream from separating.

Thus, the early universe was a cosmos without clumps, and when we try to explain how the lumps we call galaxies came to be, we need to find a mechanism responsible for such condensations. Gravitation provides just such a mechanism, since gravitational forces will make small density dif-

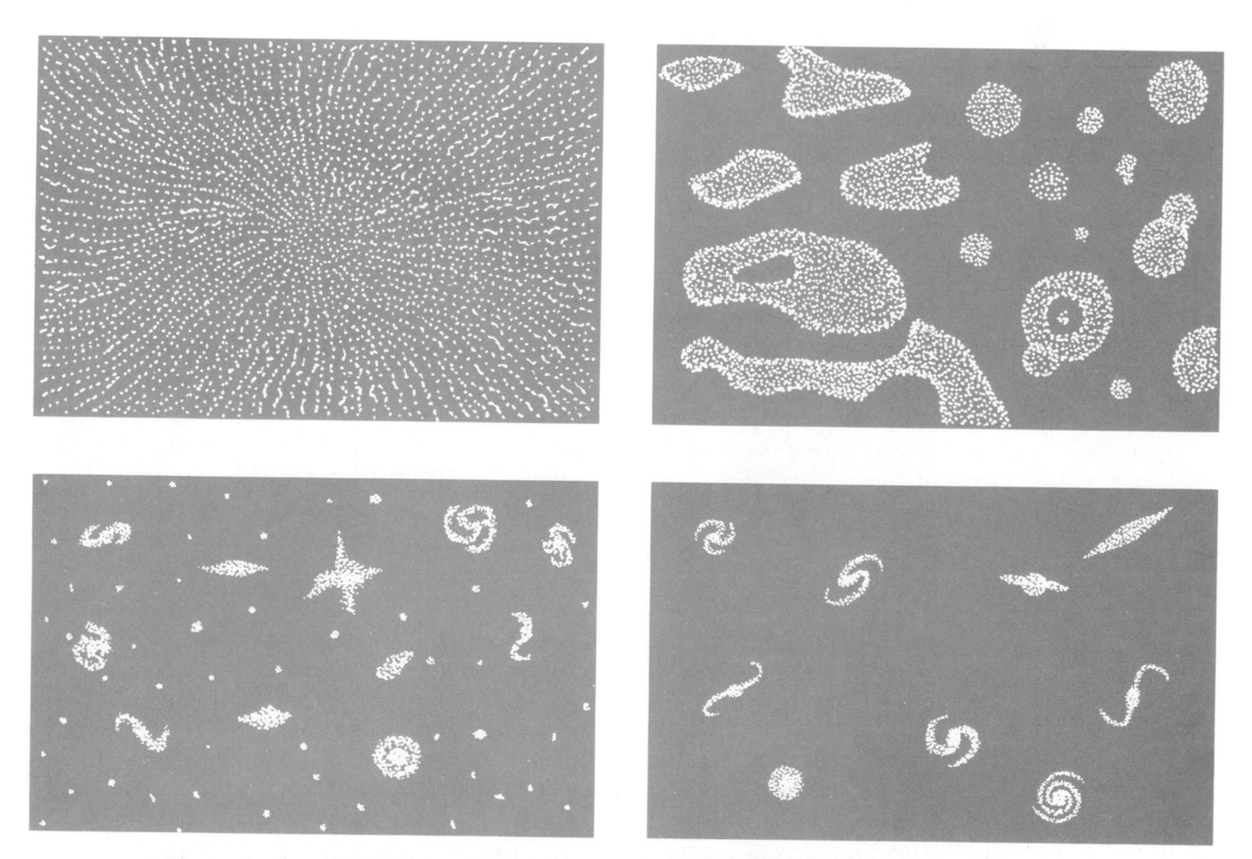

Figure 5-15 Galaxies formed from relatively small fluctuations in the density of matter from place to place. These fluctuations must have existed at the time of decoupling, about a million years after the big bang. Regions with a greater-than-average density attracted more matter and eventually coalesced to form galaxies and galaxy clusters.

ferences grow into larger ones, as the denser regions attract more matter than the less dense regions of space (Figure 5-15).

The Growth of Clumps in the Universe

But how did the universe ever acquire density differences from place to place, even small ones? We can only speculate that the differences in local density may have existed in the initial stages of the universe, shortly after the big bang, and may have then survived, at least in part, the homogenizing influence of a million years. If, at any rate, we can accept the fact that density fluctuations *did* exist in the universe after the time of decoupling, astronomers have a rather coherent scenario of how these density differences grew into galaxy clusters, galaxies, and stars. This scenario needs to be tested with more observations before we can consider it the *real* story of cosmic evolution, but for now it offers our best description of how galaxies formed between 1 million and several billion years after the big bang.

This scenario relies on calculations that show that after the time of decoupling, the clumps of matter most likely to persist and grow were those with about 100 million times the sun's mass. These clumps were initially only a bit denser than average. (Recall that density measures mass *per unit volume.*) The greater-than-average density produced a greater-than-average gravitational force to attract matter from neighboring regions of space. Thus these clumps, despite the ongoing expansion of the universe, could become steadily denser than the average density of matter. Clumps that had less than 100 million solar masses, and thus spanned a smaller volume of space at the time of decoupling, would not have enough gravitational force to survive as dense clumps. Instead, they would be smoothed away by the random motions of particles within and outside the clumps. Clumps with much more than 100 million solar masses—those that occupied a particularly large volume of space—would increase their densities relative to the average density of matter, but at a slower rate than the 100-million-solar-mass clumps, because they would have to struggle harder to overcome the universal expansion, which tends to separate all parts of a clump as it struggles to pull itself together.

Thus the clumps with about 100 million solar masses would form the basic units of the early universe, somewhere between 1 million and 1 billion years after the big bang. Individual stars eventually develop in these clumps, but first the clumps would have the chance to interact with one another through gravitational forces. Such interactions would produce galaxies (with anywhere from 100 million to 1 trillion solar masses of matter) and even clusters of galaxies (with up to several thousand galaxies and perhaps 100 trillion solar masses).

Suppose we accept, at least tentatively, the theory that 100-million-solar-mass clumps were the first to form. We still must understand why and how many galaxies with thousands of times this mass came to exist, and why some galaxies are spiral and some elliptical. To do so, we must think more carefully about the processes that occurred within the aggregates of gas that became stars and galaxies.

What Happens to a Contracting Gas Cloud?

Consider a clump of gas that has become a recognizable entity, with 100 million solar masses and a density greater than the average density of matter in the universe. Several competing processes can affect this clump. First, it may collide with other clouds, to merge with them through mutual gravitational attraction. Second, the cloud's **self-gravitation**, the attraction that each part of the gas in the clump exerts on the other parts, tends to compress the clump to a smaller size. Third, the random motions of the individual gas particles (atoms and molecules) tend to resist such gravitationally induced contraction, though not indefinitely. Finally, close encounters among clouds may give each of them some rotation (Figure 5-16).

The combination of these four factors, occurring to a different extent from clump to clump, has apparently produced the variety of galaxies and galaxy clusters we observe today. Some clouds of gas merged with many others before they formed stars: These are the giant galaxies. Some shrank through their own self-gravitation before such merging could occur, to produce dwarf galaxies. (The least massive dwarf ellipticals presumably arose from clumps that managed to survive with even less than the 100 million solar masses we have cited as the minimum for a good clump.) Some galaxies, large and small, gathered themselves into giant clusters, in which all the galaxies orbit around their common center of mass. Others, like our own Milky Way, have only a few neighboring galaxies and will not even remain within these small groups indefinitely. Some galaxies acquired a good deal of rotation as **protogalaxies** (galaxies in formation), others acquired less. Some reached a high enough gas density to form stars at relatively early stages, others only later. This last point deserves closer attention, since it promises to explain how spiral and elliptical galaxies acquired their shapes.

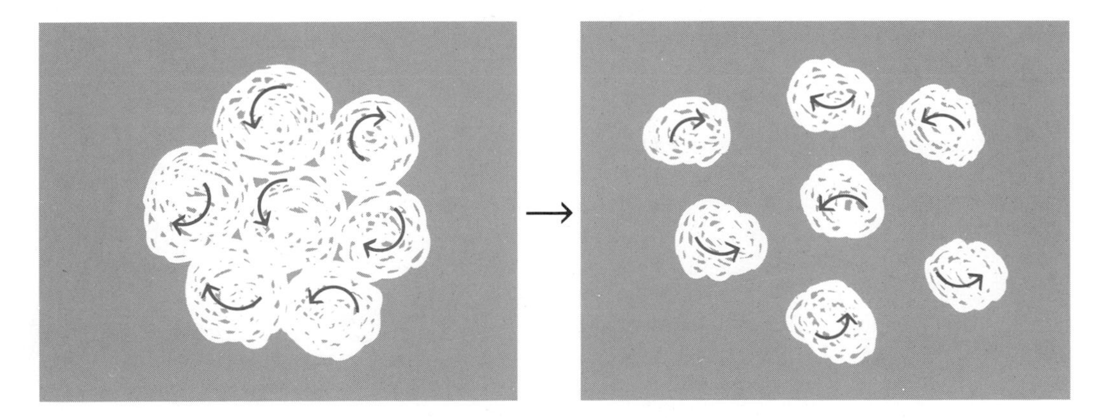

Figure 5-16 Clumps of gas attract one another through gravitational forces and may merge as a result. A large clump of gas formed in this way may later fragment into subclumps. During this fragmentation, some of the subclumps may be set in rotation.

The Conservation of Angular Momentum

Suppose that a clump of gas with somewhere between 100 million and 1 trillion solar masses begins to contract under the influence of its self-gravitation. The gas clump must obey a key rule of physics called the **conservation of angular momentum**. Any object in rotation has a nonzero **angular momentum**, defined as the object's mass times its rate of spin times the square of the object's size perpendicular to the axis of spin. According to the conservation principle, this angular momentum remains the same as the object contracts. Thus, if the object shrinks in size by a factor of two, it must spin four times more rapidly. Ice skaters and acrobatic high divers intuitively make use of this rule. They know that by contracting their bodies to a smaller size, they can increase their rates of spin (Figure 5-17).

If a clump of gas, such as a protogalaxy, has begun to spin through encounters with other clumps, it will keep on spinning, but more rapidly as it contracts. The spinning protogalaxy finds contraction easier in directions along the spin axis than in directions perpendicular to the spin axis (along its "equator"), because the rotation tends to counteract the tendency to collapse toward the axis of rotation. Therefore, if protogalaxies remained diffuse gas, we would expect them to end up as flattened, rotating disks (Figure 5-18). Indeed, this corresponds to the general shape of a spiral galaxy (Figure 5-3). Moreover, the halo population and globular clusters in spiral galaxies, as well as the overall structure of elliptical galaxies, indicate that many stars must have begun to form before the protogalaxies had all condensed into disklike shapes.

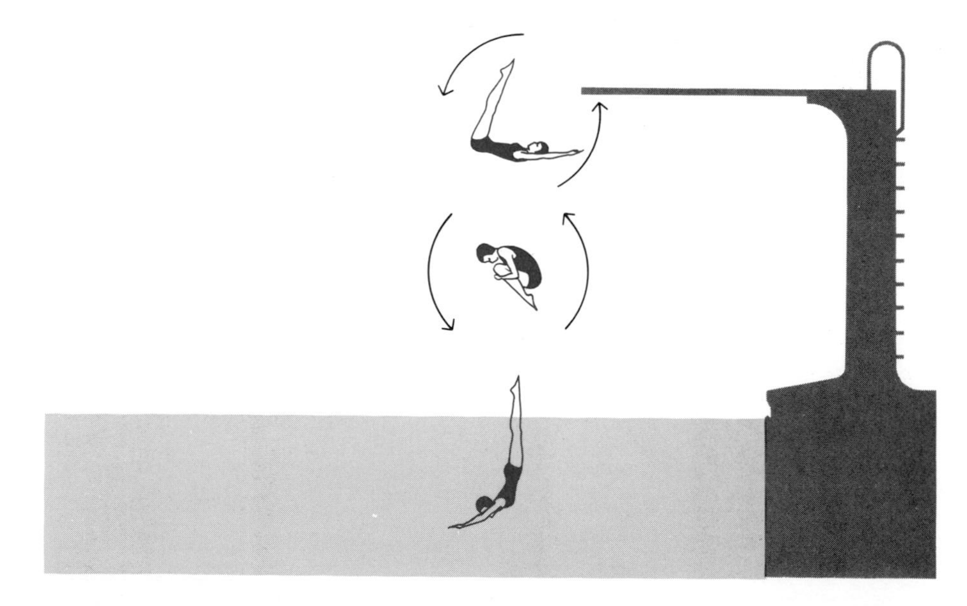

Figure 5-17 A high diver who contracts her body in the direction perpendicular to her spin axis will increase her rate of spin, because her angular momentum must stay constant as she falls.

5.4 Star Formation in Galaxies

Stars will not begin to form in a protogalaxy until the gas density rises to a certain critical value, which may vary somewhat from place to place. Thus, a crucial time in any protogalaxy's history is the moment when the density rises to the point where star formation can begin.

Star Formation in Elliptical Galaxies

In elliptical galaxies, the density apparently rose to the critical value long before the protogalaxy could assume a disklike configuration. Moreover, once the protogalaxy had formed stars, the stars did *not* continue the tendency to collapse into a disk. On the contrary, the stars occupied such a small fraction of space that they did not collide with each other, but rather could orbit freely around the center of the galaxy. In contrast, the atoms within the diffuse gas constantly collided, and only through such collisions could the gas dissipate energy, allowing the clump to assume a disklike structure (Figure 5-18). As atoms collided with a clump of gas, their electrons would jump into larger orbits. The atoms would then produce photons that carried away some energy. As a result of this energy loss, the particles in the gas would acquire new orbits around the galactic center that eventually produced the flattened disk of a spiral protogalaxy.

Star Formation in Spiral and S0 Galaxies

In spirals, we see a combination of early star formation (in the halo population) and later star formation (in the disk). Some star formation occurs even now, since clouds of gas and dust still exist, occasionally growing dense enough to contract into stars (Figure 5-19). But a great burst of star formation in spirals occurred after the halo population had formed, when the rest of

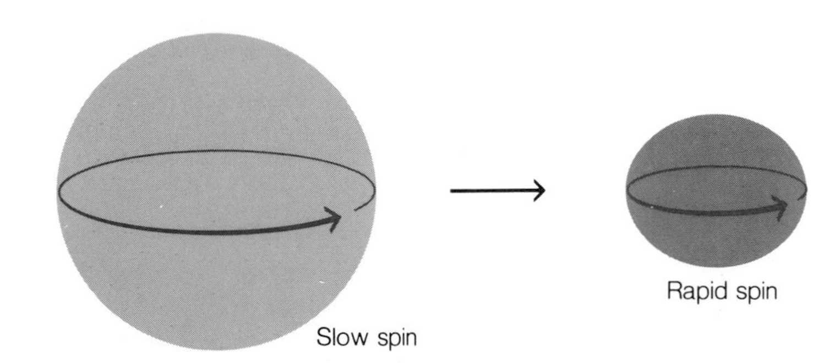

Figure 5-18 A spinning protogalaxy contracts somewhat in directions perpendicular to the spin axis, and hence spins more rapidly. But it contracts even more in directions along the spin axis and eventually becomes a disk.

Figure 5-19 This cloud of gas, lit from inside by young, hot stars, represents a region of our galaxy in which stars are now forming.

the matter had concentrated into a disk for the first time (see page 174). In S0 galaxies, which are a cross between spirals and ellipticals, this "burst" apparently made stars out of all the material just before the flat disk configuration could be reached, leaving no interstellar gas and dust for later generations of stars. Since a galaxy's spiral arms appear to need constant replenishment (see page 197), S0 galaxies have no spiral arms now, if indeed they ever did.

Star Formation in Irregular Galaxies

To complete our scenario of star formation, we note that the irregular galaxies are those in which star formation has been delayed the longest, and which still contain the largest fraction of their masses in the form of interstellar gas. In other words, irregular galaxies have not yet allowed stars to become the overwhelming feature of the galaxy or to pull the gas into a shape that conforms to their spatial distribution. Instead, with their gas and dust possessing as much mass as that in the stars, irregular galaxies still present a relatively amorphous structure, with a chaotic display of stellar motions that may be vaguely reminiscent of the earlier phases of spiral galaxies.

5.5 The Rotation of Galaxies

We have outlined a reasonable scenario of how galaxies may have formed, but we cannot feel confident that this is what really happened. We tread firmer ground when we describe the present-day features of galaxies, since here our models come from current observation, not from what may have occurred in the distant past. Galaxies certainly do rotate—so slowly that we can never "see" them turn, but rapidly enough for astronomers to measure their spins by use of the Doppler effect.

In both spiral and elliptical galaxies, stars move around the galaxy's center in elliptical orbits, ranging from almost perfect circles, such as the sun's orbit around the center of the Milky Way, to highly elongated ellipses, typified by the orbits of many globular star clusters in our own galaxy (Figure 5-20). All these orbits have a common property first explained by Isaac Newton: Any object will move in a straight line unless a force acts on it. Astronomical situations typically involve gravitational forces. In our solar system,

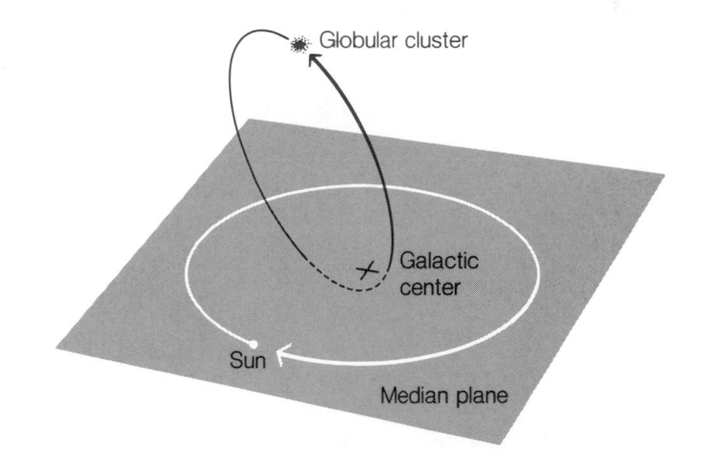

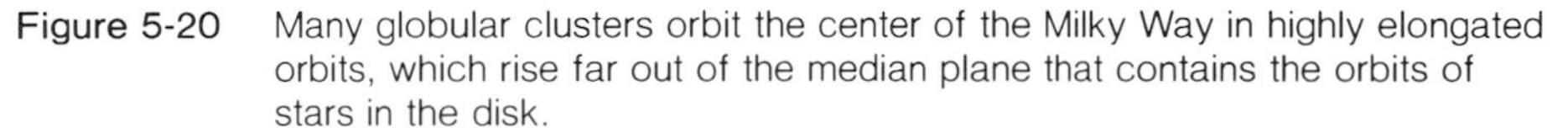
Figure 5-20 Many globular clusters orbit the center of the Milky Way in highly elongated orbits, which rise far out of the median plane that contains the orbits of stars in the disk.

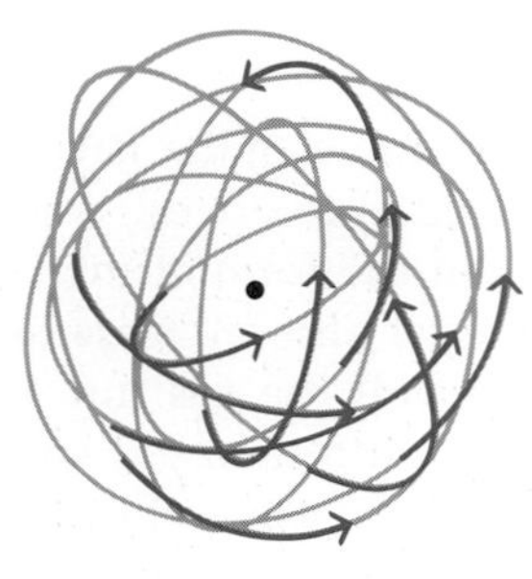

Figure 5-21
In an elliptical galaxy, stars orbit the galactic center in all directions. Nonetheless, in most elliptical galaxies, more stars orbit in one general direction than in the opposite direction, giving the entire galaxy an overall sense of rotation.

the gravitational force comes primarily from the sun; in a galaxy, the corresponding pull comes from the totality of all the stars in the galaxy, which translates roughly into a single pull directed toward the galactic center (Figure 5-20). This pull tends to change straight-line motion into repetitive, orbital motion. Objects in elliptical orbits move more rapidly as they come closer to the center, more slowly when they are farther away, in a pattern similar to that of a planet in the solar system (Figure 2-27).

The Rotation of Elliptical Galaxies

In elliptical galaxies, each star's orbit may take it in a direction different from that of other stars. However, if we look at the totality of orbits, we do find an overall rotation for the galaxy, with more stars moving in one general direction than in the opposite direction (Figure 5-21). This holds especially true for the more flattened ellipticals (E4 through E7), least true for the unflattened (E0) ones.

The Rotation of Spiral Galaxies

No doubt exists about the rotation of spiral galaxies. The stars in those galaxies' disks all move in the same direction and in almost the same plane. Our sun, relatively far from the center of the Milky Way, moves in an almost circular orbit at about 240 kilometers per second (just under one-thousandth the speed of light). Since this orbit has a radius of 9000 parsecs and a circumference of 185,000 light years, the sun orbits our galaxy once in about 240 million years. Thus mammals first appeared on Earth just about the time that the sun was last at the position along its orbit it occupies now. Stars farther from the center may take 300 or even 400 million years to complete a single orbit; those closer may complete a round trip in 100 million years, or even less.

Motions of Stars in Spiral Galaxies

In addition to their generally circular orbits around the galactic center, stars in a spiral galaxy also have a superimposed, up-and-down motion through the galaxy's median plane (Figure 5-22). These oscillations keep the galaxy's stars from contracting to form an infinitesimally thin disk. For example, our sun has an up-and-down motion that carries it about 200 parsecs above and below the median plane of the Milky Way. The sun bobs up and down perhaps four or five times during each orbit around the galactic center, reaching a maximum velocity of about 10 kilometers per second in the direction perpendicular to the median plane as it passes through it. When the sun rises "above" or falls "below" the plane of the galaxy, gravitational forces not only pull the sun toward the center, but also attract it back toward the plane (Figure 5-23). The sun's momentum carries it through the plane (without any collisions with other stars, each of which is performing a roughly similar motion) and outward on the opposite side. During the 4.6 billion years since

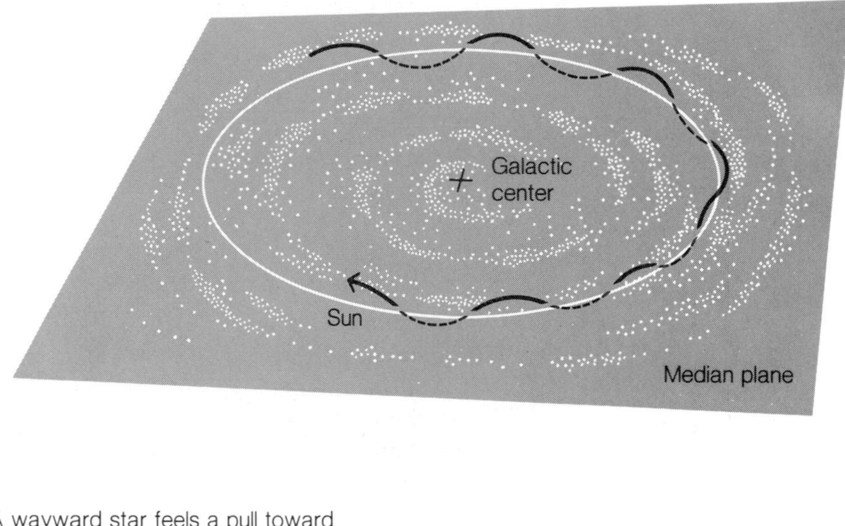

Figure 5-22
As a star in the disk of a spiral galaxy orbits the center, it oscillates up and down, through the median plane of the galaxy, to a "height" of 100 or 200 parsecs on either side of the plane.

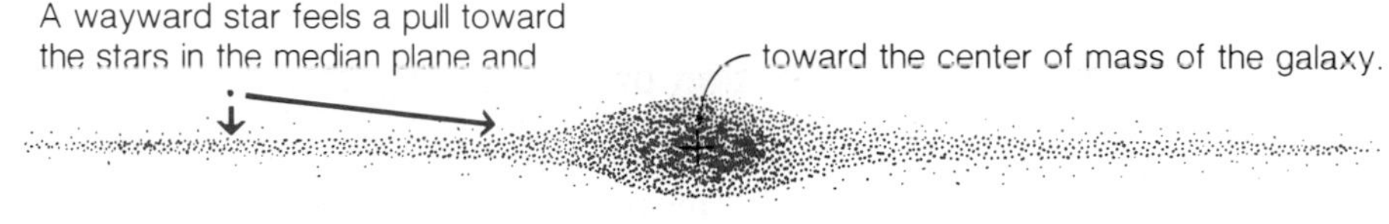

Figure 5-23 As a star rises above the galaxy's median plane, the gravitational forces from the stars in the galaxy pull the star back toward the plane, as well as toward the center of the galaxy.

the sun formed, it has made about twenty trips around the galaxy and perhaps 100 oscillations through the plane of the Milky Way. At the present time, we occupy a position within 30 parsecs of the median plane, so far as we can tell.

5.6 Clusters of Galaxies

We have seen how spiral galaxies grew flat while elliptical galaxies remained relatively spherical. If the large spirals and ellipticals did form from smaller clumps, as in the scenario we have outlined, it is intriguing to note that giant spirals and ellipticals have about the same mass (several hundred billion times the sun's mass), about the same size (some 30,000 parsecs in diameter), and occur in about equal numbers, which says something about the equal opportunity open to the smaller clumps as they collided and merged. We also ought to examine the fact that most galaxies now belong to notable clusters of galaxies, with most clusters made up of at least a few hundred members. In these clusters, the average density of matter is significantly greater than the average density throughout the universe. The clusters seem to contain enough mass to remain gravitationally bound, that is, to exert enough gravitational force on each galaxy within the cluster to keep it a member indefinitely (Figure 5-24).

Figure 5-24 A large cluster of galaxies, such as this one in Coma Berenices, has enough mass to keep its member galaxies part of the cluster for billions of years. The total gravitational force on any galaxy binds it to the cluster as the galaxy orbits the cluster's center of mass.

The Amount of Mass in Galaxy Clusters

From studying the motions of galaxies along our line of sight by means of the Doppler effect, astronomers can calculate the total mass of matter in a cluster. They can do this by the direct application of Newton's laws of motion and gravitation (which we shall discuss in Chapter 11), since these laws tell us how much acceleration a given amount of mass will produce on a given object, such as a galaxy. At the present time, the calculations indicate that most galaxy clusters contain more mass—about ten times more—than we can easily account for from the luminosity of the galaxies within the clusters. Astronomers have found that among the galaxies in a cluster, hot intergalactic gas emits x rays, but the mass of this hot gas equals no more than the mass within the galaxies themselves. Hence, at the present time we face an unresolved divergence between the mass we calculate for galaxy clusters on the basis of the motions within them, and the mass we obtain when we add up the masses of the individual galaxies. This divergence corresponds rather well to the divergence in our estimates of the average density of matter in the universe: The method based on counting galaxies gives a density ten or twenty times less than the method based on looking back into the past and seeing how the average density of matter has affected the universal expansion. We may yet find that invisible matter (burnt-out stars and galaxies, or black holes, for example) forms the bulk of matter in the universe (see page 337).

Galactic "Cannibals"

Galaxy clusters deserve special attention for one recently discovered aspect: **galactic cannibalism**. Astronomers have known for some time that rich clusters of galaxies often contain one or two extremely large and bright galaxies near their centers (Figure 5-25). These giants, called **cD galaxies** in a rather complicated naming system, may contain hundreds of trillions of stars and have masses of more than 100 trillion solar masses. Their diameters may exceed several hundred thousand parsecs, ten times the diameter of the Milky Way. If one of these galaxies replaced the Andromeda galaxy, it would spread over a tenth of the sky, and its central regions would shine with an apparent brightness greater than any star's.

In attempting to explain how cD galaxies come into existence, astronomers have found merit in the rather simple idea that these galaxies have slowly swallowed other galaxies that happened to collide with them. This process occurred only rarely in the first few billion years of the cluster, but as the one or two extremely large galaxies at the cluster's center grew in size, their gravitational force pulled the other members into the ever-larger cD "trap." Thus, we see a continuing process that may, if the cluster lasts long enough, put most of the cluster's mass into the cD galaxies.

Figure 5-25 This giant cD galaxy, at the center of the cluster called A 754, has more than ten times the mass of the Milky Way. This galaxy ranks among the largest known.

5.7 Measuring the Distances to Galaxies

The effort to make accurate determinations of the distances to galaxies has occupied astronomers for three generations, and will continue to engage their attention in the future, for we must rely on a whole chain of observations and methods of deduction, rather than on a single method that would yield the distances to all galaxies. All the methods of distance determination compare a relatively nearby object with a more distant object—a star, a gas cloud, a galaxy. If (and this will always be a key condition) we have reason to believe the nearby and faraway objects are almost identical, then the difference in distance will produce a corresponding difference in the apparent brightness and size of the objects. Measurements of this difference will, if we know how to use them, give us the ratio of the distances to the objects.

The Relationship Between Brightness and Distance

Consider, for example, the fact that any object's apparent brightness decreases in proportion to the *square* of the object's distance from us. This phenomenon is called the **inverse-square law.** If, therefore, we observe a star that we *know* has the same true brightness as the sun, but which appears 1 trillion (10^{12}) times fainter than the sun to us, we may conclude that the fainter star has 1 million (10^6)—the square root of 1 trillion—times the sun's distance from us. Notice that this method yields the *ratio* of the star's distance from us to the sun's distance from us and that it relies on our knowing that the two objects have the same *true brightness.* Astronomers usually measure the true brightness of an object by calculating the apparent brightness that the object would have at a standard distance of 10 parsecs. The object's apparent brightness at a distance of 10 parsecs provides, by definition, the luminosity (absolute magnitude) of the object (see box, page 51).

Standard Candles

Standard candles are light sources of known true brightness that have almost the same properties from one galaxy to the next. To find the distances to other galaxies, astronomers have relied heavily on two types of standard candles: pulsating variable stars and novae.

Pulsating variable stars fluctuate in brightness over a time scale of several hours to several days. By timing the period of light variations, astronomers can determine what kind of variable star they are observing. The most important of these variable stars are the **Cepheid variables,** named after the first one to be studied, Delta Cephei. The period of light variation for these stars has a direct correlation with the luminosities of the stars. A related class of stars, the RR Lyrae variables, can be recognized by characteristic features in the stars' spectra and by their relatively short periods. By finding the approximate distances to Cepheid variables and RR Lyrae variables in our own galaxy, astronomers have established "benchmark" distances. If we then assume that the same type of star will have the same true brightness in other galaxies, we are ready to find the distance ratio: The star in a distant galaxy will be farther away than a nearby star by a factor equal to the square root of the ratio of apparent brightnesses. This was the method used by Edwin Hubble to find the distance to the Andromeda galaxy (page 104) and remains the preferred method to find the distances to galaxies close enough to us for periodic variable stars to be observed.

The other type of standard candle consists of stars that flare up sporadically: the **novae,** which suddenly grow thousands of times brighter than usual, and the **supernovae,** which grow billions of times brighter. Again, if we know the true brightness of the novae and supernovae, we can measure their apparent maximum brightnesses to find their distances. The problem, however, is that the absolute magnitudes, especially of the supernovae, vary from star to star, making the distances calculated from their apparent brightnesses uncertain at best. This is regrettable, because we can see supernovae (owing to their immense true brightnesses) out to distances of hundreds of megaparsecs. In contrast, we can recognize individual pulsating variable stars only if they lie within galaxies at no more than 4 or 5 megaparsecs, and novae only at distances less than 20 or 30 megaparsecs.

Changes in Angular Size with Distance

Another kind of distance measurement relies not on apparent brightness, but on **angular size,** that is, how large an object appears to be on the sky (see box, page 23). Within our own galaxy, giant clouds of gas, in which stars have recently formed, light up when photons from the young, hot stars interact with the atoms in the gas (Figure 5-26). These glowing clouds, called **H II regions,** have various sizes and brightnesses, but astronomers have some evidence that the largest such clouds in a giant spiral galaxy may have the same size from one galaxy to the next. If this is true, presumably because gas clouds of a certain maximum size can begin to form stars, we can determine the relative distance of galaxies by measuring the angular extent—the fraction of a degree on the sky—that the largest H II regions in another galaxy cover (Figure 5-27). This angular size decreases in proportion to the cloud's distance from us. Since we can see such H II regions out to distances of 30 to 50 megaparsecs, they provide a backup method to check the distances derived from observations of novae in other galaxies and a way to extend somewhat our range of distance determinations.

Galaxies as Standard Candles

For galaxies more than 50 megaparsecs away, we can use a new kind of standard candle: the galaxies themselves. Astronomers have determined the distances to several hundred galaxies with good accuracy (meaning that they think they know the distance to within about 25%). All these galaxies lie within 20 megaparsecs of the Milky Way, and since we know their distances and their apparent brightnesses, we can calculate their true brightnesses. Suppose now that we photograph a cluster of galaxies at a distance of a few hundred megaparsecs. We cannot see any individual stars or H II regions in

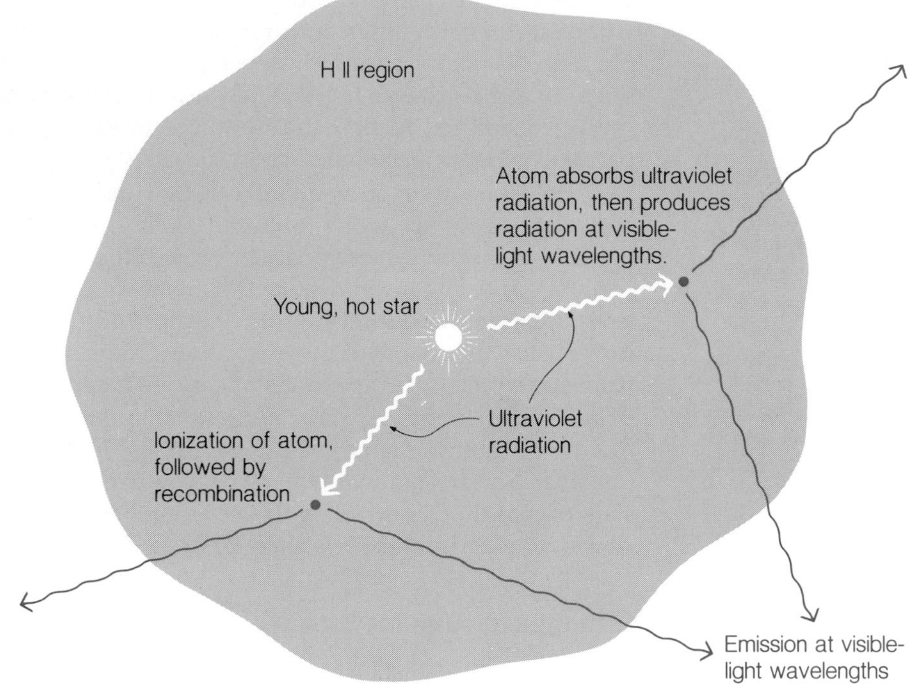

Figure 5-26
Young, hot stars inside clouds of gas will ionize atoms in the gas around them. As these atoms recombine from ions and electrons, they will emit photons of visible light, which we see as diffuse radiation from the entire gas cloud, as in Figure 5-19. Such H II regions mark the places where stars have recently been born.

the photograph, but we can see that many galaxies resemble familiar spirals and ellipticals much closer to us (Figure 5-1). If we make spectroscopic observations of these faraway galaxies and find spectra with patterns similar to those of the nearby galaxies they resemble in appearance, we can make the assumption that the nearby and faraway galaxies have about the same true brightness. Then if a spiral galaxy in the faraway cluster has 1/16 the apparent brightness of a similar-looking galaxy in the Virgo cluster, whose distance we already know, we can conclude that the faraway galaxy has 4 times the distance of the member of the Virgo cluster. This method is not as accurate as the variable-star, nova, and H II-region methods, but it has opened up the far range of the distance scale to astronomers.

The Brightest Members of Galaxy Clusters as Standard Candles

For more distant objects, astronomers have found yet another standard candle. Studies of the nearby galaxy clusters have revealed that the brightest members of a "rich" cluster—one with more than 1000 member galaxies—seem to have almost the same true brightness from cluster to cluster. Astronomers take the fifth-brightest member of a galaxy cluster (in an attempt to avoid the exceptional, tremendously bright galaxies—though why the fifth and not the fourth or seventh remains arbitrary). They then compare the apparent brightness of this fifth-ranked galaxy from cluster to cluster to determine the distance ratios, as with pulsating variable stars and novae.

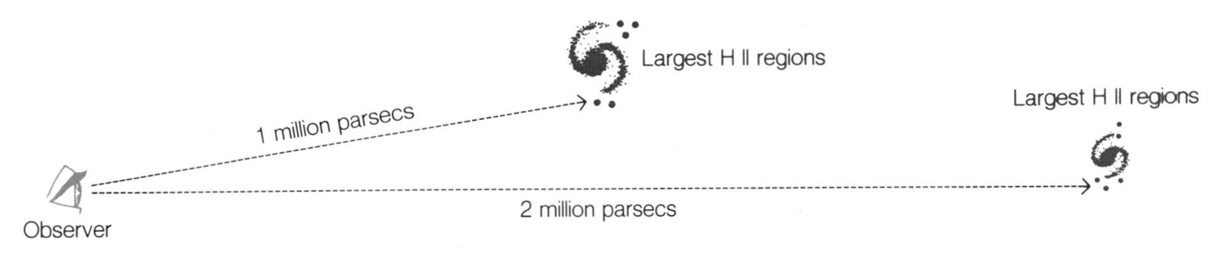

Figure 5-27 The largest H II regions in a giant galaxy are thought to have about the same size in each galaxy. Hence, the apparent *angular* size of the largest H II region will decrease in proportion to the galaxy's distance from us. We can therefore find the galaxy's distance by measuring the angular size of its largest H II regions.

The use of the luminosity, or true brightness, of the fifth-brightest member of a cluster as our standard candle has an interesting drawback. Since the light from faraway clusters takes a long time to reach us, we are looking far back into the past when we observe them. Furthermore, we are looking back by different amounts of time when we observe galaxy clusters at different distances from us. If galaxies change their luminosities over the course of billions of years—perhaps as certain types of stars disappear from them—then we may be led into error by assuming that the fifth-brightest member of a galaxy cluster has the same luminosity *regardless of the time at which its light was produced.*

Astronomers have attempted to allow for the evolutionary effects on the luminosities of galaxies, but with modest success. These evolutionary effects cannot really be accounted for without knowing the distances to the galaxy clusters, and it is just these effects that prevent us from making accurate distance determinations. Hence, for the most distant clusters of galaxies—those whose light arrives billions of years after it was emitted—the distance to the cluster suffers grave uncertainty, because we do not know whether the galaxies had the same luminosities billions of years ago that closer galaxies have *now* (or at least had less than 1 billion years ago).

Hubble's Law as a Distance Indicator

Finally, if all else fails, we can rely on Hubble's law itself, the summation of astronomers' measurements of distances and velocities. A galaxy too far away to have its structure clearly visible, or too strange to be compared directly with a known galaxy, still produces light—otherwise we would not see it. Astronomers can measure the spectrum of that light, even out to immense distances, for they have developed remarkable techniques to determine the spectra of even tremendously faint objects. If we find a familiar pattern in the light from the galaxy, we can determine the amount by which the Doppler effect has shifted the pattern and thus the relative speed of recession of the galaxy from us (Figure 4-9). If we accept Hubble's law as a universal rule, we can place the galaxy at the appropriate recession velocity

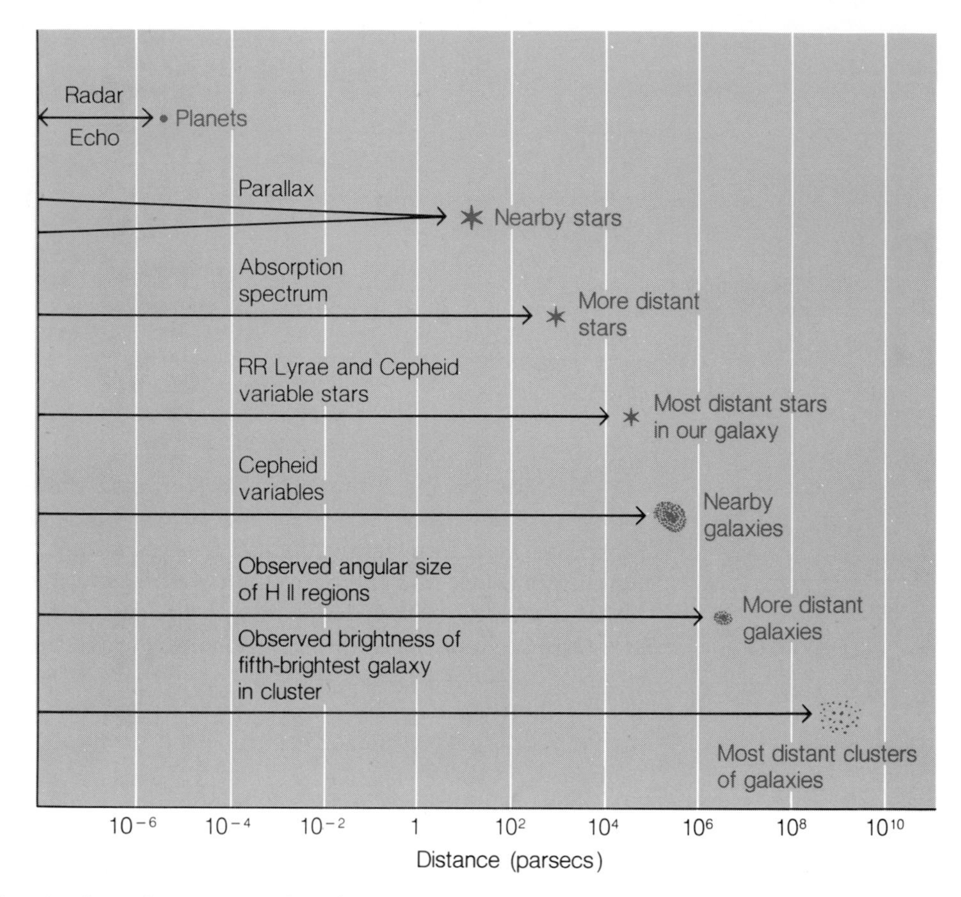

Figure 5-28

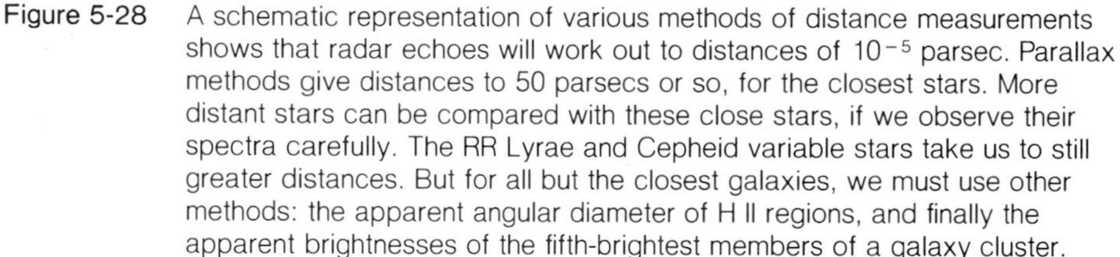
A schematic representation of various methods of distance measurements shows that radar echoes will work out to distances of 10^{-5} parsec. Parallax methods give distances to 50 parsecs or so, for the closest stars. More distant stars can be compared with these close stars, if we observe their spectra carefully. The RR Lyrae and Cepheid variable stars take us to still greater distances. But for all but the closest galaxies, we must use other methods: the apparent angular diameter of H II regions, and finally the apparent brightnesses of the fifth-brightest members of a galaxy cluster.

on the graph in Figure 4-9 and thus derive the galaxy's distance. This method, of course, can never tell us about the deviations from Hubble's law. If we could measure these deviations, we might find out whether the universe will expand forever (Figure 4-23).

The Pyramid of Distance Indicators

Figure 5-28 summarizes the methods we can employ to measure the distances to various types of objects. For the moon, sun, and nearby planets, we can use radar echoes that yield distances accurate to a tiny fraction of a percent. We encounter much greater difficulties when we attempt to measure

Figure 5-29 The Virgo cluster, about 20 megaparsecs from us, is the closest large cluster of galaxies. None of the galaxies in this cluster is close enough for us to see Cepheid variable stars, so our distance determinations rely on observations of the H II regions in the member galaxies.

the much greater distances to even the closest stars by using the parallax method (Figure 1-10). The parallax method, which works out to distances of 20 to 40 parsecs, typically includes a possible error of 10% or so. More-distant stars in our own galaxy can be compared in their spectral characteristics with nearby stars, and the method of comparing apparent brightnesses gives the distance ratio, again with an inaccuracy of perhaps 10% or more. From our vantage point inside the Milky Way, we can then employ the methods we have described for nearby galaxies, somewhat more distant galaxies, and the most distant clusters of galaxies that we can see.

Astronomers would probably agree (or hope!) that we know the distance to the Andromeda galaxy (0.68 megaparsec) with a probable error of no more than 20%, perhaps even within 10%. In other words, light from the Andromeda galaxy has taken 2.22 million years—or somewhere between 1.8 and 2.6 million years—to reach us. Such possible errors are the daily concern of astronomers who study galaxies. They seek to improve the accuracy of measurement, but console themselves with the thought that we cannot expect an easy time in measuring distances of millions of light years. Distance estimates become even more difficult for still greater distances. The galaxies in the Virgo cluster (Figure 5-29) have an average distance of 20 megaparsecs—plus or minus a few million parsecs. The distance to the Coma cluster

(Figure 5-24) is 115 megaparsecs—but might possibly be 90 or 140 megaparsecs. A galaxy cluster at a distance of 1000 megaparsecs might in fact be only 600 megaparsecs, or as much as 1500 megaparsecs, away. Astronomers, who have revised their distance estimates in the past, will probably do so in the future. This proves not their incompetence, but rather their willingness to try to understand the universe without totally accurate knowledge of such fundamental facts as the distances to galaxies. Indeed, we have made impressive strides in understanding, *despite* the uncertainties in our knowledge. Astronomers have had to accept the imprecision in their determinations and to bear them in mind as they formed and tested the suppositions based upon these determinations. To a surprising extent, they have succeeded in unraveling the secrets of the universe even as they acquire a better and better knowledge of it.

Summary

Matter in the universe is no longer spread evenly through space, but instead has formed clumps called stars, which themselves clump into galaxies and galaxy clusters. The process of clumping that produced such aggregates must have begun in earnest only after the time of decoupling, that is, only after the universal background of photons no longer exerted a homogenizing effect upon the matter.

Gravitational forces appear to be responsible for the formation of large clumps of matter, which hold together through their self-gravitational attraction. As gravity caused clouds of gas to condense the phenomenon called conservation of angular momentum made each clump rotate more rapidly as it contracted to a smaller size. Thus if the clump's rate of spin were not zero as it began to contract, the final rate of spin would be much greater than the initial spin rate.

Spiral galaxies, which contain a definite disk of stars within a larger spherical halo, appear to have arisen from protogalaxies that formed stars later in the clumping process than elliptical galaxies did. In contrast to the disk structure of spirals, and to the spiral arms within the disk, elliptical galaxies have little if any flattening. Elliptical galaxies do have an overall rotation, but the average star's motion within an elliptical galaxy has a much greater random component than within a spiral galaxy. In spiral galaxies, most of the stars in the disk orbit the galactic center in almost circular trajectories, taking from 100 to 400 million years to do so. As these stars move in their orbits, they add a smaller up-and-down motion through the median plane of the galaxy to the basic circular orbit.

Galaxies usually appear in clusters, often with many hundreds or thousands of individual galaxies in a cluster. Such "rich" clusters of galaxies often contain one or two giant galaxies (called cD galaxies), which have apparently consumed many neighboring galaxies through the gravitational forces they exert upon them. Since rich clusters of galaxies seem to resemble one another, astronomers sometimes assume that the fifth-brightest member of a rich cluster has the same luminosity in every rich cluster. With this assumption, astronomers can derive the distance to a cluster of galaxies by measuring the apparent brightness of its fifth-brightest member. Since the apparent brightness decreases with the square of the distance, if the galaxy's luminosity (absolute brightness) is known, a comparison of the apparent brightness and absolute brightness provides the galaxy's distance from us.

Key Terms

angular momentum
angular size
barred spiral galaxy
cD galaxy
Cepheid variable star
conservation of angular momentum
dwarf elliptical galaxy
elliptical galaxy
galactic cannibalism
globular star cluster
halo
H II region
inverse-square law
irregular galaxy
Magellanic Clouds
megaparsec
nova
protogalaxy
pulsating variable star
S0 galaxy
self-gravitation
spiral arms
spiral galaxy
standard candle
supernova

Questions

1. What are the major similarities between spiral and elliptical galaxies? What are the chief differences between these two types of galaxies?
2. How do the halo and disk of a spiral galaxy differ? What different populations of stars characterize these two regions? On what basis do astronomers classify stars into populations?
3. What are S0 galaxies? How are they different from irregular galaxies?
4. What prevented clumps of matter from forming during the early years of the universe, up to a time of 1 million years after the big bang?
5. What aspect of the star formation process in protogalaxies helped to determine whether a protogalaxy would become an elliptical or spiral galaxy?
6. Which types of galaxies have the largest fraction of their mass in interstellar matter? Which types have the smallest? What has happened to the interstellar matter in the latter types?
7. What does the orbit of a typical star in the disk of a spiral galaxy look like? Does this describe the sun's orbit in the Milky Way galaxy?
8. What are cD galaxies? How did they come to acquire such enormous masses?

9. The galaxies M 81 and M 83 have almost the same true brightness, but M 81 appears to be four times as bright as M 83. Which galaxy is farther from us? By how much?
10. The galaxies M 87 and NGC 7793 have almost the same apparent brightness, but M 87 is four times farther from us than NGC 7793. Which galaxy has the greater absolute brightness? By how many times?
11. Why do you think astronomers might find it useful to observe the brightest individual stars in another galaxy?
12. What is an H II region? Why are such regions important in attempts to determine the distances to other galaxies?

Projects

1. *The Andromeda Galaxy.* On a clear night in fall or winter, try to find the Andromeda galaxy by using the star maps (see endpapers). Locate the constellation Pegasus above the eastern horizon (in fall) or high above the southern horizon (in winter) at about 9 P.M. The "great square" of Pegasus—four stars of medium brightness—include one star in Andromeda at the square's northeast corner. Two lines of stars spread northeastward from this corner; when you find these stars, which are fairly dim, you can use the second star in each of the lines to point toward the Andromeda galaxy (see star maps). If no moon is out and you are well away from city lights, you should easily see the galaxy, a fuzzy object quite unlike a star in appearance. You will see it most easily by averting your vision and using the more sensitive corners of your eyes. The starlight you see from this galaxy has been traveling about 2 million years before reaching your eyes.
2. *The Distances to Galaxies.* Make a simple model of the scale of distances in the universe by representing the Earth-sun distance (150 million kilometers) with two objects on a table set one meter apart. The closest star to the sun, Alpha Centauri, lies 45 trillion kilometers away. Where should this star go in the model? The Milky Way galaxy spans 25,000 times the distance from the sun to Alpha Centauri, and the Andromeda galaxy has a distance 20 times the diameter of the Milky Way. Is there room on the Earth (diameter 12,750 kilometers) for the Milky Way in this model? The closest large cluster of galaxies has a distance from us about 25 times the distance to Andromeda, and we can see galaxy clusters that are 100 times farther away than this. In your model, how do the distances to these clusters compare to the true distance from the Earth to the moon (400,000 kilometers)? Could the farthest clusters fit within the distance to the sun?
3. *Angular Momentum.* If you are not intuitively familiar with angular momentum through exercise as a high diver or ice skater, try this experiment. Sit in a swivel chair and ask a friend to spin you around. Start the spin with a couple of heavy books held in your hands with your arms extended. Once you are spinning, bring the books in toward your body. What happens? How can you explain this result in terms of the conservation of angular momentum?

Further Reading

Arp, Halton. 1975. "The Evolution of Galaxies." In *New Frontiers in Astronomy*. Edited by Owen Gingerich. San Francisco: W. H. Freeman.

Hodge, Paul. 1966. *Galaxies and Cosmology*. New York: McGraw-Hill.

Hubble, Edwin. 1936. *The Realm of the Nebulae*. New York: Dover Books.

Meier, David, and Sunyaev, Rashid. 1979. "Primeval Galaxies." *Scientific American* 241:5, 130.

Peebles, P. James; Groth, Edward; Seldner, Michael; and Soneira, Raymond. 1977. "The Clustering of Galaxies." *Scientific American* 237:5, 76.

Sandage, Allan. 1960. *The Hubble Atlas of Galaxies*. Washington, D.C.: Carnegie Institution of Washington.

Shapley, Harlow. 1972. *Galaxies*. 3d. ed. revised by Paul Hodge. Cambridge: Harvard University Press.

Silk, Joseph. 1980. *The Big Bang*. San Francisco: W. H. Freeman.

Strom, Stephen, and Strom, Karen. 1979. "The Evolution of Disk Galaxies." *Scientific American* 240:4, 72.

Trimble, Virginia, and Rees, Martin. 1978. "Are Galaxies Here to Stay?" *Astronomy*, July 1978.

Tucker, Wallace, and Gorenstein, Paul. 1978. "Rich Clusters of Galaxies." *Scientific American* 239:5, 110.

The spiral galaxy M 33 is the third largest member of the Local Group.

6

The Milky Way and the Local Group of Galaxies

Of the myriad galaxies in the universe, we know best those closest to us. When we observe our own galaxy, however, we have difficulty in determining its overall structure because we are *inside* the Milky Way. For this reason, we know more about the total structure of nearby galaxies, such as the giant spiral in Andromeda, than we do about the spiral structure of the Milky Way. The study of our closest galactic neighbors has given us insight into far more than the structure of galaxies; much of what we know about stars and star clusters comes from studying other galaxies, those located favorably close to our own.

6.1 The Local Group

Our Milky Way belongs to a small cluster of galaxies called, as we might expect, the **Local Group.** About twenty members comprise the Local Group of galaxies, far fewer than the thousands of member galaxies in a "rich" cluster such as the Hercules or Coma cluster (Figures 5-1 and 5-24).

Properties of the Local Group

As we can see from Table 6-1, which lists some properties of the Local Group members, the Milky Way and the Andromeda galaxy contain most of the mass in our small cluster. No other member has more than 6% of the mass of either of the two giant spirals. The next most massive members are the spiral galaxy M 33 in Triangulum (page 166), the Milky Way's larger satellite, called the **Large Magellanic Cloud** (Figure 6-1), the **Small Magellanic Cloud** (Figure 6-3), the two largest satellites of the Andromeda galaxy (Figure 6-2), the irregular galaxy NGC 6822, and the elliptical galaxies NGC 185

Table 6-1 Members of the Local Group of Galaxies

Name	Galaxy Type	Distance from Milky Way (Thousands of Parsecs)	Diameter (Parsecs)	Mass (Millions of Solar Masses)
Milky Way	Sb	—	30,000	800,000
(unnamed)[a]	Irr	25	1,300	100
Large Magellanic Cloud[a]	Irr	48	10,000	25,000
Small Magellanic Cloud[a]	Irr	56	8,000	6,000
Ursa Minor system	E4	70	1,000	0.1
Sculptor system	E3	83	2,200	3
Draco system	E2	100	1,400	0.1
Carina system	E3	170	1,500	—
Fornax system	E3	250	4,500	20
Leo II system	E0	230	1,600	1
Leo I system	E4	280	1,500	4
NGC 6822	Irr	460	2,700	1,000
NGC 147	E6	570	3,000	350
NGC 185	E2	570	2,300	450
NGC 205[b]	E5	680	5,000	3,000
NGC 221 (M 32)[b]	E3	680	2,400	2,000
IC 1613	Irr	680	5,000	250
Andromeda galaxy	Sb	680	40,000	500,000
Andromeda I[b]	E0	680	500	2
Andromeda II[b]	E0	680	700	2
Andromeda III[b]	E3	680	900	2
M 33	Sc	680	17,000	30,000

[a]Satellite of the Milky Way.
[b]Satellite of the Andromeda galaxy.

Figure 6-1
The Large Magellanic Cloud, with 1/30 the mass of the Milky Way, shows a rich assortment of young stars and interstellar clouds of gas and dust.

Figure 6-2
The Andromeda galaxy's two largest satellites are elliptical galaxies called NGC 205 (top) and M 32 (bottom).

Figure 6-3
The Small Magellanic Cloud has ¼ the Large Magellanic Cloud's mass, but like our galaxy's larger satellite, contains a rich assortment of stars, gas, and dust.

and NGC 147 (Figure 6-4), which lie in the general direction of the Andromeda galaxy but are not its satellites (see Figure 1-4).

The Local Group consists basically of the Milky Way and Andromeda, plus some smaller galaxies. When we look at other galaxy clusters, we cannot expect to see galaxies so small and dim as, for example, the Fornax system (Figure 6-5). Such galaxies have so few stars—and none of the young, bright stars that outline the arms of a spiral galaxy—that they simply are too dim to be seen at distances equal to the 20 megaparsecs to the Virgo cluster, the nearest large cluster of galaxies. Hence, the Local Group represents our only chance to study these dwarf galaxies, which may be more numerous than any other type of galaxy, even though they may contain an essentially negligible fraction of the total mass in a cluster of galaxies.

Despite its lack of any giant ellipticals, our Local Group does offer an assortment of smaller and dwarf elliptical galaxies for our study. As we can see from Figure 6-4, these galaxies show an ellipsoidal distribution of stars, densest at their centers and thinning out toward their edges. Almost no gas or dust appears among the stars in these elliptical galaxies. In contrast, irregular galaxies, such as the Magellanic Clouds and NGC 6822, contain huge agglomerations of gas and dust, within which star formation continues. As a result, we find a far larger proportion of young stars in irregular galaxies than in spiral galaxies (which contain only a few percent of stars younger than a billion years) and elliptical galaxies (which contain almost no young stars).

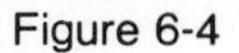

Figure 6-4 The elliptical galaxies NGC 147 (left) and NGC 185 (right) each have about 0.05% of the mass of the Milky Way. NGC 185 shows a prominent dust cloud, unusual for an elliptical galaxy.

Figure 6-5 This negative print of the dwarf elliptical galaxy in Fornax allows the loose concentration of matter to be seen in more detail than a positive print would.

The Magellanic Clouds

Our Milky Way's two large irregular satellites, the Magellanic Clouds, have contributed greatly to our understanding of stars. These galaxies, ten times closer to us than the Andromeda galaxy, cover about the same area on the sky as the full moon. From southern latitudes, the Clouds appear as patches on the clear night skies. The two Magellanic Clouds actually lie within the outer parts of the halo of the Milky Way. The Clouds may therefore be considered subunits of our own galaxy. What makes the Magellanic Clouds especially useful to astronomers is that they are close enough for us to observe many stars individually, yet far enough away that all of their stars may be judged to be at nearly the same distance from us. Thus even before the distance to the Magellanic Clouds had been determined, astronomers knew that if one of their stars appeared to be ten times as bright as another, then the first star's *absolute* brightness must also exceed that of the second star by a factor of ten. In other words, so long as we observe stars within the Magellanic Clouds, the ratio of apparent brightnesses accurately gives us the ratio of the stars' true brightnesses.

Cepheid Variable Stars as Distance Indicators

As we noted in Chapter 5, **Cepheid variable stars** change their apparent brightness in a regular way, with periods of variation that range from a few days to a few months. During the early 1900s, Henrietta Leavitt observed Cepheid variable stars in the Small Magellanic Cloud and found a correlation between the stars' apparent brightnesses and their periods of light variation: Brighter stars took longer for each complete cycle of variability. Leavitt realized that this relationship must hold true in terms of the **absolute brightnesses** of the Cepheid variable stars, since those with greater apparent brightnesses in the Small Magellanic Cloud must have greater absolute brightnesses (Figure 6-6).

The Cepheid variable stars unlocked the mystery of the distances to members of the Local Group, such as the Andromeda galaxy, NGC 6822, and M 33. For example, in 1923, Edwin Hubble discovered a Cepheid variable in the Andromeda galaxy with the same period of light variation as a Cepheid in the Small Magellanic Cloud, but with an apparent brightness 80 times smaller. If the two Cepheid variables have the same absolute brightness (averaged over their cycle of variability), then the Cepheid in the Andromeda galaxy should have a distance 9 times greater than the distance to the Small Magellanic Cloud (since the square root of 80 is close to 9).

The Cepheid variable method of distance determination forms the keystone of our estimates of the distances to other galaxies (see page 104). (Actually, this method *compares* one distance, presumably already known, with a greater distance.) Unfortunately, Cepheid variables, although intrinsically fairly bright, cannot be seen as individual stars beyond a distance of a few megaparsecs. Hence, we must use other methods to estimate distances to galaxies much farther from us than those in the Local Group (see page 157).

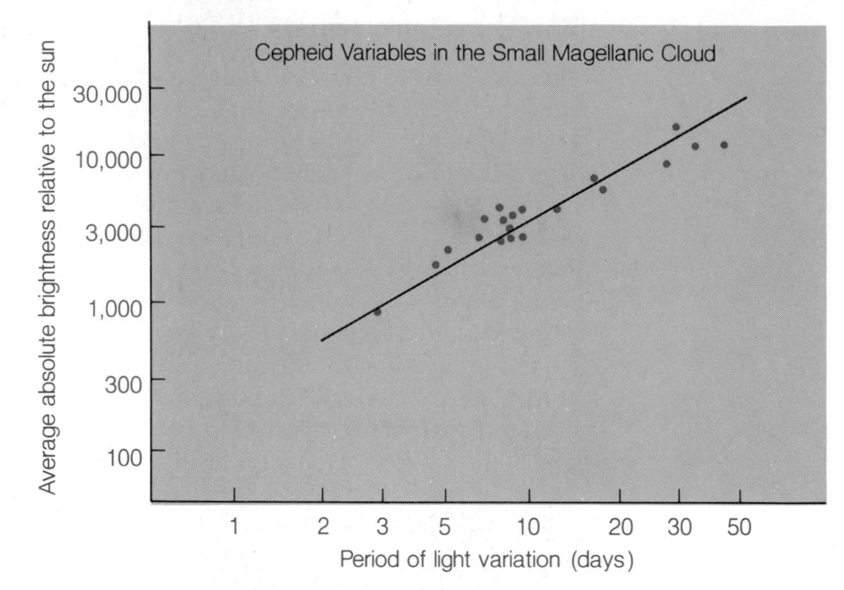

Figure 6-6 Henrietta Leavitt derived the relationship between the apparent brightness and the period of light variation for Cepheid variable stars in the Small Magellanic Cloud. The relationship has the same graphical form as the one shown here, in which the *absolute* brightness of each Cepheid variable has been calculated, using our knowledge of the stars' distance from us.

Spiral Structures in Galaxies

If the Local Group has a dearth of giant ellipticals, it cannot be faulted on spirals: Its three largest members are spiral galaxies. The relative closeness of the Andromeda galaxy (M 31) and the spiral in Triangulum (M 33) had attracted the attention of astronomers long before the Milky Way was known to be a spiral. When we study photographs of spiral galaxies, such as Figure 6-7, we are immediately struck by the regular pattern of spiral arms.

These bilaterally symmetric, rotating galactic disks call for an explanation of how the spiral pattern ever formed, and we shall examine this question on page 197. For the moment, however, we ought to recognize the fact that in terms of the distribution of mass within a spiral galaxy, the spiral arms look more spectacular than they really are. Spiral arms stand out in the photographs because they contain the youngest, hottest, and brightest stars within a galaxy, as well as the glowing gas clouds called H II regions that such stars illuminate. But the density of all stars *between* the spiral arms hardly differs from the density of all stars *within* the arms. The basic structure is that of an entire rotating disk of stars. Something, however, produces young, bright stars in a spiral pattern—stars that burn themselves out within a few million years, far less time than it takes for the galaxy to rotate once. Later we shall consider what that mechanism to produce young stars in spiral arms may be (see page 200).

Figure 6-7 The spiral galaxy NGC 6744 in Pavo, about 15 million parsecs from the Milky Way, shows a well-developed pattern of spiral arms.

6.2 Stellar Populations

Through many spectroscopic observations of individual stars, astronomers have determined, with good precision, the fraction of various stars' masses that consists of elements heavier than hydrogen and helium. On the basis of this *heavy element fraction,* astronomers group almost all stars into two populations, named Population I and Population II.

Population I and Population II Stars

The **Population I stars** have about 1% of their masses in the form of elements heavier than helium. These stars, astronomers believe, represent the

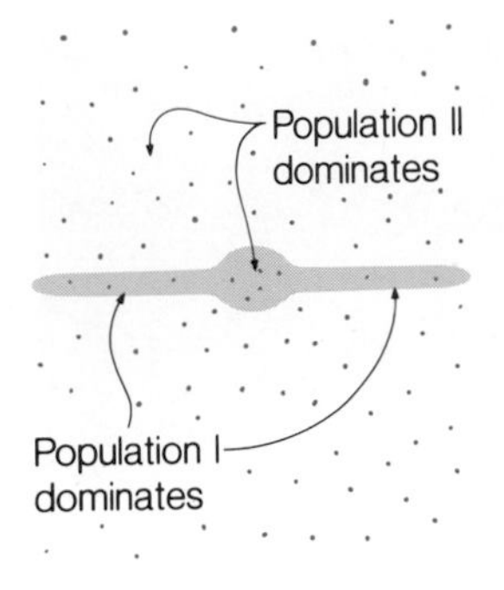

Figure 6-8
Population I stars dominate the disk of the Milky Way galaxy, while Population II stars appear mostly in the nucleus and halo. Globular clusters consist almost entirely of Population II stars.

younger stars. Certainly, all extremely young stars (less than 1 billion years old) appear to be Population I stars. In contrast to the "large" fraction of heavy elements in Population I stars, the **Population II stars** have only 0.1%, or even less, of their mass in the form of elements heavier than hydrogen and helium. These stars apparently formed more than 10 billion years ago, when the clouds of gas and dust that condensed into stars contained a smaller fraction of heavy elements (see page 323).

Most of the stars in the disk of a spiral galaxy are Population I stars, of which our sun provides a good example. Although some of these Population I stars are the youngest, brightest stars that outline the spiral arms, most of them, like our sun, are several billion years old. Most of the stars in the halo that surrounds the galactic disk are Population II stars, typically 10 billion years old and somewhat less massive than the sun. The nucleus of the Milky Way, the region within a kiloparsec of the galactic center, also contains large numbers of Population II stars (Figure 6-8). Population II stars include the most mature stars, those as old, or almost as old, as the galaxy itself.

Population I Stars in Spiral Arms

Within the local regions of our galaxy, the bright stars Rigel (in Orion's foot) and Bellatrix (in Orion's dimmer shoulder), as well as the stars in Orion's belt and most of the other stars in Orion and to the south, represent the

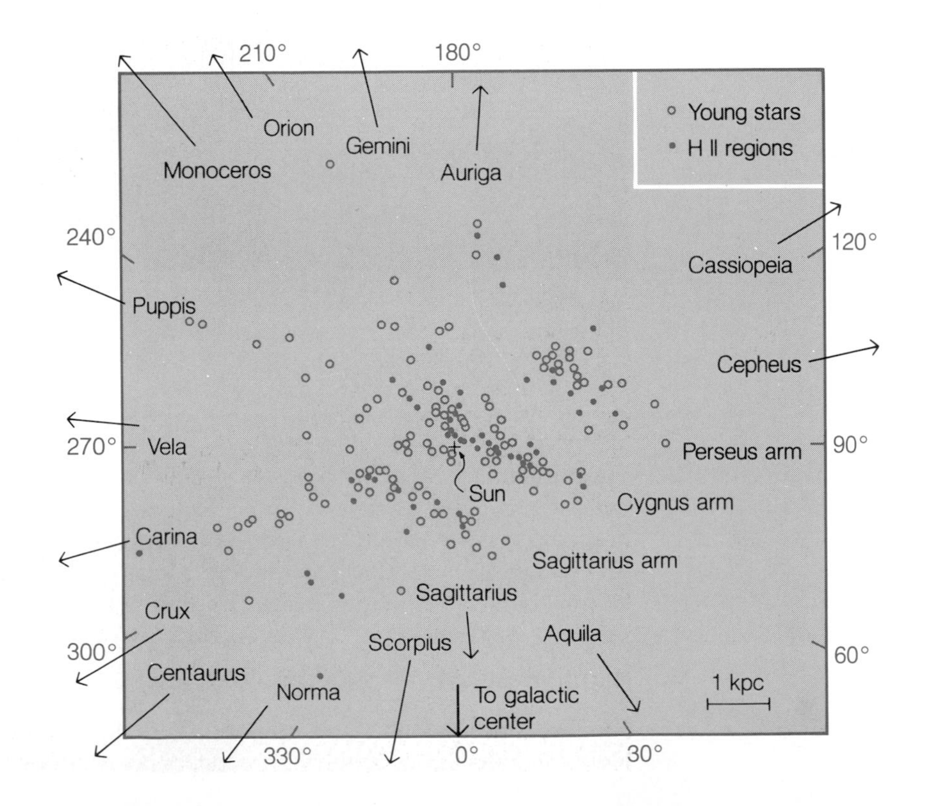

Figure 6-9
We can map the spiral arms of our galaxy out to distances of a few kiloparsecs from the sun by determining the distances to young, hot stars and to H II regions. These objects appear primarily within a galaxy's spiral arms.

extremely young sort of Population I stars, those with ages of a few tens of millions of years or less. These stars each have an absolute brightness thousands of times greater than the sun's, so although they rank among the most distant stars visible to the unaided eye (150 to 600 parsecs away), they have apparent brightnesses equal to those of many stars a hundred times closer. Such stars outline the spiral arms of our galaxy (and of other spiral galaxies), making the arms stand out in brightness from the background of the disk. Measurements of the distances to young, hot stars in the Milky Way provided the first method of determining its spiral structure (Figure 6-9). Using this technique, astronomers found that the sun appears to lie within a spiral arm of the Milky Way, originally called the Orion arm, since it contains the bright stars of Orion, but now called the Cygnus arm.

When we examine nearby galaxies, we find that there, too, stars divide into Population I and Population II. Only a small fraction of the total have abundances of heavy elements intermediate between those of Population I and Population II stars. Elliptical galaxies apparently turned almost all of their material into stars long ago, so although they contain both Population I and Population II stars, they have almost none of the young Population I stars, the sort that outline spiral arms and light up H II regions (Figure 6-10). In spiral galaxies, Population II stars seem to have formed as the protogalaxy contracted toward a disklike configuration. Thus Population II stars occupy either the halo or the central regions of spiral galaxies. Population I stars, on

Figure 6-10
The spiral arms in a galaxy such as M 101 are defined by the bright H II regions they contain. Young, hot stars illuminate the H II regions with ultraviolet radiation, which the gas around the stars converts into visible light (see page 157).

the other hand, formed (and continue to form) after the galaxy had assumed a disklike structure. Even the irregular galaxies, which contain predominantly Population I stars, do have some Population II stars. At the other extreme, some dwarf elliptical galaxies appear to consist entirely of Population II stars and may have completed their star formation some 10 billion years ago.

6.3 Star Clusters

Surveys of the sky show that stars in our galaxy, and indeed in other galaxies, often appear in groups, or clusters. Of these, two distinct types exist: the **open clusters** (sometimes called by the older name of **galactic clusters**) and the **globular clusters.** Open clusters each contain up to several thousand stars (Color Plate Q), while globular clusters contain hundreds of thousands or even millions of stars (Figure 6-11). Open clusters lie in the plane of the Milky Way galaxy, whereas globular clusters form a basically spherical distribution around the galactic center (Figure 6-12). Open clusters consist primarily of Population I stars, whereas globular clusters consist exclusively of Population II stars. Finally, open clusters are loose agglomerations of stars that may not persist as definite clusters for billions of years, whereas the stars

Figure 6-11 The globular cluster M 13 in Hercules has a distance of 8000 parsecs from the sun.

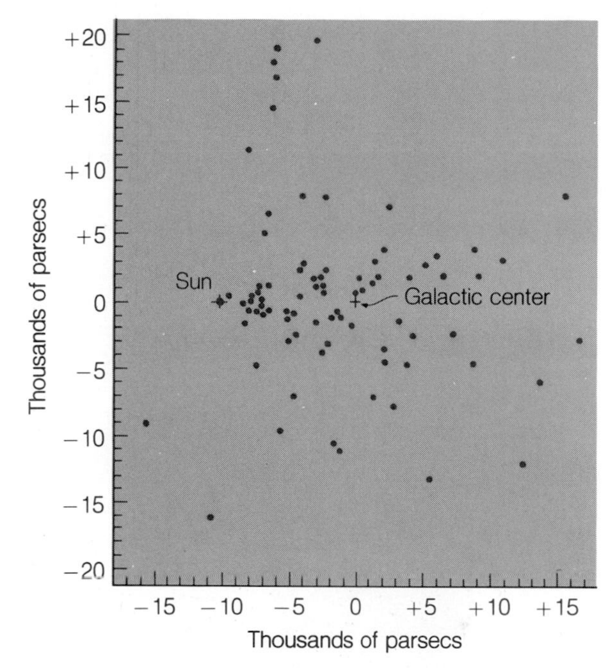

Figure 6-12 A plot of the distribution of globular clusters with respect to the plane of the Milky Way shows that some of them are 15,000 or 20,000 parsecs above or below the plane.

in globular clusters will remain gravitationally bound to the cluster by their stronger mutual attraction.

From these distinctions between the two types of clusters, which reappear when we examine other spiral galaxies, astronomers can draw some definite conclusions about the formation of star clusters. Globular clusters condensed as our protogalaxy contracted into its present form, many billion years ago. Measurements of the ages of stars in globular clusters confirm this (see page 297). Open clusters, however, must have formed after the bulk of the galaxy had assumed a disklike configuration. Some have existed for billions of years, although others are no more than a few million years old.

The Distribution of Globular Clusters

As Figure 6-12 shows, the globular clusters, although not particularly concentrated toward the galactic plane, do tend to appear close to the center of the galaxy. Since the sun's position within the Milky Way puts it rather close to the edge of the galaxy, about one-third of all the globular clusters we can see lie within a relatively small area of the sky (Figure 6-13). This fact provided the first definite indication that the center of the galaxy lies far from the sun, as Harlow Shapley realized in 1918 (see page 102). By now about 150 globular clusters have been found within the Milky Way. Several dozen

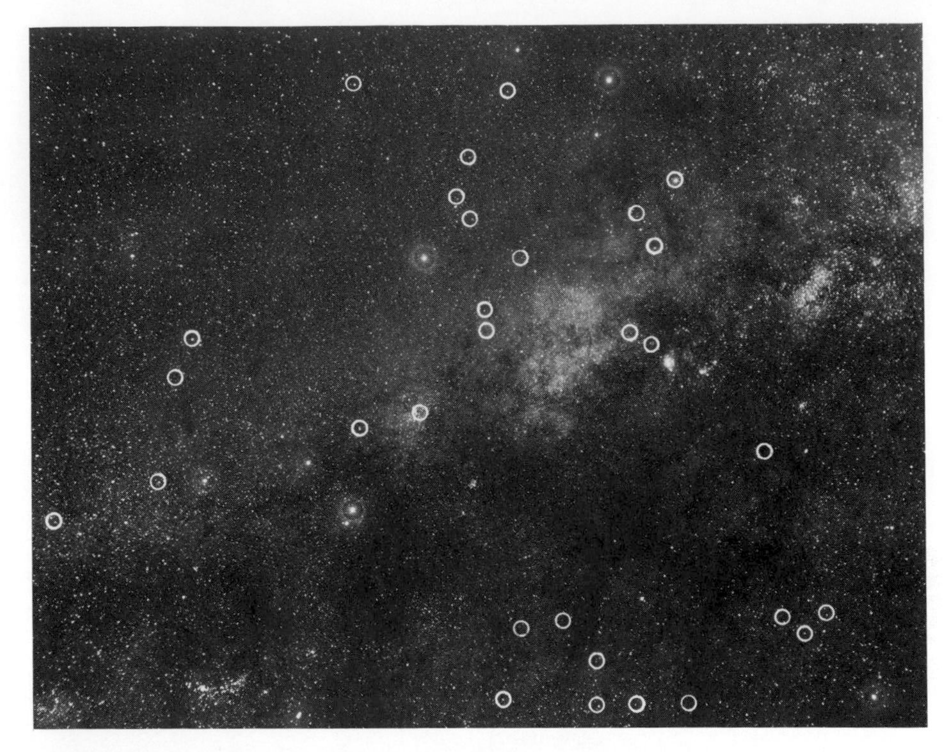

Figure 6-13
The central regions of the Milky Way show about four dozen globular clusters, of which the brightest have been circled in this photograph.

more may lie on the far side of the galactic center, obscured from our view by interstellar dust.

A typical globular cluster such as that shown in Figure 6-11 may span 10 or 20 parsecs and contain a million stars. If the sun were a member of such a cluster, our night skies would contain many stars with greater apparent brightness than the full moon! On the other hand, the stars move around their common center of mass in such a cluster with almost no danger of collision—a reminder of how empty space must be. The compact nature of globular clusters, along with their relatively great true brightnesses (since they each contain many thousands of stars), makes them easy to spot in other galaxies. Astronomers have mapped the distribution of globular clusters in such faraway galaxies as those in the Virgo cluster and even beyond. From the ubiquitous appearance of globular clusters in both spiral and elliptical galaxies, we may conclude that as protogalaxies condensed, subunits with masses close to a million solar masses had a definite tendency to form compact groups.

Open Clusters

Open clusters, unlike globular clusters, clearly belong to the disk of a spiral galaxy (Figure 6-14). The most important open clusters (from *our* point of view) are the Hyades, the Pleiades (Color Plate Q), and the Double Cluster in Perseus (Figure 6-15). Open clusters consist of Population I stars, whereas

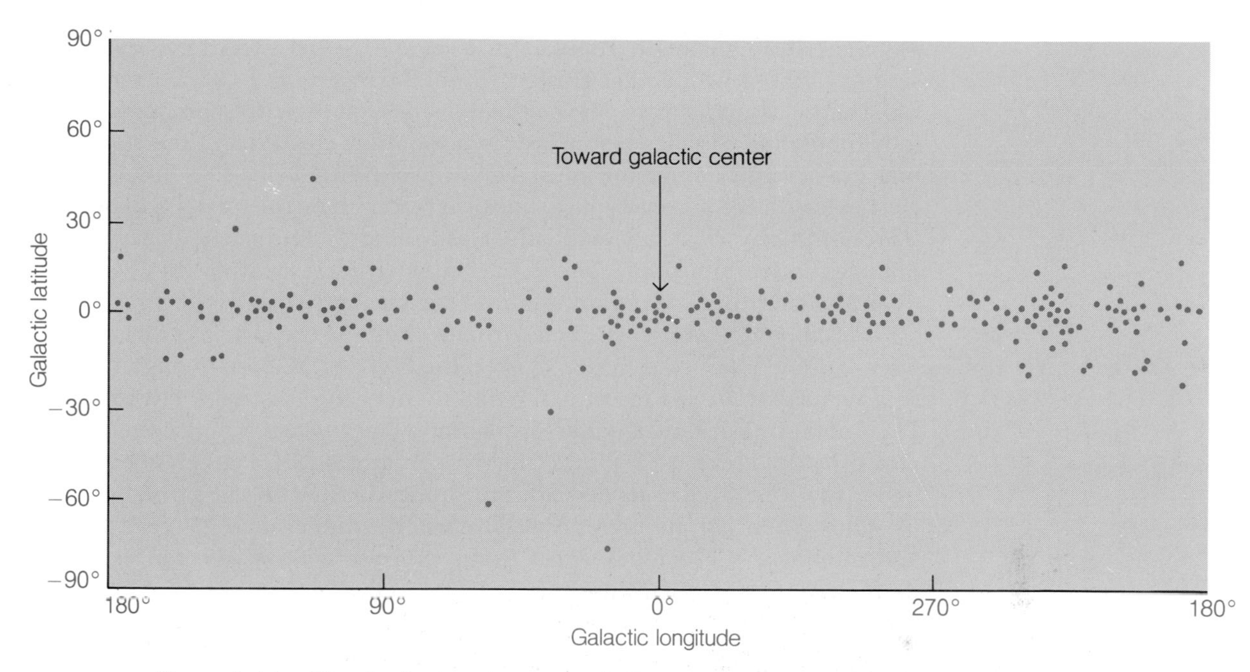

Figure 6-14 The distribution of selected open clusters (dots) shows a strong concentration toward the plane of the Milky Way (galactic latitude = 0°).

Figure 6-15 The Double Cluster in Perseus, a rich open cluster, contains several thousand stars at a distance of about 2000 parsecs from the sun.

globular clusters contain Population II stars. Open clusters have diameters similar to those of globular clusters (5 to 20 parsecs), but contain only one-thousandth as many stars. Hence the density of stars within an open cluster falls thousands of times below that in a globular cluster, and the stars' mutual gravitational attraction must be correspondingly less. A "loose" open cluster may have a density just slightly greater than the average density of stars not in any cluster. Nevertheless, astronomers find such a similarity in the ages and compositions of the stars in such a cluster that they feel confident in assigning a common origin to the cluster's stars.

Additional reassurance comes from a study of the *motions* of stars in an open cluster. The nearest open cluster, the Hyades, is close enough (50 parsecs away) that we can measure the motion of each of its stars across the sky. This motion, the stars' angular displacement per year, is called the **proper motion.** As Figure 6-16 shows, the proper motion of a given star with a given speed relative to us decreases in proportion to the star's distance from us. Thus when we photograph a star cluster at intervals of several decades and eliminate the parallax effect from our observations, we can expect to detect a change in a star's position (relative to the background of much more distant stars) only for stars within 100 parsecs or so of the sun. If the Hyades were ten times as far from us, we would have to wait ten times as long to observe the same angular displacement on the sky, provided that their motions in space were the same as before.

Measuring the Distance to the Hyades

The measurements of the proper motions of the stars in the Hyades reveal that these stars all move in almost the same direction (Figure 6-17). The stars indeed move almost parallel to one another. The effect of our perspective, however, directs their apparent motions toward a single **convergent point.**

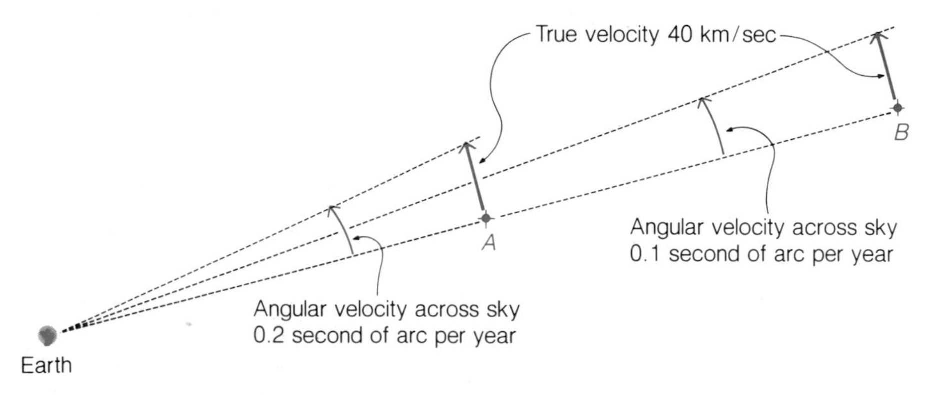

Figure 6-16 For a star with a given velocity through space (in kilometers per second), the proper motion (angular displacement per year) that we observe depends on the star's distance from us. More distant stars have smaller proper motions, even though they may have the same velocity through space as less distant stars.

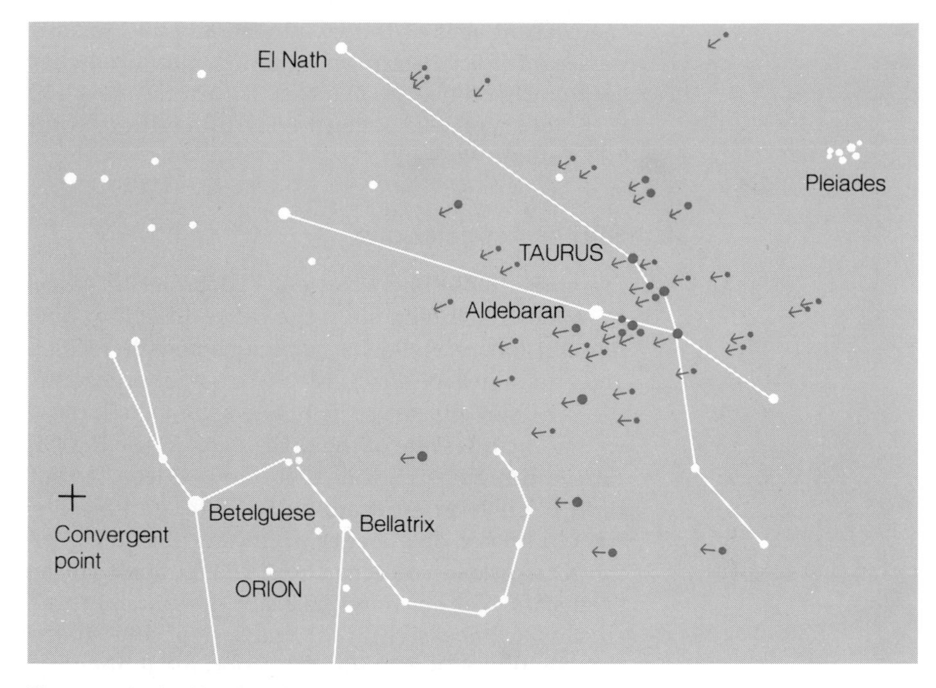

Figure 6-17 The stars in the Hyades cluster have almost the same direction and amount of proper motion. In fact, the stars' motions seem to head to a single convergent point because of our perspective on their motions, much as railroad tracks appear to converge in the distance.

This effect is analogous to what we would see if the stars moved along parallel railroad tracks that seemed to come together in the distance. We can use our knowledge of trigonometry to find the *distance* to the stars in the Hyades by locating the convergent point on the sky. Once we use the Doppler effect to determine the stars' motions along our lines of sight, we can combine this information with the stars' proper motions and their angular distances from the convergent point to derive their actual distances. This approach could also be used to find the distance of a railroad train if we measured its Doppler shift, its proper motion, and its angular distance from the convergent point where the tracks appear to vanish. Though the method may seem roundabout, it has provided us with the distances to the stars in the Hyades cluster. These stars in turn serve as benchmarks against which the entire scale of distances in the universe has been calibrated. Thus when we speak of a galaxy as being, say, 50 million parsecs away, we really mean a million times the distance to the Hyades.

The Dissolution of Open Clusters

Careful studies of the motions of the stars in the Hyades show that the slight differences in the stars' proper motions should lead to the dissolution of the cluster within the next few hundred million years, a period about equal to the present ages of the Hyades stars. The Pleiades, too far away (about 125 parsecs) for accurate proper-motion measurements, appear headed for even swifter dissolution as a cluster, within the next 100 million years or so. This may not surprise us when we learn that these young, hot stars are only 50 to 100 million years old.

Stellar Associations

Extremely loose open clusters, called **stellar associations,** consist of young stars moving in generally the same direction, but separating within only a few million years. Such associations may be recognized by the common ages and common motions of the stars that compose them.

The sun appears to be inside an association (without being a member of it) that includes five of the seven stars in the Big Dipper, as well as Sirius and Procyon in quite another part of our skies. The sun may actually have been born in such an association. If so, the gentle divergence of stellar motions has eliminated any chance of finding the sun's "brothers and sisters."

Our galaxy contains about 20,000 open clusters, each with a few hundred stars. Neither these stars nor the few hundred million stars in globular clusters form a significant fraction of the total mass of the Milky Way, though the clusters provide important guideposts to the formation of the galaxy. Most stars, including our own sun, belong to no cluster, but simply orbit the center of mass of their galaxy with unhurried calm.

6.4 Motions of Stars in Our Galaxy

Most of the stars and gas clouds within the disk of the Milky Way move in almost circular orbits around the galactic center, taking somewhere between 100 million and 400 million years for each orbit, as we described on page 152. The exceptions to this rule are called **high-velocity stars,** because their noncircular orbits give them a large velocity relative to the stars that move in circular orbits (Figure 6-18).

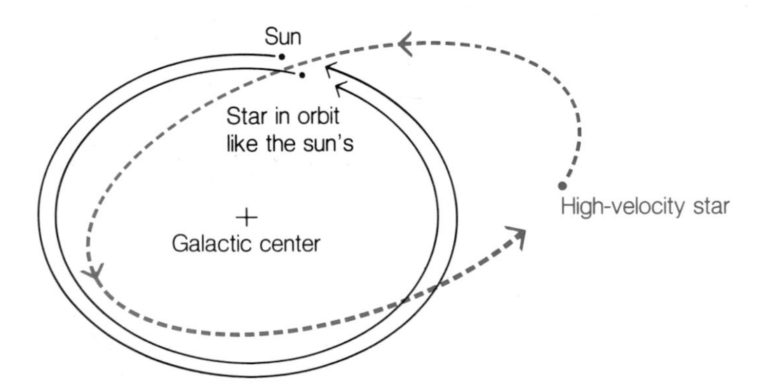

Figure 6-18 High-velocity stars are those that have large velocities with respect to the sun. Such stars are actually moving around the galactic center more *slowly* than the sun, and in rather elliptical orbits.

From calculations based on Newton's law of gravitation (see page 372) and from observations of the motions of stars toward us or away from us, astronomers have achieved a fairly detailed understanding of the orbital speeds that most stars have at various distances from the galactic center. Our sun's speed around the center, 240 kilometers per second, typifies the orbital velocities of stars circling at 9000 parsecs from the center. Stars in circular orbits at other distances have other velocities, each reflecting the interplay of the gravitational attraction on the star and the momentum it has in its orbit. The knowledge of what these orbital velocities must be comes in handy when astronomers want to assign an object a position within the galaxy solely on the basis of its motion around the galactic center.

6.5 Interstellar Matter

Among its 400 billion or so stars, the Milky Way, like other spiral galaxies (and like irregular galaxies), contains a quantity of gas and dust, spread out among the stars like smoke. This interstellar matter represents both the residue of the material that once formed stars and potential material for new stars, which form even now, though not at the high rate that characterized the early protogalaxy. In total, the interstellar matter amounts to perhaps a few percent of the galaxy's mass in stars, with **interstellar gas** (individual atoms, or molecules made of no more than a few atoms) far more abundant than **interstellar dust** (atoms joined by the millions into small clumps).

The Absorption of Starlight by Interstellar Dust

Although interstellar dust is not as abundant as interstellar gas, it can nonetheless be seen far more easily than interstellar gas, because dust absorbs starlight, and gas does not. (More precisely, gas absorbs only certain frequencies of starlight, while most frequencies penetrate unaffected.) As Figure 6-19 shows, if we use special techniques to highlight the **absorption** by dust, the Milky Way suddenly looks like an edge-on spiral galaxy, even though we are inside it (Figure 5-3). The absorption by dust becomes especially noticeable when we look toward the galactic center, 9000 parsecs away, and discover so much dust that we cannot see the center at all. An interesting "window" in the dust does exist a few degrees away from the center of the galaxy, allowing us to see globular clusters that lie almost directly behind the center (Figure 6-20).

Interstellar Reddening

We have described the absorption of starlight by interstellar dust grains. A subtler effect produced by the dust consists of **interstellar reddening.** Dust particles absorb and scatter visible-light photons of all wavelengths and frequencies, but they affect photons of blue light more than photons of red light. Hence, light that passes among dust grains emerges *redder* (if it emerges at all) than when it entered, because more blue than red photons

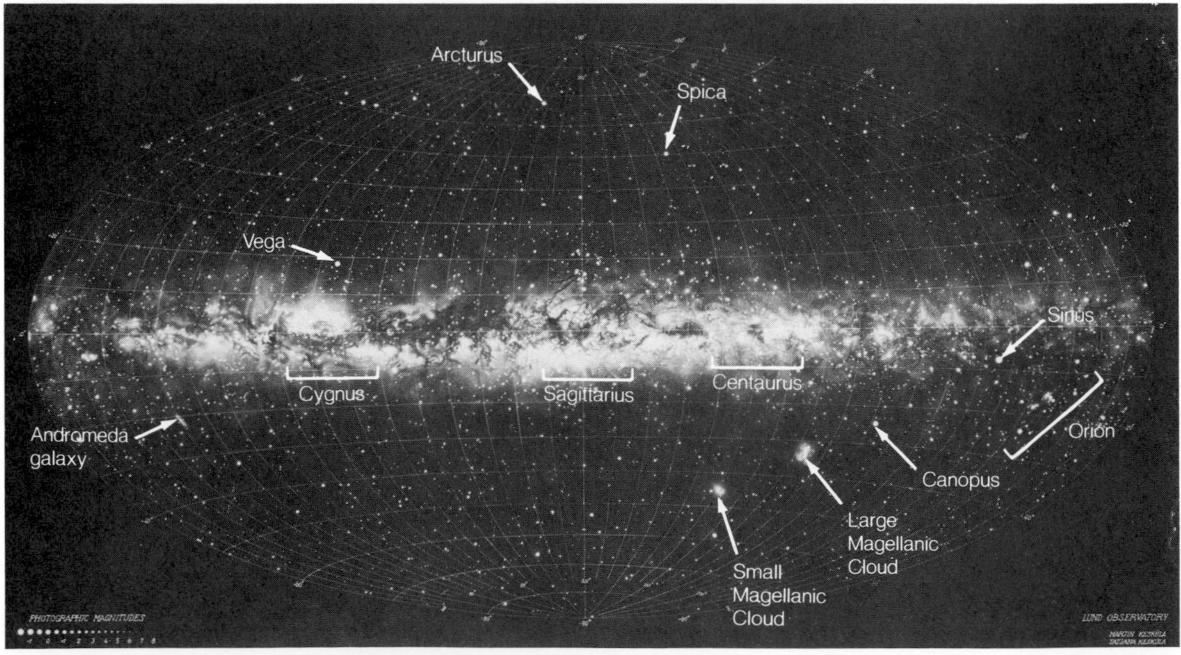

Figure 6-19
A wide-exposure photograph of the Milky Way, showing almost 180° of the galactic disk, highlights the dust and reveals that we live inside the disk of a spiral galaxy (top). (The three dark lines are the camera supports.) A mosaic drawing based on photographs gives a full 360° view of the Milky Way, as seen from the sun (bottom). Some familiar objects are marked on the drawing for reference.

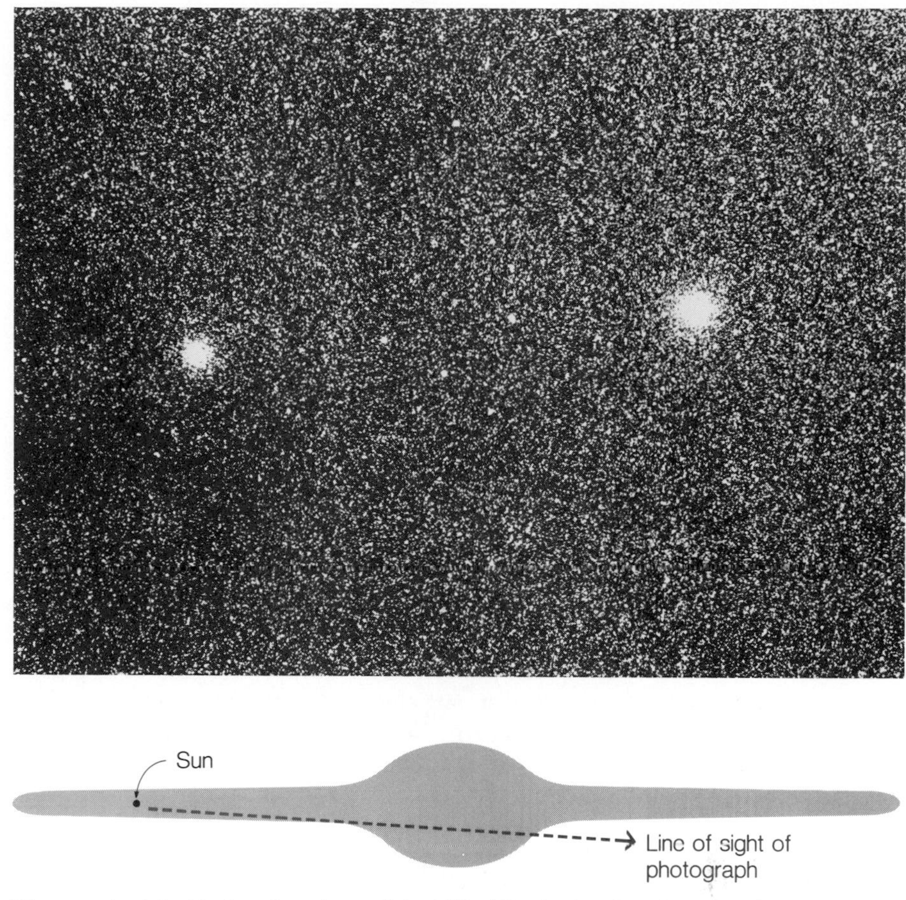

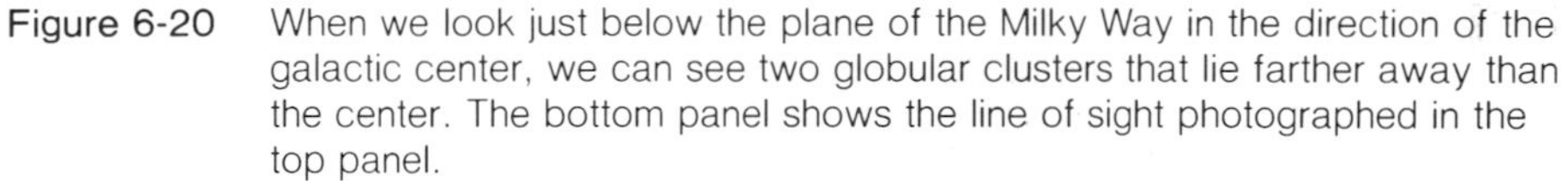

Figure 6-20 When we look just below the plane of the Milky Way in the direction of the galactic center, we can see two globular clusters that lie farther away than the center. The bottom panel shows the line of sight photographed in the top panel.

have been removed from the beam. (A similar phenomenon is responsible for the redness of sunrises and sunsets, when the sun's light must pass through a large amount of the Earth's atmosphere—which scatters out most of the blue light—before it reaches our eyes.)

This interstellar reddening becomes more pronounced for longer paths through the disk of the galaxy, where the dust is concentrated (Figure 6-21). Light from nearby stars suffers almost no reddening, while the light from stars 10,000 parsecs away may be completely changed in color, if it can penetrate at all the dust along the line of sight. Astronomers have used this reddening effect to their advantage. They compare the color a star "ought" to have (from its spectral resemblances to nearby stars) with the color it does have. From this they can estimate, in a rough way, the distance to faraway stars in the Milky Way, simply by assuming a constant number of dust particles in each thousand parsecs of distance.

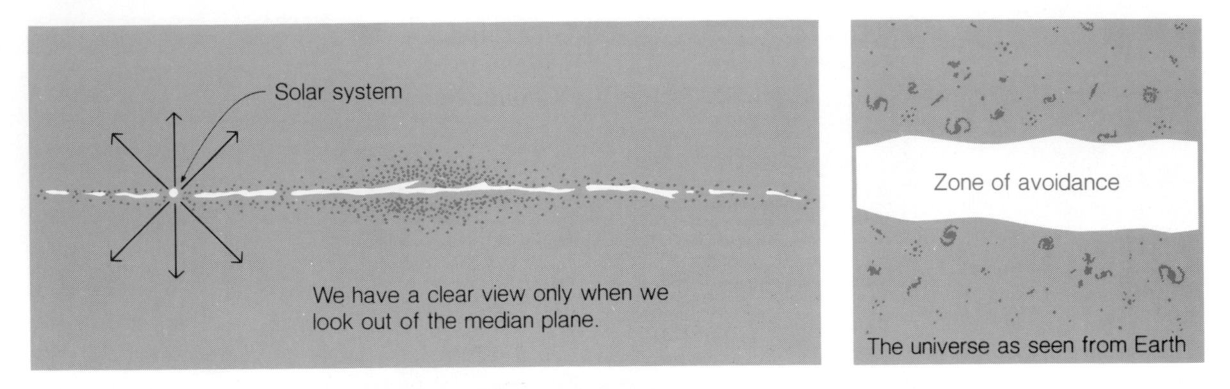

Figure 6-21 The dust particles that absorb and redden starlight concentrate heavily toward the plane of the Milky Way galaxy. Therefore, we cannot see out of our galaxy when we look in directions along the plane, the zone of avoidance.

The Zone of Avoidance

The extinction of light by interstellar dust particles becomes important when we search for galaxies. When we try to look outward in visible light through the plane of the Milky Way, we cannot see any galaxies at all! The concentration of dust particles within the plane of our galaxy (as within other spirals) produces a **zone of avoidance**—not a zone the galaxies avoid, but one where dust voids any chance of seeing galaxies outside the Milky Way (Figure 6-21).

The Composition of Interstellar Dust

Most interstellar dust grains apparently consist of carbon (graphite), perhaps with an outer layer of water ice. However, silicon-oxygen grains exist as well.

All dust grains may have been made in the outer atmospheres of cool stars, then expelled into interstellar space as the stars aged (see page 311). The dust does more than interfere with our observations, though. Dust grains appear to be the sites where the most abundant interstellar molecules, hydrogen molecules, manage to form.

Interstellar Clouds

Basic Elements. Hydrogen molecules, each made of two hydrogen atoms, form one of the two basic constituents of interstellar gas. The other is none other than the hydrogen atoms themselves, not joined into molecules. Where the interstellar gas has densities between 0.1 and a few hundred particles per cubic centimeter, atoms predominate and few molecules exist. Such densities typify the bulk of interstellar space, as well as the less dense of the **interstellar clouds,** regions where interstellar matter has clumped together.

In contrast, the dense interstellar clouds, where more than a few hundred particles occupy each cubic centimeter, are primarily molecular. The atoms have linked together to form diatomic hydrogen (H_2) or more complex molecules.

In addition to hydrogen, interstellar matter consists of helium (an amount equal to about 10% of the hydrogen total) and much smaller amounts of carbon, nitrogen, oxygen, neon, iron, sulfur, silicon, and other elements found in stars. The elements heavier than hydrogen appear in about the same proportion as they do in Population I stars (the younger stars), suggesting again that these stars did condense out of interstellar gas within relatively recent times. In the less dense interstellar clouds and in the intercloud regions of interstellar space, almost no molecules exist. The individual atoms, often ionized by the background starlight or by fast-moving electrons and protons, occasionally collide but rarely combine. The situation is completely the opposite in molecular clouds, those clouds with densities of several hundred to several million particles per cubic centimeter. Here the temperatures tend to be lower—about 20 K, in contrast to the 50 to 100 K of less dense clouds, and the 10,000 K or more of the intercloud regions. Lower temperatures and higher densities promote molecular formation, and almost all the atoms combine into various types of molecules. A ring of dense molecular clouds lies in the plane of the Milky Way, about 5000 parsecs from the center. Other such clouds dot the Milky Way at various distances from us and from the galactic nucleus.

The Variety of Interstellar Molecules. The situation just described, almost unsuspected until the 1960s, has allowed astronomers to discover more than 50 different molecular types within dense interstellar clouds. A favorite hunting ground for new molecular species has been the denser parts of the Orion Nebula (Figure 6-22), but a still more fertile area has been the giant molecular cloud complexes close to the galactic center (Figure 6-23). By making detailed radio measurements of these and similar regions, astronomers have detected radio emission (or, in a few cases, absorption of the radio waves in the cosmic microwave background) caused by different sorts of molecules.

6.6 The Galactic Center

The central region of the Milky Way long remained shrouded in mystery, because the concentration of dust particles there prevents any direct observation in visible light (Figure 6-19). However, as astronomers acquired the ability to observe the universe at longer wavelengths than those of visible light, they began to probe the galactic center. The center turns out to be a place of mystery, dominated by a single massive object.

Observations of the Galactic Center

Interstellar dust near the galactic center absorbs visible light with such high efficiency that only one photon of visible light in every 100 billion passes

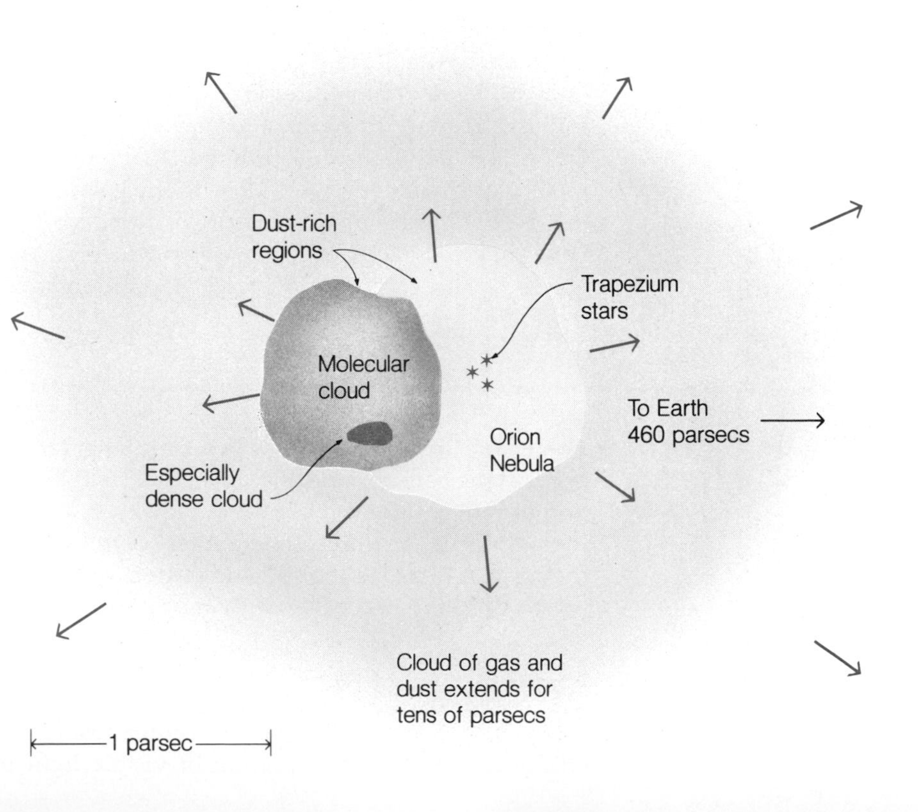

Figure 6-22
The Orion Nebula (top left) is an H II region lit by several dozen young, hot stars. Radio maps of the nebula show great amounts of short-wavelength radio emission, as well as of infrared emission (top right), a sign of contracting clouds of gas and dust. Around the visible Orion Nebula is a huge cloud of interstellar matter. The densest parts of this cloud lie behind the nebula we see lit by young stars (bottom).

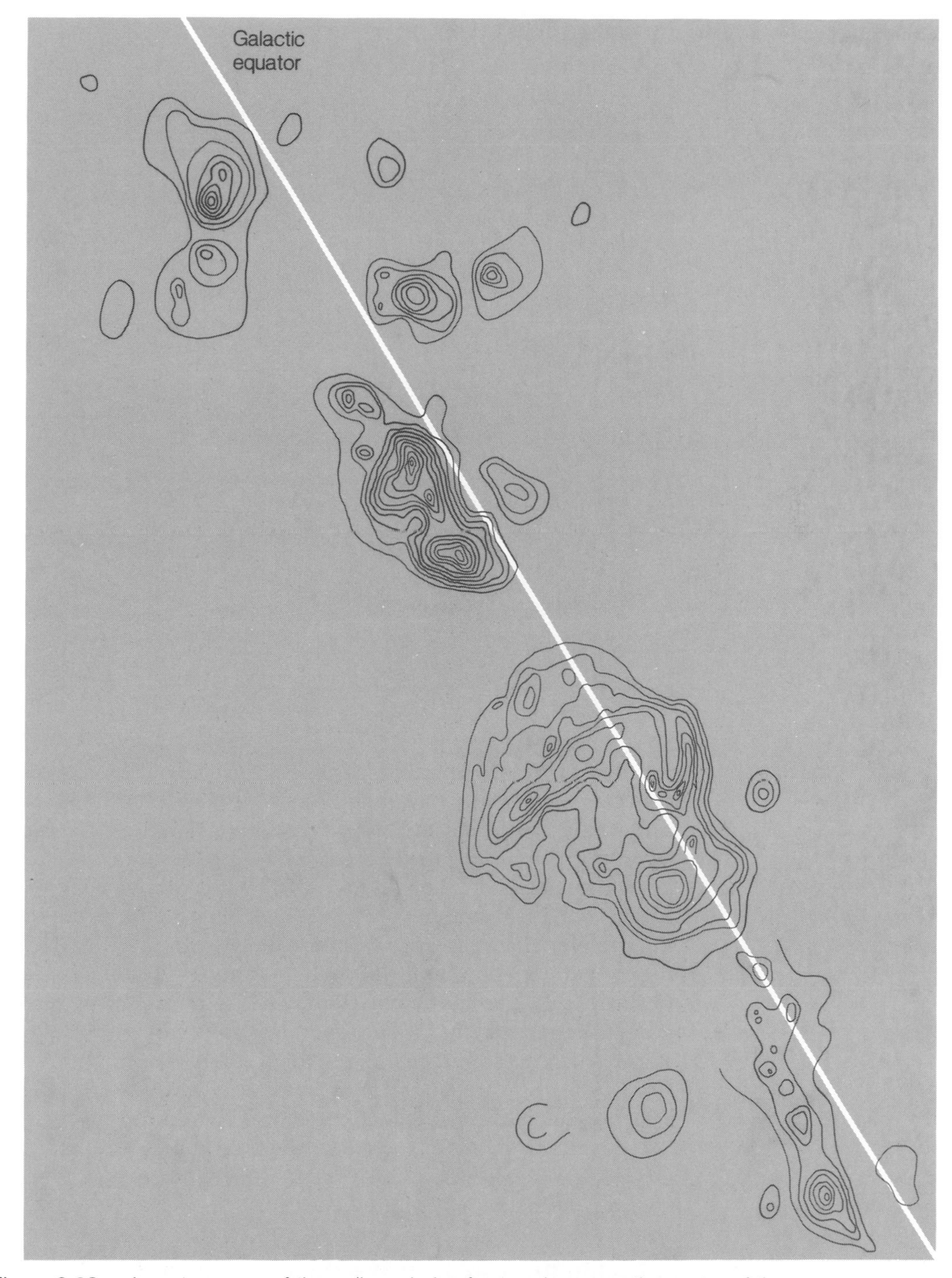

Figure 6-23 A contour map of the radio emission from regions near the center of the Milky Way shows observations at a radio wavelength of 3.75 centimeters. Radio waves can penetrate the dust that absorbs visible light and infrared radiation. An intense source of radio emission, Sagittarius B, lies at the point where the contour lines are most dense.

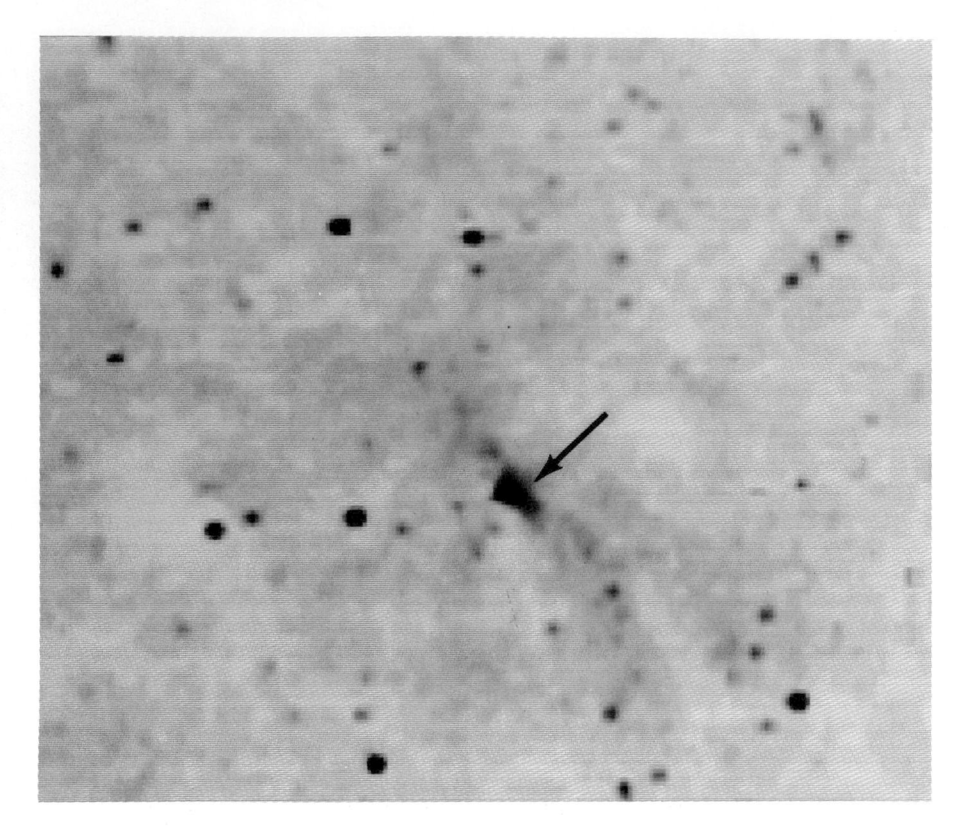

Figure 6-24
A contour map of the emission observed from the galactic center at a wavelength of 2.2 microns (2.2×10^{-4} centimeter) shows the most intense emission from regions along the galactic plane, and in particular from the galactic center (arrow). The map covers a distance of about 200 parsecs along the galactic plane.

unabsorbed among the dust particles. Since the efficiency of absorption declines as the photon wavelength increases (see page 183), infrared photons suffer much less absorption as they pass through the dust. Figure 6-24 shows a contour map of the intensity of infrared emission from the galactic center at a wavelength of 2.2 microns (2.2×10^{-4} centimeter). An extended source of emission spreads along the galactic plane, peaking at the center itself. This intense emission may arise from matter swirling near a black hole. Observing longer wavelengths of infrared radiation, astronomers see emission both from the dust particles themselves, which are heated by the radiation from young, hot Population I stars, and also from vast numbers of Population II stars, which predominate close to the galactic center.

Observations of the galactic center at *radio* wavelengths give a different picture than those made at infrared wavelengths. One reason for this difference is that the radio waves are almost completely unaffected by dust absorption. A more important reason, however, is that different processes produce the radio and infrared emission.

Emission Processes at the Galactic Center

Continuum Emission. Infrared radiation typically arises from matter that has been heated to temperatures of a few hundred to a few thousand degrees above absolute zero. We find such warm matter in the dust particles near the galactic center or in the outer layers of red supergiant stars. Radio emission can also occur from warm particles (as in the case of the early universe), but this sort of radiation is usually swamped by other radio emission processes. For example, near the galactic center, some of the radio emission arises from the synchrotron emission process when fast-moving electrons accelerate in a magnetic field (see page 208). Such emission reveals the existence of electrons that have reached velocities close to the speed of light. This implies the sudden, violent release of energy at the center of our galaxy. In addition, regions of hot, ionized hydrogen gas, the H II regions we see throughout the Milky Way, emit radio waves as well as visible-light radiation. This radio emission arises from the close interaction of electrons moving rapidly by one another (though not at speeds close to the speed of light).

Figure 6-23 shows a contour map of the galactic center at a wavelength of 3.75 centimeters, more than 10,000 times the wavelength of the infrared radiation used to produce the map in Figure 6-24. As we can see from Figure 6-23, the radio emission from the galactic center appears to spread along the median plane of the galaxy. Both the infrared and radio emission maps show a peak emission at the same spot, which is thought to be the very center of the Milky Way.

The infrared and radio observations we have described refer to **continuum emission,** that is, to the emission of photons over a relatively wide range of frequencies and wavelengths. Likewise, the emission processes we have described—synchrotron emission and the radio emission from H II regions—are the ones that can produce continuum emission.

Line Emission. In addition to the processes just described, we must also consider those that result in **line emission,** the emission of photons over only a narrow range of frequencies and wavelengths (see page 81). Line emission in the radio domain typically arises from the interaction of the elementary particles that form an atom or molecule. Each type of atom or molecule can emit radio waves at certain frequencies. Hence, by studying the emission at these particular frequencies, astronomers have been able to identify the many molecular types found in interstellar clouds.

An advantage of studying line emission from the galactic center comes from measuring the Doppler shift of the emission. Observations of carbon monoxide, the best-studied type of molecule near the galactic center, have revealed that a violent explosion within the last few million years has expelled vast quantities of carbon monoxide gas from the center at velocities of several hundred kilometers per second (Figure 6-25). Our galaxy has puffed out a sort of smoke ring, perhaps just one of a series.

The Central Parsec of the Milky Way

Within the last few years, astronomers have made detailed maps of the very center of the galaxy, called *Sagittarius A West,* at many infrared and radio wavelengths. These observations have shown that the *central single parsec* of the Milky Way, which spans one ten-thousandth of the distance from the sun to the center of the galaxy, contains 5 to 8 million times the sun's mass.

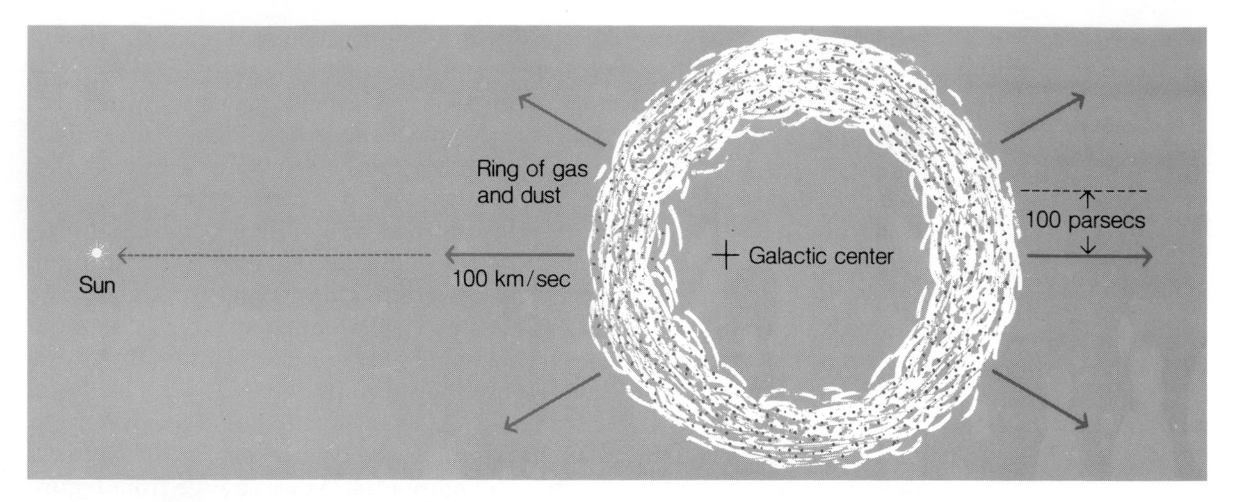

Figure 6-25 Observations of carbon monoxide molecules toward the galactic center have shown that a ring of material in the plane of the Milky Way is expanding away from the center. Apparently, some sudden release of energy at or close to the center has expelled this gas within the last few million years.

(An equal volume centered at the sun would contain 1 solar mass—the sun's.) Infrared-emission studies imply that only about 2 million solar masses of this material consist of stars, and a much smaller amount of dust and gas. If the tentative conclusions drawn from this account stand up to further study, some 3 to 6 million solar masses of matter at the galactic center form a nonstellar, nongaseous object, quite possibly a black hole (see page 337).

Whatever the truth may be about the distribution of matter at the galactic center, we must expect to find something that can suddenly release large amounts of energy. Only then can we explain the ejection of material from this region as seen in the outward motion of carbon monoxide molecules. This ejection resembles, though on a much smaller scale, the violent release of energy in quasars, Seyfert galaxies, and radio galaxies (see Chapter 7).

6.7 Mapping the Milky Way with Radio Waves

If we turn our attention from the central region of our galaxy back to its overall structure, we can admire the fact that even though we cannot see this structure in visible light, we can use radio emission from interstellar gas to map out the spiral arms. To understand the method of mapping that astronomers have developed during the past few decades, we must first understand why hydrogen gas, even when far from any star, emits radio photons of a particular frequency and wavelength.

Every hydrogen atom consists of a single electron in orbit around a nucleus, which, for most hydrogen atoms, consists of a single proton. These

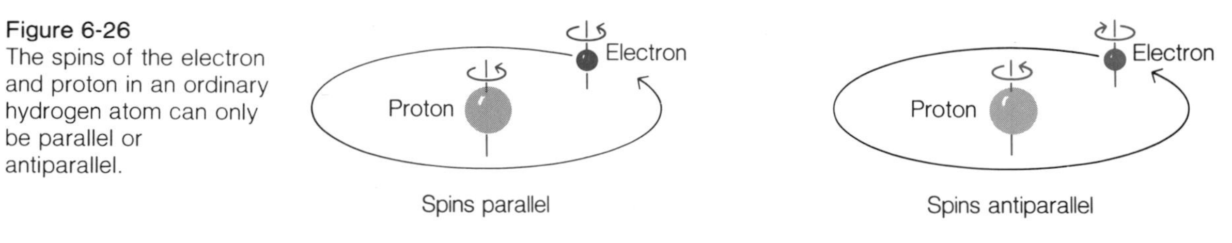

Figure 6-26
The spins of the electron and proton in an ordinary hydrogen atom can only be parallel or antiparallel.

atoms, which make up most of the interstellar gas, possess the ability to produce radio waves. They can do this because of a subtle interaction between the electron and proton in each atom, an interaction that depends on the "spin" of these two elementary particles.

The Spins of Protons and Electrons

We have described the mass and electric charge of the electron and proton on page 77. In addition to these two properties, these particles have a third distinguishing characteristic, their spins. We can think of an electron or proton as a tiny spinning magnet, something like a common bar magnet in rotation. If we measure what corresponds to the strength of the magnet and the magnet's rate of spin, we find that an electron and a proton have the same amount of **spin,** the combination of the strength of their magnetism and their rotation rate.

When an electron orbits a proton in a hydrogen atom, the electron's spin direction can only be *parallel* or *antiparallel* to the proton's spin direction (Figure 6-26). That is, the electron and proton spins must line up in parallel directions, but the two spins can be either in the same direction or in opposite directions. These two possibilities define the *hyperfine structure* of a hydrogen atom.

Spin-Flip Radio Emission

When the electron and proton spins of a hydrogen atom are parallel, the atom has a tiny amount more energy than it does when the electron and proton spins are antiparallel. As a result, an atom with parallel spins will eventually flip the electron spin over to reach a lower-energy state. This **spin flip** into the antiparallel position must be accompanied by the emission of a photon that carries away just the energy difference between the two spin-state possibilities (Figure 6-27).

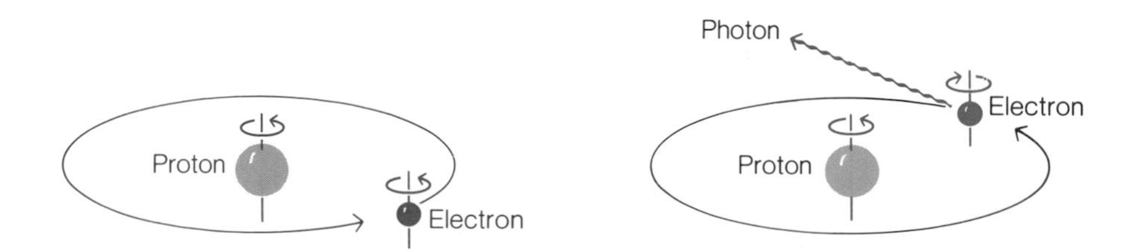

Figure 6-27
When the electron in a hydrogen atom flips its spin from the parallel into the antiparallel position, the atom will emit a photon of wavelength 21.10611 centimeters and frequency 1420.4058 megahertz.

An average hydrogen atom will spend 10 million years in the parallel-spin state before the electron spin flips over, but our galaxy contains so many hydrogen atoms that about 10^{54} spin flips, and accompanying photon emissions, occur each second. Collisions among atoms tend to flip some of the antiparallel-spin atoms back into the parallel-spin configuration. At any given time, three-quarters of the hydrogen atoms have parallel spins, while the remaining quarter have antiparallel spins.

The 21-Centimeter Radio Line

Because the difference in energy between the parallel-spin and antiparallel-spin configurations is so small (9.5×10^{-18} erg), the photons produced by the spin flips in hydrogen atoms have equally small energies. Hence, as we discussed on page 74, these photons must have small frequencies and large wavelengths: They are radio photons. Spin flips in hydrogen atoms produce radio waves of a wavelength and frequency that rank among the most accurately known parameters of the physical universe. To astronomers, the basic frequency of hydrogen spin-flip photons, 1420.4058 megahertz, and the corresponding photon wavelength, 21.10611 centimeters, set a universal standard. This is called the **21-centimeter emission** from hydrogen atoms. By measuring the number of photons of this frequency and wavelength, observed from a given direction, astronomers can determine the number of hydrogen atoms in that direction. Best of all, by measuring the Doppler shift away from the standard frequency, they can determine the relative velocity toward us or away from us that characterizes a particular group of hydrogen atoms. With this method we can map the spiral arms, rich in interstellar hydrogen atoms, of the Milky Way galaxy.

Observing the Galaxy at 21 Centimeters

When we look in directions close to the plane of the Milky Way, we observe large numbers of photons from the spin-flip process in hydrogen atoms, but with wavelengths slightly changed by the Doppler effect. We can combine our observations with our knowledge of the way the galaxy rotates (see page 152). If we assume that all the hydrogen atoms move in circular orbits around the galactic center, we can assign a distance along our line of sight to each particular Doppler shift (Figure 6-28). The assumption of perfectly circular orbits introduces some imprecision, but the distances derived from this assumption are thought to be rather accurate. Each circular orbit, including the sun's, has a characteristic velocity. Therefore, the velocity *difference* that we observe through the Doppler shift tells us which orbit carries a group of hydrogen atoms whose 21-centimeter radio emission shows a particular Doppler shift, hence a particular velocity difference along our line of sight (Figure 6-28).

Using this method of observing the emission from spin-flip transitions in hydrogen atoms, astronomers have derived the map of the hydrogen dis-

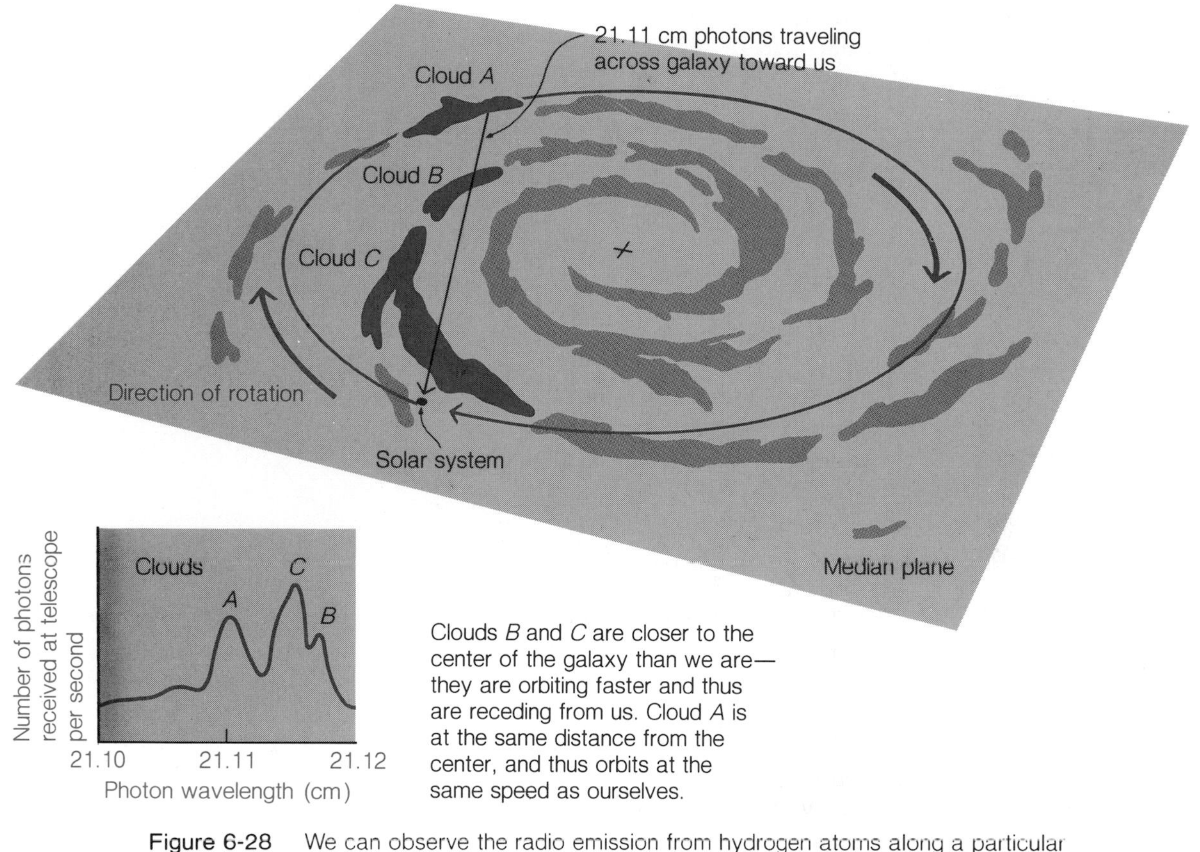

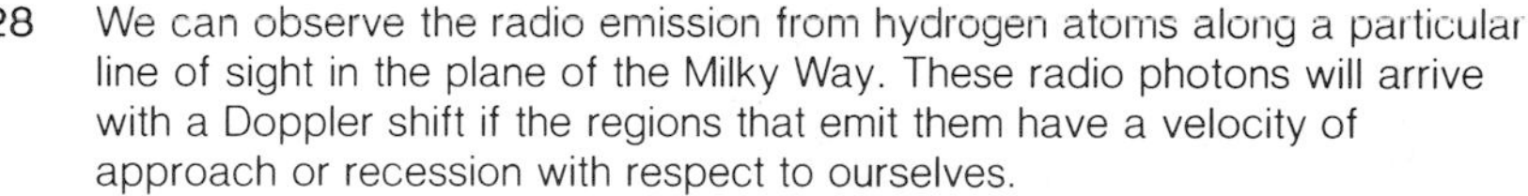

Figure 6-28 We can observe the radio emission from hydrogen atoms along a particular line of sight in the plane of the Milky Way. These radio photons will arrive with a Doppler shift if the regions that emit them have a velocity of approach or recession with respect to ourselves.

tribution shown in Figure 6-29. We can see that the hydrogen gas does concentrate in particular "arms" that show a vaguely spiral pattern, although our overall understanding of the distribution of gas in our own galaxy does not equal that for nearby spiral galaxies (Figure 6-30). The spiral arms mapped out by radio observations correspond fairly well, though by no means perfectly, to the spiral arms determined by observations of young, hot stars (Figure 6-9). In both methods of tracing the spiral arms, uncertainties in the *distance* determinations, either of young, hot stars or of groups of hydrogen atoms, inevitably make our maps less reliable than we would like. The method of observing hydrogen gas, which does concentrate in spiral arms, has the advantage that we can see almost the entire galaxy through the 21-centimeter emission. The method of observing bright, young stars works only out to distances of about 4000 parsecs from the sun and thus covers only a small fraction of the Milky Way.

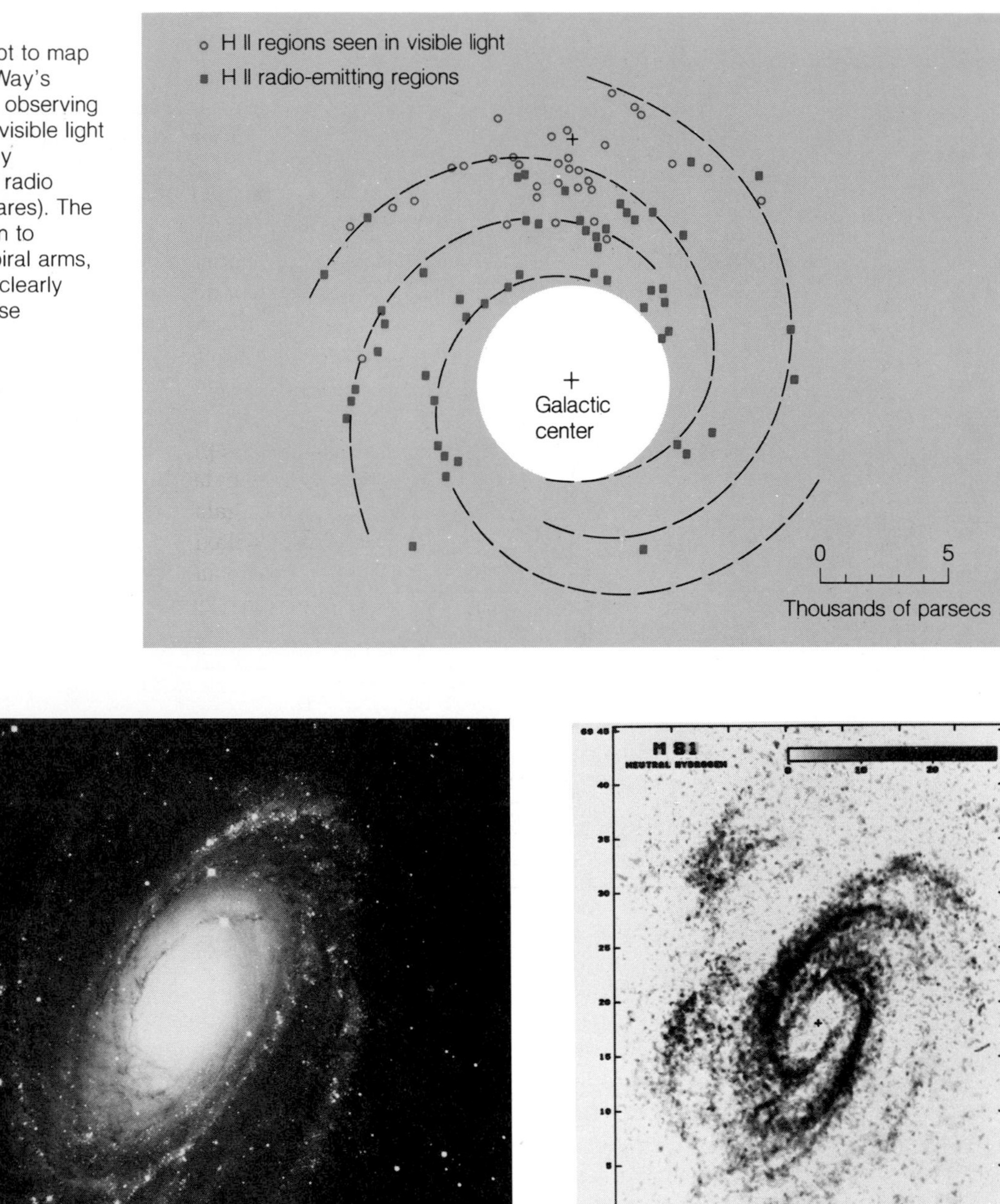

Figure 6-29
We can attempt to map out the Milky Way's spiral arms by observing H II regions in visible light (circles) and by detecting their radio emission (squares). The lines are drawn to suggest the spiral arms, which are not clearly defined by these observations.

Figure 6-30 A map of the intensity of radio emission at 21-centimeter wavelength (right) from the galaxy M 81 (left) shows the darkest areas at the points of most intense radio emission and hence of the greatest number of hydrogen atoms. Notice that hydrogen atoms do not appear in great numbers at the galaxy's center, where we see the greatest amount of visible light.

6.8 The Persistence of the Spiral Structure

Almost half the galaxies we know are spiral galaxies like the Milky Way, with a spiral-arm pattern of young, bright stars and interstellar hydrogen gas. When we consider the fact that the inner parts of these spiral galaxies rotate in less time than the outer parts, we face the question, why does the spiral pattern persist? A galaxy like our Milky Way takes 250 million years to rotate in its outer parts (400 million years in its farthest reaches), but less than 100 million years close to its center. Thus, after a few billion years—less than the age of most galaxies—the spiral pattern should be wound up tight (Figure 6-31). We may compare this situation to a marching band that makes a pinwheel maneuver. If the inner band members complete a turn in less time than the outer members, the pattern will be completely distorted after a few turns (Figure 6-32).

Theoreticians, however, know that the spiral pattern *does* persist, for we would not otherwise see so many spiral galaxies. Thus, astronomers have attempted to construct models of spiral galaxies that explain the *pattern*. Whatever it is that causes spiral arms in galaxies does so by collecting interstellar gas and forming young stars, which last for only a few tens of millions of years before fading away (see page 294). What process can produce such a pattern?

The Density-Wave Theory

Current theories of spiral structure propose that somehow—we really don't know how—a **density-wave pattern** begins in the disk of a spiral galaxy as

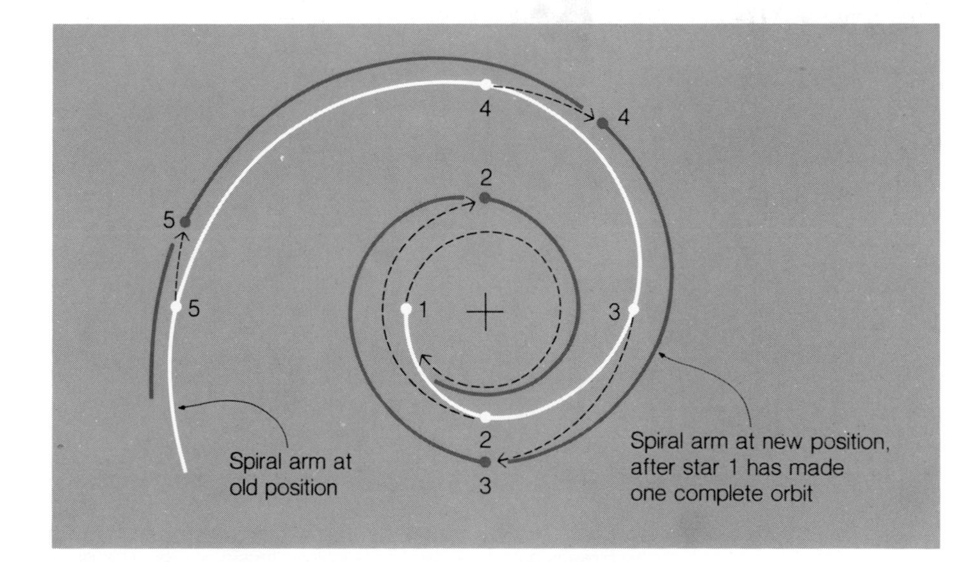

Figure 6-31 Because the inner parts of the Milky Way complete an orbit in less time than the outer parts, we would expect that the galaxy's spiral arms would be wound up tight after a few rotations.

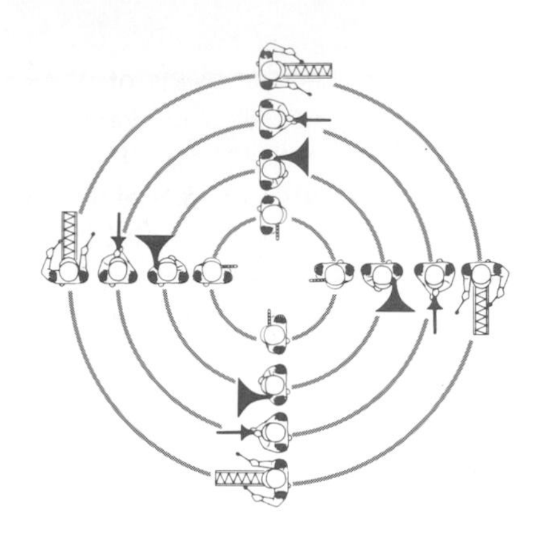

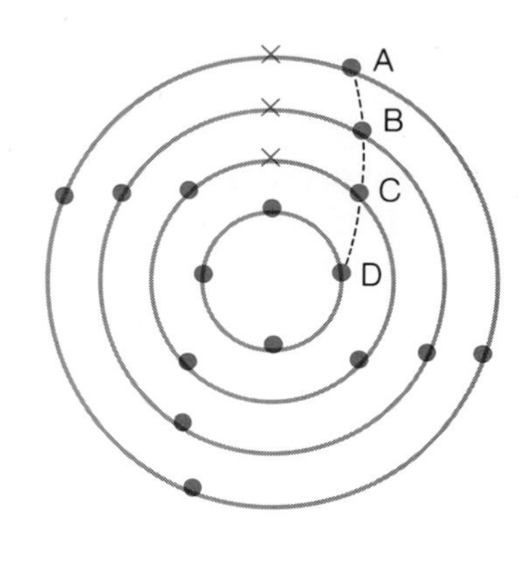

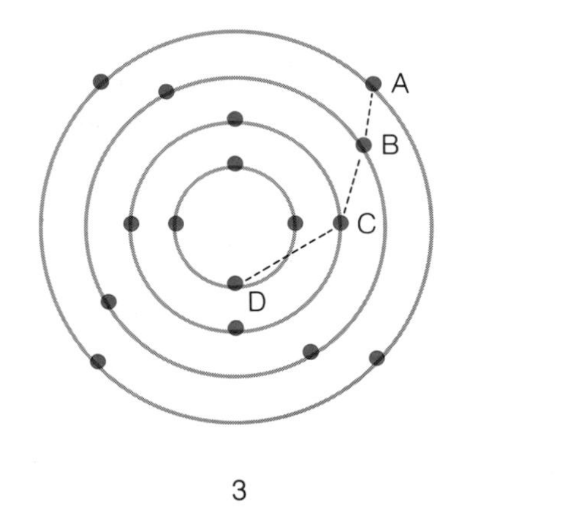

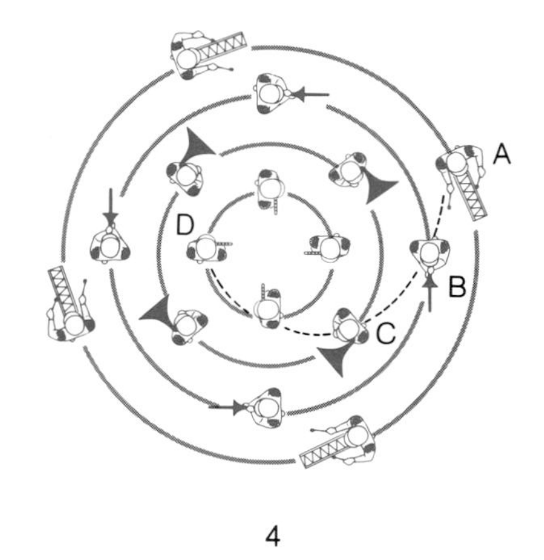

Figure 6-32
If a marching band makes a circle and the inner band members complete each turn in less time than the outer band members, the initial alignment of the band will be destroyed after a few turns.

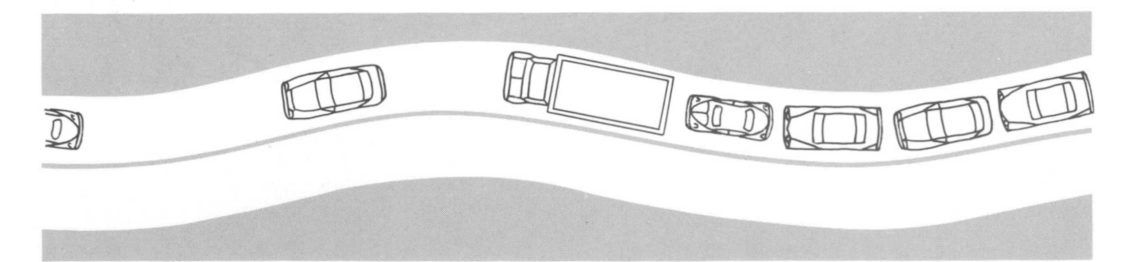

Figure 6-33 When a slow truck heads uphill, with cars passing as they can, we have a one-dimensional analogue to a spiral density wave. The pattern of bunched cars persists, even though a given car passes into and then out of the denser part of the pattern. Furthermore, the pattern moves at a slower speed than that of the average car.

it forms. The density-wave pattern consists of regions of alternately higher and lower gas densities. It resembles the water-wave pattern we see in a pond when we drop in a stone: Regions of alternately higher and lower water surface appear. Both the density-wave and the water-wave patterns move in a way that is different from the individual particles' motions. In the water wave, the drops of water bob up and down while the pattern sends ripples outward. Similarly, in the disk of a spiral galaxy, the density-wave pattern rotates around the center in the same direction as, but more slowly than, the individual elements of the galaxy, the stars and gas clouds.

Another analogy to the density-wave pattern in spiral galaxies occurs when traffic backs up behind a slow-moving vehicle on a highway. Individual cars may pass this vehicle and thus occupy a different part of the pattern, but the pattern persists, moving down the highway at a speed different from that of the average car (see Figure 6-33).

Star Formation in the Density-Wave Pattern

In a spiral galaxy, the density-wave pattern, like the slow-moving vehicle in the highway analogy, tends to "block" the passage of particles. Stars and gas overtake the spiral pattern, and as they enter it, the gas density increases to five or ten times its density outside (Figure 6-34). This increase in gas density has an important effect. It triggers the formation, and the collapse, of large gas

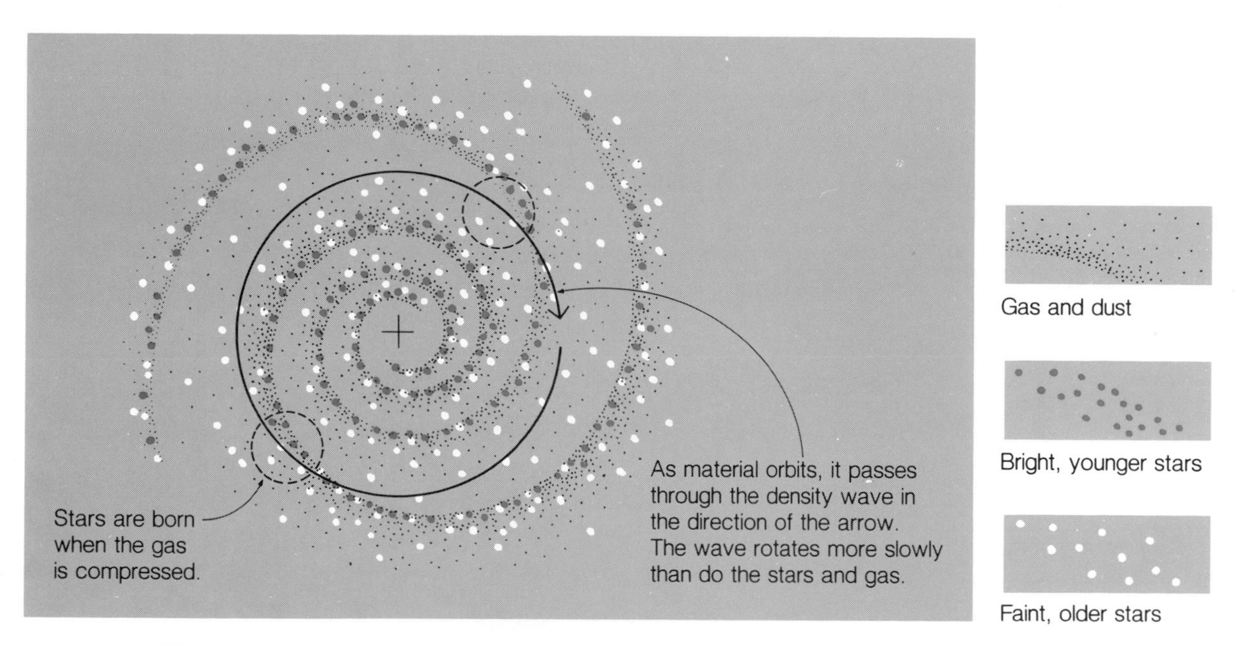

Figure 6-34 A schematic representation of the spiral structure in our galaxy shows new stars being born soon after clouds of gas and dust enter the denser part of the spiral density wave. As the density wave rotates, the location of young, bright stars and H II regions echoes the recent passage of the dense part of the wave past that part of the galaxy.

clouds, which condense into protostars within a million years or so, producing clumps of young, bright stars. Thus within a few million years after the gas enters the dense part of the pattern, some of it becomes the sort of bright stars that outline spiral arms for a few million years before fading away into obscurity. After a few million (or tens of millions) of years, these stars burn themselves out, because they burn their nuclear fuel so prodigiously (see page 294). By the time the pattern has moved a significant distance around the galaxy, relative to the stars, the stars that once outlined a spiral arm will be fading away, while a new group of stars forms from the gas that entered the dense part of the pattern more recently (Figure 6-34). Thus the density-wave pattern persists even as the individual bright stars perish.

The Origin of the Density-Wave Pattern

Calculations show that if a galaxy can start a density-wave pattern with a two-armed spiral structure, the pattern will be stable and last for billions of years. But how does such a spiral density-wave pattern get started? One theory assigns the origin of the pattern to close encounters between galaxies. Another theory suggests that the rotating central bulge of the spiral galaxy somehow generates the density-wave pattern.

In support of the first theory, we know that galaxies in clusters often pass close to one another, though actual collisions are rare. (Here "often" means every few billion years or less). Computer simulation of these near misses shows that the close encounters can generate galactic "tides" that become spiral density-wave patterns (Figure 6-35). In this theory, the Milky

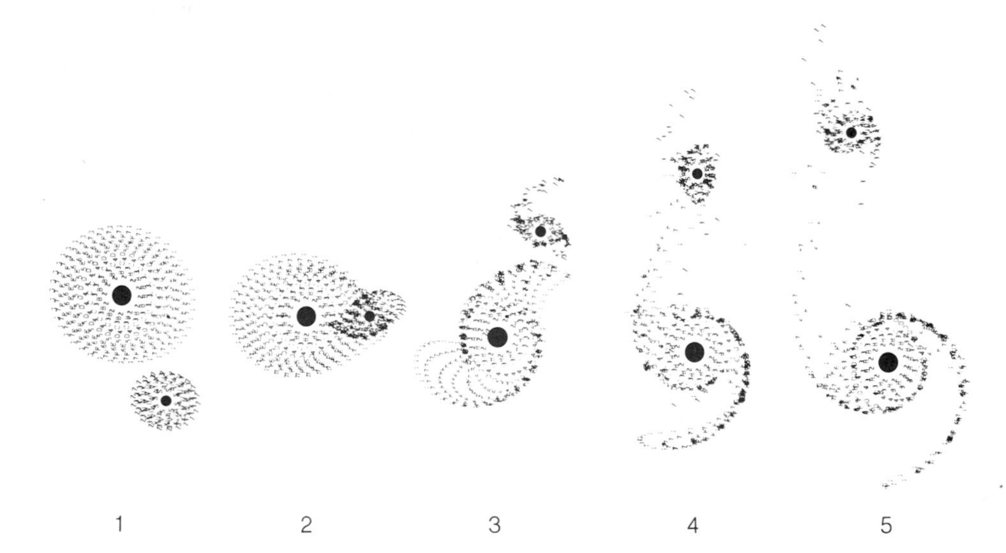

Figure 6-35 A computer simulation of a close encounter between two galaxies shows how the galaxies' gravitational forces tend to pull bunches of stars into spiral-arm patterns. This simulation suggests that close encounters can produce spiral arms, at least temporarily, and may start a spiral density-wave pattern that persists long after the encounter.

Way's density-wave pattern arose most recently from a close encounter with the Large Magellanic Cloud.

The second theory suggests that spiral galaxies generate their own density-wave patterns. Spiral galaxies possess a central rotating spheroid of stars, or a bar-shaped rotating spindle of stars in the case of barred spiral galaxies. The stars in the central part of the galaxy, 500 to 1000 parsecs across, tend to orbit the center with the same orbital period. This coherent motion of the innermost stars may be able to generate spiral density waves that propagate outward from the central agglomeration. Barred spirals seem better candidates for this mechanism than ordinary spiral galaxies, because their rotating bars of stars provide natural sources for a two-armed, symmetric density-wave pattern.

Both theories for how the spiral pattern begins have their supporters among astronomers. We shall have to make more detailed observations of the motions within galaxies to see which theory (if either one) more closely corresponds to reality.

Summary

Our Milky Way galaxy belongs to a small cluster of galaxies called the Local Group. In addition to our own galaxy, the Local Group contains the large spiral galaxy in Andromeda, the smaller spiral M 33 in Triangulum, and about twenty still less massive galaxies. Some of these smaller galaxies form satellite galaxies to our own and to the Andromeda spiral. Others, either dwarf or irregular galaxies, occupy positions within the 1-megaparsec extent of the Local Group, but not particularly close to any large galaxy.

Spiral galaxies dominate the Local Group, since the Milky Way and the Andromeda galaxy each has more mass than all the other members combined. The spiral arms of a spiral galaxy represent regions in which the density of interstellar gas and dust, and to some extent of stars, rises above the average in the disk of the galaxy. Spiral arms appear preeminent because they contain all the young, bright stars within the disk, stars that last for only a few tens of millions of years before fading away. These stars represent the extreme of the younger Population I stars. Population I stars contain a larger fraction of elements heavier than hydrogen and helium than do the older Population II stars. Our sun is a typical Population I star, not particularly young at 4.6 billion years of age, but younger than the bulk of the Population II stars. Within the disk of spiral galaxies, Population I stars predominate. In the spherical halo of stars that surrounds the disk, Population II stars form the dominant component, as they do in the innermost regions of spiral galaxies. Elliptical galaxies have both Population I and Population II stars, but they contain none of the young, bright Population I stars that outline spiral arms.

Among the stars in a spiral galaxy such as the Milky Way, interstellar matter, consisting of both gas and dust, forms perhaps 10% of the total mass of the galaxy. Irregular galaxies contain a larger fraction of their mass in interstellar matter, while elliptical galaxies have almost none. The dust particles may be graphite (carbon) or silicon-oxygen mixtures, probably overlain by a mantle of ice. This dust absorbs and scatters starlight well, but red light suffers somewhat less extinction than blue light. Thus starlight passing among the dust particles emerges redder than before. The interstellar gas consists mostly of hydrogen atoms and hydrogen molecules, with a small amount of helium and other elements. In dense interstellar clouds, more than 50 types of molecules have been detected, proof that interstellar chemistry plays an active role in the denser clouds.

Hydrogen atoms emit radio waves of 21.1-centimeter wavelength. Astronomers who study the amount of radio emission at each precise wavelength from a given direction can use the Doppler effect to determine the number of hydrogen atoms with each particular velocity relative to ourselves. Using their knowledge of how the Milky Way rotates, they can then make a map of the distribution of hydrogen in the galaxy from these observations, locating, with some accuracy, the various spiral arms of our Milky Way.

Key Terms

absorption
Cepheid variable stars
continuum emission
convergent point
density-wave pattern
globular cluster
high-velocity stars
interstellar clouds
interstellar dust
interstellar gas
interstellar reddening
Large Magellanic Cloud
line emission
Local Group
open cluster
Population I stars
Population II stars
proper motion
Small Magellanic Cloud
spin flip
stellar associations
21-centimeter emission
zone of avoidance

Questions

1. What is the Local Group? What are its chief constituents?
2. How does the galaxy NGC 6822 compare in size with the Milky Way? How does it compare in mass? In terms of mass, does NGC 6822 rank in the upper half or the lower half of the Local Group galaxies?
3. What are spiral arms? Why do they "look" more prominent than they really are in terms of the distribution of stars within a galaxy?
4. Why do young stars appear predominantly inside the arms of a spiral galaxy?

5. What are the chief differences between globular star clusters and open star clusters? Which type of cluster, on the average, has older stars?
6. What are stellar associations? How do they differ from star clusters?
7. What does interstellar matter consist of? How can we observe interstellar gas? How can we detect interstellar dust particles?
8. In what regions of interstellar space do complex molecules appear? How does the density within these regions affect the chances that such molecules will form?
9. By what process do hydrogen atoms emit photons with a 21.1-centimeter wavelength? How does this emission process allow astronomers to map the spiral arms of the Milky Way and of nearby galaxies?
10. Why are we prevented from observing the central regions of our galaxy in visible light? How do astronomers manage to make observations of this region?
11. Why might we expect that the spiral arms of a galaxy such as our Milky Way will be wound up tight after a few billion years?
12. How does the density-wave theory of spiral structure explain the persistence of the spiral arms of galaxies? How might a density-wave pattern begin within the disk of a galaxy?

Projects

1. *The Galactic Merry-Go-Round.* Take a ride on a merry-go-round to determine how many similarities you can find between your motion and the sun's motion in the Milky Way galaxy. How many up-and-down oscillations do you make in one complete circle? How many does the sun make? Why do you have difficulty in seeing the other side of the merry-go-round? How does this compare with the situation in the Milky Way galaxy? What is the relationship between the riders' speeds and their distances from the center of the merry-go-round? What are the riders' speeds relative to yourself? How are the motions of stars around the Milky Way more complex than those of riders on a merry-go-round? In particular, why do none of the riders on a merry-go-round have a velocity relative to yourself, while stars closer to the center of the galaxy or farther from it have velocities relative to our sun?
2. *The Galactic Center.* In summer, you can observe the region of the center of the Milky Way at the boundary between the constellations Sagittarius and Scorpius (see star maps). Pick a clear, moonless night and look for the Milky Way on the sky in these constellations. Does the center of the Milky Way seem especially bright, in comparison with parts of the Milky Way 10° or 20° on either side of the center? How can you explain this fact, in terms of the wavelength of the photons your eyes detect? In what wavelengths would the center of the Milky Way seem especially bright?
3. *The Spiral Galaxy M 33.* On a clear, dark night in fall or winter, try to use binoculars to find the spiral galaxy M 33 in Triangulum. The galaxy is

about 4° to the west of the sharp tip of the triangle that forms Triangulum (see star map for September). How does the apparent brightness of M 33 compare with that of M 31, the Andromeda galaxy? M 33 and M 31 have about the same distance from us, but M 31 has about twenty times the mass of M 33. Does this seem to offer a reasonable explanation for the observed ratio of the galaxies' apparent brightnesses? What assumptions are involved in judging the acceptability of this explanation?

4. *The Magellanic Clouds.* If you have a chance to travel to latitudes south of the equator, seize the opportunity to observe our galaxy's two large irregular satellite galaxies, the Large and Small Magellanic Clouds. If you look at the evening sky between October and March, you should be able to find the Large Magellanic Cloud with no difficulty (it is directly south of the bright star Rigel in Orion). The Small Magellanic Cloud is in the same general area of the sky, but about 55° toward the west from the Large Cloud. These two galaxies each contain only a few percent of the mass in the Andromeda galaxy, yet their apparent brightnesses exceed that of the Andromeda galaxy. How can you explain this fact? How can you use it to derive a maximum distance to the Magellanic Clouds in terms of the distance to the Andromeda galaxy?

Further Reading

Bok, Bart, and Bok, Priscilla. 1981. *The Milky Way.* 5th ed. Cambridge: Harvard University Press.

Gammon, Richard. 1977. *Chemistry Between the Stars.* Washington, D.C.: NASA, U.S. Government Printing Office.

Geballe, Thomas. 1979. "The Central Parsec of the Galaxy." *Scientific American* 241:1, 60.

Heiles, Carl. 1978. "The Structure of the Interstellar Medium." *Scientific American* 238:1, 74.

Sanders, Robert, and Wrixon, Gerald. 1975. "The Center of Our Galaxy." In *New Frontiers in Astronomy.* Edited by Owen Gingerich. San Francisco: W. H. Freeman.

Turner, Barry. 1975. "Interstellar Molecules." In *New Frontiers in Astronomy.* Edited by Owen Gingerich. San Francisco: W. H. Freeman.

Verschuur, Gerrit. 1974. *The Invisible Universe.* New York: Springer-Verlag.

The peculiar galaxy NGC 5128 is the source of tremendous amounts of radio emission.

7

Radio Galaxies, Exploding Galaxies, and Quasars

Most of the matter we see in the universe resides in galaxies, and most belong to the familiar types described in Chapter 5, which are undistinguished by peculiar events. However, an important minority of galaxies, discovered only within the past few decades, has attracted attention through the violent events that occur within them. The precise nature of this cosmic violence still eludes explanation, though astronomers do not lack theories to fill the gaps in our knowledge. At present, however, we cannot be certain which of these theories corresponds most closely to reality. Before examining such theories, we must turn our attention to the various types of galaxies in which the explosive release of energy has recently taken place. Perhaps then we can find a relationship among the various forms of violent outbursts.

7.1 Radio Astronomy: Thermal and Nonthermal Emission

Most of what we know about the sudden, violent release of energy within certain galaxies comes from the science of **radio astronomy.** Radio observations of the cosmos began during the 1930s as a hobby of two American radio engineers, Karl Jansky and Grote Reber (Figure 7-1). Only during the late 1940s did astronomers begin to make radio observations in earnest, as new techniques in detecting radio waves, developed during the second World War, became available. As they gained experience with radio observations, astronomers learned how to measure the spectra of sources of radio waves (number of photons emitted at each wavelength or frequency), just as they had learned to measure the spectra of visible-light sources. Astronomers could then classify radio spectra into various categories, based on the general shapes of the spectra they observed.

Astronomers have long known that every object in the universe with a temperature above absolute zero emits some radio waves (and some infrared, visible-light, ultraviolet, and x and gamma radiation as well). This **thermal emission** typically appears in the form of an ideal-radiator spectrum, or close to it in shape (Figure 4-18). However, the thermal emission from any object is unlikely to produce much radio emission. The low level of radio output from thermal emission appears strikingly in the cosmic microwave background that arises from the entire universe. The low temperature of the universe, now characterized by 2.7 K, makes the photon emission appear mostly in the form of radio waves, but the total signal strength of the cosmic microwave background falls far below that of many individual radio sources.

Hence, except for the microwave background and a few objects relatively close to us (the sun, moon, planets, and H II regions), astronomers

Figure 7-1 Radio astronomy began with observations made by Karl Jansky (left), shown with his array of antennas, and Grote Reber (right), who built a steerable-dish antenna to study the sky in radio waves. Jansky and Reber worked quite independently of each other.

cannot observe thermal radiation from cosmic sources in the radio domain. Stars and galaxies emit plenty of thermal radiation, but because they are hot, their radiation emerges primarily as ultraviolet, visible-light, and infrared photons (see page 72). The radio emission that astronomers study largely consists of **nonthermal radiation,** that is, radio waves produced by processes other than the natural emission from an object at a temperature above absolute zero. Nonthermal radiation has two distinct categories: line emission and continuum emission.

Line Emission

Line emission, as its name implies, consists of emission concentrated within a limited range of frequencies or wavelengths (Figure 7-2). Such emission typically occurs when an electron in an atom jumps into a smaller orbit (Figure 3-17). In this case the line emission usually appears at ultraviolet, visible-light, or infrared wavelengths. The most important process that produces line emission at radio wavelengths, the famous 21-centimeter emission from hydrogen atoms, occurs wherever cool hydrogen gas exists (see page 194). This process commands our attention when we consider the distribution of hydrogen gas within galaxies, and in fact it forms the best way to map out the presence of hydrogen. However, it has little to do with the cosmic violence that is the focus of this chapter. Here the major emission of importance consists of nonthermal continuum radio emission.

Continuum Emission

Continuum emission describes the emission of photons over a broad band of frequencies or wavelengths, called a continuum (Figure 7-2). The objects we investigate in this chapter have spectra whose shapes differ noticeably from the ideal-radiator spectra that typify sources of thermal emission. As Figure 7-3 shows, ideal-radiator spectra have one basic shape and differ only

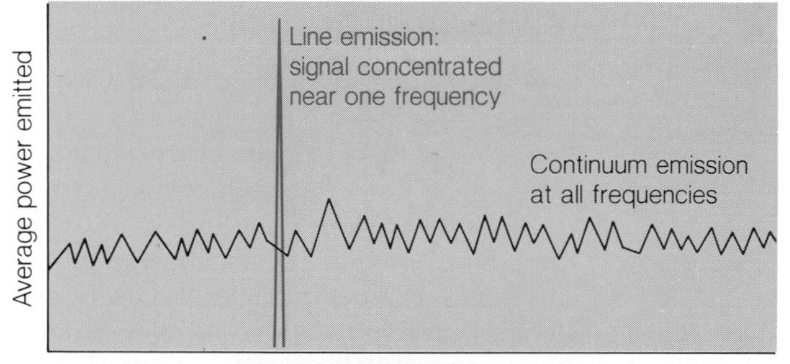

Figure 7-2 Line emission consists of the production of photons within a narrow range of wavelengths and frequencies. Continuum emission consists of the production of photons over a wide range of frequencies and wavelengths. The continuum emission drawn here is nonthermal.

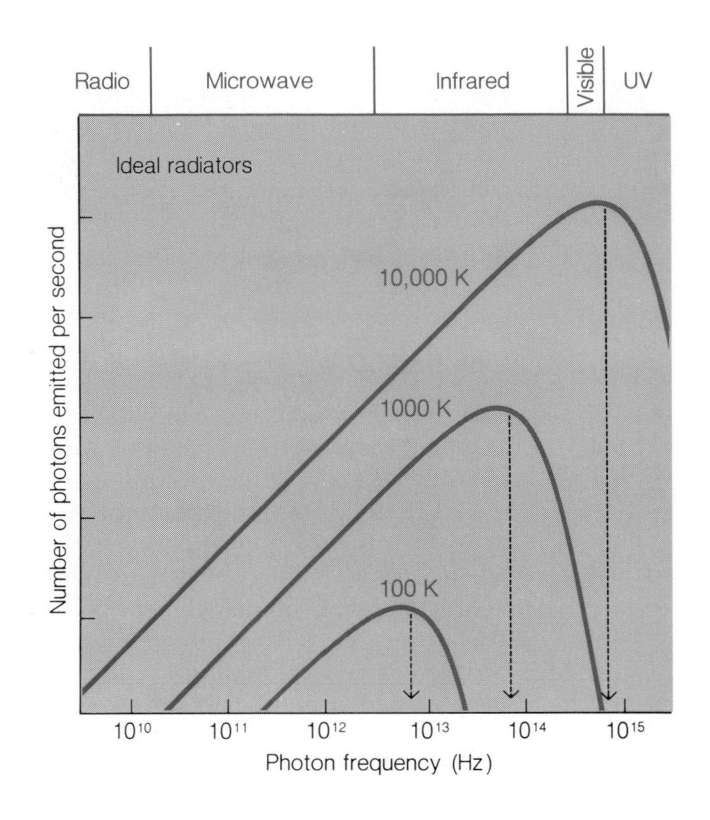

Figure 7-3
Continuum emission can arise from the natural emission from any object hotter than 0 K. Such thermal emission has a characteristically shaped spectrum, in which the peak frequency increases in proportion to the object's temperature.

in the temperature of the object producing the thermal emission. Nonthermal continuum sources of emission, however, have spectra with various shapes, all distinctly different from those of ideal radiators. The most important process we must consider in nonthermal continuum emission, the key process involved in cosmic violence, has the intriguing, if slightly misleading, name *synchrotron emission.*

Synchrotron Emission

We have discussed several ways in which photons can appear in the universe. In addition to the emission processes we have just discussed, scientists have found that charged particles that accelerate in magnetic fields will emit photons as they change their speed, their direction, or both. This process is called **synchrotron radiation,** or **synchrotron emission,** because physicists first discovered it with synchrotron particle accelerators (Figure 7-4). In such machines, charged particles (typically protons or electrons) accelerate to velocities close to the speed of light. Since the particles follow a circular path, they constantly accelerate.[1] They change the direction of their motion at each instant, whether or not they change their speed. Because these accelera-

[1]Acceleration is defined as any change in the speed or in the direction of motion of an object.

Figure 7-4 Particle accelerators called synchrotrons use magnetic fields to make the charged particles follow circular orbits. Particles accelerated to almost the speed of light emit photons as they change their speeds or their directions of motion.

tors use magnetic fields to impose forces on the particles, accelerating them to larger and larger velocities and making them move in circles, the particles accelerate in magnetic fields. Scientists found that this acceleration produced photons that carried energy away from the particles. In fact, this energy had to be resupplied to keep the particles traveling at a constant speed.

The theory of this emission states that at speeds much less than the speed of light, the number of photons and the energy they carry amounts to little. (This particular process has the name *cyclotron emission* after an earlier, less powerful particle accelerator.) Only when charged particles moving at nearly the speed of light accelerate in the presence of a magnetic field do the amount of energy and the total number of photons emitted reach substantial figures. This occurs in synchrotron emission. The change in the amount of energy radiated as the particles' velocity nears the speed of light forms an interesting sidelight to the theory of relativity, which points to many strange effects that occur at velocities close to the light velocity.

The key fact is that the synchrotron emission process can indeed produce copious amounts of radio emission, with a spectrum that extends over a

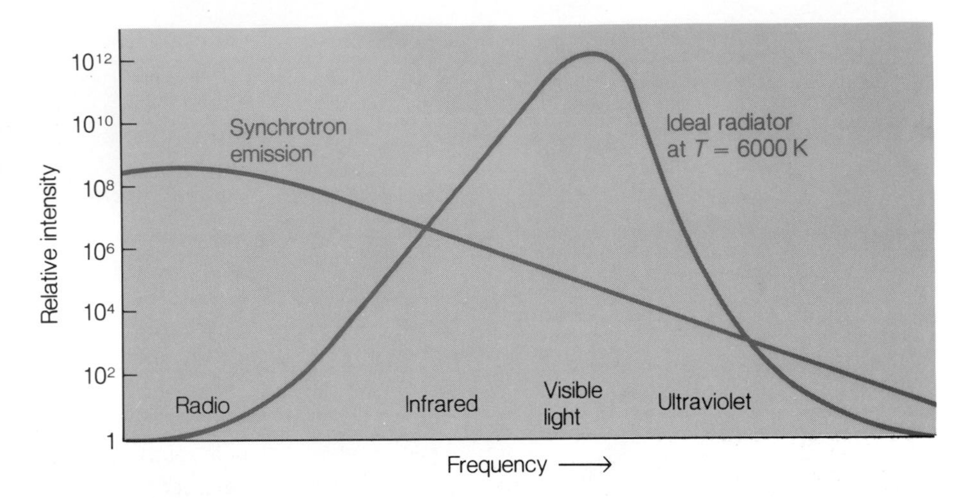

Figure 7-5 The synchrotron emission process, like the process of thermal emission, produces photons with a wide range of frequencies and wavelengths. The two types of spectra, however, differ noticeably in their shapes.

broad range of frequencies and wavelengths (Figure 7-5). If we can identify the characteristic shape of this spectrum from observations, we have a definite sign that the synchrotron process is at work. Furthermore, the fact that the synchrotron process "works" only if charged particles move at nearly the speed of light provides a valuable clue. If we observe photons made by the synchrotron process, then we know that large numbers of protons, electrons, or other charge particles must somehow have been accelerated to nearly the speed of light. Just as the cosmic microwave background provides a "fingerprint"—a clue to cosmic violence soon after the big bang—so does the synchrotron spectrum indicate a later, localized scene of violence, one that has boosted particles' velocities to close to 300,000 kilometers per second. We must now examine the evidence that such violence has occurred and see how the synchrotron process helps explain what we observe.

7.2 Radio Galaxies and Exploding Galaxies

Most galaxies, such as our own Milky Way, produce radio waves through a process that occurs within hydrogen gas. This process involves not cosmic violence but rather the flipping of electron spins within hydrogen atoms (Figure 6-27). Still more important, the total radio emission from a "normal" galaxy such as ours amounts to only a minuscule fraction of the total energy emitted. Most of the energy appears as ultraviolet, visible-light, and infrared waves. In other words, the radio emission from most galaxies forms a tiny part of the bulk of the emission, a sign that the most important events (in energy terms) have little to do with radio emission.

In contrast, some galaxies emit as much as, or even more than, their "normal" energy output in the form of radio waves. Astronomers define a

radio galaxy as one in which the energy released each second in the radio domain of the spectrum equals or exceeds the energy released at shorter wavelengths. In radio galaxies, the contribution of hydrogen gas to the total emission cannot account for the much-greater-than-usual radio output. Instead we must look to violent events within the galaxy, for spectral analysis tells us that most of the radio emission comes from the synchrotron process.

Radio Emission from Cygnus A

Consider, for example, the radio galaxy Cygnus A (Figure 7-6). This galaxy has such a great distance from us (300 megaparsecs) that we cannot get a clear picture of its structure. However, we can tell that we are not dealing with an ordinary galaxy simply from the shape that we see, which is neither spiral nor elliptical. Cygnus A has a relatively low apparent brightness in visible light, as indeed we would expect from its great distance. Nevertheless, its intense radio output makes it the second strongest source of radio emission that we observe, despite its being 300 megaparsecs away from us. This galaxy emits far more energy in radio waves than in visible light, in contrast to the Milky Way, which emits one-millionth as much energy each second in radio emission as in visible-light emission.

Figure 7-6 The radio galaxy Cygnus A presents an unusual, fuzzy appearance, shining with light that has taken a billion years to reach us.

When we make a careful study of the radio emission from Cygnus A, however, we find that most of the radio waves come not from the galaxy that we see but from its surroundings. This is typical of most radio galaxies. As in the case of Cygnus A, we usually observe the most intense emission to come from two regions symmetrically located on either side of the visible galaxy (Figure 7-7). The regions of radio emission may have diameters of hundreds of thousands of parsecs and may be separated by millions of parsecs in extreme cases (Figure 7-8). The galaxy itself, located just between the two "clouds" of radio emission, has a diameter of perhaps 30,000 parsecs. How, then, can we explain the vast sizes of the regions that emit radio waves? What processes have accelerated great numbers of particles to almost the speed of light while apparently ejecting them in opposite directions from the galaxy?

The Ejection Hypothesis

In recent years, astronomers have accumulated mounting evidence that explosions within certain galaxies do produce twin jets of material in opposite directions. These galaxies are called **exploding galaxies,** not because an entire galaxy is exploding, but because explosive events occur in its core. Although we have only the vaguest notion of what causes such explosions, the presence of synchrotron emission tells us that particles must have been accelerated to almost the speed of light, and the presence of clouds of radio

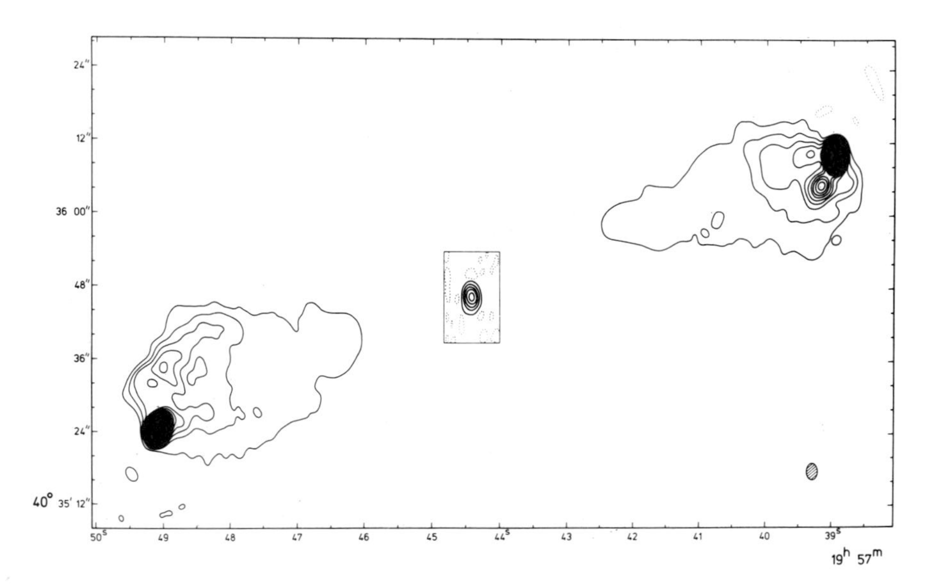

Figure 7-7 When we map the radio emission from the vicinity of Cygnus A, we can draw contour lines that show the rising levels of radio emission. The most intense emission arises from the dark ellipses, symmetrically located on either side of the visible galaxy. The visible galaxy would fit snugly within the contour lines at the center.

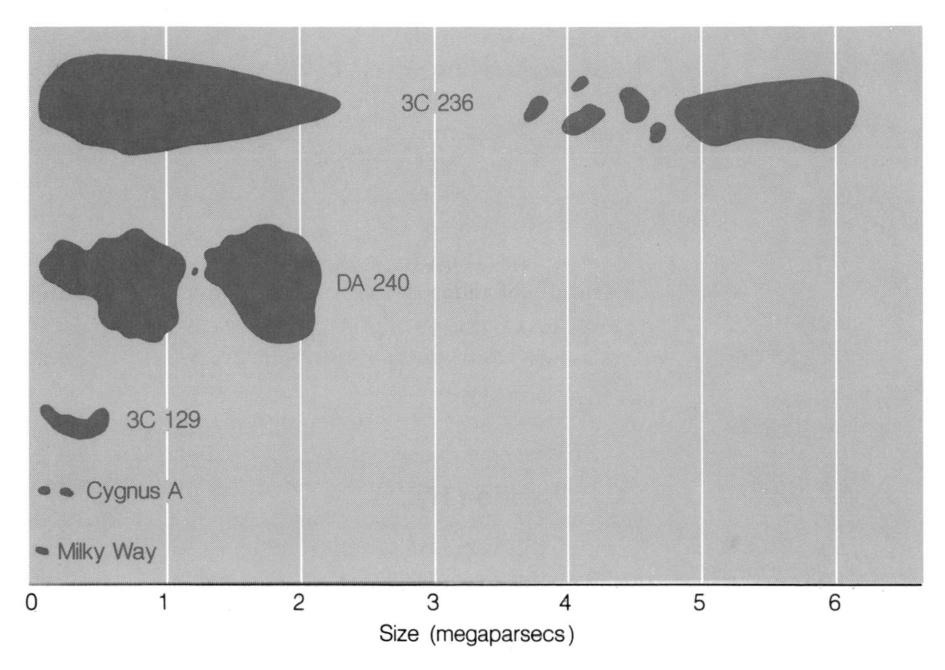

Figure 7-8 A comparison of the sizes of four radio-emitting regions around radio galaxies of the same size as the Milky Way shows the tremendous dimensions of the largest radio galaxies. Typical radio galaxies have sizes comparable to that of Cygnus A.

emission surrounding the galaxy with precise symmetry tells us that something within the galaxy itself should be responsible for the outburst that has led to the radio waves we observe. According to this model of radio galaxies, the initial outburst should appear as two narrowly confined jets, relatively close to the galaxy itself. As the outburst continues, the particles ejected from the galaxy move outward, pushing aside the gas of the intergalactic medium, the material that lies between galaxies (Figure 7-9). This deflection of intergalactic gas causes a "snowplow" effect, which piles up material at the heads of the two expanding jets, making the radio emission most intense at these points. In addition, the snowplow effect slows the speed with which the jets move outward, so that the clouds of radio emission—the regions within which particles are moving at almost the speed of light—eventually produce less energy in the form of radio waves as they expand to merge with the general intergalactic medium.

The Outburst in M 87

Although the picture we have described remains somewhat speculative, we can see what appears to be a good example of the early stages of a galactic outburst in the giant elliptical galaxy M 87 (Figure 5-12). The innermost

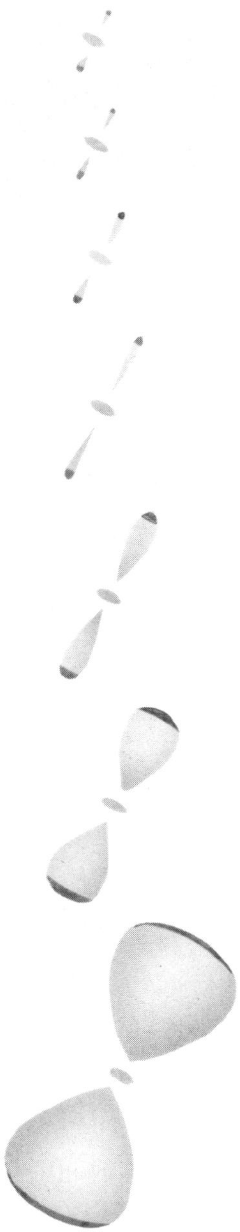

Figure 7-9
A radio galaxy with two symmetrically located regions of radio emission might arise from the ejection of particles from the galaxy itself. As the particles ejected from the galaxy move through the intergalactic gas, they pile up this matter in a "snowplow" effect. This slows down the ejected particles and produces a lateral expansion in their distribution.

parts of this galaxy show a jet of material, divided into smaller condensations, arrowing out from the galaxy's center (Figure 7-10). Still closer study reveals a "counter-jet," much harder to see, in exactly the opposite direction. We may presume that the counter-jet emerges in the direction away from the Milky Way, so that a greater amount of material within the central regions of M 87 obscures our view of it. The jet and counter-jet emit radio waves by the synchrotron process, in agreement with the model we have suggested. Other galaxies, not so close to us as M 87, show larger jets by the radio waves we detect (Figure 7-11). These galaxies should be transitional cases between M 87 and the giant radio galaxies whose schematic development appears in Figure 7-10.

Infrared Emission by Radio Galaxies

The science of radio astronomy developed before astronomers had the ability to make good observations in the infrared region of the spectrum. When reliable infrared detectors were developed during the 1960s, however, astronomers realized that most radio galaxies emit great amounts of infrared radiation—as much as, or more than, the great amount emitted in the form of radio waves. These infrared photons arise within the galaxy itself, presumably as part of the same violence that leads to the emission of radio waves, but not as the result of the synchrotron process itself. A good example of a radio galaxy that emits infrared radiation is M 82 (Figure 7-12), where a vast outpouring of energy within the past few million years has led to the emission of ten times more energy per second in infrared radiation than in visible light. The infrared radiation apparently comes from a complicated interaction of the explosive release of energy with dust particles within the galaxy. The photons from the galaxy heat the dust particles to several hundred K through impact, at which temperature the dust particles produce infrared radiation. Unfortunately, these dust particles obscure our view of the central regions of M 82, so we cannot get a good look at the regions where the violent release of energy has occurred and may be continuing.

7.3 Seyfert Galaxies

We have described the currently accepted model for the emission of radio waves by explosive events within some galaxies. Closely related to the radio galaxies we have described are the **Seyfert galaxies,** named after their dis-

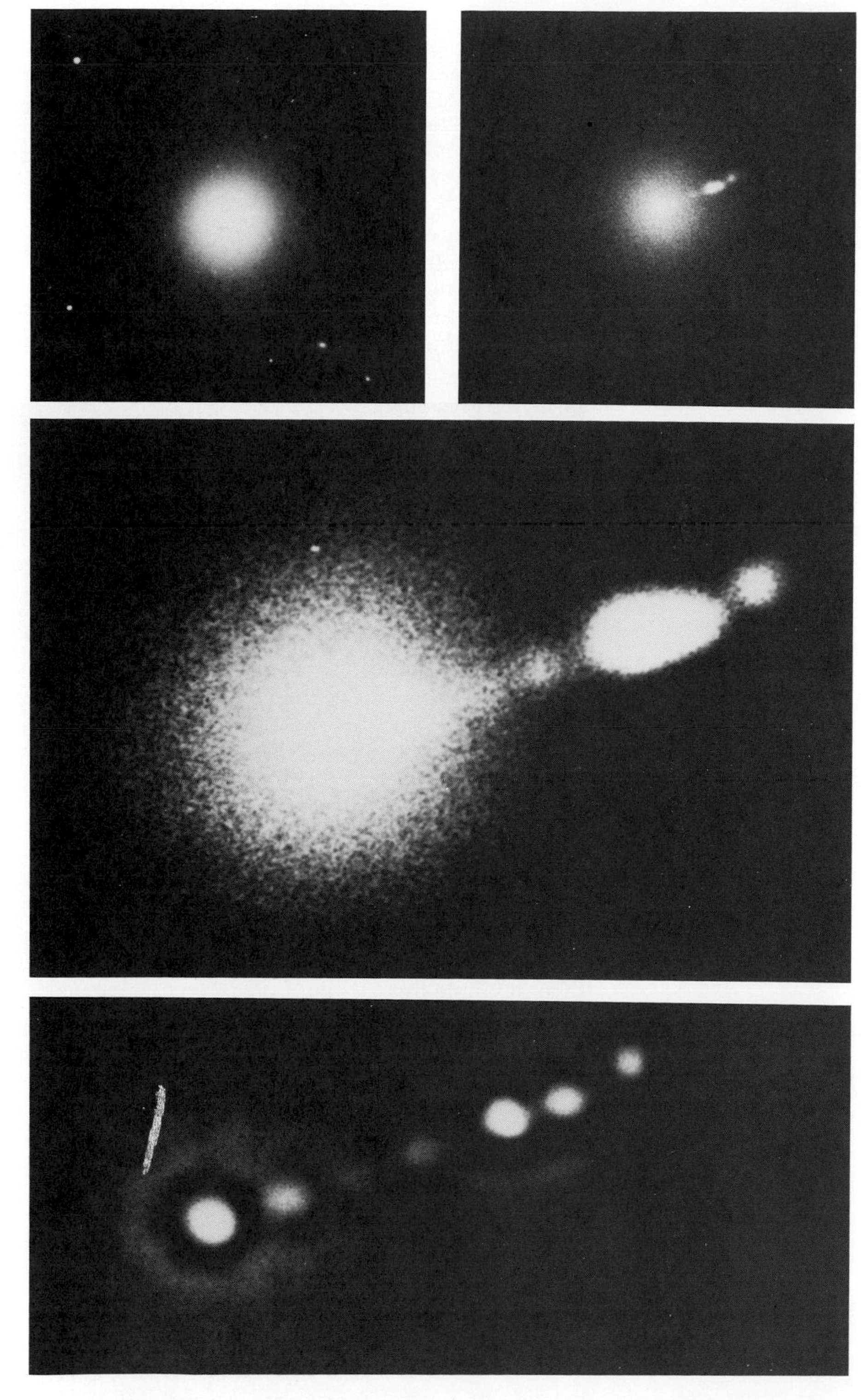

Figure 7-10
Careful, short-exposure photographs of the central regions of M 87 (Figure 5-12) reveal the jet of matter from the galaxy's center (middle) and that this jet consists of several individual knots (bottom).

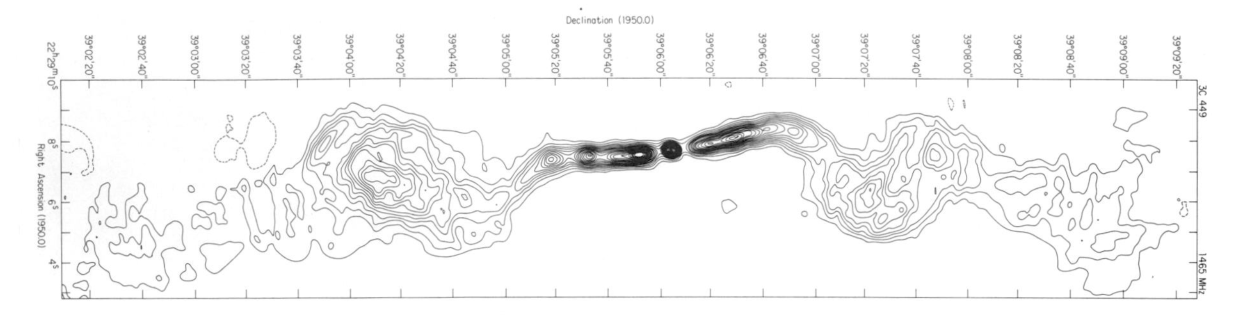

Figure 7-11 A contour map of the radio emission from the region of the radio galaxy 3C 449 shows jets of radio emission on either side of the visible galaxy, which lies close to the region of most intense radio emission.

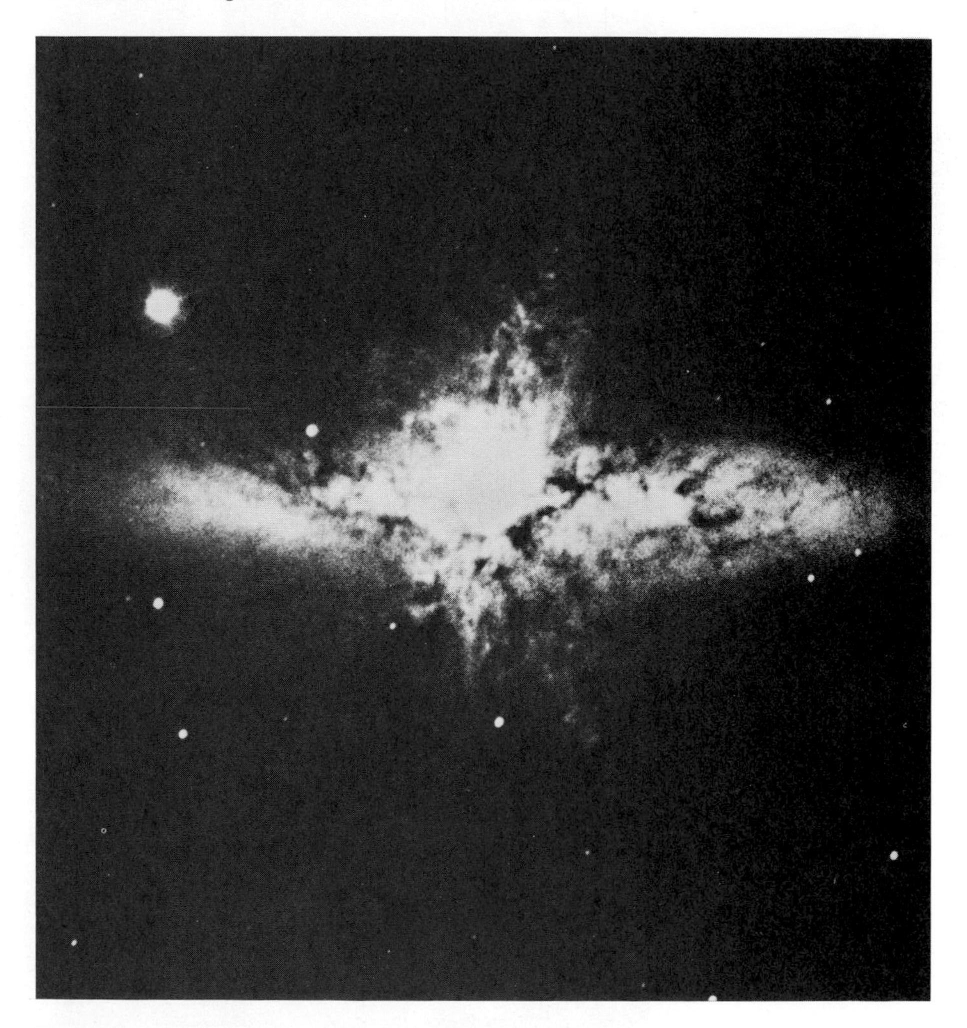

Figure 7-12 The irregular galaxy M 82, about 3 million parsecs away, is a strong source of radio emission, and an even stronger source of infrared radiation. This galaxy contains great amounts of interstellar dust and gas.

coverer, Carl Seyfert (1911–1960). Seyfert galaxies resemble ordinary spirals, but their innermost regions emit tremendous amounts of energy in infrared radiation and radio waves (Figure 7-13). The region that produces this extraordinary emission—often more energy per second than the entire galaxy emits in the form of visible light—has a diameter only a few tens of parsecs across, less than 1% of the diameter of the entire galaxy.

With Seyfert galaxies we again encounter an apparently violent release of energy, in this case localized within the galaxy's center. Our own Milky Way, and perhaps other spirals as well, have strange emission processes within their innermost parts (see page 192), but Seyfert galaxies have a much greater release of energy in their nuclei than any "ordinary" spiral galaxy. Among the thousands of galaxies examined by astronomers, only a few dozen are Seyfert galaxies. Their strangeness, however, is dwarfed by a still more spectacular class of objects, the quasi-stellar objects, or **quasars.**

Figure 7-13 The Seyfert galaxy NGC 4151 (center) emits great amounts of radio and infrared power from its innermost 100 parsecs.

7.4 Quasars

The development of radio astronomy during the 1950s opened a new spectral domain. Astronomers quickly found entire new classes of objects, unfamiliar to them from visible-light studies, but strong sources of radio emission. They saw the need to make complete surveys of the *radio sky*, that is, the cosmos as it appears in radio emission at various frequencies. (Astronomers would later make similar surveys at infrared, ultraviolet, and x-ray frequencies.)

The radio surveys showed pointlike sources of radio emission, which could be compared with the sources of visible light known from extensive photography of the heavens. Upon making this comparison, astronomers found some sources of radio emission that matched nicely in position with peculiar objects in our own galaxy (see page 321) and with peculiar galaxies, such as M 82 and M 87. Some of the locations of radio sources, though, appeared to correspond to quite ordinary-looking stars (Figure 7-14). But when astronomers made careful spectroscopic measurements of these "stars," they reeled in amazement. The spectra looked like no stellar spectra ever seen before.

The Spectra of Quasars

Although the spectra showed some absorption lines reminiscent of most stellar spectra, the frequencies at which the lines appeared differed from those that characterize stars. Moreover, the spectra showed prominent emission lines, the result of intense photon emission at particular energies. Again, the frequencies of the emission lines did not correspond to any frequencies familiar from the study of stellar spectra. Finally, the "stars" emitted far more blue light than yellow light, while most stars show a lower ratio of blue-to-yellow light emission. Astronomers use this ratio, called the **color index,** as a convenient way to classify stars' spectra without making detailed examination. On this basis, the peculiar stars had a much bluer color index than usual.

In 1963, two astronomers at the California Institute of Technology, Jesse Greenstein and Maarten Schmidt, unlocked the secret of the "blue stellar objects." But their discovery led to a host of greater mysteries. What Greenstein and Schmidt realized was that the spectra of these objects do indeed show a familiar pattern of absorption and emission lines, which in fact arise from such common elements as hydrogen, magnesium, sulfur, and carbon, all well known in stellar atmospheres. But the pattern shows a temendous shift to lower frequencies and longer wavelengths. For the first two objects examined in detail, the shifts amounted to 16% and 37%. That is, all the wavelengths had increased by these percentages, relative to their values as measured in terrestrial laboratories. If we interpret these increases as the result of the Doppler effect (see page 85), then we must conclude that the objects have a tremendous velocity of recession relative to ourselves, equal to 14.5% and 30% of the speed of light, respectively. (See box, page 87.)

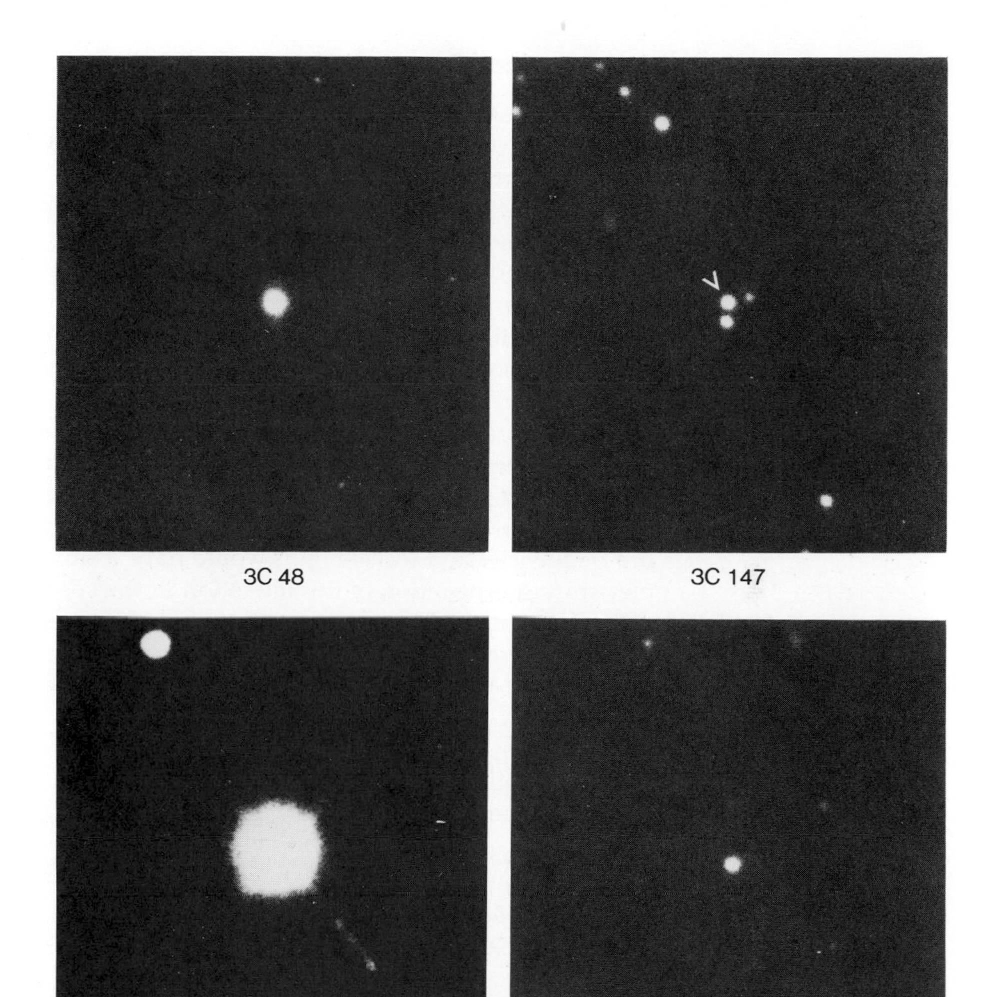

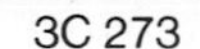

Figure 7-14 Four of the earliest quasars discovered show basically starlike images, although one of them, 3C 273, does have an accompanying jet reminiscent of the galaxy M 87. The designation by 3C and a number comes from the *Third Cambridge Catalogue of Radio Sources.* Since the compilation of the 3C catalogue, more complete 4C and 5C surveys of the sky at radio wavelengths have been made.

Greenstein and Schmidt coined the name *quasi-stellar radio sources* for these peculiar objects. The *quasi-stellar* (like a star) part indicated that these objects look like stars, but are *not* stars. The *radio sources* part indicated that the objects emit great amounts of radio waves. Shortened to QSRS or *quasars,* the name had immediate appeal.

What Causes Quasars' Red Shifts?

The enormous Doppler shifts—if indeed that is the explanation for the shifts to longer wavelengths (or **red shifts**) observed in quasars—imply one of two things about these peculiar objects. Either they owe their great recession velocities to the expansion of the universe, in which case they must have immense distances from us (see page 109), or else they receive tremendous velocities, and always of recession, from some other source.

Five hundred quasars have been discovered in the two decades since Greenstein and Schmidt deciphered their spectra, and every one of them has a spectral red shift, never a **blue shift** to shorter wavelengths and higher frequencies. The largest red shift yet found for any quasar reaches an amazing 3.53, which means that all the wavelengths have had 3.53 times their original values added to them and thus have 4.53 times their original values. Correspondingly, all the frequencies have been decreased to 1/4.53, or 22% of their original values. (We use the word *original* to denote the values that the frequencies and wavelengths would have if no shift occurred.) The Doppler-shift interpretation of this alteration in the spectrum implies a recession velocity equal to 92% of the speed of light! If we assume that the universal expansion produces this Doppler shift, the quasar must be some 5 billion parsecs away—the most distant object yet discovered.

Since quasars have immense distances from us, we are looking far back in time when we observe them—1 billion years for a quasar at 300 million parsecs, 16 billion years for a distance of 5 billion parsecs. Hence, astronomers have speculated that whatever quasars may be, they represent a type of object much more common in the early universe than now. This seems reasonable, but we should note that 1 billion years represents only a small fraction of the age of the universe. Thus, a quasar one billion light years away cannot be described as part of the "early" universe, even though quasars more than 10 billion light years away certainly can be.

If quasars were simply the most distant objects known, with a starlike appearance and tremendous Doppler shifts in their spectra, astronomers would be impressed but hardly mystified. In addition to their strange spectra, however, quasars present still more startling puzzles, not as yet explained: their small sizes and tremendous energy outputs.

The Sizes of Quasars

To understand why we conclude that quasars have small sizes, we must consider the variations in apparent brightness on time scales ranging from a few hours to a few weeks observed for at least some quasars. These changes imply that the primary light-emitting regions within quasars must be millions of times smaller than galaxies such as our own.

Detailed observations show that some quasars can double their visible-light emission within one day! This indicates that whatever region produces this light cannot be more than 1 light day across (the distance light travels in a day, or about 26 billion kilometers). If the light-emitting region had a diameter greater than 1 light day, then even if the brightening occurred in-

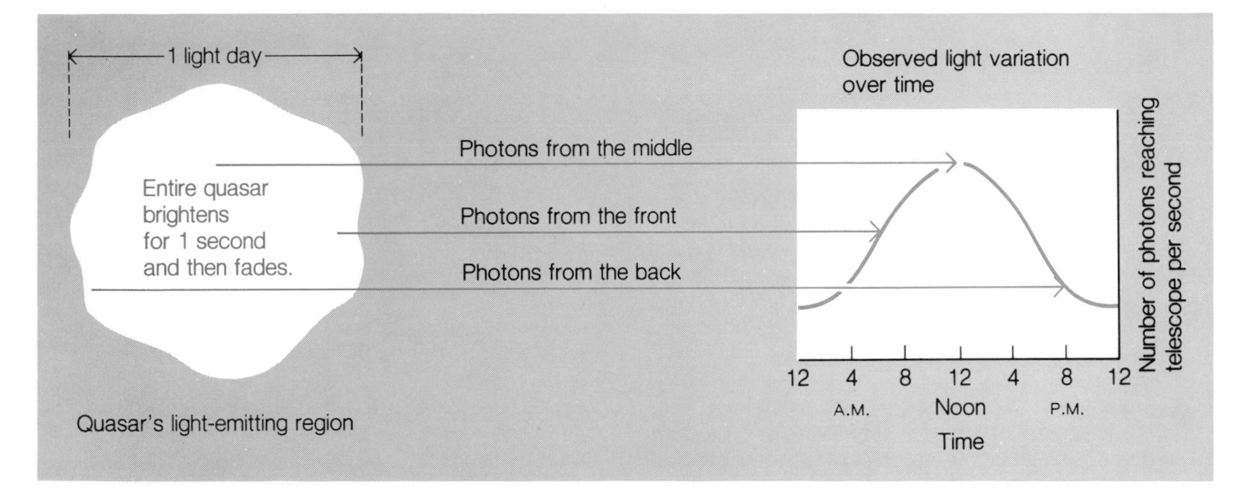

Figure 7-15 If we observe a quasar to vary in brightness in less than a day, we conclude that whatever produces the quasar's emission must be less than 1 light day across. Even if the quasar changed its brightness instantaneously (say, in 1 second), the news of the brightening would take different times to reach us from different parts of the quasar. A size larger than 1 light day would prevent us from seeing any significant brightness change in just one day.

stantaneously, we would see a change in brightness spanning more than one day's time because of the differences in travel time for the light to reach us from various parts of the quasar (Figure 7-15). Since 1 light day equals 1/365 of a light year, or 1/1191 of a parsec, we can see that the quasar's light-emitting region must be fantastically small in comparison with the tens of thousands of parsecs that span a galaxy. Light takes about 6 hours to travel from the sun to Pluto, the farthest planet. Thus, some quasars apparently produce more energy each second than a bright galaxy, and from a region that is at *most* only a few times the size of the solar system!

Energy Output from Quasars

Clearly, the radio and visible-light observations of quasars produced a dilemma for astronomers. They had apparently detected the most distant objects known, with intrinsic brightnesses greater than any galaxy's. But at least some of these objects released their visible-light energy within a tiny fraction of a galaxy's volume.[2] The next problem came when astronomers measured the infrared output of quasars. They discovered several quasars emitting even *more* energy in infrared radiation than in visible light or radio waves! (One such quasar is 3C 273, shown in Figure 7-14.)

[2]The disk of our galaxy, for example, has a volume 10^{21} (or a billion trillion) times larger than the maximum size of the energy-emitting region of some quasars.

Energy and Power

Scientists define **energy** as the capacity to do work, and they define **work** as force times the distance through which the force acts. (See box on forces, page 250.) To lift a 1-kilogram mass requires a certain force; to lift the mass 2 meters high requires twice as much work as to lift it 1 meter. To perform twice as much work requires twice as much energy. The rate at which energy is generated (energy per second) is called **power.**

The scientific world measures energy in joules and ergs (1 joule = 10^7 ergs). One joule is the energy needed to accelerate a 2-kilogram mass from zero to a velocity of 1 meter per second. This acceleration gives the mass a kinetic energy, or energy of motion, equal to 1 joule. But kinetic energy is not the only form of energy. A 1-kilogram mass, for example, has an enormous energy of mass: 9×10^{16} joules. This energy of mass represents the kinetic energy that would be liberated upon the complete annihilation of the 1-kilogram mass.

The sun and other stars produce kinetic energy from energy of mass. The *rate* at which this conversion occurs gives the power of the star. In the sun, for example, 4.4×10^9 kilograms of mass disappear each second, producing a kinetic energy of 4×10^{26} joules. One joule per second is called a watt, so the sun's power is 4×10^{26} watts.

Astronomers often give the power of a given object in ergs per second, rather than in watts. (There is no special name for 1 erg per second, as there is for 1 joule per second.) In this system, the sun has a power of 4×10^{33} ergs per second. A 40-watt light bulb nominally has a power output of 40 joules per second, or 4×10^8 ergs per second. Thus, the light bulb produces just one part in 10^{25} of the sun's power output. At the other end of the power scale, the most impressive quasars apparently produce 10^{50} ergs per second, or 10^{43} watts, and thus have a power output that is more than 10^{16} times the sun's.

For the past fifteen years, astronomers have worked toward understanding how quasars emit so much energy, and from such small volumes of space, but they have fallen short of success. No theory exists that explains in any detail how a tiny volume can produce so much energy. Exploding stars by the millions have been suggested (see page 315); black holes with matter falling in have been suggested (see page 346); matter-antimatter annihilation has been suggested. But put to the hard test of calculations, these suggestions fail to produce completely viable models for quasars.

Similarities Among Quasars, Seyfert Galaxies, and Exploding Galaxies

If we cannot explain how quasars work, at least we can take some comfort from the general resemblance among Seyfert galaxies, exploding galaxies, and quasars. Like Seyfert galaxies, quasars emit great amounts of infrared and radio waves from a small volume. In fact, in some ways quasars resemble superactive, extremely small Seyfert nuclei. Like exploding galaxies, quasars emit radio waves. In at least one case (3C 273), we can see a jet connected with the quasar, presumably emerging from it, which itself produces radio emission (Figure 7-14).

Are Quasars Really the Most Distant Objects?

We have assumed that we can derive quasars' distances from the red shifts we measure in their spectra. This presupposes that the Doppler effect arising from the universal expansion has increased all wavelengths in these spectra by the same fractional amount. Even though this assumption results in immense distances to quasars, we have no better way to explain the spectral red shifts. Most astronomers favor the expanding universe Doppler red shift as the explanation for the quasars' spectra, but a minority of astronomers disagree and assign much smaller distances to quasars, placing them no farther away than many "ordinary" galaxies.

Arp's Theory of Peculiar Galaxies. An American astronomer from the Mount Palomar Observatory, Halton Arp, has provided most of the evidence suggesting that quasars are not billions of parsecs away. In examining many peculiar galaxies and their immediate surroundings, Arp found that such galaxies appear to have more quasars located close to them on the sky than we would expect by the workings of pure chance. For example, the galaxy NGC 1073 has three quasars within 30 seconds of arc of its center (Figure 7-16). This alignment on the sky of the galaxy and the quasars could be accidental, however. In that case, the quasars' distances from us should be about ten times the galaxy's distance, since the quasars' spectra show red shifts ten times that of NGC 1073.

Arp has found many other examples of quasars lying close to galaxies on the sky. He argues that chance alignments cannot explain the large number of cases he finds, and concludes that the quasars must be close to peculiar galaxies in space and thus at approximately the same distance as the galaxies from us. If this is true, then the quasars' spectral red shifts cannot arise from the expansion of the universe. They would be unlikely to arise from the Doppler effect at all, since we would expect to see almost as many blue shifts as red shifts if quasars are somehow shot from galaxies like popcorn. Besides, what process could accelerate quasars to almost the speed of light (in some cases) while keeping them so small?

Arp's hypothesis does have one advantage. If quasars are 10 or 100 times closer than we think, then their absolute brightnesses must be 100 to 10,000 times less than we think, because a given object appears 100 times brighter if it moves 10 times closer to us. This helps with the problem of explaining how quasars radiate energy (though hardly explaining all of it), but leaves a still greater mystery to be resolved, the quasars' red shifts.

Reaction to Arp's Theory. Why do most astronomers reject Arp's hypothesis of relatively small distances to quasars? Statistical arguments such as Arp's must rely on an "unbiased sample," a set of cases not selected for any particular characteristic, in order to be convincing. (Notice that we are describing the sample, not the scientist, as unbiased! The scientist is free to have a bias, so long as the scientific part of the investigation follows well-defined rules.) Many astronomers argue that Arp has concentrated on find-

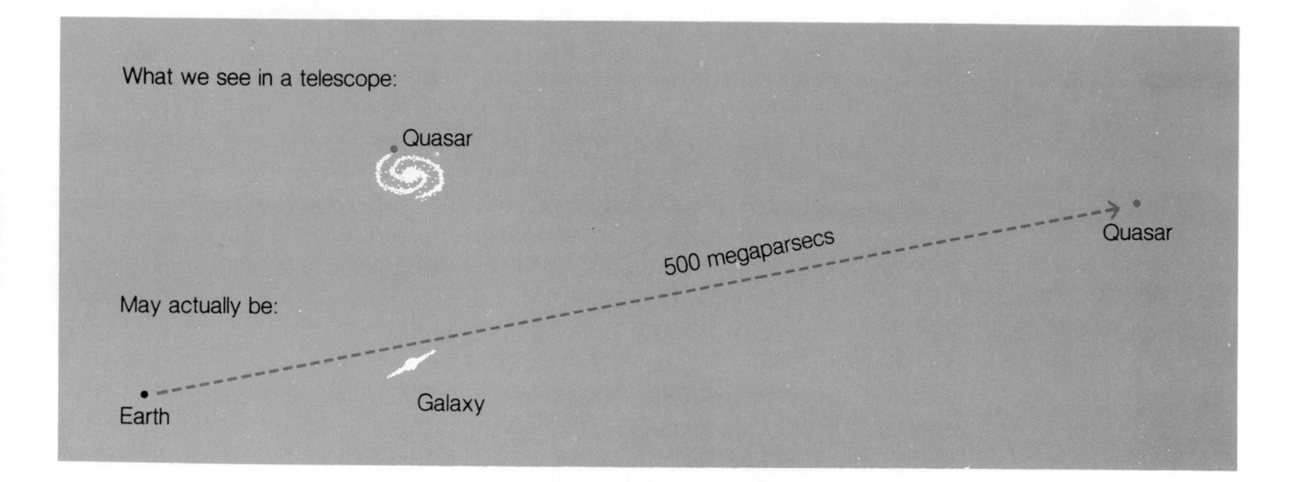

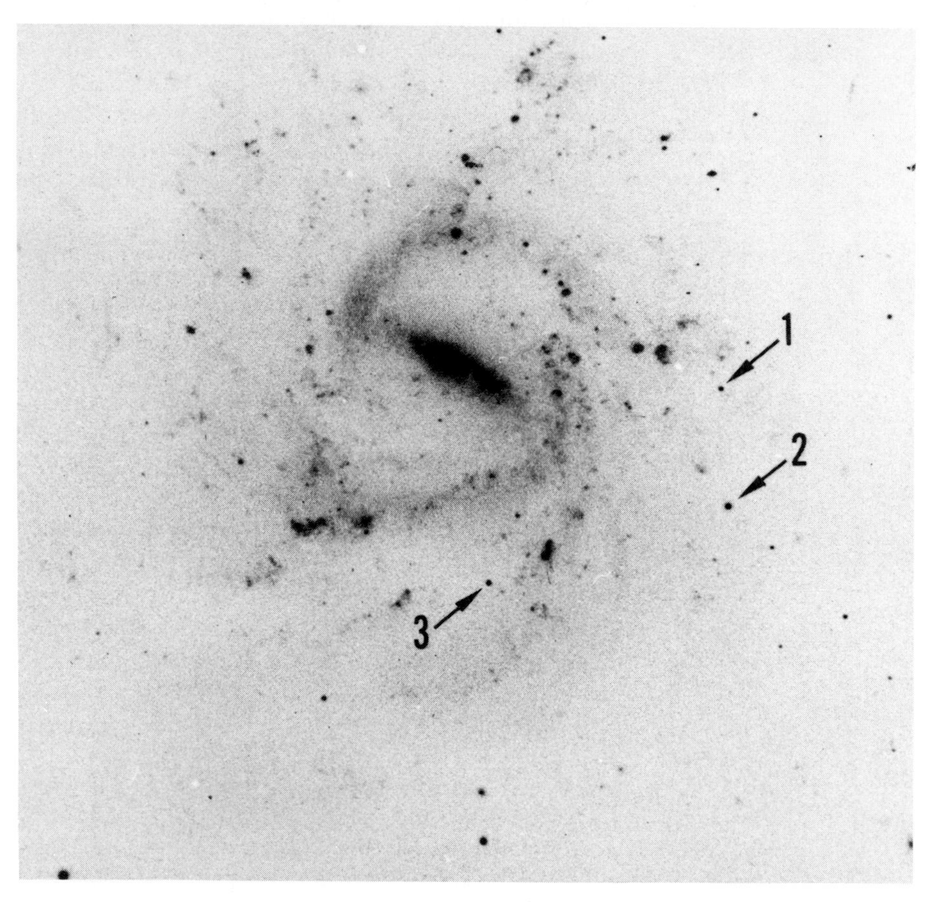

Figure 7-16 This negative print of a barred spiral galaxy shows three quasars (arrows) quite close to the galaxy. The top part of the illustration shows that we might be seeing a chance line-up of the galaxy and much more distant quasars.

ing quasars close to peculiar galaxies. They contend that if he made an equal effort for other galaxies, or for areas of the sky selected at random, he would find quasars there too, in about the same numbers as he finds close to peculiar galaxies. Arp still maintains, however, that he can prove statistically that more quasars appear close to peculiar galaxies than can reasonably be accounted for on the basis of chance.

There the disagreement rests, unresolved for more than ten years. Arp may be right, but if he is, quasars' red shifts represent a puzzle in astrophysics so striking as to be capable of shaking the pillars of modern science. He may be wrong (and most say that he is), but if so, quasars still present the enigma of a huge energy output from a small region. Improved observations of quasars and their surroundings offer the best hope of resolving the quasar dilemma. Here, too, astronomers look forward to markedly better observations from the Space Telescope, which will be launched in the mid-1980s and which will make more detailed observations of quasars than ever before possible.

The Double Quasar

In 1979, astronomers found an intriguing **double quasar** that may help clear away some of the mystery that surrounds quasars. This object, with the name 0957 + 561 A,B (little more than a specification of the object's position on the sky and a reference to two sources of radio emission), in fact consists of two

Figure 7-17 The double quasar 0957 + 561 A, B appears as two points of light less than 6 seconds of arc apart.

objects of almost equal brightness, separated by 5.7 seconds of arc on the sky (Figure 7-17). Spectral analysis of the objects revealed that they have almost identical spectra, with the same pattern of emission and absorption lines, and—most important of all—with almost the same red shift. The increase in wavelengths for objects A and B equals 139.13% and 139.12%, respectively. If we interpret these red shifts as the result of the Doppler effect, then the difference in recession velocity of the two objects amounts to no more than 15 kilometers per second, for objects that are moving away from us at 70% of the speed of light, or 210,000 kilometers per second!

What can we make of such a pair of matched quasi-stellar objects? We shall have difficulty explaining them as two separate objects unless we can imagine some process that makes near-perfect duplicates. The most favored explanation among astronomers, at least at the present time, turns out to be a mechanism that partakes of Einstein's general theory of relativity: a gravitational lens!

Gravitational Bending of Light. In 1916, Albert Einstein published a theory of space, time, and matter that predicted a *gravitational bending of light*. According to this theory, light rays passing close by a massive object would be bent toward it (Figure 7-18). In other words, gravity affects light, which

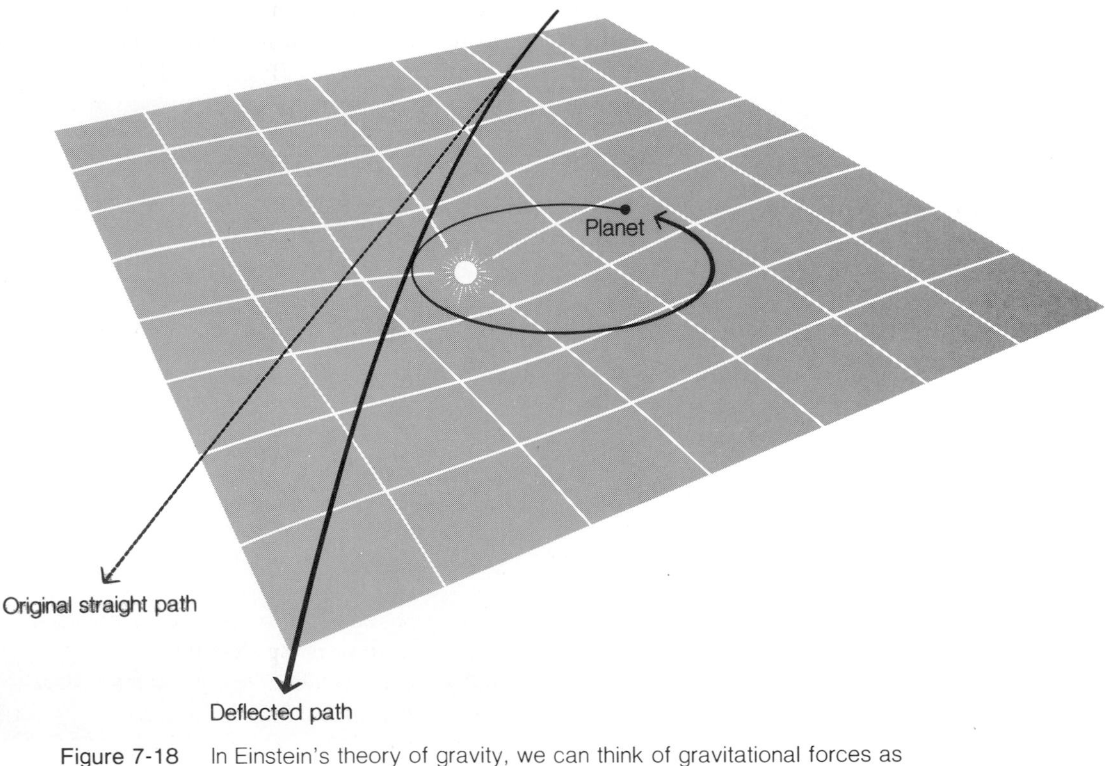

Figure 7-18 In Einstein's theory of gravity, we can think of gravitational forces as producing a curvature of space. The sun's gravitational force thus makes light rays passing close by it deviate from straight-line trajectories.

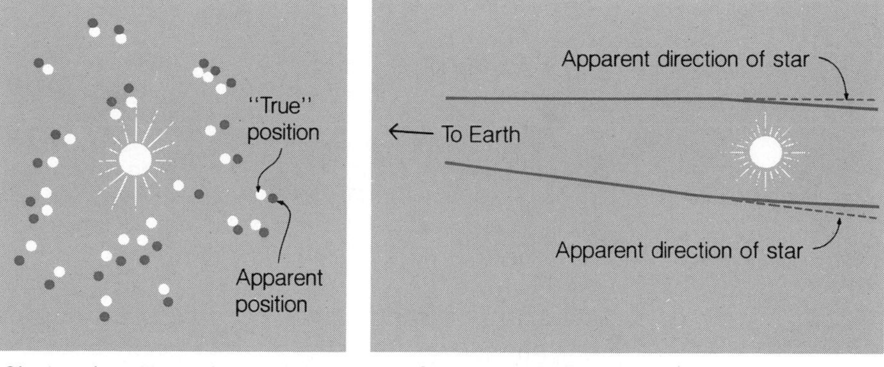

Observed positions of stars during a total eclipse of the sun

Side view: bending and scale exaggerated

Figure 7-19 Confirmation of Einstein's theory came when astronomers observed the stars near the sun at the time of a total solar eclipse. These stars had slightly different positions, relative to one another, than they did at a time when the sun no longer lay near our line of sight to the stars. The difference arises from the gravitational deflection of starlight passing by the sun.

consists of photons that have no mass, as well as particles that do have mass. Einstein showed that we may indeed think of gravitational forces as able to curve space, producing a greater curvature in regions of stronger force and a lesser curvature where the force decreases (Figure 7-18). Einstein's theory received dramatic confirmation during an eclipse of the sun in 1919, when astronomers found that the positions of stars visible during the total eclipse differed from the stars' relative positions when the sun no longer appeared among them (Figure 7-19). Today, much more accurate measurements of the bending of radio waves from quasars provide still better confirmation of Einstein's theory.

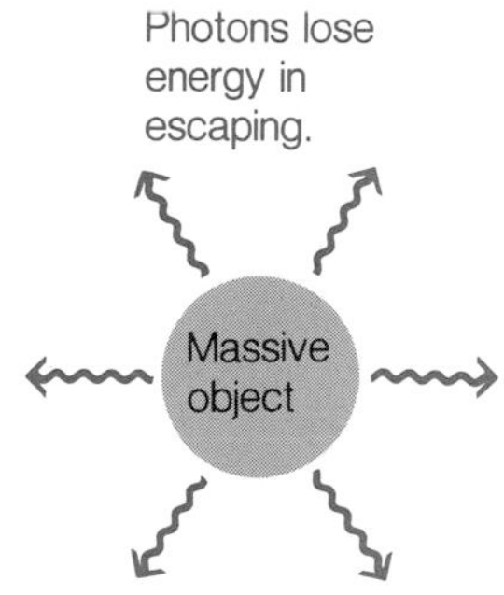

Figure 7-20
Although photons always escape from an object at the speed of light, they lose some energy in escaping. This energy loss appears as a decrease in the photons' frequencies and a corresponding increase in the photons' wavelengths.

Another part of the same theory predicts that photons must lose some energy in escaping from a massive object and that stronger gravitational fields will cause a greater fractional decrease in the energy of escaping photons (Figure 7-20). In extreme cases, the photons cannot escape at all, in which case we have a black hole (see page 340).

The decrease in photon energy caused by gravity, or **gravitational red shift,** might seem to offer a resolution to the puzzle of quasars' spectra. A sufficiently massive object could mimic the effects of the Doppler red shift, since all photons lose energy (and thus lower their frequencies and increase their wavelengths) by the same fractional amount as they escape from the same object. Calculations show, however, that an object capable of producing red shifts as large as those found in quasars' spectra while radiating as much energy as quasars do (even if they are as close to us as galaxies) must collapse to form a black hole in a matter of seconds. Hence, the gravitational-red shift explanation seems untenable, although we cannot rule it out completely. What the effect of gravity on light can explain, however, is a double source such as that found in the double quasar.

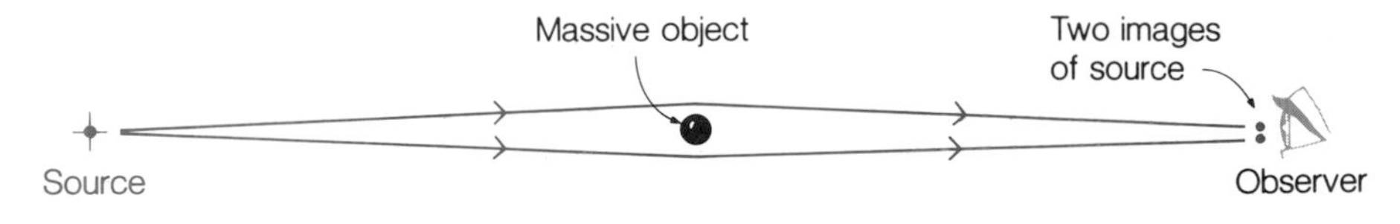

Figure 7-21 If a small, extremely massive object lies almost directly between ourselves and a source of photons, the object can bend the light waves and radio waves around either side. Then we may see a double image, as if the single source had become two almost identical sources.

The Gravitational Lens. Consider the situation shown in Figure 7-21. A small, massive object lies almost directly between ourselves and a source of light and radio waves. If we look past the object, which itself may be entirely invisible, to see the source of radiation, we may observe *two* sources, even though only one actually exists. Light and radio waves from the source passing on either side of the small object will be focused—bent back toward us from both sides of the object. If the intervening object lies extremely close to the straight line between ourselves and the source, we may see two sources of almost equal intensity, and the spectra of the "two sources" will be identical, since in fact we observe the same source twice.

Figure 7-22
The triple quasar shows three images within a region 3 seconds of arc in diameter.

This **gravitational lens** hypothesis seems to be the best way to explain the double quasar. Astronomers think they have found a dim galaxy between the two components, the presumed source of gravitational focusing.

One year after the discovery of the double quasar, astronomers found a *triple* quasar composed of three closely spaced images. Like the double quasar, the triple quasar apparently arises from a gravitational lens. Light from a single object is bent by a lens along three separate trajectories to produce three images in our field of view (Figure 7-22).

Summary

Some galaxies emit far more radio waves than most, apparently the result of violent events within the centers of these objects. Much of this radio emission arises from the synchrotron process, which produces photons when charged particles accelerate at almost the speed of light in magnetic fields. The detection of synchrotron emission from radio galaxies signals the presence of magnetic fields, as well as the existence of something capable of accelerating vast numbers of charged particles to near-light velocities.

Many radio galaxies (those that emit at least as much power in the form of radio waves as in visible light) produce much of their radio emission in two regions far outside the visible galaxy, but symmetrically located on either side of the galactic center. Other radio galaxies show more centralized outbursts, such as the jet in the middle of the elliptical galaxy M 87. Radio galaxies often produce huge numbers of infrared photons, a characteristic they share with Seyfert galaxies, which show intense emission of visible-light, infrared, and radio photons from a small central region.

Quasars resemble radio galaxies in the tremendous output of radio waves that many of them produce; they resemble Seyfert galaxies in the vast amounts of infrared emission they produce. Quasars differ from galaxies, however, in their pointlike appearance and smaller size. Furthermore, the visible-light spectra of quasars show the largest Doppler shifts yet observed. If these Doppler shifts arise from the overall expansion of the universe, then quasars are the most distant family of objects known and the most powerful sources of photon energy. We cannot satisfactorily explain how quasars can produce so much energy, and we have still more difficulty understanding their rapid variations in energy output. These variations imply that the energy-producing regions of at least some quasars are no greater in size than our solar system, a minuscule fraction of a galaxy's extent.

Many quasars appear on the sky close to peculiar-looking galaxies. The galaxies have much smaller Doppler red shifts than the quasars, so if the quasars have a physical association with the galaxies, their enormous red shifts cannot arise from the universal expansion. If we accept this theory, then we can no longer explain what causes the large Doppler red shifts observed in quasars' spectra. However, this hypothesis has the advantage that the distances to quasars would then be hundreds of times less than we estimated, and the energy outputs of quasars would be tens of thousands of times smaller than those derived on the assumption that the red shifts arise from the expansion of the universe. Most astronomers, however, believe that the preponderance of evidence favors the hypothesis that quasars indeed have the immense distances and fantastic power outputs implied by their red shifts, taken as the result of the universal expansion.

Key Terms

blue shift
color index
continuum emission
double quasar
energy
exploding galaxy
gravitational lens
gravitational red shift
line emission
nonthermal radiation
power
quasar
radio astronomy
radio galaxy
red shift
Seyfert galaxy
synchrotron radiation (emission)
thermal emission
work

Questions

1. What is a radio galaxy?
2. Why don't most galaxies produce large amounts of radio emission?

3. What is thermal emission? In what way does this kind of emission depend on the temperature of the emitting object?
4. What is synchrotron emission? How does it differ from thermal emission?
5. What can we conclude about an astronomical object that is emitting synchrotron radiation?
6. How can we explain the fact that radio galaxies often show much of their radio emission from regions located far outside the visible galaxy?
7. What is a Seyfert galaxy? How does it resemble a radio galaxy? How is it different from a radio galaxy?
8. What is a quasar? In what ways do quasars resemble stars? In what ways do they differ from stars?
9. What evidence suggests that quasars are the most distant objects yet observed? What evidence suggests that quasars are no more distant than certain peculiar galaxies?
10. Which interpretation of quasar distances best explains their visible-light spectra? What does the competing interpretation say about these spectra?
11. Which interpretation of quasar distances raises greater problems in explaining the energy output of quasars? Why?
12. What is a gravitational lens? How does this phenomenon help explain the double quasar?
13. The largest known radio galaxy, 3C 236, has two main components of radio emission separated by 5 megaparsecs. If this galaxy's radio emission arises from particles that were shot out from the galaxy between the components at almost the speed of light, what is the minimum time that must have elapsed since the particles were ejected?
14. The quasars 3C 147 and 3C 196 show red shifts in their spectra of 0.55 and 0.87, respectively. All the wavelengths have been increased by 55% and 87%, respectively, over the values they would have if no red shift existed. How have the frequencies of the spectra changed? If these red shifts arise from the Doppler effect caused by the universal expansion, which quasar is farther away? By how much?

Projects

1. *Radio Waves.* Examine the dials of an AM and an FM radio to determine which frequencies of radio waves they receive. Why is it so hard to find out from the radio itself? Does this tell us anything about popular understanding of technology? Consider FM radio waves with a frequency of 100 megahertz and AM radio waves with a frequency of 1000 kilohertz (10^8 and 10^6 cycles per second, respectively). Calculate the wavelengths of these radio waves, using the fact that the frequency times the wavelength always equals the speed of light (3×10^{10} centimeters per second). How do these wavelengths compare with the sizes of radio and television antennas? Is this simply a coincidence?

2. *Angular Resolution.* If we observe a source of photons that produces visible light, radio waves, and other kinds of photons, we will obtain a different "picture" or image when we observe at different photon frequencies and wavelengths. This is partly because the source emits different amounts of photons from different parts of the source at various frequencies. But even if the source emits the same number of photons at all frequencies, our image would change as we changed our observing frequency, because our ability to "see" the source changes with frequency. Any detector of photons—the human eye, a telescope and photographic plate, a radio antenna, an x-ray detector—faces an inherent "best" limit on its ability to resolve small angles, that is, its ability to detect the existence of two closely spaced sources as individual points of emission rather than as one single source. This limit varies in direct proportion to the wavelength of the photons being detected, and inversely as the size of the antenna or detector. Compare the theoretical "best limit" of the human eye, with a diameter of 1 centimeter for its detector, with a radio antenna of 100-meter diameter, if the eye detects photons of 5×10^{-5}-centimeter wavelength and the radio antenna and detector observe photons of 5-centimeter wavelength. Which theoretical "best limit" gives better angular resolution? What does this tell us about the ability of radio telescopes to "see" fine details on the sky? What solutions can you suggest to improve the ability of radio telescopes to "resolve" smaller angles on the sky, that is, to discriminate among closely spaced sources of radio emission?

3. *The Energy of a Snowflake.* A "strong" source of cosmic radio emission produces huge amounts of radio energy, but only a tiny fraction reaches the Earth, and a still smaller fraction enters the radio telescopes we point to the skies. The most intense radio galaxy, Cygnus A, emits enough radio power that on Earth, a billion parsecs away, each square meter of surface pointed at Cygnus A receives about 10^{-7} erg per second of radio energy. Suppose that the largest radio-astronomy antenna, the 300-meter-diameter dish at Arecibo, Puerto Rico, observes Cygnus A. How much time must pass before the antenna collects as much kinetic energy from Cygnus A as it would from a falling snowflake (1 erg)? (Of course, Puerto Rican snowflakes are quite rare.) What does this tell us about the difficulty of observing radio sources much weaker than Cygnus A?

4. *Arguments by Probabilities.* Part of the argument about the distances to quasars hinges on the question of whether certain observed "line-ups" of quasars and peculiar galaxies could be the chance result of our particular line of sight to a galaxy and a much more distant quasar, or whether such line-ups are much too improbable. Discuss the use of our assessment of probability and improbability in daily life. For example, if the chance of an earthquake in California is 1 in 100 per year, what effect does this have on the people who live in California? If you were told that the chance of a serious earthquake *this* year is 1 in 4, and you

believed this, would it change your attitude toward living in California? If you were told that the chance of an earthquake in Maine is 1 in 1000, and an earthquake occurred, could you *prove* that the prediction was wrong? Would you accept *any* such statement of probabilities (1 in 10, 1 in a million, and so on) as valid once an earthquake had occurred?

Further Reading

Burbidge, Geoffrey, and Hoyle, Fred. 1970. "The Problem of the Quasi-stellar objects." In *New Frontiers in Astronomy*. Edited by Owen Gingerich. San Francisco: W. H. Freeman.

Disney, Michael, and Veron, Philippe. 1977. "BL Lacertae Objects." *Scientific American* 237:2, 32.

Field, George; Arp, Halton; and Bahcall, John. 1973. *The Redshift Controversy*. Menlo Park, Ca: Benjamin/Cummings.

Golden, Fred. 1976. *Quasars, Pulsars, and Black Holes.* New York: Charles Scribner's Sons.

Strom, Robert; Miley, George; and Oort, Jan. 1975. "Giant Radio Galaxies." *Scientific American* 233:2, 26.

Weymann, Ray. 1975. "Seyfert Galaxies." In *New Frontiers in Astronomy*. Edited by Owen Gingerich. San Francisco: W. H. Freeman.

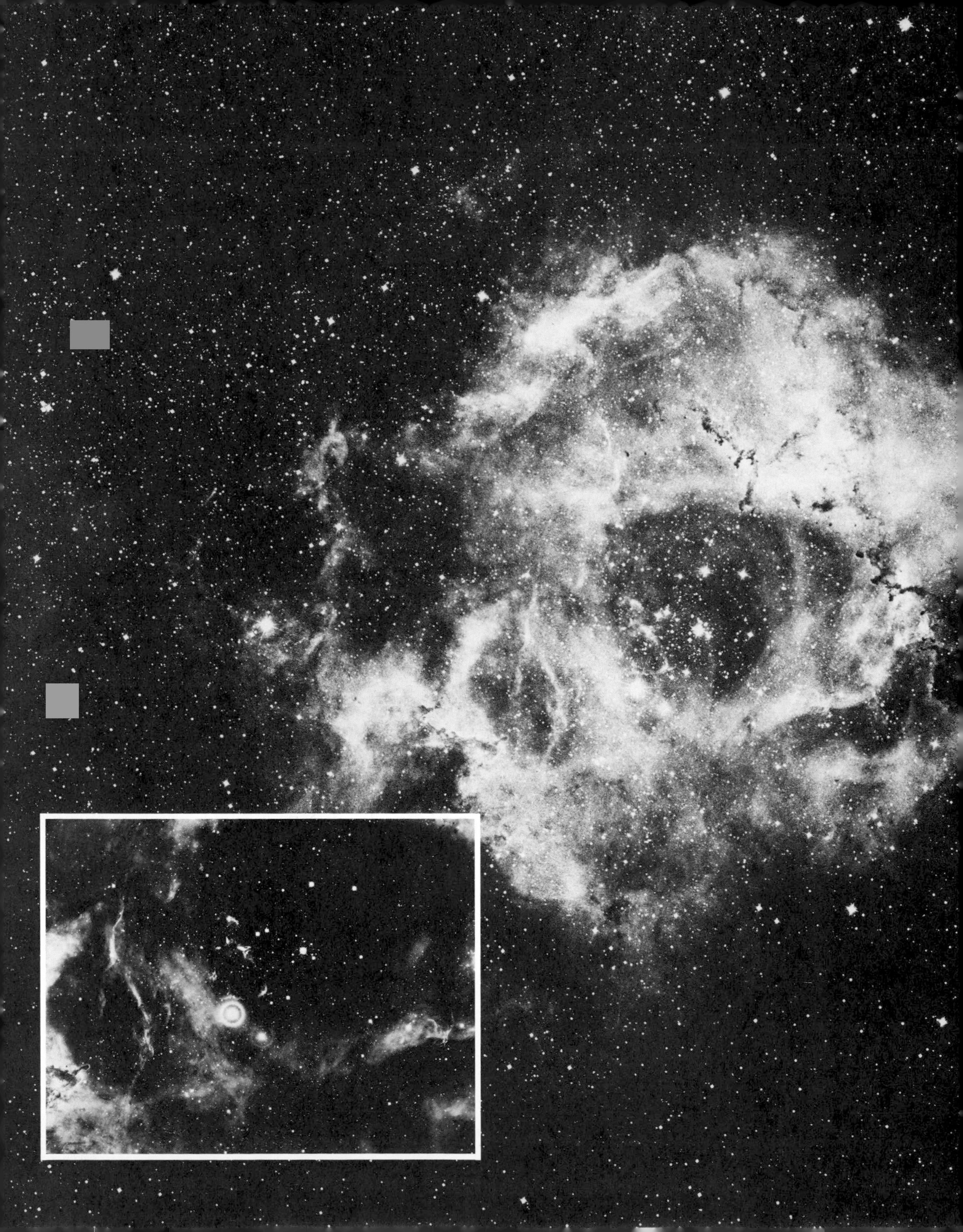

PART 3

Stellar Evolution

Our star, the life-giving sun, has been shining for 4.6 billion years, and will continue to shine for another 5 billion years. During this time, some stars will die while others are born in the Milky Way. The Rosette Nebula, a cloud of gas and dust millions of times larger than the Earth's orbit around the sun, is turning into a small cluster of stars. This process will take only a few hundred million years, a small part of the lifetime of the sun and Earth. Some of the stars formed in the Rosette Nebula will last for 20 million years, others for 20 billion. What causes this difference in stellar lifetime? The answer lies in the mass of each star, which determines how rapidly the star will consume its nuclear fuel. Why do some stars form with more mass than others? As yet we do not know. Perhaps further study of regions like the Rosette Nebula will help us find the answer.

Color Plate K The "diamond ring" effect appears just before a total solar eclipse begins. *Light travel time: 8 minutes, 20 seconds*

Color Plate L During totality, the sun's pearl-white corona shines as brightly as the full moon. *Light travel time: 8 minutes, 20 seconds*

Color Plate M
The Lagoon Nebula in Sagittarius is a place where stars are now forming. *Light travel time: 4000 years*

Color Plate N
Light that reflects from a grating of finely ruled parallel lines reveals the spectrum of component colors (see page 84).

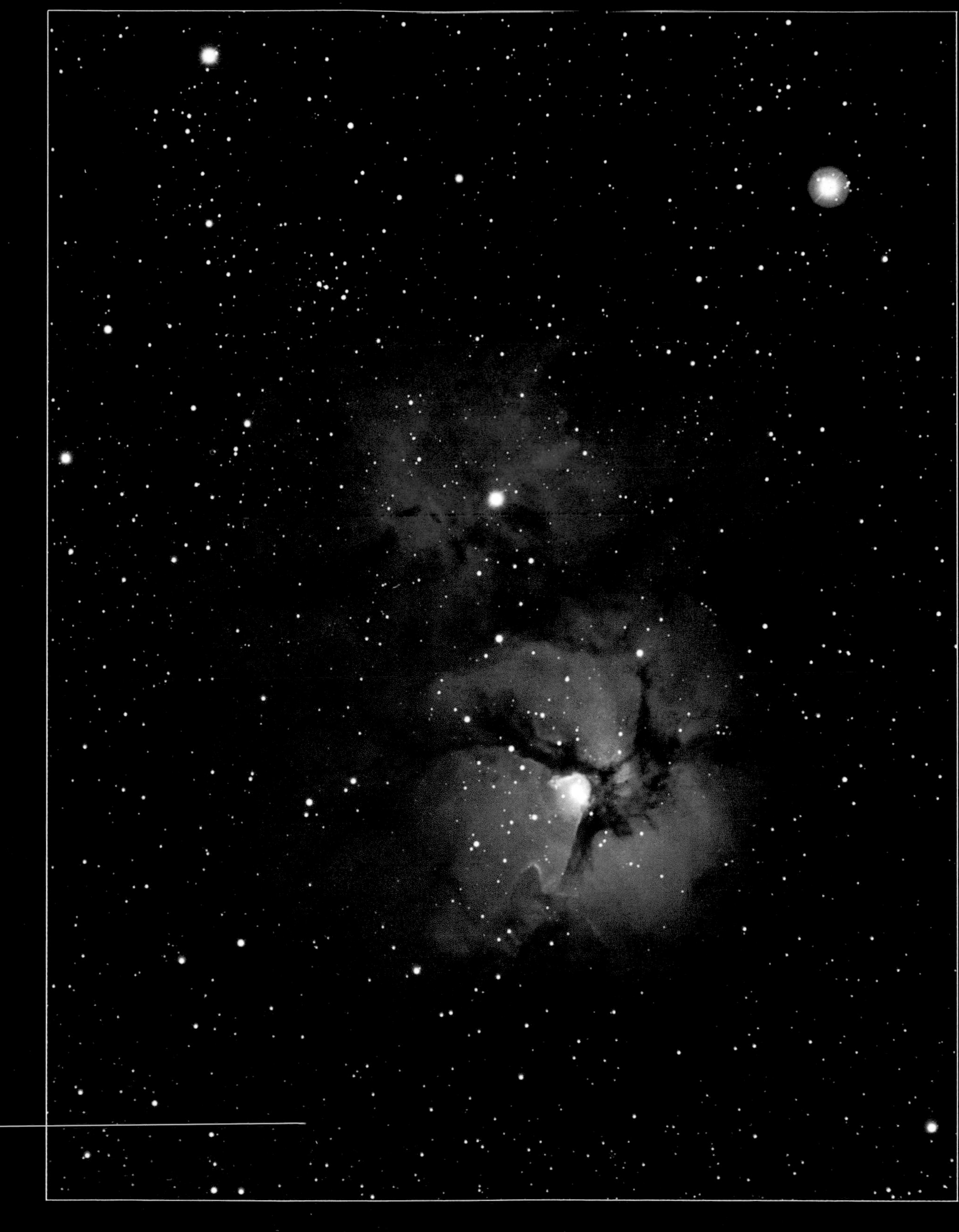

Color Plate O The Trifid Nebula is lit from within by recently born stars. *Light travel time: 3300 years*

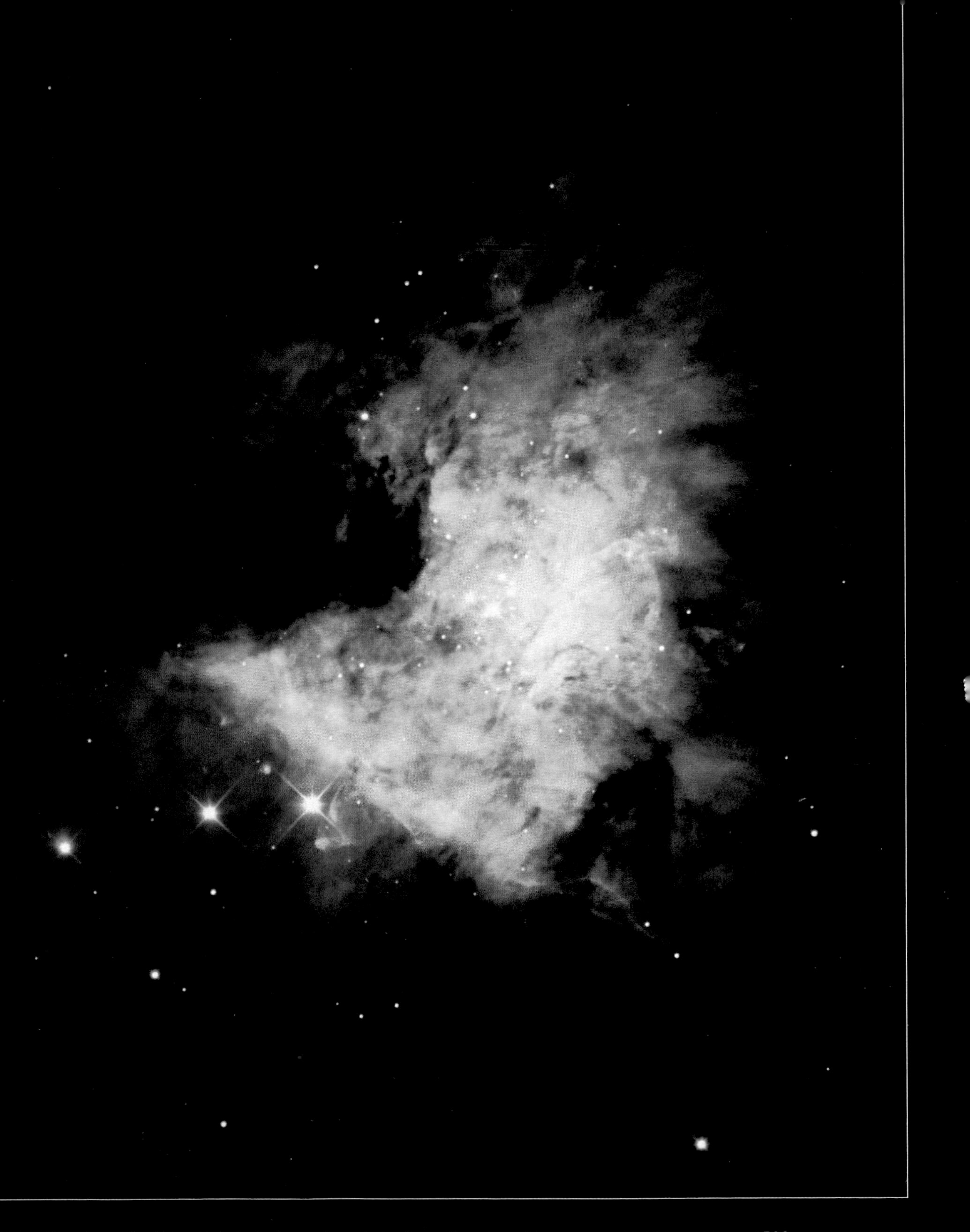

Color Plate P The Orion Nebula is the region of star formation closest to our sun. *Light travel time: 1500 years*

Color Plate Q
The Pleiades form a relatively young star cluster, with some stars as old as 50 million years. *Light travel time: 400 years*

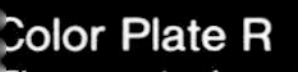

Color Plate R
The central regions of the Milky Way spread across the sky in Sagittarius. *Light travel time: 30,000 years*

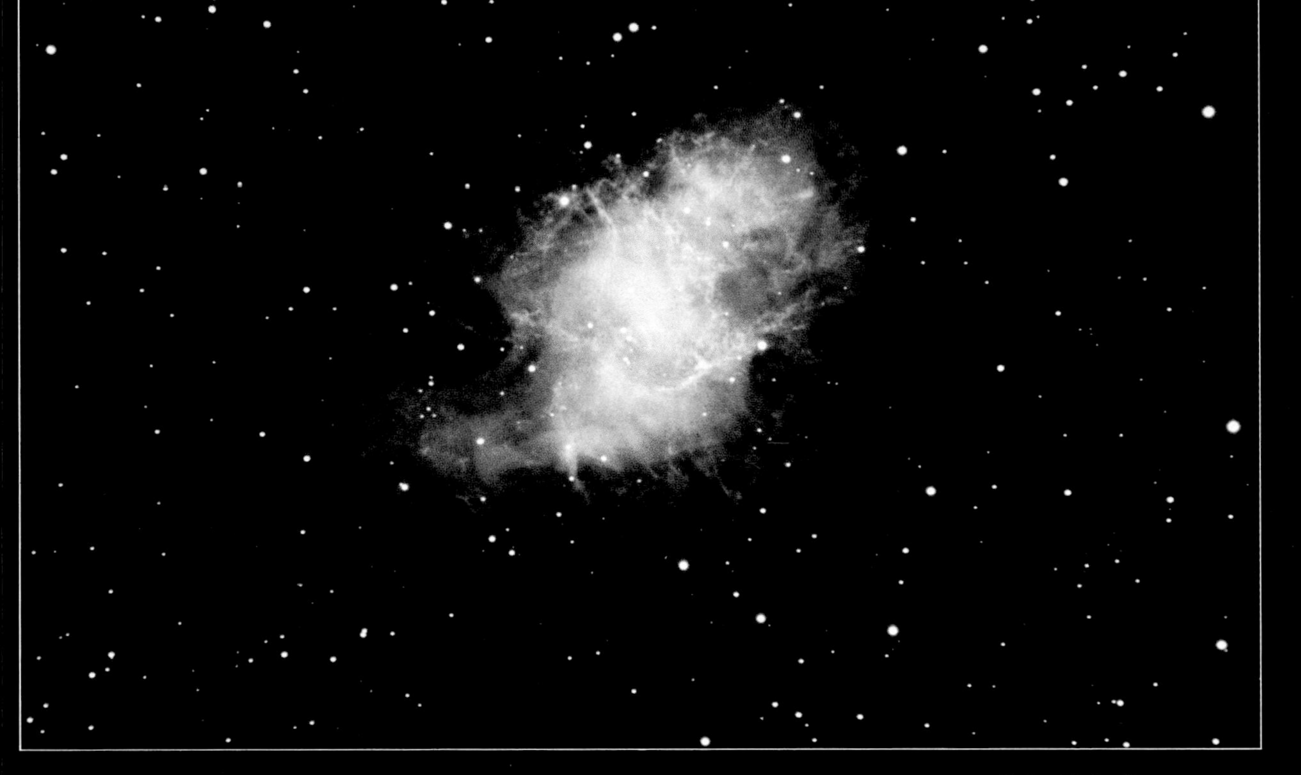

Color Plate S
The Crab Nebula is the remnant of an exploded star. *Light travel time: 5500 years*

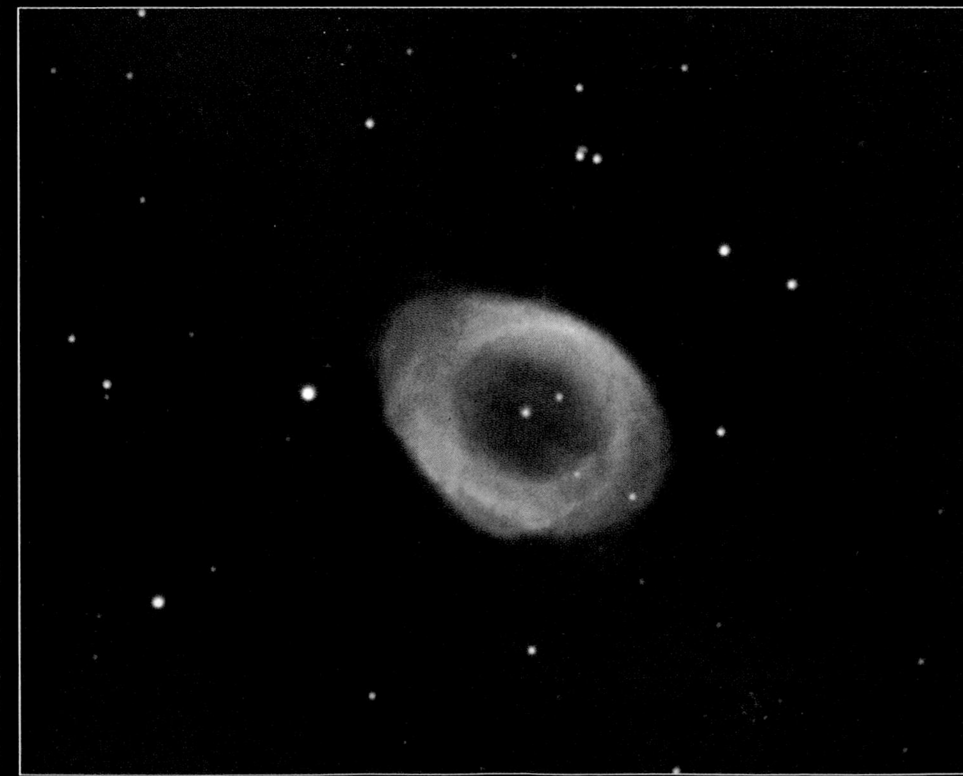

Color Plate T
The Ring Nebula in Lyra surrounds an aging star. *Light travel time: 2300 years*

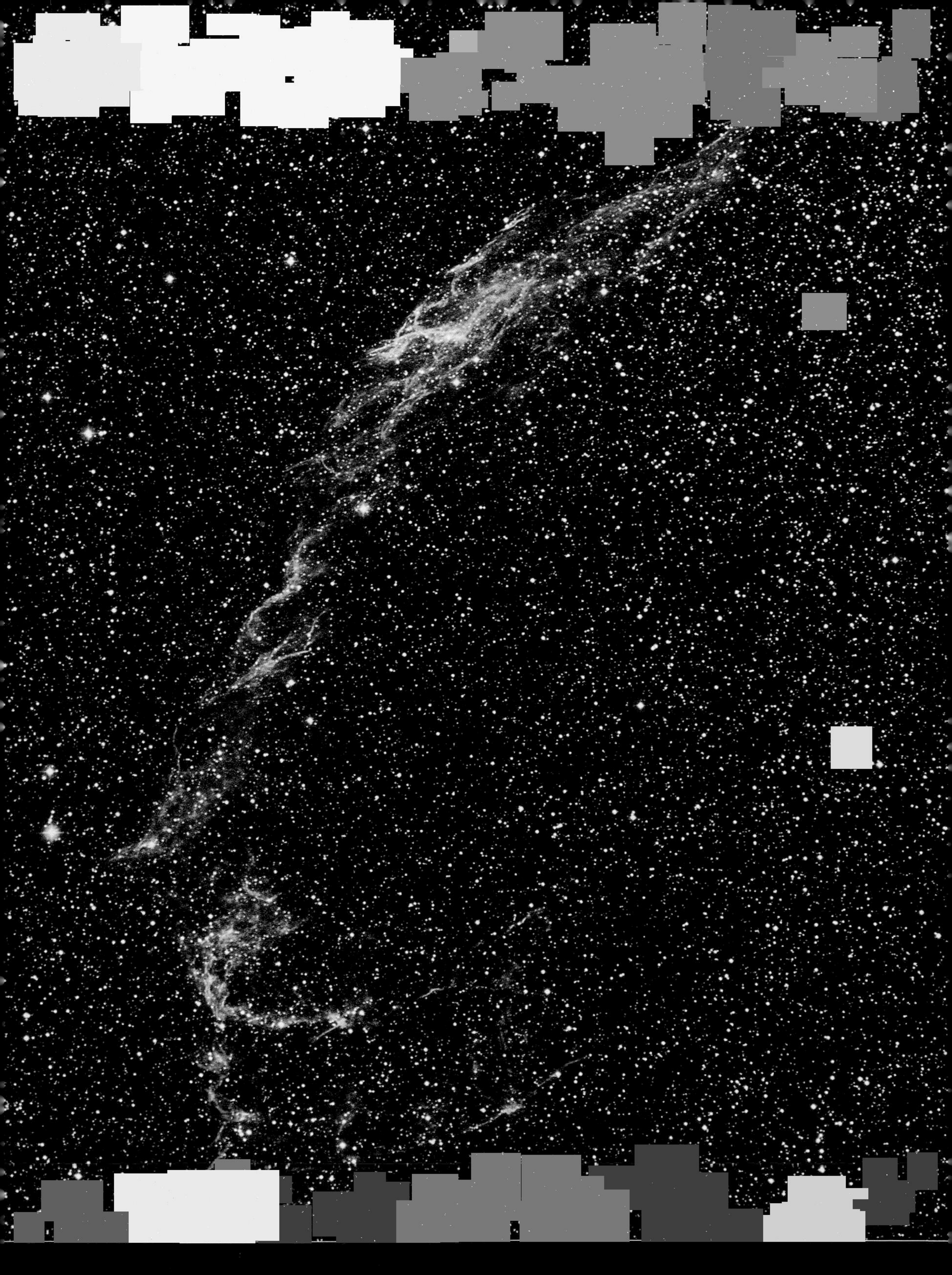

Color Plate U The Veil Nebula in Cygnus represents part of a supernova explosion. *Light travel time: 1600 years.*

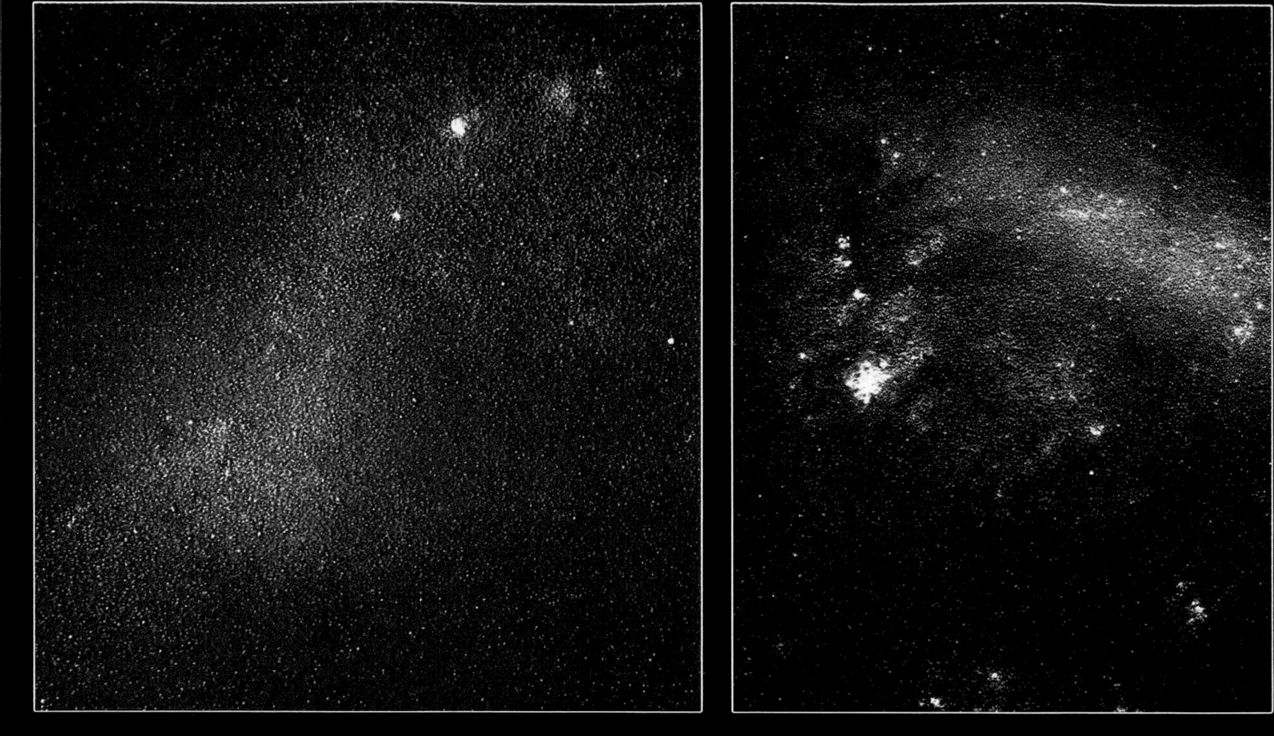

Color Plates V, W The Milky Way's two satellite galaxies, the Large (right) and Small (left) Magellanic Clouds, each contain more than a billion solar masses. *Light travel time: 160,000 years*

Color Plates X, Y The barred spiral galaxy M 83 (left) contains more than a hundred billion stars. *Light travel time: 10 million years.* The peculiar galaxy NGC 5128 (right) is a powerful source of radio emission. *Light travel time: 15 million years*

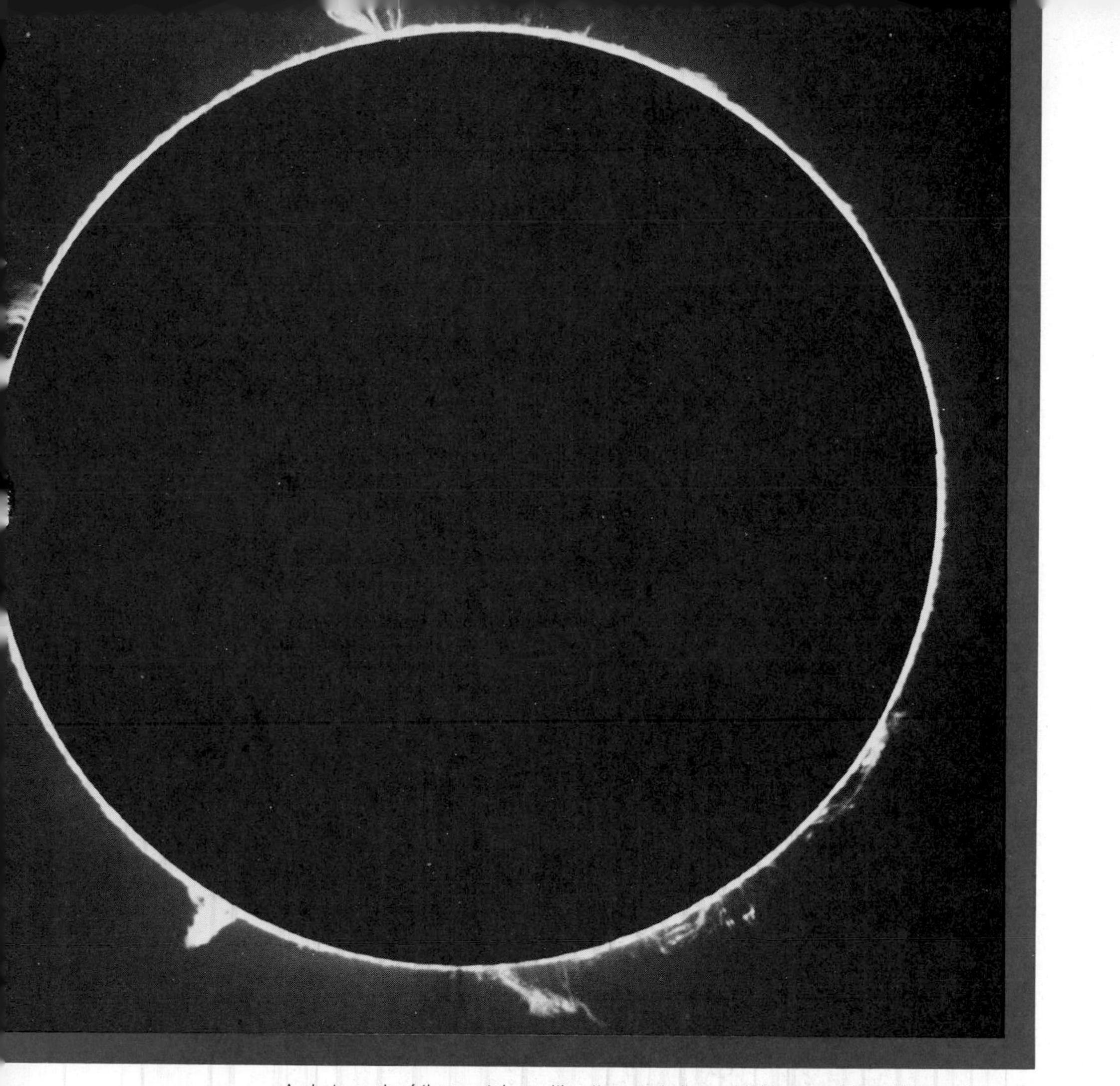

A photograph of the sun taken with a "spectroheliograph" blocks out the light from the sun's disk to show the prominences rising above the surface.

8

Stars: From Gas Clouds to Energy Liberation

When we think of astronomy, we usually think of stars and planets. Our affection for planets has understandable roots: We live on one. Stars, too, deserve respect. Our sun, a typical star, provides us with the energy we need to live. When humans realized that their planetary home is merely a dust speck in the enormous universe, they began to examine the distribution of matter around them and to see that most of the visible material in the universe resides in stars. These compact spheres of gas, each capable of liberating energy by natural processes, form the basic units of the astronomical universe. Stars occasionally group into clusters, and most stars belong to much larger groups called galaxies, but the stars themselves merit direct attention. If we ask what most of the matter in the universe has been doing for the past 10 or 20 billion years, the best reply is that matter has been assembling itself into stars, which then convert some of the matter into new elements. Among other things, this makes our existence possible.

8.1 The Formation of Stars

The clumps of matter that became galaxies in galaxy clusters themselves fragmented into billions of individual pieces that formed stars. As we discussed on page 146, we do not yet know how galaxies and galaxy clusters began to form, and it should come as no surprise that the initial stages of star formation likewise remain shrouded in mystery. But since we can observe stars forming even today, much closer to us than any galaxy (Figure 8-1), we have some well-developed scenarios for the stage of cosmic evolution that produces stars.

The Masses of Stars

We start with the remarkable fact that all stars have about the same mass. The most massive star known has about 80 times the sun's mass, and the least massive has about 1/50 of the sun's mass. This range in mass may seem large, but when we consider how much greater the range *could* be, and the vastly greater range in the luminosities and densities of stars, we may conclude that masses so close to that of the sun must have special properties. These properties turn out to be *enough mass* to produce nuclear fusion reactions (see page 256), but *not so much mass* that radiation pressure ejects matter from the star.

All stars have apparently formed from interstellar clouds of gas and dust, each of which had a mass ranging from a few hundred to many thousand times the mass of the sun (2×10^{33} grams). As these clouds contracted, they somehow fragmented into subclumps with starlike masses (Figure 8-2). We don't know much about the details of this fragmentation, but its results—stars with masses in the range mentioned above—light the skies around us.

Contracting Interstellar Clouds

What causes interstellar clouds to contract in the first place? Calculations show that clouds with masses of a few thousand times the sun's mass tend to be unstable, ready to collapse under their own gravitational forces if they receive a relatively slight perturbation. In spiral galaxies, such perturbations arise periodically from the passage of the spiral density wave (see page 199). The density wave triggers star formation by a rapid "squeeze" on the interstellar gas clouds, which increases their density through the increase of the pressure upon them (Figure 8-2). As a result of this squeeze, some of the clouds will cross the brink of instability and begin to contract. Supernova explosions (page 368) may also trigger star formation, as apparently happened in the case of our sun, and there may be still other mechanisms that make interstellar clouds start to contract.

Within a few million years, an interstellar cloud that has a density of perhaps a few thousand atoms per cubic centimeter may produce subclumps with densities of trillions of molecules per cubic centimeter. Just a million

Figure 8-1 Stars are forming today inside these dark Bok globules, contracting clouds of gas and dust in the Milky Way.

years or so after that, these subclumps, now called **protostars** (stars in formation), will have shrunk under their self-gravitational forces to the point where the density at their centers equals the density of water (10^{22} molecules per cubic centimeter) (Figure 8-3). Soon after, nuclear fusion begins.

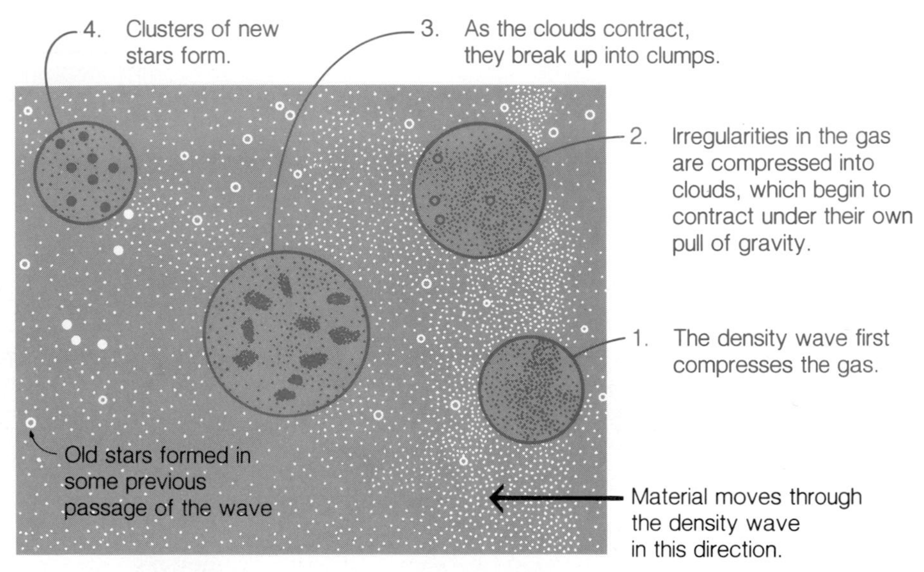

Figure 8-2 The gas compressed to higher density by the passage of a density wave can begin to form clumps which later fragment into protostars.

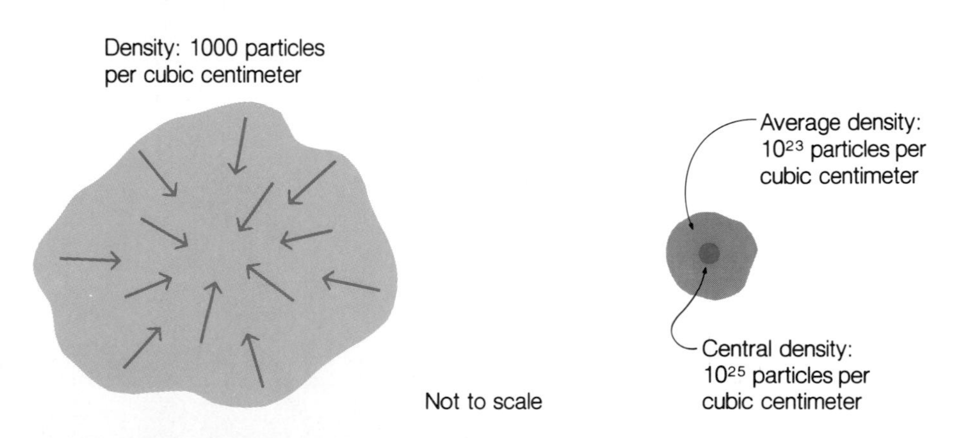

Figure 8-3 Stars form from clouds of gas and dust that begin contracting with a density of about a thousand particles per cubic centimeter. The density at the center of a protostar must rise enormously before nuclear fusion begins.

The Start of Nuclear Fusion

The onset of nuclear fusion occurs because the temperature at the center of the protostar rises along with the density, as all of the particles move more rapidly under the influence of the increasing pressure there. The tremendous increases in density, pressure, and temperature all come from the same cause, the enormous increase in the weight of the overlying layers. These layers press down upon the center of the protostar with ever greater force as the protostar contracts. Thus the contraction of a protostar under its own gravity must lead to enormous temperatures in its center, large enough to begin nuclear fusion reactions. This happened for our sun 4.6 billion years ago, but only a few tens of millions of years ago for most of the stars in Orion (Figure 8-4).

Figure 8-4
A photograph of the constellation Orion shows many young stars, one older star (Betelgeuse), and the Orion Nebula, a cluster of stars still in the process of formation. The inset above is a negative print of the Horsehead Nebula.

8.2 Double and Multiple Stars

As we mentioned earlier, the interstellar clouds most likely to begin contraction and thus to form stars may have masses many thousand times greater than the sun's. Some of these clouds have enough mass to remain bound together as a globular star cluster (see page 176); others, with less mass, persist for a time as much smaller open star clusters or as loose stellar associations (page 178). The vast majority of stars, however, appear in neither form of star cluster. Although they may have been born together with several hundred other stars, they have gradually drifted apart from their stellar brothers and sisters. Our sun typifies this case. We have no idea which stars formed along with the sun, or where we might find them along the sun's orbit around the galactic center (see page 152).

Great numbers of stars *have* managed to stay together in pairs, triplets, or higher-multiple star systems. These stars formed in close proximity to one another, astronomically speaking. As the interstellar cloud fragmented into protostars, the protostars' mutual gravitational attraction got them into orbits around their common center of mass (Figure 8-5). As the other stars in the association that formed from the interstellar cloud drifted away, the double- and multiple-star systems persisted, and can persist indefinitely. More than half of all the stars in our galaxy are members of such systems. In this sense, the sun cannot be regarded as completely typical.

Visual Binary Stars

We can detect double and multiple stars in a variety of ways. By far the simplest method consists of looking through a telescope and seeing the individual stars (Figure 8-6). The best-known stars often turn out to be **visual binary stars:** Sirius, the brightest star in the sky, has a much fainter companion; so does Polaris, the pole star, and Aldebaran, the brightest star in Taurus. Castor, the second-brightest star in Gemini, turns triple in even a modest telescope, but this fails to complete Castor's story (page 246).

Visual binary stars furnish a good way to measure the masses of stars. Two stars close to one another will move in accordance with Kepler's laws: Their orbits will be ellipses, with their common center of mass at one focus

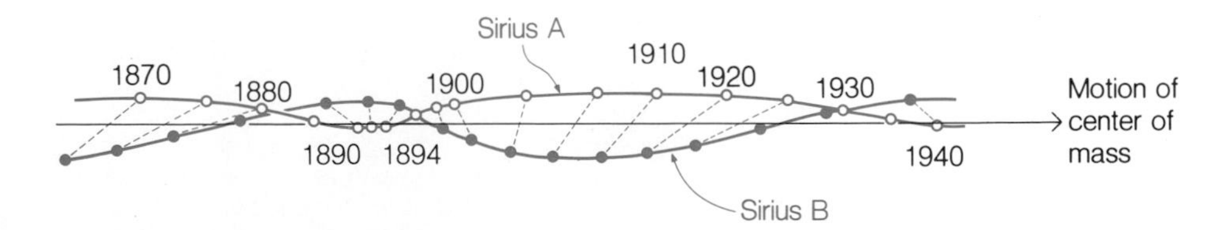

Figure 8-5 Stars that form close to one another will end up in orbits around their common center of mass. In a double-star system, the two stars must always be on opposite sides of the center of mass.

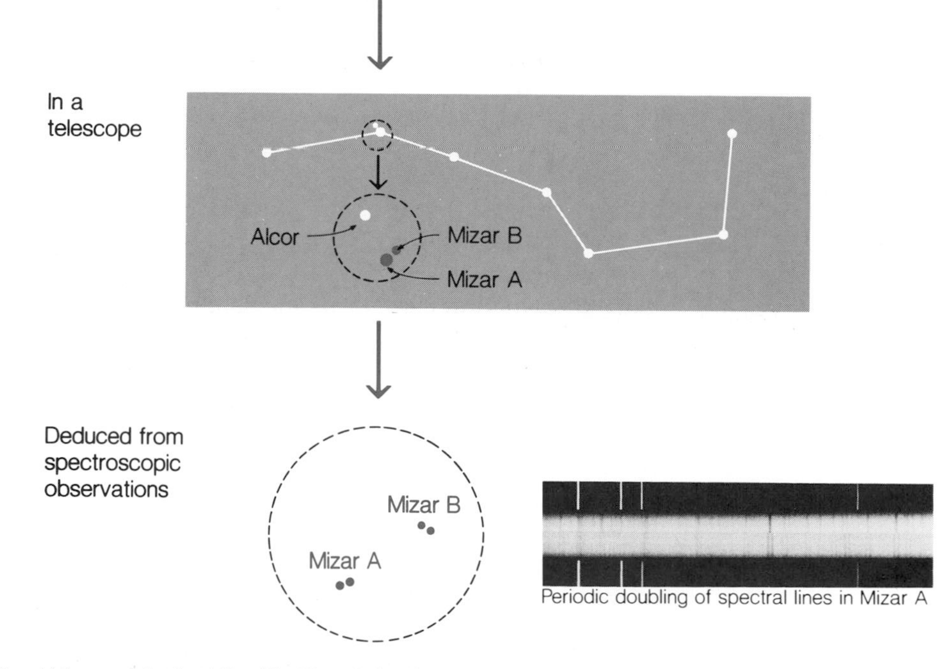

Figure 8-6 When we look at the Big Dipper (top), we can see a bright star in the middle of the handle, Mizar, with a fainter star, Alcor, close to it. This apparently represents a chance line-up, since Mizar and Alcor have different distances from us. In a telescope, Mizar turns out to be two stars, Mizar A and Mizar B. Each of these is itself double, as shown by spectroscopic observations.

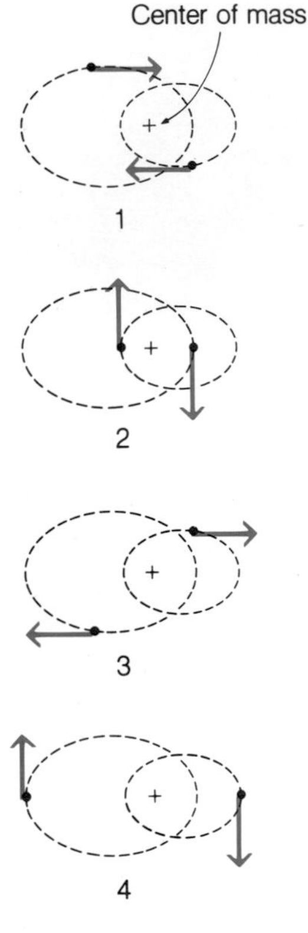

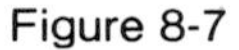
Figure 8-7
A pair of stars in mutual orbit will move around their common center of mass. The more massive star will have an orbit smaller than that of the less massive star, and the two stars will always be on opposite sides of the center of mass.

of each ellipse (Figure 8-7). The two elliptical orbits will have the same shape, but the sizes of the orbits will have a ratio in inverse proportion to the ratio of the stars' masses. The two stars will always be on directly opposite sides of their center of mass as they move. Finally, from Kepler's third law as modified by Isaac Newton (see page 375), we can determine the masses of the two stars if we know the distance between them and the time they take to complete an orbit around the center of mass. We can find the distance between the stars by measuring the angular separation of the stars on the sky. Knowledge of the distance to the stars (if we can measure the parallax shift for each of them) then allows us to use trigonometry to calculate the stars' separation in kilometers. The time it takes for the stars to complete one orbit comes from careful, repeated observations of the pair (Figure 8-5).

Table 8-1 lists several visual binaries, along with the masses and separation of each pair. We can see that a typical separation equals about twenty times the Earth's average distance from the sun, called the **astronomical unit** (A.U.). Since the sun's giant planets, and Jupiter in particular, may be thought of as would-be stars with too little mass to become real stars (see page 449), we may regard the solar system as a near-miss multiple-star system with dimensions typical of visual double-star systems. From this we may infer that many apparently single stars may have planets, low-mass companions that we can detect indirectly by their effect on the motion of the primary star.

Proper Motion of Stars

Each star's random motion, in addition to its general orbit around the center of our galaxy, gives it a particular **proper motion,** or motion across our line of sight. By photographing a given region of the sky year after year, we can determine an individual star's proper motion, which typically turns out to be a straight line. For some stars with large proper motions, we can detect apparent wiggles around the basic straight-line motion (Figure 8-8).[1] The simplest explanation for these deviations is the presence of an unseen companion, a planet or star too dim to be seen whose gravitational pull makes the visible star move a bit first one way and then another as the two objects orbit around their common center of mass. Barnard's star, the star with the largest proper motion, seems to have two unseen companions with masses about equal to Jupiter's. These may be the first planets detected outside the solar system.

Spectroscopic Binary Stars

In addition to the visual binary stars and the binaries found by their gravitational effects, we must also consider the **spectroscopic binary stars,** those that can be detected only by spectroscopic techniques. If two stars are too

[1]We must allow for the effects of parallax caused by the Earth's motion before we can detect such wiggles.

Table 8-1 Selected Binary Star Systems Compared with the Solar System

System Name	Distance from Sun (Parsecs)	Component A Mass (Sun = 1)	Component B Mass (Sun = 1)	Separation (Astronomical Units)
Visual Binaries				
Alpha Centauri[a]	1.3	1.1	0.9	23
Sirius	2.6	2.2	1.0	20
Procyon	3.5	1.8	0.7	16
Krüger 60	4.0	0.3	0.2	10
Spectroscopic Binaries				
Mizar	27	~8[b]	~8[b]	?
Epsilon Aurigae	1350	~35	~22	35
Eclipsing Binaries				
Algol	27	5.2	1.0	0.07
Alpha Coronae Borealis	22	2.5	0.9	0.2
Stars with Unseen Companions				
Barnard's star	1.8	0.2?	0.002?	5?
Luyten 726-8	2.6	0.1?	?	?
Solar System				
Sun and Jupiter		1.0	0.001	5.2
Sun and Saturn		1.0	0.0003	9.5

[a]The Alpha Centauri system is in fact triple; the table gives data on the two close components.
[b]These masses are highly uncertain, but in any case exceed 1.67 solar masses.

close together to be seen as individual points of light in a telescope, we may still detect their individual existences from the changes that occur in their combined spectra. As the stars orbit their common center of mass, one star will be approaching us in orbit while the other is receding, and we may see a periodic doubling of all the features in the spectrum (Figure 8-9). This doubling arises from the different Doppler shifts that the light from the two stars undergoes. We can detect spectroscopic binaries if the stars have fairly large velocities in orbit, at least 10 kilometers per second. Only then can we see the spectral lines grow double and then merge again as the stars cross our line of sight.

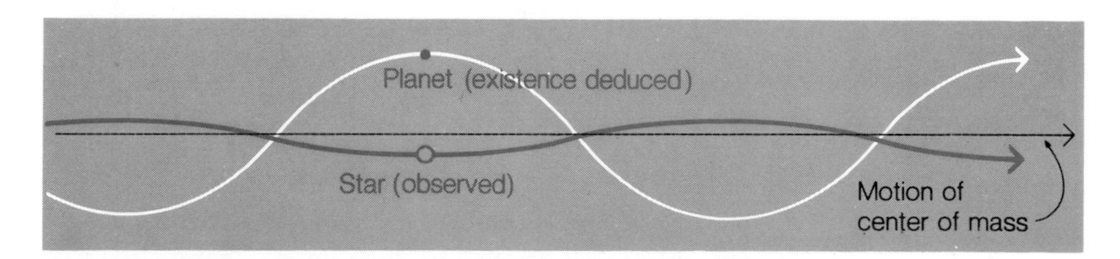

Figure 8-8 Some stars show periodic wiggles in their proper motion, even though we can see no companion star. One or more planets or dim stars may be in orbit with the star and may therefore perturb the star from straight-line motion. The amount of the star's wiggle has been highly exaggerated in this drawing.

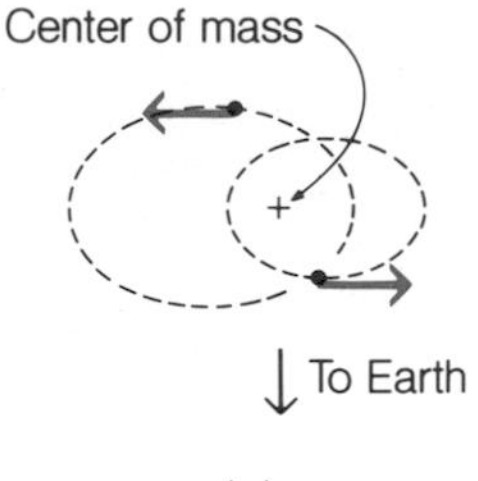

(a)

(b)

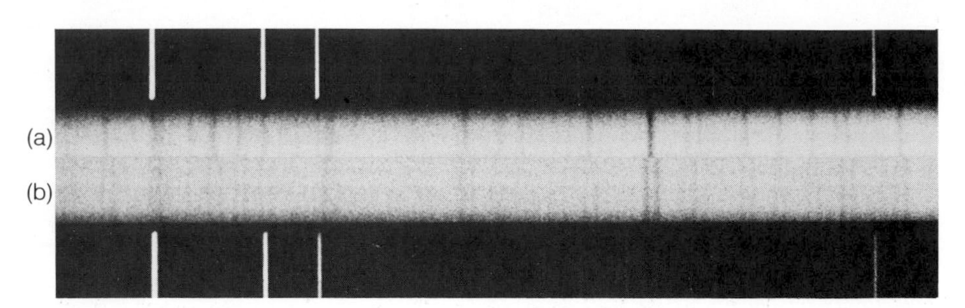

Figure 8-9
The spectrum of the star Mizar A shows a periodic doubling of the spectral lines (b) from the single-lined spectrum (a). The doubling occurs when the two stars are moving parallel to our line of sight (in different directions), rather than across our line of sight.

Luckily, however, the same thing that makes the stars difficult to *see* as double—their closeness in space—makes their velocities large, since the closer together two objects are, the faster they must orbit their center of mass (see page 372). The greater velocity in orbit comes from the greater amount of force with which they attract one another. Spectroscopic binary stars typically have a separation of 1 A.U. or less and orbit their center of mass in several days. In contrast, visual binary stars, with much greater separations, take many years to complete a single orbit. An exceptional spectroscopic binary, Epsilon Aurigae, consists of two extremely massive stars spaced 35 A.U. apart.

Castor, the triple star we mentioned earlier, turns out to be sextuple. Each of the three stars we see in a telescope is a spectroscopic binary! Castor thus consists of three closely bound pairs, all moving around the system's center of gravity in rather elongated orbits (Figure 8-10).

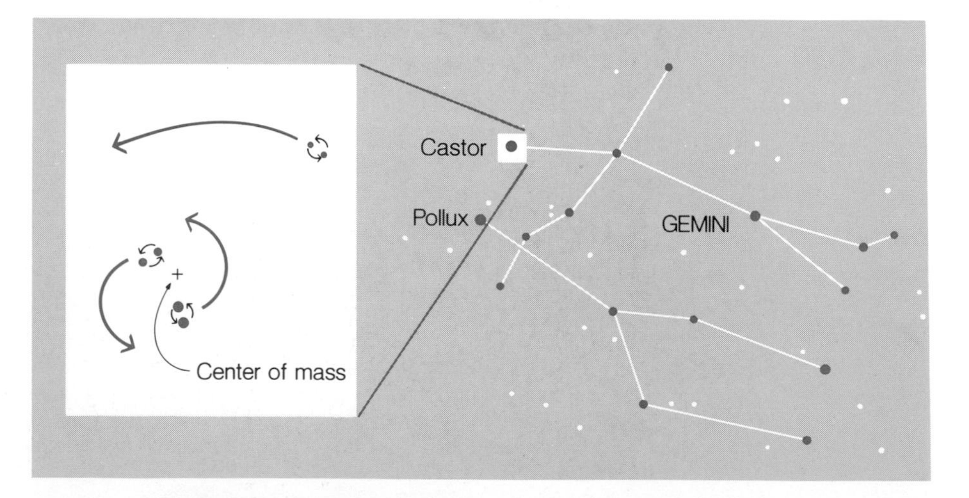

Figure 8-10
The star Castor actually consists of six stars, three spectroscopic binary pairs, which all move in elliptical orbits around their common center of mass.

Eclipsing Binary Stars

A final class of binary stars consists of the **eclipsing binaries.** These systems happen to have their orbital plane in, or nearly in, our line of sight. Thus, as the stars move around their center of mass, we can see one star eclipse the other, and then the second star moves in front of the first (Figure 8-11). Whether or not we see a given pair as an eclipsing binary is a matter of luck. What appears as an eclipsing binary to us would not be seen as such from a planet around Sirius or Aldebaran. But when we do see a double system in periodic eclipse, the effect can be spectacular.

The first eclipsing system to attract attention was Algol, named the *demon star* by Islamic astronomers because it shows a sudden decrease in brightness every two days. Generations of astronomers marveled at these

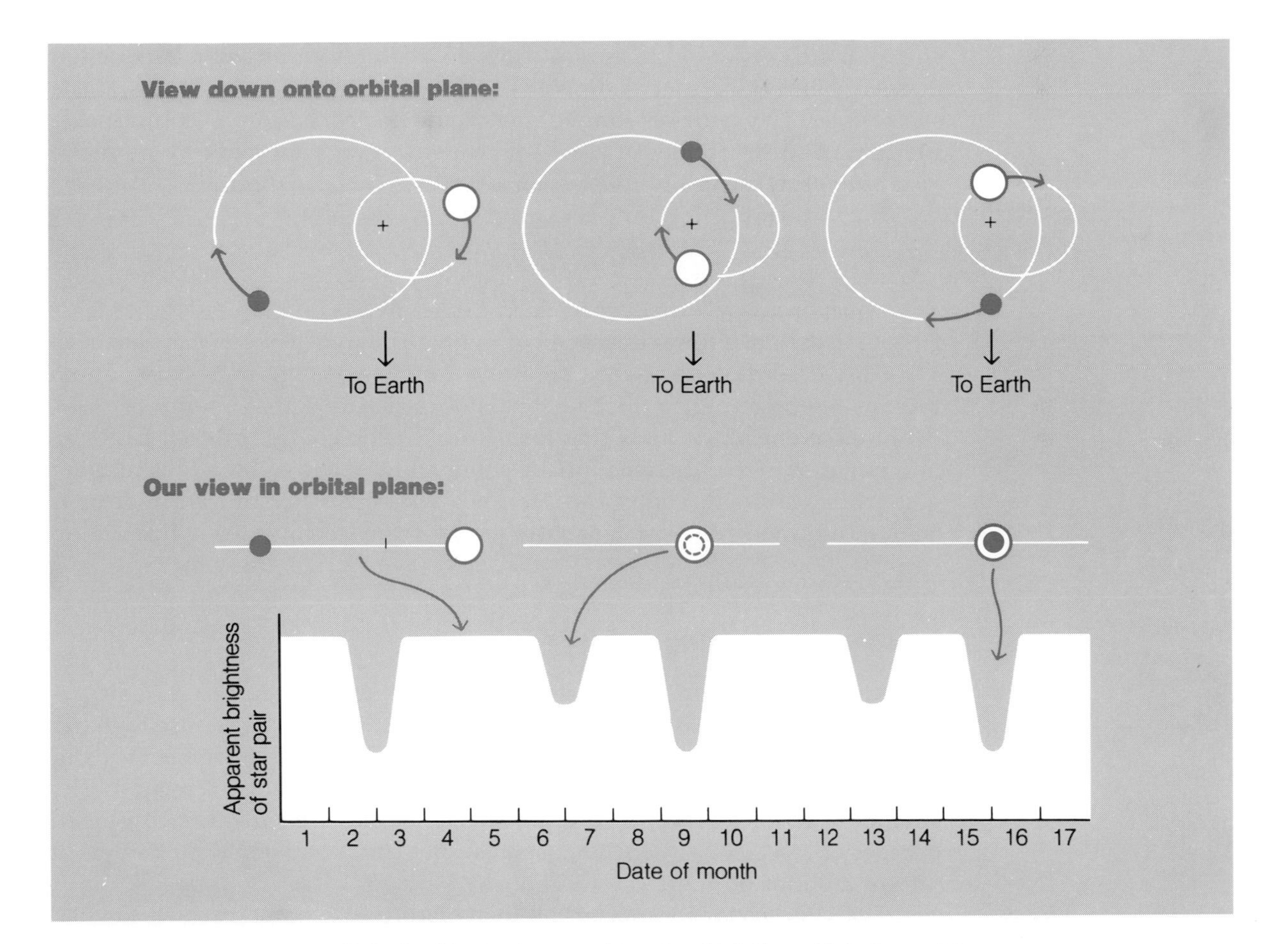

Figure 8-11 Eclipsing binary stars are those in which the orbital plane of the two stars almost coincides with our line of sight. As the two stars orbit around their center of mass, we see two eclipses in each complete orbit.

changes without a satisfactory explanation. Today, though, we know that the decrease occurs when one member of the pair passes behind its partner for a few hours.

8.3 Energy Liberation in Stars

Double- and multiple-star systems have provided astronomers with much of their information about stellar masses, thanks to the gravitational forces that result in observable motions and allow the calculation of the masses producing the attraction. With their knowledge of stellar masses and their ability to derive the surface temperatures and absolute brightnesses of stars, astronomers have made detailed models of how stars live and die. These models have been checked against reality and confirm the conjecture about how stars shine first made in the 1930s.

Stars shine by emitting photons from their surfaces. Every photon carries away from the star a certain kinetic energy that departs forever. Thus, to shine for billions of years, as most stars do, requires an ongoing production of energy of motion, not at the star's surface, but buried deep inside, at the star's center. The source of kinetic energy, the nuclear fusion reactions inside the star, converts energy of mass into kinetic energy moment by moment, year after year, according to Einstein's most famous equation,

$$E = mc^2.$$

This equation tells us that every object with a mass m has an energy of mass E equal to the mass times the speed of light, c, squared. The nuclear fusion in a star involves protons, each with a tiny 1.67×10^{-24} gram of mass. These protons fuse together to liberate about 1% of their original energy of mass. Because the energy of mass that turns into kinetic energy in each fusion is minuscule, so is the liberated kinetic energy. But with a tremendous number of nuclear fusion reactions each second, a star can liberate an enormous amount of kinetic energy, far more energy than we could ever liberate on Earth.

Temperatures Inside Stars

Why do nuclear fusion reactions occur in the centers of stars? High temperatures and densities give the answer. In particular, for nuclei to fuse together requires temperatures of about 10 million K. On Earth, we have never maintained such high temperatures for more than a fraction of a second. The attempt to achieve "controlled" nuclear fusion like the sun's has consumed decades of research without final success. But stars can keep their centers at tens of millions of degrees for billions of years.

The same factor that raises the temperature—the immense self-gravitational force of the stars—also allows the stars to release huge quantities of energy of motion without blowing themselves apart. As the matter that forms a protostar weighs down on its central regions, the gravitationally induced pressure rises. The increased pressure makes all the particles near the center move more rapidly, and they collide with greater and greater force, breaking molecules into atoms and stripping the electrons from the atoms. The rise in the average velocity of each particle occurs in any gas as the pressure increases, and corresponds to an increase in the gas temperature, since temperature simply measures the average kinetic energy per particle. We can see this effect whenever we use an air pump to compress air into a tire by increasing the pressure on it. The air grows warmer as the pressure increases.

When the central region of a protostar reaches a density equal to that of water (1 gram per cubic centimeter), its temperature approaches 1 million K, much hotter than any furnace on Earth. But the protostar continues to contract until the central density increases another 10 to 100 times and the temperature climbs to 10 or 20 million K. Then nuclear fusion begins, and the flood of kinetic energy released successfully opposes the contraction and eventually maintains the star at a constant size. The protostar has then become a true star.

The Proton-Proton Cycle

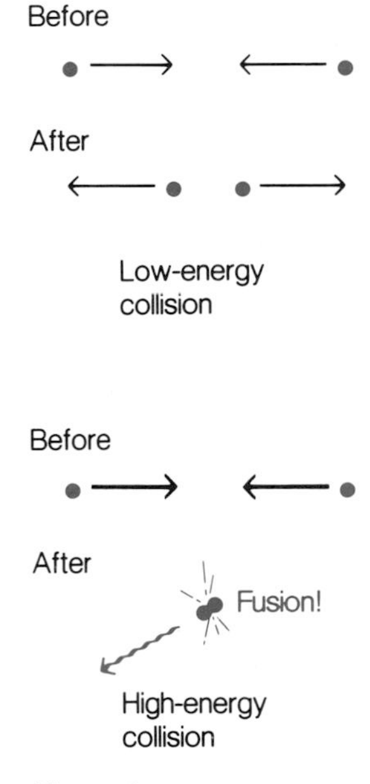

Figure 8-12
Protons tend to repel each other with electromagnetic forces. Hence only at high temperatures, when the protons have large kinetic energies, can they come close enough to fuse together.

The interstellar gas from which a protostar forms consists mainly of hydrogen (90% by number, 75% by mass) and helium (10% by number, 25% by mass). All elements heavier than hydrogen and helium provide at most 1% of the mass of the protostar. As the hydrogen and helium atoms become ionized within the contracting protostar, the composition of the protostar becomes a mixture of hydrogen nuclei (protons), helium nuclei, and electrons, with a sprinkling of nuclei heavier than helium. The protons have 75% of the protostar's mass and provide the source of energy for the star.

At temperatures of about 10 million K, protons can fuse together upon collision. At lower temperatures, any protons that happen to collide do so with too little energy to overcome their mutual repulsion from electromagnetic forces (Figure 8-12). If protons move rapidly enough, however, they may collide with enough energy of motion to fuse together despite the repulsion that two positive charges exert on each other. The **strong forces** that hold atomic nuclei together and make protons fuse act only at distances of about 10^{-13} centimeter or less (see box on the four kinds of forces). Hence, protons must be able to approach one another to within this distance if they are ever to fuse together.

The fusion of two protons represents the first step in the **proton-proton cycle**, the basic energy-liberating process in stars. This fusion produces a deuteron, a neutrino, and a positron, as shown in Figure 8-13. A **positron** is the electron's antiparticle, and a **deuteron** may be thought of as a proton and a neutron bound together by strong forces. We might think that each deuteron should have more than twice the mass of a proton, since a neutron has slightly more than a proton's mass. Our intuition misleads us in the realm of elementary particles. In fact, a deuteron has *less* mass than two protons. This fact has crucial importance in a star's liberation of energy by nuclear fusion.

Table 8-2 Change in Mass and Energy of Mass During the First Step of the Proton-Proton Cycle

Before			After		
Mass of proton	=	1.6724×10^{-24} gram	Mass of deuteron	=	3.3432×10^{-24} gram
Mass of proton	=	1.6724×10^{-24} gram	Mass of positron	=	0.0009×10^{-24} gram
			Mass of neutrino	=	0
Total mass before	=	3.3448×10^{-24} gram	Total mass after	=	3.3441×10^{-24} gram
Energy of mass before	=	$(3.3448 \times 10^{-24}$ gram$) \times c^2$	Energy of mass after	=	$(3.3441 \times 10^{-24}$ gram$) \times c^2$

Energy of mass before − Energy of mass after = $(0.0007 \times 10^{-24}$ gram$) \times c^2$
= 6.3×10^{-7} erg

As Table 8-2 shows, the two protons that fuse in the first step of the proton-proton cycle have a total mass of 3.3448×10^{-24} gram. The deuteron mass plus the positron mass equals only 3.3441×10^{-24} gram, and the neutrino has no mass at all. Thus, the products of the fusion reaction have 0.0007×10^{-24} gram *less* mass than the particles that began the fusion.

The Four Kinds of Forces

Force, according to the dictionary, is the capacity to do work or to cause changes to occur. In scientific language, a force on an object is something that can *accelerate* the object. If we pick an object of a standard mass—1 kilogram—then we can define a unit of force, called the *newton,* as the amount of force capable of accelerating a 1-kilogram mass by 1 meter per second of velocity every second. A force equal to 10 newtons can accelerate a 1-kilogram mass by 10 meters per second every second or can accelerate a 10-kilogram mass by 1 meter per second every second.

Four basic types of forces exist in nature. **Gravitational forces** make every object in the universe attract every other object with an amount of force that varies in proportion to the product of the objects' masses and in inverse proportion to the square of the distance between the objects' centers. **Electromagnetic forces** act only between particles with nonzero electric charge. The amount of electromagnetic force between two charged objects varies in proportion to the product of the objects' charges and in inverse proportion to the square of the distance between the objects' centers. Unlike gravitational forces, electromagnetic forces can be either attractive (between particles with opposite electric charges) or repulsive (between particles with the same sign of electric charge).

The two additional types of forces, strong forces and weak forces, act only over distances no larger than the size of an atomic nucleus (no more than a few times 10^{-13} centimeter). **Strong forces** hold together atomic nuclei. They are always attractive, and they act between any two *nucleons* (protons or neutrons). **Weak forces** have effects that appear when one kind of elementary particle turns into other kinds. An example occurs when a neutron turns into a proton, an electron, and an antineutrino. As their name implies, weak forces have less strength than strong forces. However, they play an important role in the interactions among elementary particles.

Recent theoretical and experimental work on elementary particles has shown that electromagnetic forces and weak forces basically represent the same force, working in different ways at different levels of distance and energy. This "unification" of weak and electromagnetic forces suggests that we may yet discover that all four types of force are manifestations of a single kind of force in the universe. The quest for such a grand "unified field theory" occupied the last twenty years of Albert Einstein's career. It remains one of the exciting frontiers of modern physics.

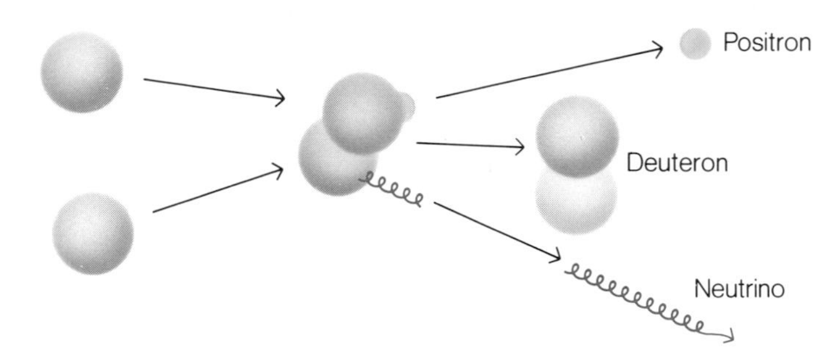

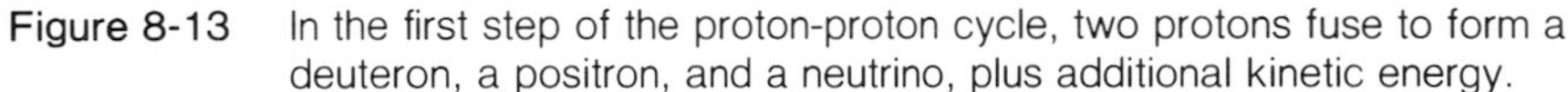
Figure 8-13 In the first step of the proton-proton cycle, two protons fuse to form a deuteron, a positron, and a neutrino, plus additional kinetic energy.

We can find the amount of kinetic energy liberated in the proton-proton fusion by using Einstein's equation, $E = mc^2$. If we multiply the decrease in mass, 0.0007×10^{-24} gram, by the square of the speed of light, $c^2 = 9 \times 10^{20}$ centimeters squared per second squared, we find a decrease in energy of mass and a corresponding increase in kinetic energy of 6.3×10^{-7} erg. The new kinetic energy appears as extra energy of motion in the deuteron, positron, and neutrino that emerge from the fusion (Figure 8-13).

A new kinetic energy of 6.3×10^{-7} erg may not seem a great amount, since we need a million times more energy of motion to blink our eyes. On the other hand, in stars like the sun, about 10^{38} proton-proton fusions occur each second. Thus the total effect can be seen quite easily. To complete our knowledge of how stars liberate energy of motion, we must examine the two final steps of the proton-proton cycle, which actually liberate more kinetic energy than the first.

In the second step of the cycle, a proton collides with a deuteron at speeds high enough to produce fusion (Figure 8-14). The collision produces a nucleus of helium-3 and a photon, plus some extra energy of motion. Helium-3 (He^3) has a nucleus of two protons and one neutron, while ordinary helium, called helium-4 (He^4), has a nucleus of two protons and two neutrons. Again, the particles that appear after the fusion (a photon and a nucleus of helium-3) together have slightly more energy of motion, and

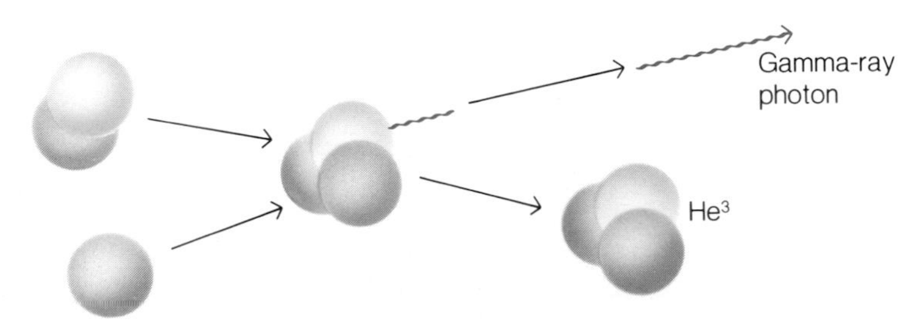

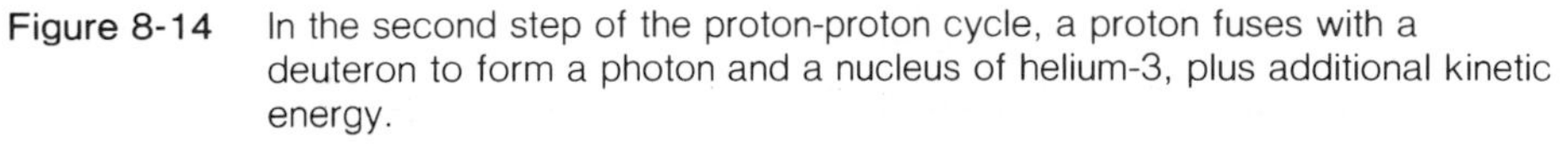
Figure 8-14 In the second step of the proton-proton cycle, a proton fuses with a deuteron to form a photon and a nucleus of helium-3, plus additional kinetic energy.

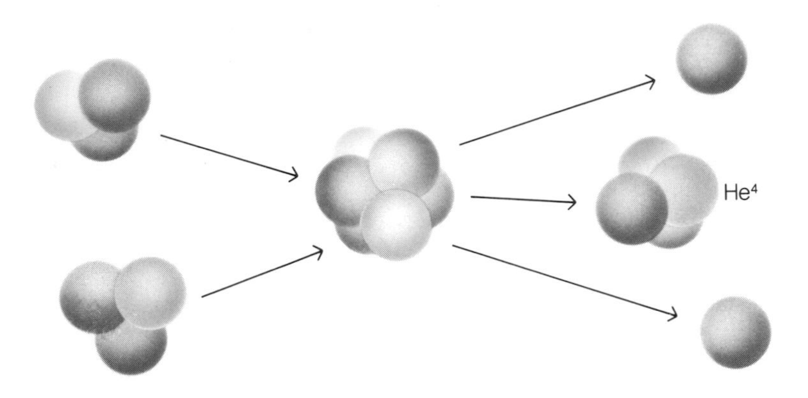

Figure 8-15 In the third step of the proton-proton cycle, two nuclei of helium-3 fuse together, producing a nucleus of helium-4, two protons, and still more kinetic energy.

slightly less energy of mass, than the particles that collided. Again, some energy of mass has turned into additional kinetic energy.

The third step of the proton-proton cycle liberates the most energy of motion. Once helium-3 nuclei have been created, they, too, can collide (Figure 8-15). Helium-3 nuclei are likely to fuse together upon collision, and when they do, they produce a nucleus of helium-4 and two protons, plus additional energy of motion.

The positrons that appear when protons fuse with protons have an extremely short lifetime. They soon encounter electrons and undergo mutual annihilation with them, producing photons, neutrinos, and antineutrinos as they turn *all* of their energy of mass into energy of motion (see page 115). We may rightfully count the energy of motion released by the two positron-electron annihilations in our total of the kinetic energy liberated from energy of mass during the three steps of the proton-proton cycle. This total comes to 426×10^{-7} erg. The new kinetic energy appears in the motions of the helium-4 nucleus, protons, photons, and neutrinos that emerge from the fusion reactions, hundreds of thousands of kilometers deep inside the star.

The neutrinos escape directly from the center of the star, taking a small part of the energy of motion with them (see page 276). The photons travel only a tiny distance before they collide with some other particle—an electron, a proton, or a helium nucleus—and pass some of their energy of motion to that particle. The protons collide with other particles and share their energy of motion with them.

Figure 8-16 summarizes the three steps of the proton-proton cycle. We can see that the cumulative effect of the three kinds of fusion consists of converting four protons into one helium-4 nucleus, two positrons, two photons, two neutrinos, and additional energy of motion. We need two each of steps one and two to have one of step three, which involves the collision of *two* helium-3 nuclei. Table 8-3 compares the power output from proton-proton fusion on various scales.

The Balance Between Gravitation and Energy Liberation

Let us now consider the state of affairs throughout a star once nuclear fusion has begun in its central regions. The weight of the overlying layers, which raised the central temperature to 10 million K, continues to press down upon

Table 8-3 Mass Lost in Fusion and Power Output from Various Sources

Source	Mass Lost in Fusion Per Second (Kilograms)	Power Output (Kinetic Energy Per Second) in Watts
Single proton-proton fusion (in one second)	7×10^{-31}	6.3×10^{-14}
Possible fusion reactor	10^{-4}	9×10^{12}
Thermonuclear (hydrogen) bomb (assumed one-second duration)	1	9×10^{16}
Barnard's star	5×10^{6}	4.5×10^{23}
Sun	4.4×10^{9}	4×10^{26}
Spica	10^{13}	9×10^{29}

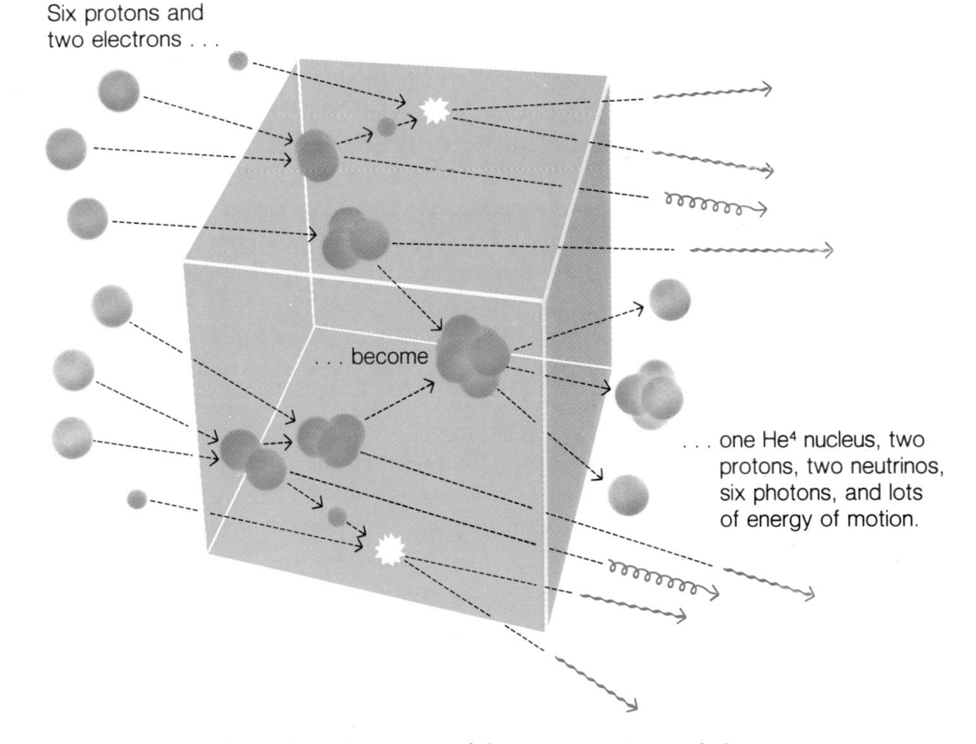

Figure 8-16 The combined effect of the three steps of the proton-proton cycle is to turn four protons into one helium-4 nucleus, two neutrons, six photons, and additional kinetic energy.

the center. But now nuclear fusion produces new kinetic energy each second, which appears as additional energy of the moving particles. This new energy of motion provides the exact amount needed to maintain the star at a constant size, neither expanding nor contracting. The exactness comes not by accident but by a natural, self-regulation characteristic of every star.

Suppose, for example, that a star should happen to liberate a bit more kinetic energy than is needed to keep the star at a constant size. The additional kinetic energy would expand the central regions slightly, pushing the various parts of the star to slightly greater separations. Such readjustments occur within a star soon after it has begun nuclear fusion. As a result, the pressure on the center, and the temperature there, would decrease slightly, and so would the rate at which particles fuse together. The rate of nuclear fusion increases tremendously as the temperature rises slightly, and decreases enormously if the temperature falls only a bit, because the ability of protons to overcome their mutual repulsion depends critically on the average kinetic energy per proton. If the rate of nuclear fusion decreases, so does the amount of kinetic energy liberated each second. With less energy of motion liberated, the star's central regions must contract under the pressure induced by the star's self-gravitation. This contraction will bring the star back to its initial size and thus to its initial rate of nuclear fusion.

The Carbon Cycle

We have discussed the proton-proton cycle of nuclear fusion, by which most stars turn energy of mass into kinetic energy. However, a minority of stars employ a different series of fusion reactions, called the carbon cycle, to achieve the same end. In the **carbon cycle,** a nucleus of carbon-12 (C^{12}) serves as a catalyst to fuse four protons into one helium-4 nucleus, converting energy of mass into kinetic energy of motion (Figure 8-17). In this process, protons combine with the carbon nucleus and then with heavier nuclei, until four protons have fused with the original nucleus to make an unstable nucleus of oxygen. This, in turn, splits into a nucleus of carbon-12, just like the original one, plus a nucleus of helium-4. In each of the six steps of the carbon cycle, some energy of mass turns into energy of motion.

The carbon cycle requires not only the presence of carbon nuclei, but also a higher temperature than that needed for the proton-proton cycle. A star must have a central temperature close to 20 million K before the carbon cycle predominates over the proton-proton cycle in the liberation of kinetic energy. Like most other stars, the sun relies primarily on the proton-proton cycle, since its central temperature reaches "only" 16 million K. Why, then, do some stars have higher central temperatures? The answer lies in the basic difference among stars: their masses.

The Importance of Mass

Each star's mass determines how heavily its outer layers press inward on the center and thus how much energy the star must liberate each second to oppose this pressure. More massive stars have a greater self-gravitation and

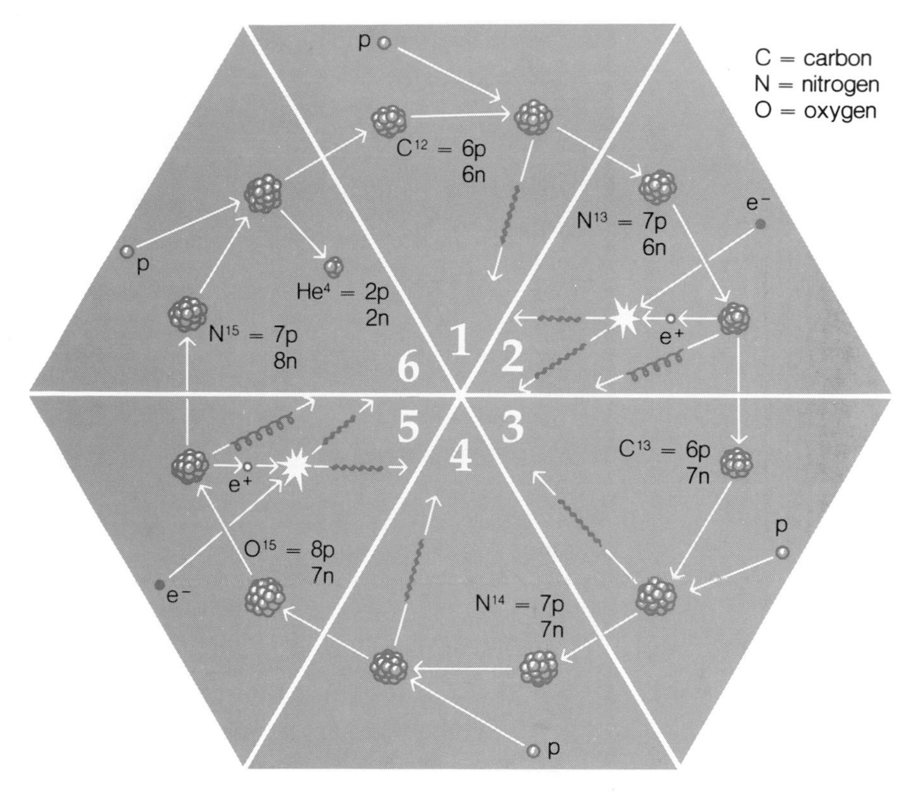

Figure 8-17 In the carbon cycle, carbon nuclei act as a catalyst to help turn protons into nuclei of helium-4. The six stages of the cycle involve the fusion of protons with successively heavier nuclei, until finally a helium-4 nucleus and a new carbon-12 nucleus appear. Each fusion liberates some new kinetic energy.

thus a greater weight that squeezes their central regions with more force. They need a greater rate of energy liberation to oppose this force, and they get it from the higher temperature produced by the greater pressure at their centers.

When we consider a particular star, we must ask whether that star will keep the same mass, and thus the same rate of energy liberation, throughout its lifetime. It turns out that for *most* of its energy-liberating existence, any star will have a nearly constant mass. The conversion of energy of mass into kinetic energy would use up less than 1% of the original mass, even if the star subjected every proton it contains to the proton-proton cycle of nuclear fusion. But exceptions to the rule of constant mass do occur at the start and at the finish of energy liberation. We shall discuss the death of stars in the next chapter, but for now let us look at the beginnings of nuclear fusion and its effect upon a newly formed star.

Radiation Pressure in Young Stars

We have considered the gravitationally induced contraction of a protostar, which raises the star's central temperature so much that nuclear fusion be-

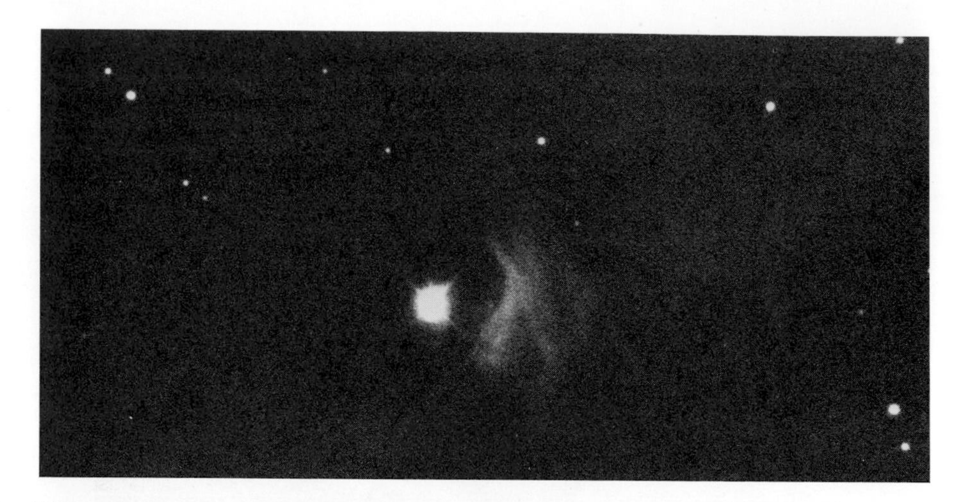

Figure 8-18 The variable star T Tauri has changed its brightness by a large factor during the past century. Around the star, wisps of gas that reflect the star's light are visible. These wisps are probably gas left over from the formation process.

gins. However, the protostar's mass must equal at least 2% of the sun's mass, or else the gravitational force will not suffice to produce a temperature of 7 million K, hot enough for nuclear fusion to begin. Less massive "protostars" will become planets, not stars. When we look around our part of the Milky Way, we can see regions where stars seem to be forming now (Figure 8-18) and where young stars appear to be wrapped in the "cocoon" of gas and dust from which they formed. As the young stars begin to shine, the light they emit pushes on the gas that surrounds them. This **radiation pressure** moves the gas outward from the star. This early phase of the star's lifetime, called the T Tauri phase after the first example studied, produces a measurable Doppler shift in the layers of gas that move toward us.

In a massive protostar, the radiation pressure on the material around the star can become so great that much of the would-be star ends up pushed back into the interstellar medium. In fact, radiation pressure places a limit on the mass any star can have. Above this limit, the star would create so much kinetic energy per second, and thus so much radiation pressure, that the outer layers of the star would quickly be blown away. This upper limit equals about 80 times the sun's mass.

The Mass-Luminosity Relationship

Stars can therefore exist only if they have between about 1/50 the sun's mass and 80 times the sun's mass. Within this range, as we have seen, a star's mass determines the rate at which nuclear fusion occurs and thus the rate at which the star liberates kinetic energy. The greater the liberation of kinetic energy, the greater the star's luminosity (Figure 8-19). More massive stars

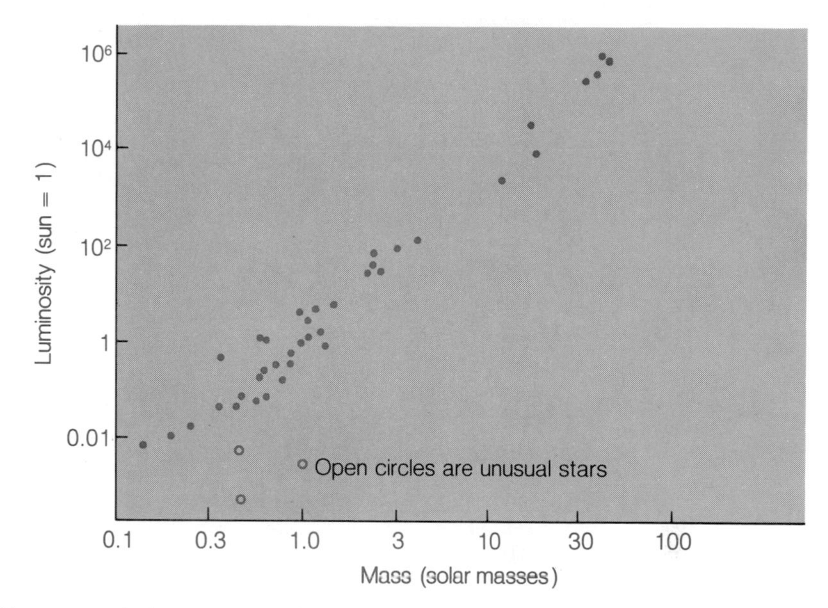

Figure 8-19 We can graph the masses and luminosities of those stars whose distances and masses have been well determined. This graph shows a definite relationship between mass and true brightness. More massive stars have greater luminosities.

liberate far more kinetic energy each second than less massive stars do. The rate of energy liberation varies approximately as the cube of the mass.

This sharp increase in the rate of nuclear fusion with the star's mass serves as a tribute to the importance of mass in determining how a star behaves. As we shall see later (page 294), stars with different masses have noticeably different lifetimes. But before we can deal with the end of energy liberation, we must examine what goes on between the stellar center, where energy is liberated, and the stellar surface, where energy is radiated into space.

8.4 Surface Conditions of Stars

The power of the human mind has enabled us to penetrate the interiors of stars and to determine what processes occur to liberate energy of motion there. But since we can see only the surfaces of stars, we must learn to link what we see with the more essential stellar facts—the star's composition, mass, and structure. By so doing we can look (mentally!) at the seething interior of the star, starting with the 15-million-K region near the center (Figure 8-20). Here particles zip to and fro, often colliding and sometimes fusing together. The trillions upon trillions of fusions generate kinetic energy as well as photons, which themselves collide with particles. The result of all these collisions is that photons of every energy exist, but not in equal num-

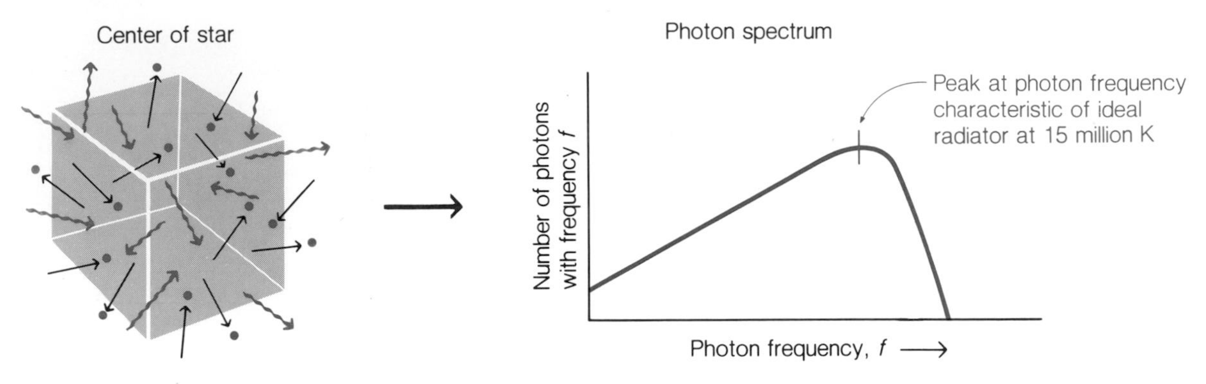

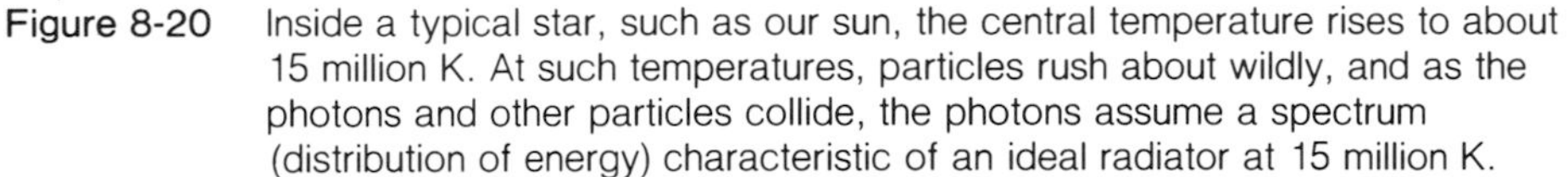
Figure 8-20 Inside a typical star, such as our sun, the central temperature rises to about 15 million K. At such temperatures, particles rush about wildly, and as the photons and other particles collide, the photons assume a spectrum (distribution of energy) characteristic of an ideal radiator at 15 million K.

bers. The star's center resembles the ideal radiator we described on page 120, typified by a temperature of 7 to 35 million K. Thus, the spectrum of photons there has the characteristic ideal-radiator shape, with a peak number of photons at the energy specified by this temperature (Figure 8-20).

The Flow of Energy in Stars

Why does the interior of a star resemble an ideal radiator? Inside each small volume of the star, photons constantly collide with other particles and thus exchange energy back and forth with them. Furthermore, the amount of kinetic energy that enters each small volume per second almost exactly equals the amount that leaves. Almost, but not quite. Even though particles rush in and out in all directions, a steady net flow of energy outward from the center occurs (Figure 8-21).

The central regions, where kinetic energy is liberated, have the greatest energy of motion per particle and so the highest temperature. Surrounding this region are progressively larger spherical shells of matter. As the liberated energy flows outward into and through these shells, each shell farther from the center acquires progressively less energy per particle and so assumes a lower temperature (Figure 8-21). If we could measure the spectrum of photons within any volume of the star, we would continue to find an ideal-radiator spectrum at each point, but characterized by steadily lower temperatures as we moved outward from the star's center.

What is a Star's "Surface"?

Since stars consist entirely of gas—free-moving particles, rather than the electromagnetically linked particles in a liquid or solid—we cannot expect to find a definite surface like the Earth's. Hence, we can best define a star's surface as those layers from which a photon has an even chance of escaping directly into space if it is headed outward (Figure 8-22). These layers are what we see when we look at a star, but they may have a thickness of several hundred, or even several thousand, kilometers (Figure 8-23), still only a small percentage of the star's total radius. Below the surface, a photon will collide with a particle rather than escape. Such collisions change the photon's energy and help maintain the ideal-radiator conditions we described.

Stellar surface layers have temperatures that range from a few thousand degrees above absolute zero to 50,000 degrees or more. The temperature of the surface layers determines the peak in the photon spectrum and hence the star's color. Hotter stars are bluer, cooler stars redder. Astronomers use a

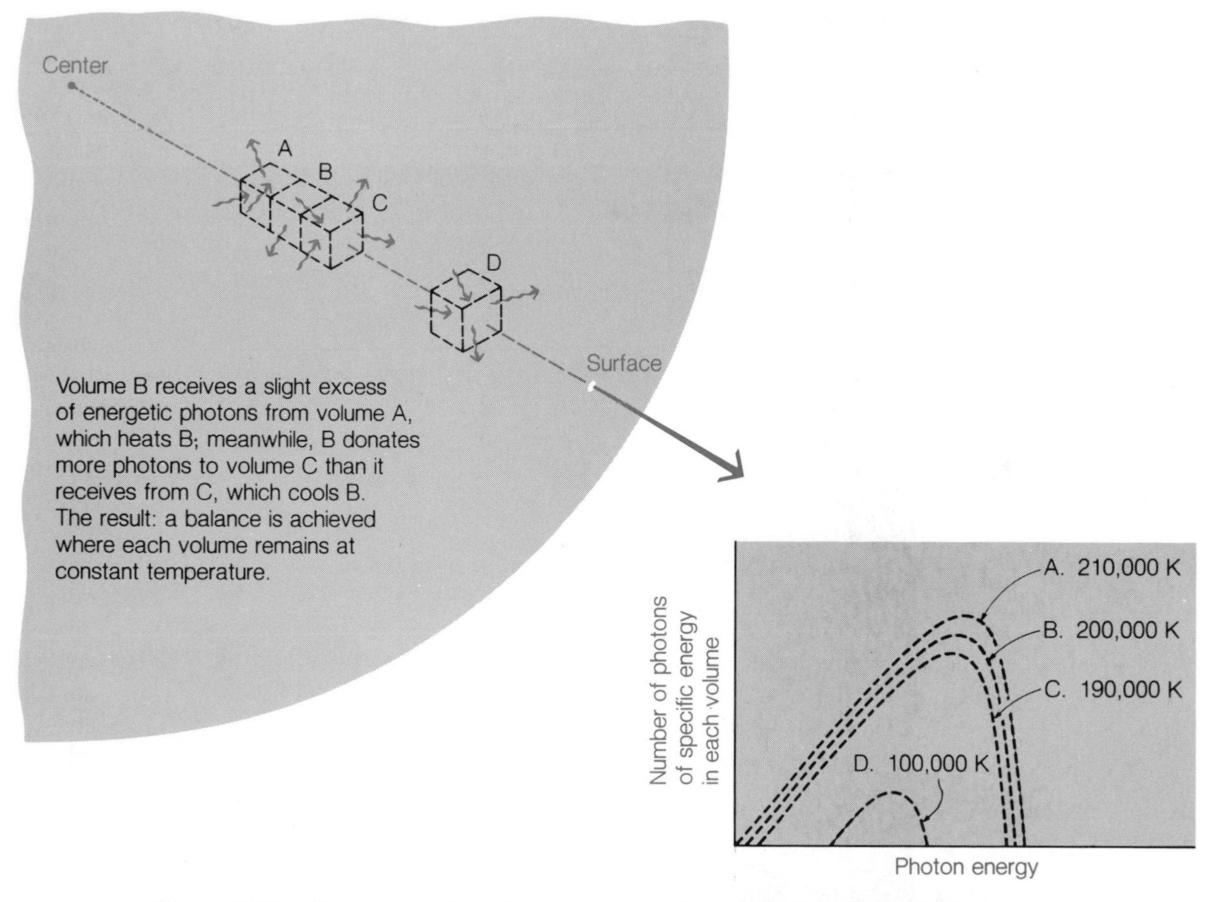

Figure 8-21 Each small volume inside a star receives photons from all adjacent regions and emits photons in all directions, but on balance a net flow of energy outward exists. As a result, the spectrum of photons that characterizes each small volume will be that of an ideal radiator, but the characteristic temperature will decrease outward from the star's center.

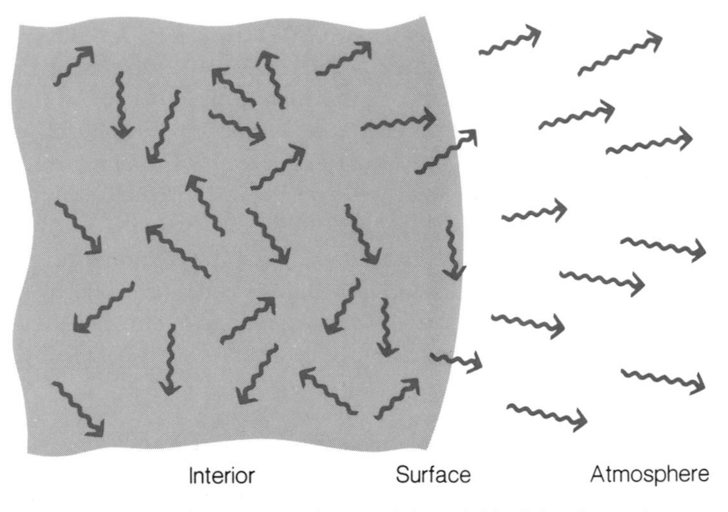

Figure 8-22 We define a star's surface as the level from which a visible-light photon has an even chance of escaping if it is headed outward, rather than being absorbed by an atom or ion or colliding with an electron.

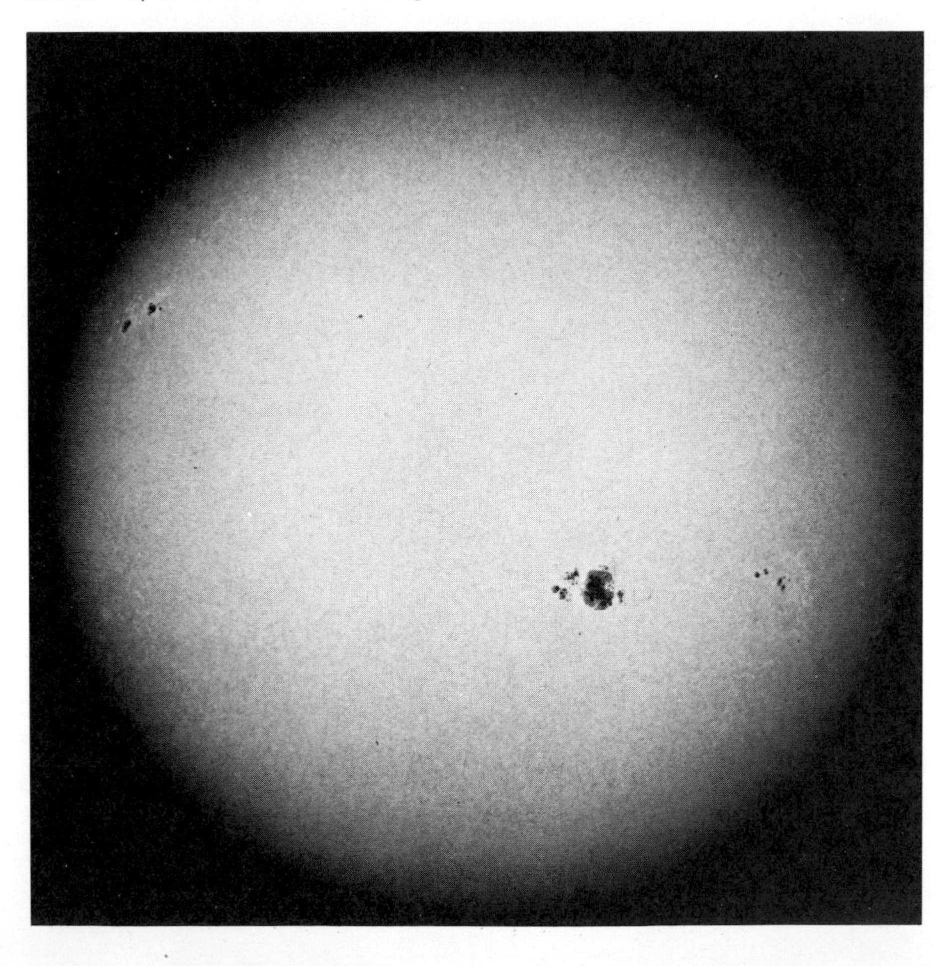

Figure 8-23
When we observe the closest star, the sun, we see its surface—about 500 kilometers (in thickness) of gas that has a density of matter 1000 times less than the density of air. The dark sunspots are regions of cooler-than-average gas.

Table 8-4 Surface Temperatures and Spectral Types of Selected Stars

Star	Distance from Sun (parsecs)	Surface Temperature (K)	Color Index[a]	Spectral Type	Luminosity Class[b]	Absolute Magnitude
Mintaka	460	28,000	−0.21	09.5	II	−6.1
Spica	80	28,000	−0.23	B1	I	−3.4
Rigel	250	15,000	−0.03	B8	I	−7.0
Vega	8	10,000	0.00	A0	V	+0.5
Sirius	3	10,000	0.00	A0	V	+1.4
Deneb	500	9,500	+0.09	A2	I	−7.3
Denebola	13	9,500	+0.09	A3	V	+1.6
Altair	5	8,000	+0.22	A7	V	+2.3
Canopus	60	7,500	+0.16	F0	I	−4.7
Procyon	4	6,600	+0.41	F5	IV	+2.7
Polaris	240	6,000	+0.6	F8	I	−4.6
Sun		5,800	+0.65	G2	V	+4.8
Alpha Centauri A	1.3	5,700	+0.7	G2	V	+4.3
Pollux	11	4,500	+1.00	K0	III	+1.0
Arcturus	11	4,200	+1.23	K2	III	−0.2
Aldebaran	21	3,800	+1.53	K5	III	−0.7
Antares	130	3,000	+1.81	M1	I	−4.7
Betelgeuse	200	3,000	+1.86	M2	I	−6.0
Barnard's star	2	2,800	+1.74	M5	V	+9.5
Wolf 359	2	2,400	+1.8	M8	V	+13.5

[a]The color index, as astronomers define it, equals the star's magnitude in blue light minus the magnitude in yellow light (either apparent or absolute magnitudes may be used). Thus, the color index is given by

$$B - V = -0.4 \log_{10}\left(\frac{L_{blue}}{L_{yellow}}\right)$$

where L_{blue} and L_{yellow} are the star's luminosity in blue and yellow light, respectively. B stands for the star's magnitude in blue light, and V stands for the star's magnitude in yellow ("visible") light.

[b]See page 273.

simple **color index,** the ratio of the number of blue photons to the number of yellow ones, to gauge the color of a star. This color index can be found by using a photomultiplier tube to measure the brightness of a star first in blue light, then in yellow light. The key advantage of using a ratio of blue to yellow light is that this index does not depend on the star's distance (except for interstellar reddening) or on its apparent brightness.

Table 8-4 gives the surface temperatures of some well-known stars. Such temperatures can be estimated directly from the color index of each star, but we can find out much more about the star's surface layers if we examine its spectrum in detail.

8.5 Stellar Spectra

If a star (or galaxy) has a faint apparent brightness, we may be able to do no more than measure its relative brightness at a few standard wavelengths and thus form the color index of that object. But for many stars, we can study the distribution of photons at various wavelengths and frequencies and thus analyze the star's spectrum. These spectra have characteristic absorption features, which represent the removal of photons of certain definite energies as they rush outward (see page 82). The details of the absorption lines in a stellar spectrum provide numerous clues to the nature of that star.

The Balmer Series of Absorption Lines

In a star such as Sirius, hydrogen atoms in the outer layers absorb photons with a particular series of wavelengths called the **Balmer series,** named after the man who first studied it. The wavelengths in the Balmer series of absorp-

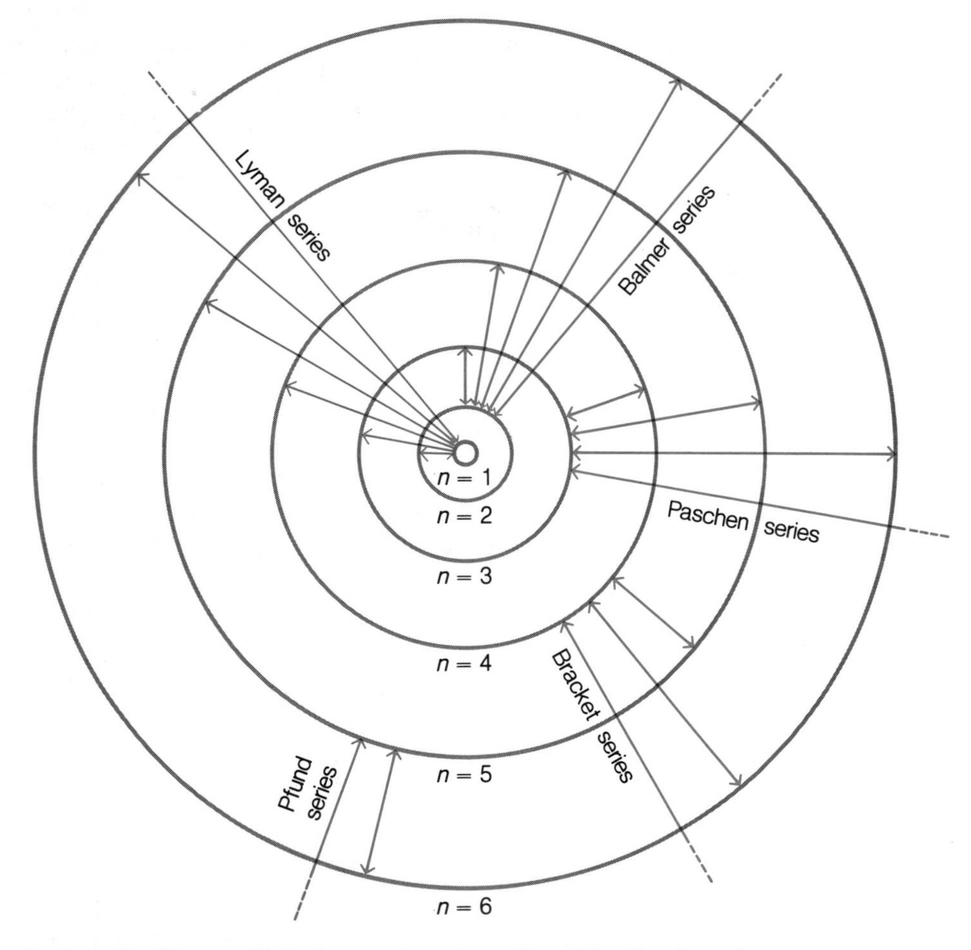

Figure 8-24 A schematic diagram of a hydrogen atom shows the orbits of various sizes that the electron can take around the proton. The Balmer series of emission and absorption lines consists of the frequencies and wavelengths of photons that are emitted when an electron jumps into the second-smallest orbit, or that are absorbed in making an electron jump out of the second-smallest orbit.

tion lines characterize the photons that have just the right energies to make an electron in the second-smallest orbit of a hydrogen atom jump into the third-smallest, fourth-smallest, or any other orbit larger than the second-smallest (Figure 8-24). Likewise, the Balmer series of emission lines arises when a hydrogen atom's electron jumps from a larger orbit into the second-smallest orbit. The spectrum of light from Sirius shows the Balmer series of absorption lines as its predominant feature. Hence, Sirius's outer layers must contain many hydrogen atoms. Furthermore, since the Balmer series of absorption lines arises from hydrogen atoms with their electrons in the second-smallest orbit, the atoms in the star's outer layers must collide with enough speed to knock a fraction of the electrons from the smallest into the second-smallest orbit, but not so much speed that collisions destroy the atoms by knocking the electrons loose. Such speeds occur among hydrogen atoms when the temperature is between about 5000 and 15,000 K.

Temperature and Density of the Surface Layers

As we have seen, the Balmer lines provide a useful guide to the temperature in the surface layers of a star, as well as to the number of hydrogen atoms there. By studying how the Balmer absorption lines are formed and how much of a spread in wavelength each line covers, astronomers can even determine the force of gravitation in the star's outer layers; that is, they can find out how compact a given star is for its mass.

The Balmer lines are merely the most useful of a tremendous set of possible absorption lines from hydrogen, helium, nitrogen, oxygen, and other atoms, as well as from molecules such as carbon dioxide and titanium oxide in cooler stars. When we stop to consider that each species of atom or molecule can produce various sets of absorption lines, depending on how the atom's or molecule's electrons happen to orbit when a photon is being absorbed, it might appear that unraveling just which species produced which absorption lines goes beyond our abilities. But astronomers have risen to this challenge. During the past 80 years, they have sorted through the spectra of star after star. They now have a thorough understanding of the abundances of different atomic and molecular species in various stars, of the temperatures in the stars' outer layers, and even of the force of gravity within these layers—all from studying the light of various colors they detect from stars.

Classification of Stellar Spectra

In this analysis of starlight, the Balmer lines came first, for the great abundance of hydrogen gives them special prominence. To this day, astronomers classify stars with a system that emphasizes the stars' Balmer absorption lines. Figure 8-25 shows seven stellar spectra, of which type A exhibits the most prominent Balmer lines. Astronomers named these stars type A before they understood *why* a given star may or may not have prominent Balmer lines in its spectrum. Today we know that if a star has too high a temperature

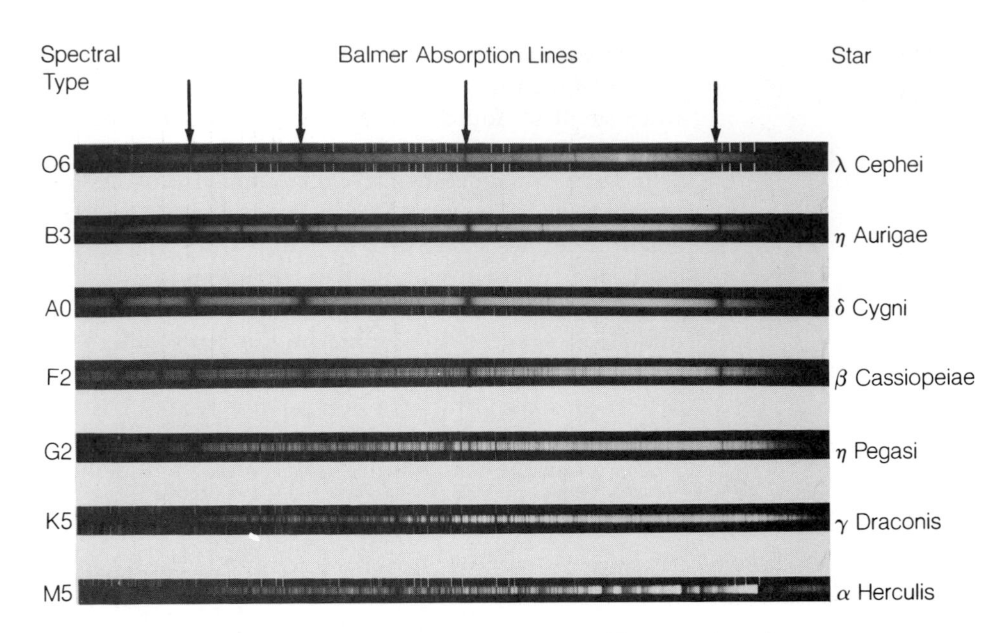

Figure 8-25 The seven principal classes of stellar spectra show different series of absorption lines produced by different atoms and ions. The Balmer absorption lines from hydrogen atoms dominate the spectra of type A stars and, to a lesser degree, those of type B and type F stars. The hottest stars (type O) show absorption lines from helium ions, while the coolest surface temperatures (type M) lead to absorption lines from various kinds of molecules.

in its outer layers, it will ionize all its hydrogen atoms by collisions and show no Balmer lines. Conversely, with too low a temperature, the star's outer layers will have no hydrogen atoms with electrons bumped into the second-smallest orbit by collisions. Again, no Balmer lines will appear in the spectrum. Thus, when astronomers went on to name stars with less visible Balmer lines B stars, F stars, K stars, O stars, and so on (some classes were soon dropped), they were naming stars both hotter and cooler than the A stars.

The names stuck, though, and are still used today. Now that we know the processes that produce the Balmer absorption lines, we can classify stars on the basis of the temperature in their outer layers. The stars range from high to low surface temperatures in the following order of classification:

O B A F G K M

This makes no particular sense unless you know the history of the classification. To remember this order, generations of young astronomers have memorized the rather sexist mnemonic phrase, "Oh Be A Fine Girl, Kiss Me." Many other suggestions for such a memory device have been made, including "Oh Be a Fairy Godmother, King Me" and "Orchids Blossom Among Fresh Green Kale, Mister." You may enjoy creating your own

Table 8-5 Surface Temperatures and Spectral Features of Different Types of Stars

Spectral Type	Average Surface Temperature (K)	Color	Outstanding Spectral Features	Examples
O	30,000	Blue	Absorption lines from ionized helium; weak hydrogen absorption lines	Mintaka
B	20,000	Blue	Absorption lines from un-ionized helium; hydrogen absorption lines stronger than in O stars	Rigel Spica
A	9,000	Blue	Hydrogen absorption lines strongest; very weak absorption from neutral helium; some absorption lines from singly ionized magnesium, calcium, and titanium	Sirius Vega
F	6,600	Blue to white	Hydrogen absorption lines still dominate spectrum; stronger lines from "heavy" elements	Procyon Canopus
G	5,500	White to yellow	Hydrogen lines weak; many absorption lines from "heavy" elements singly ionized; most prominent absorption from calcium ions	Sun Alpha Centauri A
K	4,000	Orange to red	Increasing number of absorption llines from un-ionized "heavy" elements	Arcturus Aldebaran Alpha Centauri B
M	2,800	Red	Absorption lines from un-ionized "heavy" atoms and from simple molecules; absorption from titanium oxide dominates spectrum	Betelgeuse Barnard's Star

mnemonic slogan if you plan to memorize the order of stars' spectral classes from hottest to coolest.

The basic spectral types were further divided into subcategories, designated (luckily) by simple numbers. Thus a G2 star such as our sun has a slightly hotter surface than a G3 star. The G9 class merges into the K0 class and so on. Each basic classification has ten subdivisions, running from 0 for the hottest members of each class to 9 for the coolest. Table 8-5 lists the seven basic classes, and Figure 8-26 gives a visual representation of what the table describes. Some special classifications for cool stars, classes R, N, and S, are not included in the basic table and indeed have interest only to specialists in stellar spectra. But the seven types of stars shown in Figure 8-26 deserve closer attention. From these spectra astronomers have unraveled the mysteries of how stars are born, age, and die.

8.6 The Temperature-Luminosity Diagram

Astronomers began to use the spectral classification scheme during the early years of this century. Before long, they realized that its greatest importance lay in combination with information on the absolute brightness of stars. Since the determination of any star's absolute brightness depends on the star's distance, astronomers looked at star clusters, all of whose members have about the same distance from us. As in the case of the stars in the galaxy called the Small Magellanic Cloud (page 171), if we compare the apparent brightnesses of stars we believe to have the same distance from us, we are automatically comparing the stars' luminosities. The trick is to put this comparison together with the information on the stars' surface temperatures (from their spectra) and see what we can find out about the stars in a cluster.

The result was the **temperature-luminosity diagram,** a graph of stars' relative brightnesses (or absolute brightnesses, if we can determine the distance to a star cluster) and surface temperatures. Figure 8-26 shows such a diagram for the stars closest to the sun. Notice that the temperature scale runs backward: The temperature *decreases* from left to right. This anomaly arises from the use of the spectral class letters at a time when the role of surface temperature was just beginning to be understood. Astronomers often refer to diagrams such as the one shown in Figure 8-26 as **Hertzsprung-Russell (H-R) diagrams,** because they were first made independently by two outstanding astronomers of the early twentieth century, Ejnar Hertzsprung and Henry Norris Russell.

The Main Sequence

When astronomers constructed the temperature-luminosity (H-R) diagrams for stars in clusters, they found that the chief feature of all such diagrams is the appearance of most stars along a certain part of the diagram: the **main sequence** (Figure 8-27). Most stars have surface temperatures and absolute

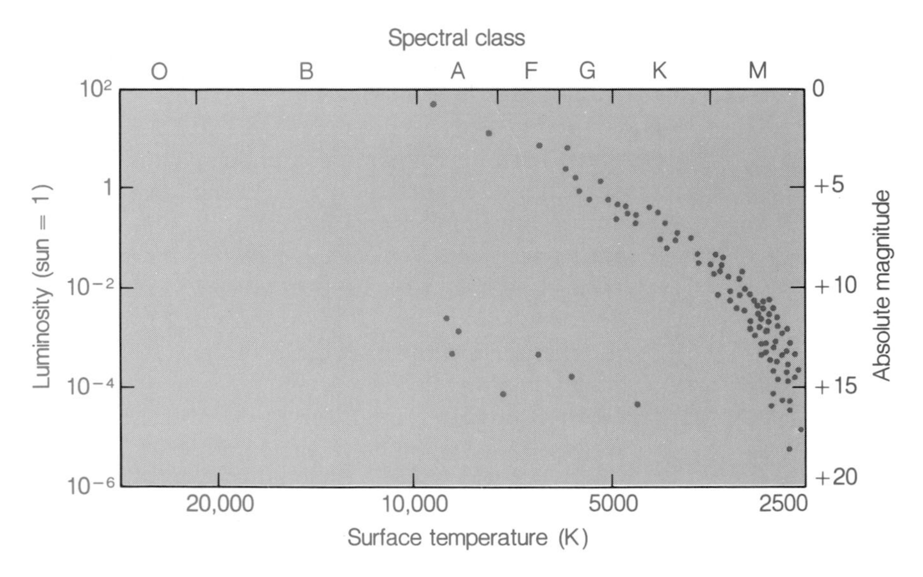

Figure 8-26 The temperature-luminosity diagram for the stars closest to the sun shows that most of these stars have lower luminosities and lower surface temperatures than the sun. Note that the surface temperature scale decreases from left to right.

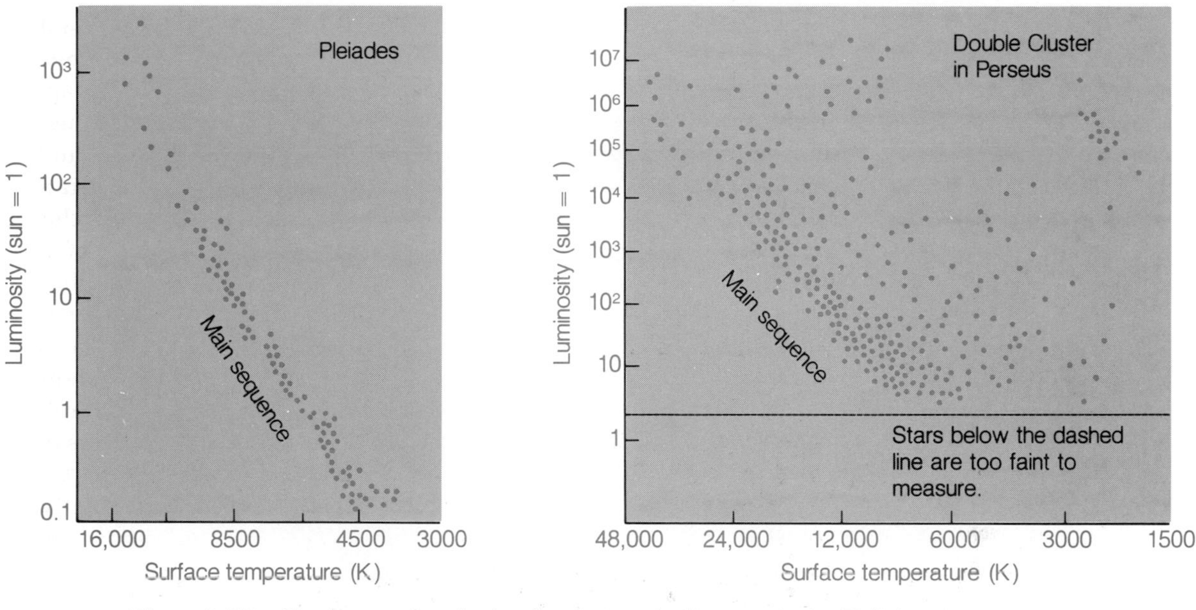

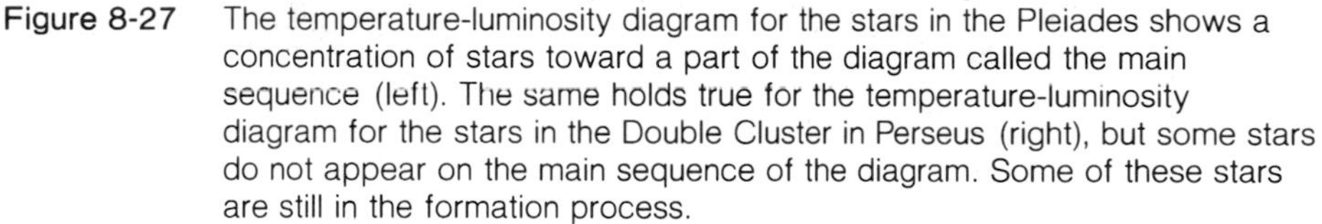
Figure 8-27 The temperature-luminosity diagram for the stars in the Pleiades shows a concentration of stars toward a part of the diagram called the main sequence (left). The same holds true for the temperature-luminosity diagram for the stars in the Double Cluster in Perseus (right), but some stars do not appear on the main sequence of the diagram. Some of these stars are still in the formation process.

luminosities that place them on this particular portion of the temperature-luminosity diagram. Hotter surface temperatures correspond to greater luminosities, and vice versa. We now know that the hotter stars on the main sequence are those with the larger masses and that low-mass stars have low surface temperatures and low luminosities. Our knowledge of how stars liberate kinetic energy, developed during the 1940s and '50s, has provided us with a fairly good understanding of the principal events in a star's lifetime. As a star contracts, it passes through a stage of infrared emission, making its way in a few million years to a given position on the main sequence (Figure 8-28). This position, which a star will occupy during most of its lifetime, depends only on the star's mass (Figure 8-29).[2] Stars do not move up or down the main sequence as they grow older. So long as a star can balance its self-gravitation with the internal release of kinetic energy, it will maintain the same position on the main sequence of the temperature-luminosity diagram.

---

[2]This statement is completely accurate only if all stars have the same chemical composition. In fact, the chemical composition varies, but only slightly, from star to star (page 323).

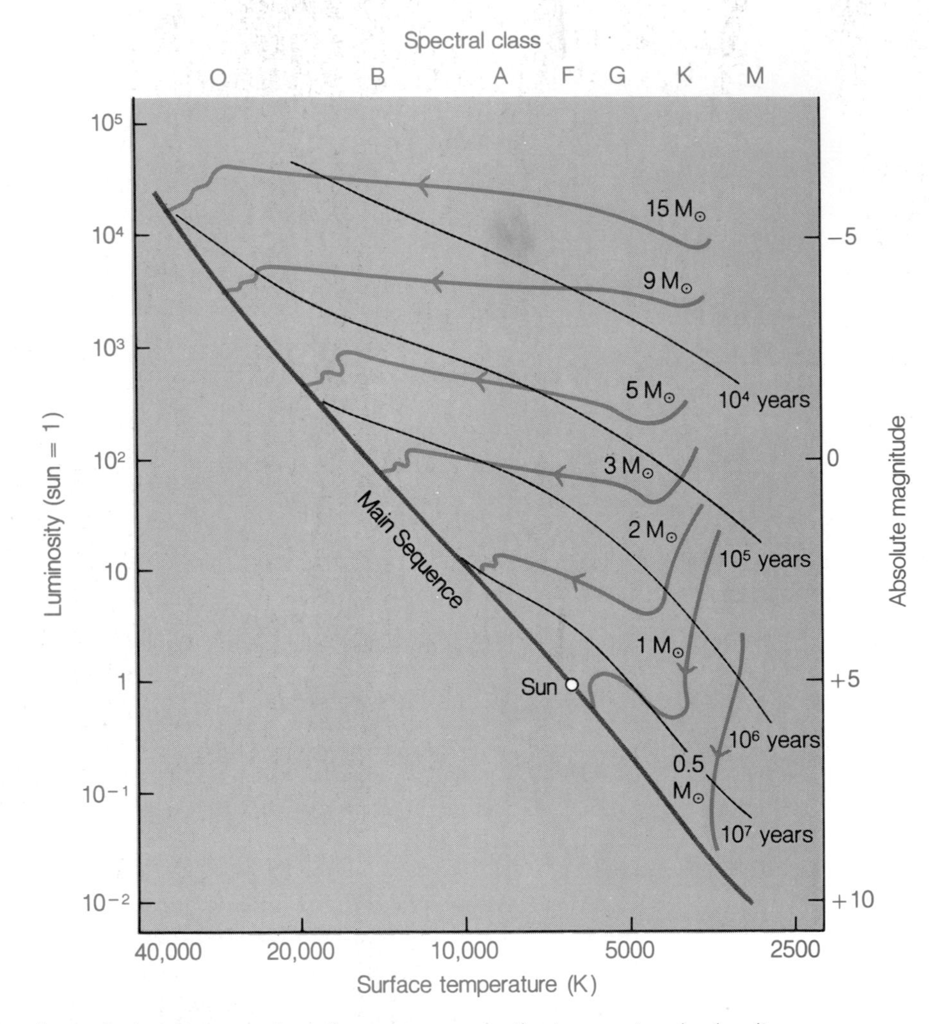

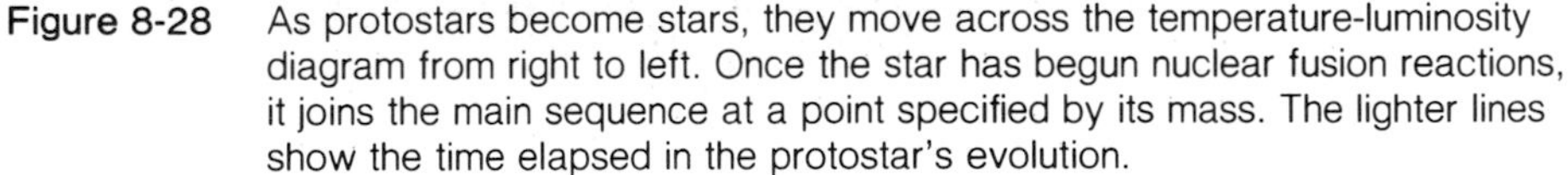

Figure 8-28 As protostars become stars, they move across the temperature-luminosity diagram from right to left. Once the star has begun nuclear fusion reactions, it joins the main sequence at a point specified by its mass. The lighter lines show the time elapsed in the protostar's evolution.

The "Usefulness" of the Main Sequence

The main sequence has the same location on the H-R diagram for cluster after cluster of stars. In other words, stars behave in the same way throughout our galaxy, and presumably in other galaxies as well. Thus if astronomers observe the spectrum of a star and find that it looks like a typical main-sequence spectrum at a given temperature, they can assume that the star *is* in fact typical. Then they can assign its absolute brightness (or luminosity) by giving the star its proper location on the main sequence, based on the star's temperature in its outer layers. Having determined the star's absolute bright-

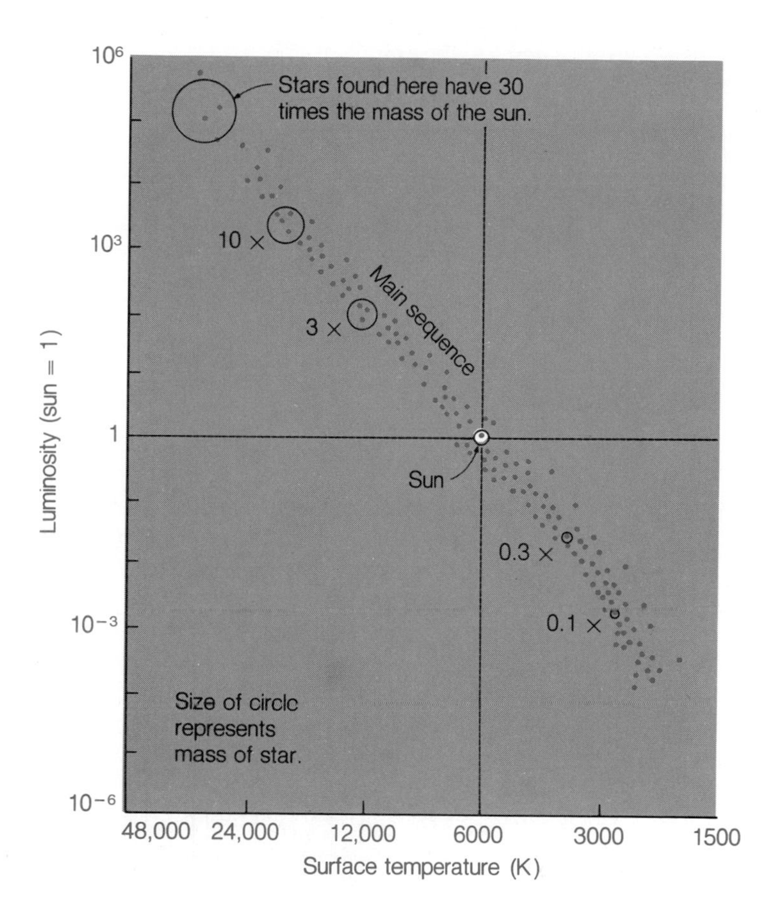

Figure 8-29
Stars on the main sequence with lower masses have lower surface temperatures and smaller luminosities than more massive stars.

ness in this way, astronomers can compare the apparent and absolute brightnesses to find the star's distance (Figure 8-30). This spectroscopic distance estimate (based on observations of the star's spectrum), sometimes misleadingly called the star's **spectroscopic parallax,** does not have the same reliability as the true, trigonometric parallax determination. Spectroscopic distance estimates do have great usefulness, however, for stars too far away to have their parallaxes measured accurately (page 25). With careful work, astronomers have even been able to apply this spectroscopic method to the stars that do *not* fall on the main sequence of the temperature-luminosity diagram.

8.7 Red Giants and White Dwarfs

What are these stars that do not appear on the main sequence? These exceptions fall into one of two categories. Some stars, far above the main sequence, have much larger true brightnesses than main-sequence stars with the same surface temperature, whereas other stars have far smaller true brightnesses (Figure 8-31). Astronomers realized 60 years ago that these stars must be ei-

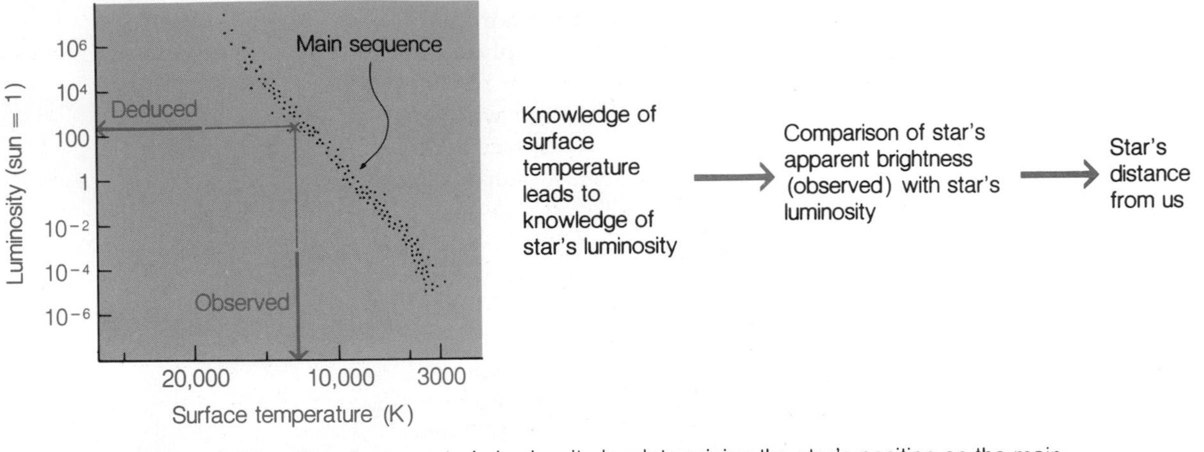

Figure 8-30 If we know a star's luminosity by determining the star's position on the main sequence, we can compare the star's apparent brightness with its true brightness to determine the star's distance.

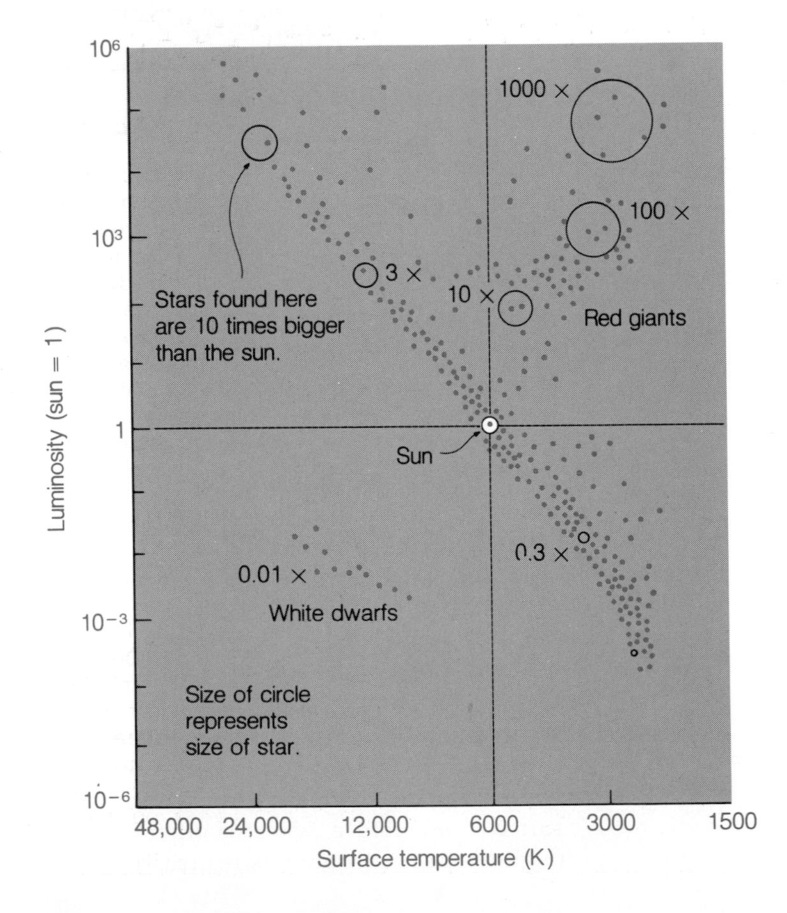

Figure 8-31
Some stars have much larger luminosities than main-sequence stars with the same surface temperature. These stars are much larger than main-sequence stars of the same temperature and are called red-giant stars. The stars whose luminosities are much smaller than main-sequence stars with the same surface temperature are much smaller in size and are called white-dwarf stars.

ther extraordinarily large or extraordinarily small in comparison with normal stars. They knew this by applying the laws of physics that deal with ideal radiators, which stars closely resemble. Experiments show that the luminosity of a given ideal radiator will vary in proportion to its surface area times the fourth power of its surface temperature (see box). Thus, stars with the same surface temperature should have absolute brightnesses that vary in proportion to the stars' surface areas. These areas in turn vary in proportion to the square of each star's radius. Hence if we determine from our study of stellar spectra that two stars have the same surface temperature, but then find that one star has 100 times the true brightness of the other, we can conclude that the brighter star must have 10 times the radius of the fainter star. But what astronomers found was that some stars with the same surface temperature differed in absolute brightness by a factor of 10 billion (10^{10}). The brighter star had a radius 100,000 times the radius of the fainter star! In that case, the star with the larger absolute brightness must be immense—as large as the sun would be if it swelled beyond the Earth's orbit.

Red Giants and Red Supergiants

Stars of immense size and relatively low surface temperature are called **red giants** and (in the largest and brightest cases) **red supergiants.** The red su-

The Surface Temperature, Surface Area, and Absolute Brightness of a Star

Any object that radiates photons with a spectral distribution that follows the ideal-radiator distribution (page 120) will emit a total photon energy per second that varies in proportion to the object's surface area times the fourth power of its surface temperature. The mathematical expression of this fact, called the **Stefan-Boltzmann law,** states that

$$L = A\sigma T^4$$

where L is the star's absolute luminosity (in ergs per second), A is the star's surface area (in square centimeters), T is the surface temperature (in Kelvins), and σ is a universal constant, called the **Stefan-Boltzmann constant,** equal to 5.67×10^{-5} erg/cm^2 K^4 sec.

Thus if two stars have the same surface area, but one has twice the surface temperature of the other, the hotter star will radiate sixteen times more photon energy per second. This assumes that both stars have ideal-radiator spectral distributions, which is usually a fairly good approximation. Conversely, if two stars have the same temperature, but one has sixteen times the absolute luminosity of the other, the intrinsically brighter star must have sixteen times the surface area (hence four times the radius) of the fainter star.

The ideal-radiator spectral distribution has a characteristic shape, with a maximum number of photons emitted at a wavelength and frequency that depend on the temperature of the ideal radiator. If we measure temperature in degrees Kelvin and wavelength in centimeters, the relationship between the surface temperature T and the wavelength of peak photon emission, λ_{max}, follows the **Wien displacement law,** which gives

$$\lambda_{max} T = 0.28978$$

Thus if a star has a surface temperature of T = 10,000 K, the wavelength of maximum photon emission equals about 2.9×10^{-5} centimeter or 2900 angstroms. If the surface temperature equals 5000 K, the value of λ_{max} doubles, to 5.8×10^{-5} centimeter or 5800 angstroms. This shifts the location of λ_{max} to the yellow region of the visible-light spectrum.

Betelgeuse

Equal areas at the same surface temperature radiate the same amount of energy each second.

Barnard's star

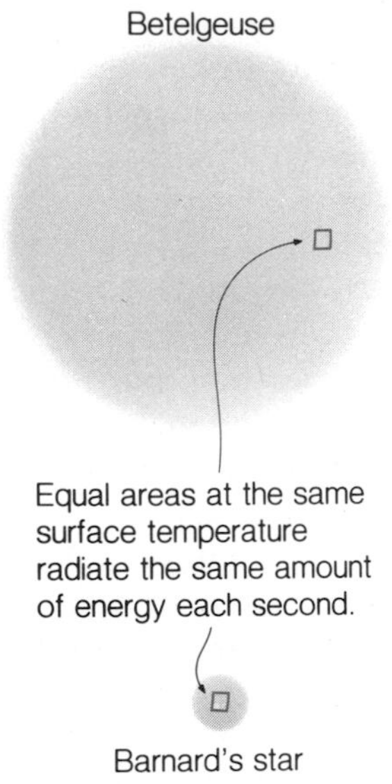

Figure 8-32
Betelgeuse and Barnard's star have almost the same surface temperature (3000 K), but Betelgeuse's luminosity is 100 million times that of Barnard's star. Since each square centimeter of Betelgeuse radiates the same amount of energy per second as a square centimeter at the surface of Barnard's star, Betelgeuse must have a radius 10,000 times that of Barnard's star.

pergiant Betelgeuse has about the same surface temperature as the main-sequence red star we call Barnard's star (see page 244). Betelgeuse's absolute brightness, however, exceeds that of Barnard's star by 100 million (10^8) times. Thus Betelgeuse must have a radius greater than that of Barnard's star by the square root of 100 million, or 10,000 times (Figure 8-32).

Red supergiants such as Betelgeuse may have radii of hundreds of millions of kilometers, but their *masses* do not reach such impressive figures. Ten or twenty times the sun's mass is normal for a red supergiant. Therefore the average density of matter inside a red giant or red supergiant must fall to extremely low values, because a relatively ordinary mass (for a star) must spread through an enormous volume. Betelgeuse has a radius about 200 times the sun's and a volume about 8 million times the sun's volume. But with only ten times the sun's mass, the material in Betelgeuse has an average density a million times less than the sun's average density. Most of any red supergiant consists of a relatively cool, fantastically rarefied cover that surrounds a small, energy-liberating core. But the area of the star's outer layers is so large that this cool envelope of gas produces a greater luminosity than the sun's.

White Dwarfs

What about the stars that lie below the main sequence in the temperature-luminosity (H-R) diagram? These stars have much lower absolute brightnesses than main-sequence stars at the same temperature and thus must have much smaller surface areas. They are neither particularly red nor particularly blue and have the name **white dwarfs.** White dwarf stars turn out to be quite common, but their low luminosities make them hard to see. For example, the white-dwarf companion to Sirius A, called Sirius B, has about the same surface temperature as its bright neighbor, but only 1/100 its radius and hence 1/10,000 its true brightness (Figure 9-18). Inside a white dwarf, the density rises enormously high, in sharp contrast to the red giants. Sirius B has a mass almost equal to the sun's within about a millionth of the sun's volume. As a result, the density of matter inside Sirius B equals about a million times the average density in the sun. Sirius B is far denser than any substance known on Earth.

Four major categories of stars—main-sequence stars, red giants, red supergiants, and white dwarfs—include almost all of the stars found in the cosmos. We use the terms *red* giants and *red* supergiants to stress the fact that most of them have low surface temperatures and red colors. These stars, like the white dwarfs, represent later stages of stellar evolution, which we shall describe in the next chapter.

How to Recognize Red Giants and White Dwarfs

By studying the details of the absorption lines observed in a star's spectrum, astronomers have learned how to tell the strength of gravity within the star's surface layers. The strength of gravity describes the pull toward the star's center. Higher gravitational forces tend to go with higher densities, which in turn means a greater tendency for ions to re-form atoms. This tendency in turn produces stronger absorption lines from the atoms. Thus Sirius A, a main-sequence star of spectral type A1, will show a weaker pattern of Balmer series absorption lines than Sirius B, a white dwarf of almost the same surface temperature. Similarly, a red-giant star exhibits a noticeably different spectrum than a main-sequence star of the same temperature (Figure 8-33).

Luminosity Classes

Astronomers add **luminosity classes** to the basic spectral types to specify giants, supergiants, or main-sequence stars. Roman numeral I denotes the largest giants, while V specifies a main-sequence star. Thus the sun is a G2V star, while Betelgeuse is type M2I. Aldebaran, the red giant in Taurus, is a K5III star. White dwarfs have a capital D before the spectral type, so Sirius B becomes a DA star, and the white-dwarf companion to Procyon is a DF star. With such notation we can classify stars in detail even before we think about how one type of star turns into another. But now we must pause to consider the one star we can examine in full detail: our own star, the sun.

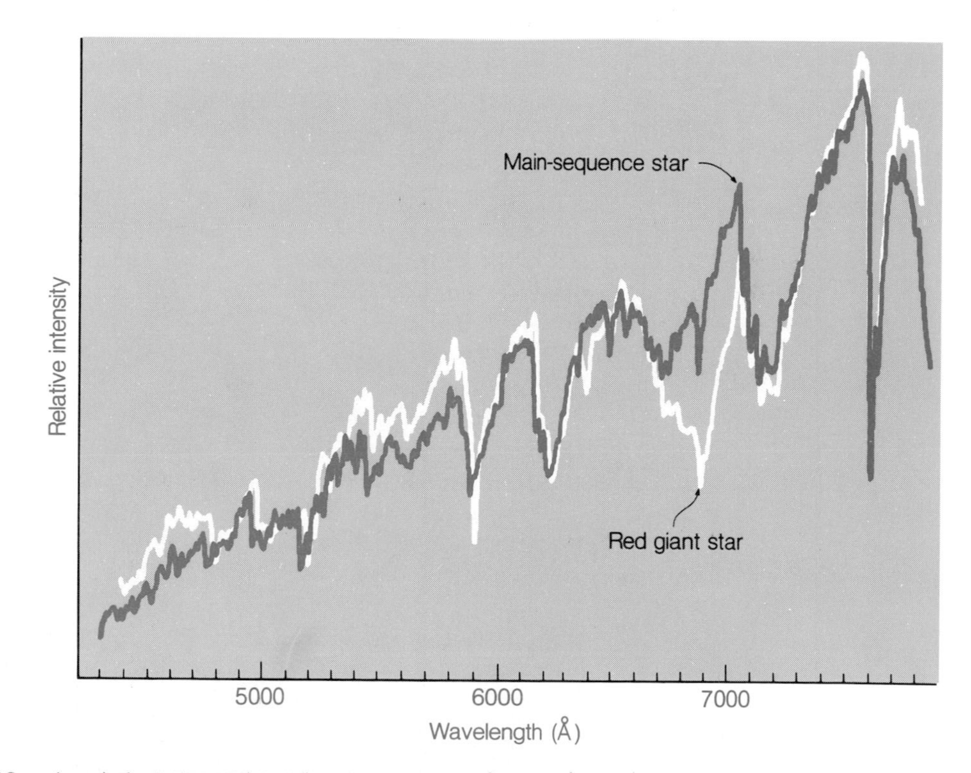

Figure 8-33 A red giant star and a red main-sequence star may have the same surface temperature. However, the main-sequence star has a much greater density of matter in its surface layers. This makes the spectra of the two stars noticeably different.

8.8 The Sun

The sun, our source of light and heat, was worshiped for countless ages by our ancestors, who sensed quite correctly that all life on Earth depends on our own star. The sunlight portrayed in Figure 8-34 as the hands of the

Figure 8-34 This Egyptian carving, more than 3500 years old, shows the sun's rays as hands reaching toward the Earth.

Egyptian sun god, Ra, in reality reaches the Earth from the solar surface, or photosphere, which glows in yellow-green light at a temperature of 5800 K. Because our atmosphere scatters more blue than red light, we see the sun as yellow-orange (Figure 8-35), and our eyes have evolved to be most sensitive to light of about that color and wavelength.

The sun's diameter of 1,400,000 kilometers exceeds the Earth's by 110 times, so it has a volume 1.3 million times greater than the Earth's. Since the sun's mass, 2×10^{33} grams, is "only" 333,000 times the Earth's, the sun has an average density of matter four times less than the Earth's: 1.4 times the density of water rather than the Earth's 5.5. In its size, mass, and density, the sun stands as a typical star, a bit more massive than most, but a bit less massive than the bright stars that typify our night skies.

In its luminosity, 4×10^{33} ergs per second, the sun likewise has an intermediate place between the most luminous stars, which have a million times the sun's true brightness, and the least luminous red main-sequence stars, with 10,000 times less true brightness than the sun. Our planet Earth, 150 million kilometers from the sun, intercepts one part in every 2.2 billion of the sun's energy output. Nevertheless, each square meter of the top of

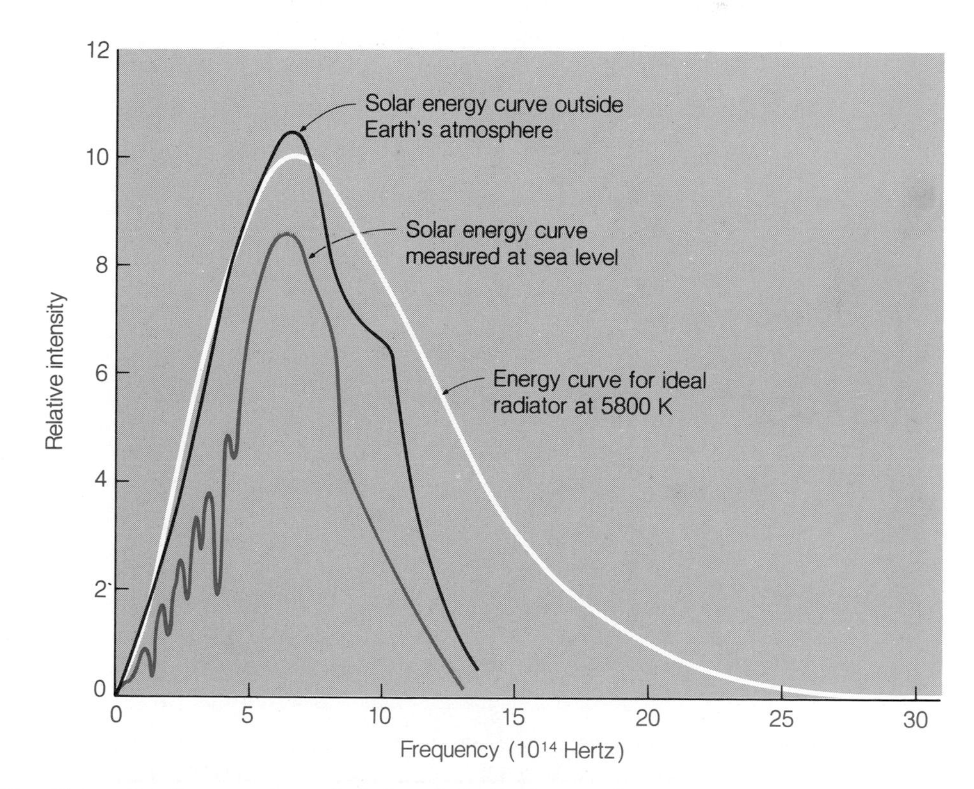

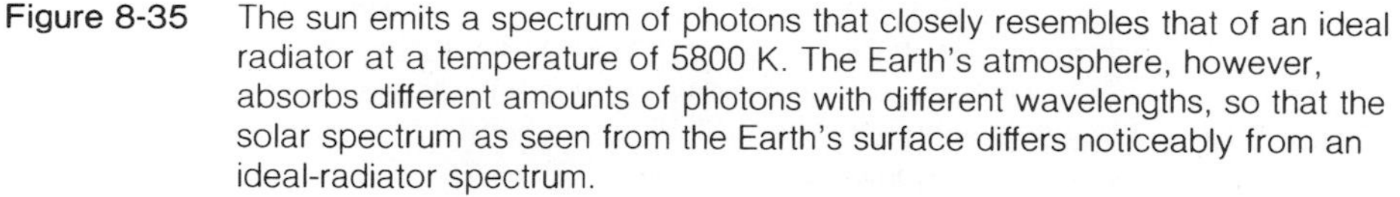
Figure 8-35 The sun emits a spectrum of photons that closely resembles that of an ideal radiator at a temperature of 5800 K. The Earth's atmosphere, however, absorbs different amounts of photons with different wavelengths, so that the solar spectrum as seen from the Earth's surface differs noticeably from an ideal-radiator spectrum.

Earth's atmosphere receives 1.39 kilowatts of power (1.39×10^{10} ergs per second) when exposed to the sun. (By the time the radiation reaches the ground, our atmosphere has reduced this figure by about half.) The sunlight that reaches the Earth every day could provide 10,000 times the energy the human race now consumes each day if we knew how to use it efficiently. We *do* use solar energy, of course, when we grow crops, bask in the warmth of the sun, feel the wind in our faces, or ride the surf. Solar radiation provides the ultimate driving force for all of these and more.

Neutrinos from the Sun's Interior

We have seen that all stars liberate kinetic energy through nuclear fusion, and the sun, a typical star, does so through the series of reactions called the proton-proton cycle (page 249). Astronomers can now calculate how the sun turns some of its energy of mass into energy of motion each second, deep in its interior at a temperature of 16 million K (Figure 8-36).

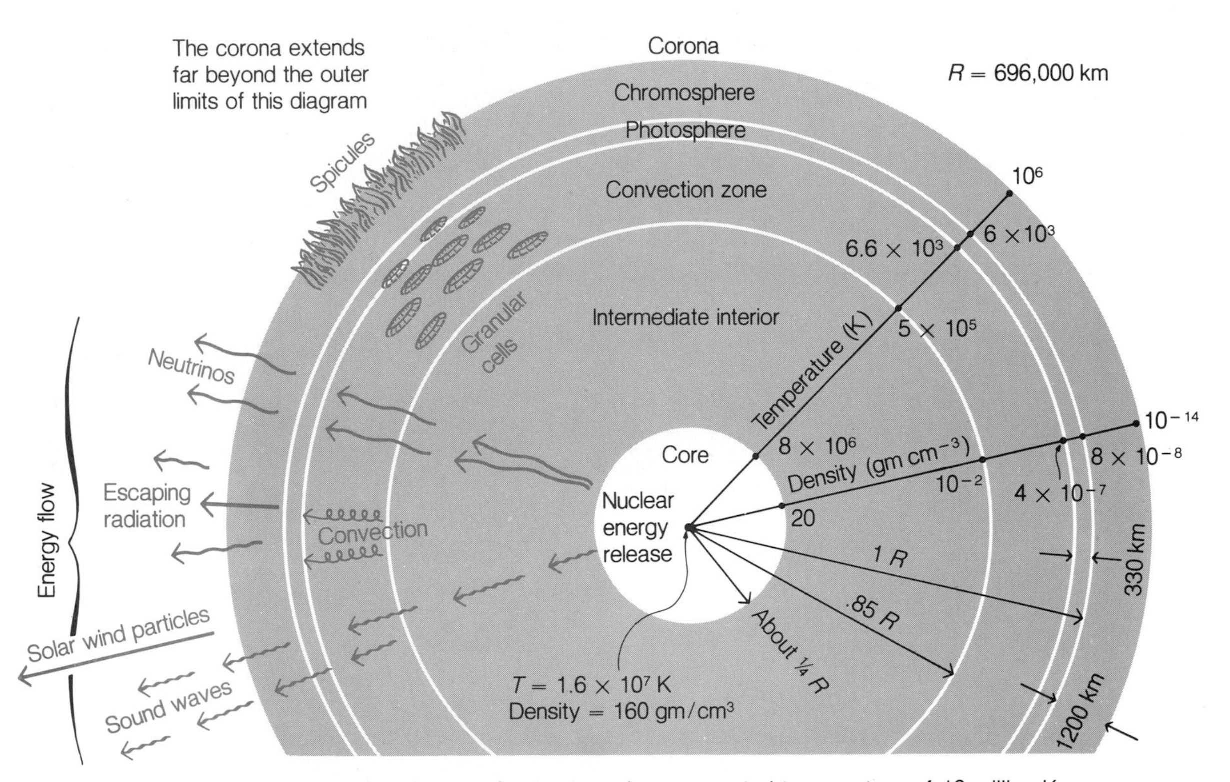

Figure 8-36 A model of the sun's structure shows a central temperature of 16 million K and a central density 160 times the density of water. Neutrinos produced in the sun's core usually escape directly into space with no further interaction with other particles in the sun. In contrast, the bulk of the kinetic energy released in the core takes millions of years to diffuse outward through collisions among particles.

Figure 8-37
The experiment designed to detect neutrinos from the sun holds 400,000 liters of perchloroethylene (C_2Cl_4) in a tank deep underground. Neutrinos from the sun have a tiny chance to strike one of the chlorine nuclei in the tank and turn it into an argon nucleus. The experiment can detect these argon nuclei one by one—a good thing, since the expected rate of chlorine-to-argon conversion for the entire tank is about one per day!

These fusion reactions also produce neutrinos (see page 249), which can escape directly from the center of the sun. Astronomers would like to detect these neutrinos and thus make the first direct observation of the sun's interior. To do this, they must build a neutrino detector—a tank with 400,000 liters of fluid that might react with neutrinos—and place it one kilometer underground, to shield it from other kinds of nuclear reactions that might confuse their instruments. For several years, such a neutrino detector has been in operation in (or under) South Dakota (Figure 8-37). With this detector, Professor Raymond Davis and his collaborators have detected neutrinos believed to come directly from the center of the sun. An odd result of the neutrino experiment is that far fewer neutrinos have been detected than astronomers expected—about one-sixth as many.

Consider how the energy of motion from nuclear fusion reactions makes its way to the sun's surface. As we saw for stars in general (page 258), the new kinetic energy (except for the neutrinos) spreads outward through huge numbers of collisions among photons and other particles in the sun. About 10 million years is needed for the energy liberated at the center of the sun to reach the surface. Thus it is possible that the sun's interior *now* (as represented by the neutrinos) may be doing something a bit different from what it was doing 10 million years ago (as represented by the energy now leaving the sun's surface). Although geological records on Earth testify to the approximate constancy of the sun's output over the past millions of years, some variation might have occurred. In fact, some theories suggest that ice ages on Earth arise from changes in the sun's energy output. Thus the neutrino puzzle and the ice-age patterns on Earth just might have a common explanation, though this theory remains far from proven.

The Solar Surface and Above

The Photosphere. High-resolution photographs of the sun's surface, or **photosphere,** reveal a seething, granulated set of gases in constant rippling motion. Each of the cells, or granules, in Figure 8-38 is a thousand or so kilometers across and lasts for only a few minutes as a distinguishable entity. The material in these granules consists mostly of hydrogen and helium gas, emitting radiation with a basically ideal-radiator spectrum. As the photons make their dash for freedom, some of them are absorbed, resulting in the characteristic solar spectrum (Figure 3-22).

The Chromosphere. Just above the solar photosphere lies the **chromosphere,** so called because of its pinkish color, visible during a total solar eclipse (see Chapter 11). The chromosphere extends upward for about 10,000 kilometers, and is hundreds of times less dense than the photosphere that forms the sun's visible surface. Although the chromosphere has far less density than the photosphere, its temperature far *exceeds* that of the photosphere. Energy transported into the chromosphere by shock waves gives the chromosphere a temperature of up to 40,000 K. Much of the upward transport of energy comes from jets of hot gas called **spicules,** which carry hot material from the photosphere into the chromosphere.

The Corona. The same principle that heats the chromosphere to tens of thousands of degrees heats the outermost layers of the sun to *millions* of degrees! These outermost layers, the sun's **corona,** extend outward for millions of kilometers (Color Plate L). Just a few hundred kilometers above the chromosphere, the temperature rises above 500,000 K! We know this from observing which kinds of ions appear in the coronal spectrum. When we find ions of iron with 16 of the atom's 26 electrons missing, we know that collisions

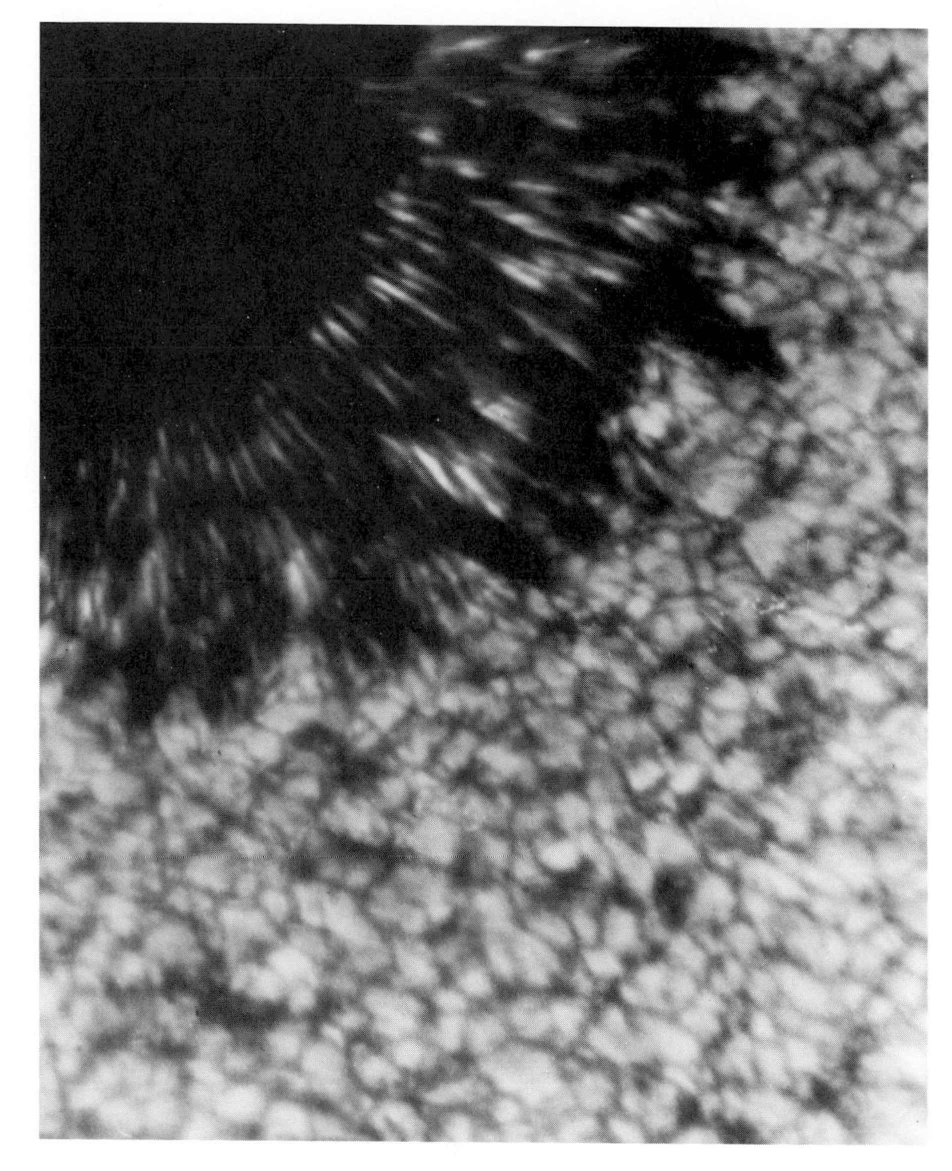

Figure 8-38
The sun's surface, or photosphere, shows a rippling pattern of granules, each 1000 to 2000 kilometers across. Part of a sunspot appears at the top left.

must occur at temperatures close to 2 million K. Such temperatures characterize the corona at distances more than 10,000 kilometers above the surface (Figure 8-39).

The corona achieves its high temperature by obtaining some of the kinetic energy from the sun's surface and distributing this energy among a tiny number of particles, in comparison to the number in the sun's surface layers. The energy reaches the corona through wave motion in the lower solar atmosphere, although astronomers do not know precisely how this occurs.

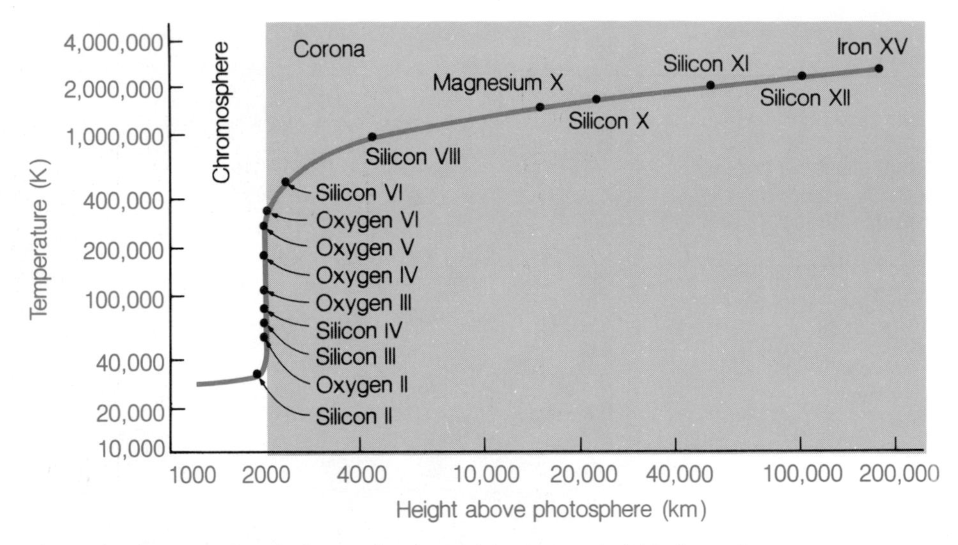

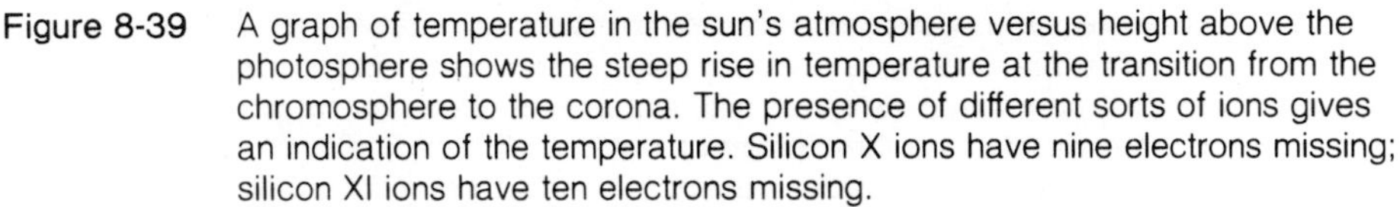
Figure 8-39 A graph of temperature in the sun's atmosphere versus height above the photosphere shows the steep rise in temperature at the transition from the chromosphere to the corona. The presence of different sorts of ions gives an indication of the temperature. Silicon X ions have nine electrons missing; silicon XI ions have ten electrons missing.

The Solar Wind

As a result of the corona's great temperature, some of its particles constantly escape from the sun. Massive though the sun may be, and strong as its gravitational force may become, the sun cannot retain particles with kinetic energies characteristic of 2 million K. As some particles continuously escape from the sun's gravity, they form a flow of gas called the **solar wind.** The solar wind flows faster and faster as it moves outward, because the same number of particles fill an ever-increasing volume as we look outward from the sun. By the time the solar wind reaches the Earth's orbit, it moves at a speed of 500 kilometers per second, still only 1/600 the speed of light.

Most of the solar wind consists of electrons and light nuclei, such as protons and helium nuclei, along with some heavier ions only partially stripped of electrons. Some parts of the corona are cooler than average and have a lower-than-average density of particles. The solar wind apparently escapes preferentially through these regions, called **coronal holes** (Figure 8-40).

Sunspots

Photographs of the sun such as those shown in Figure 8-41 often reveal **sunspots,** darker regions on the photosphere that can sometimes be seen with a carefully protected eye. (*Never* try to look at the sun directly—especially with a telescope! You can damage your eyes before you feel any pain.) Galileo claimed first discovery of the sunspots (and promptly encountered an opposing claim from the Jesuit astronomer Christoph Scheiner). With sun-

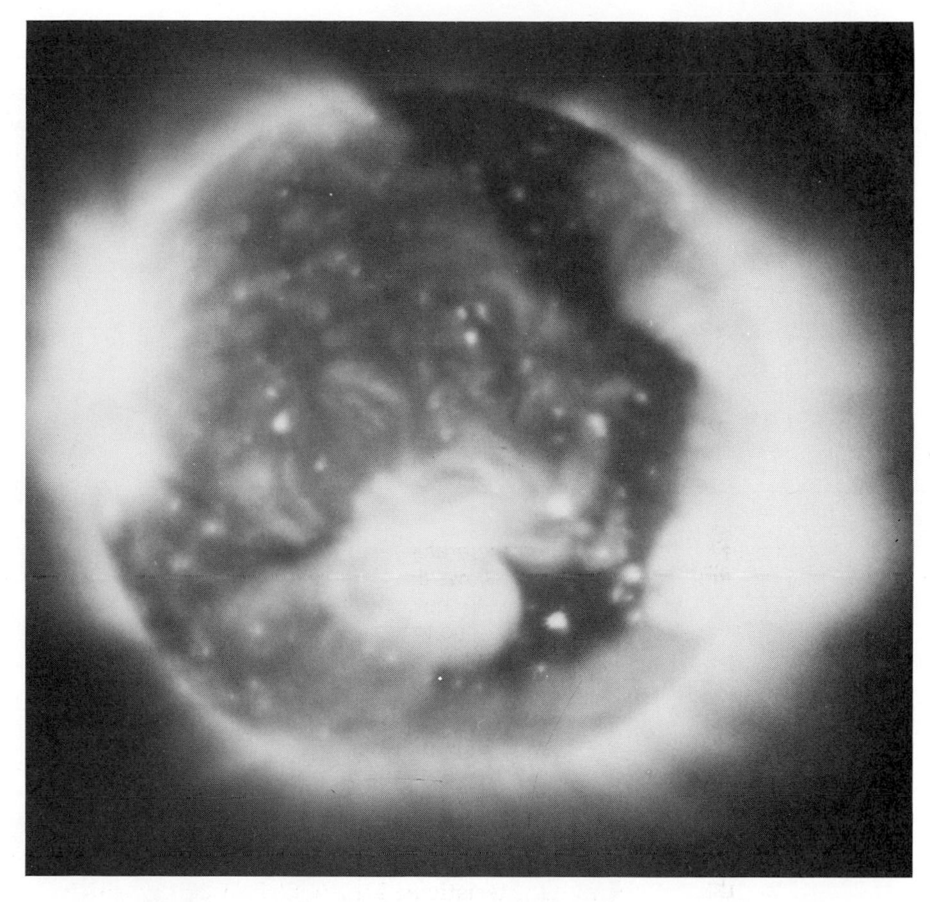

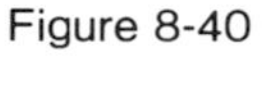

Figure 8-40 A photograph of the sun in x-ray emission, taken from a satellite in orbit above the Earth's atmosphere, shows the sun's hot corona. The dark region running from the top of the sun down the right side of the image is a coronal hole. Loops and streamers of gas seen above the solar disk reflect the orientation of the sun's magnetic field.

spots as markers, Galileo did make the first measurement of the sun's rotation: Our star rotates about once each month. More careful measurements based on sunspots, as well as on the Doppler effect for the gases at the rim of the sun, reveal that the sun rotates more slowly at regions farther from the solar equator (Figure 8-41).

Sunspots appear darker than the rest of the photosphere because they are cooler than average: 4500 K instead of the 5800 K for the rest of the sun's surface. In fact, sunspots radiate plenty of light, but less per square centimeter than other regions of the photosphere.

Sunspots have an intimate connection with the sun's magnetic field. Measurements show that the strength of the magnetic field within sunspots is much greater than that of the field outside. The sun's overall magnetic field has a strength of about 1 gauss, similar to the strength of the Earth's magnetic field. In sunspots, however, the magnetic field can easily be 2000 or 3000 gauss. On the other hand, no complete theory of how the magnetic field helps to form sunspots has yet been developed. The average sunspot forms over a period of a few days, lasts a few days or a few weeks, and merges back into the photosphere over a few days more.

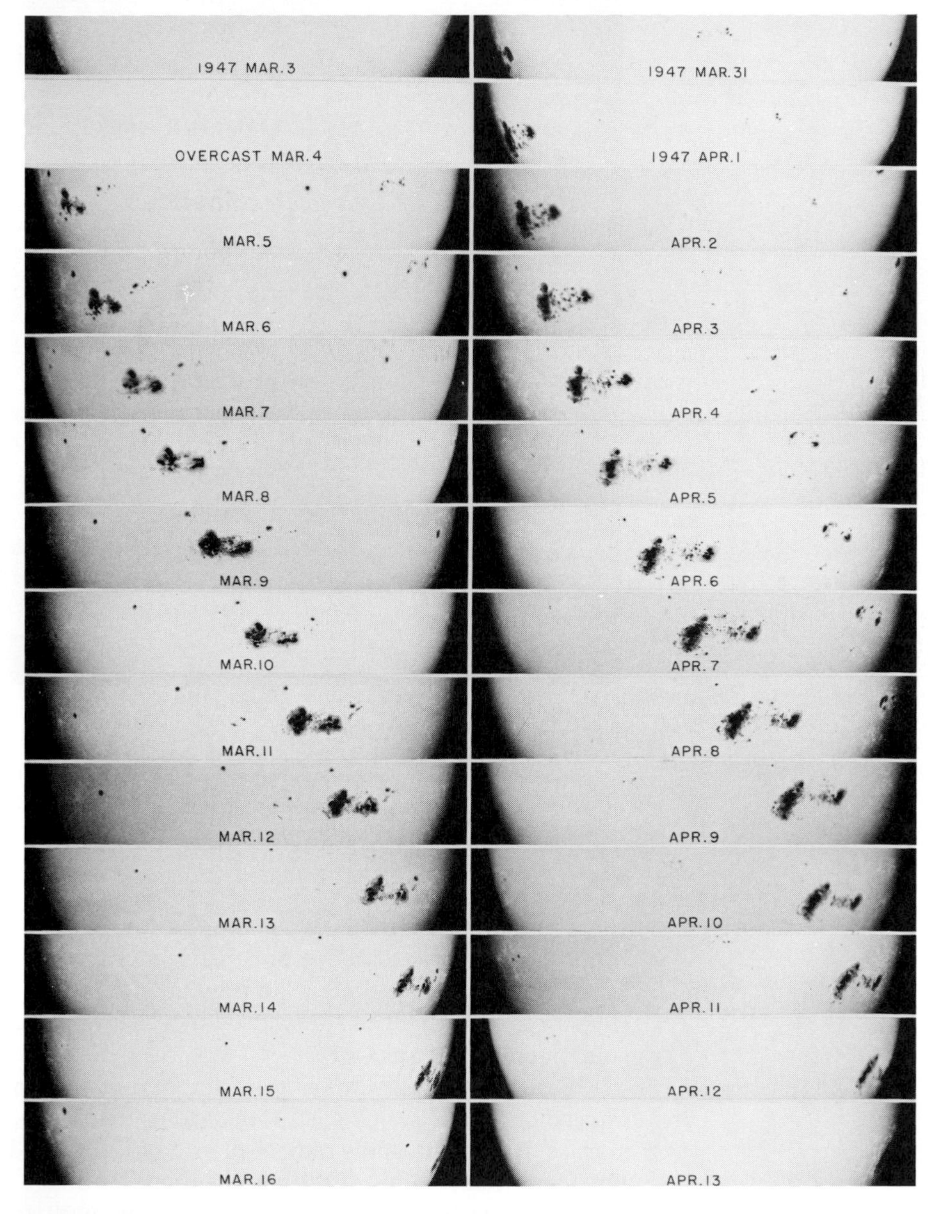

Figure 8-41 Photographs of the giant sunspot group of March–April 1947 show the sun's rotation. If we measure the rotation at various solar latitudes, we find that the lower latitudes (closer to the sun's equator) rotate in less time than the higher-latitude regions.

The sun's magnetic field varies in overall intensity over an eleven-year cycle, during which the average number of sunspots seen at any time varies too. Figure 8-42 shows the number of sunspots by year over the past few hundred years. During the period from about 1640 to 1715, far fewer sunspots than usual appeared, even during years of sunspot maximum. This absence of sunspots, first noted by E. W. Maunder and later verified by John Eddy, may have had important repercussions on Earth: The period coincided with the "little ice age" in Europe and with a pronounced drought in the southwestern United States. As we shall now see, this is not the only instance in which activities near the sun's surface affect our lives on Earth.

Flares, Filaments, and Prominences

We may think of sunspots as temporary "storms" on the sun's surface, somehow associated with increases in the magnetic field at those points. These storms are often accompanied by **solar flares,** the sudden release of energy stored in sunspot regions. Flares often bridge two or more sunspots above the photosphere. A flare, or solar eruption, can begin within a few seconds and can last a few hours, sending streams of electrons and ions tens of thousands of kilometers, or more, above the solar surface (Figure 8-43). The temperature within a flare can rise to 3, 4, or even 5 million K. Flares typically add to the stream of particles in the solar wind, and especially to the high-energy particles. These particles interact with the charged particles trapped

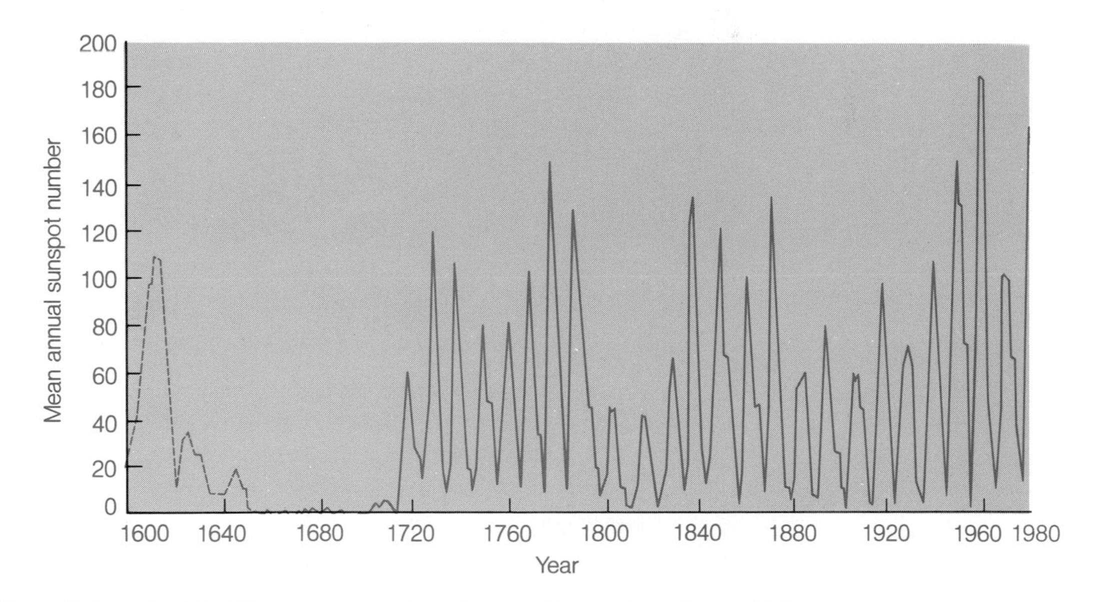

Figure 8-42 A plot of the average number of sunspots seen in a standard telescope each year shows the eleven-year cycle of variation. Also discernible on the graph is the Maunder minimum, or lack of sunspots between the years 1640 and 1715.

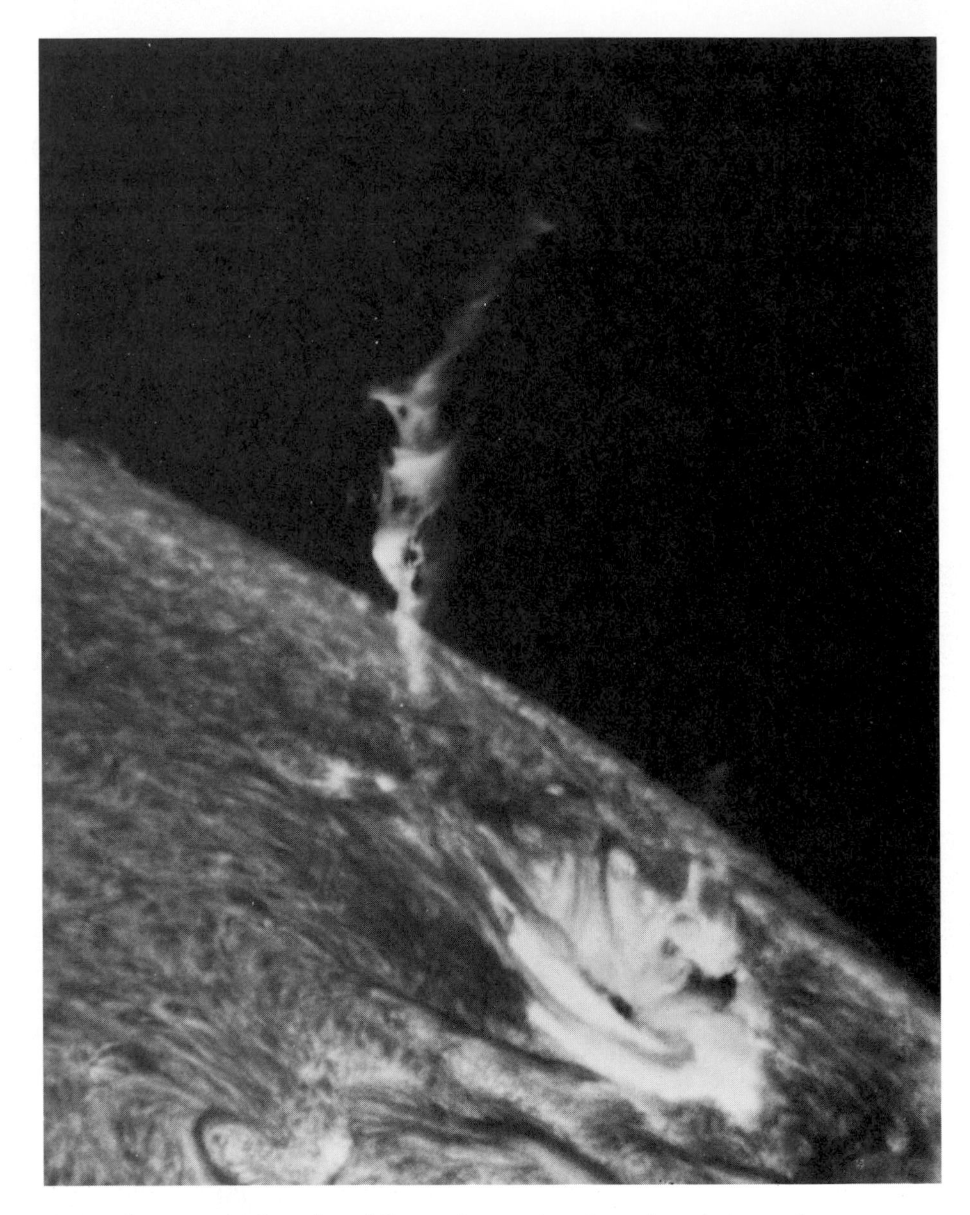

Figure 8-43 A giant flare seen at the edge of the sun has sent matter outward at speeds of several hundred kilometers per second. The area underneath the flare could contain a few hundred Earths.

by the Earth's magnetic field in the Van Allen belts (Figure 8-44). The result of this interaction is to make charged particles from the Van Allen belts spiral in toward the Earth's magnetic poles. Collisions of the charged particles with nitrogen and oxygen ions high in the Earth's atmosphere then produce the spectacular northern lights (technically, the *aurora borealis* and *aurora australis,* or aurora for short).

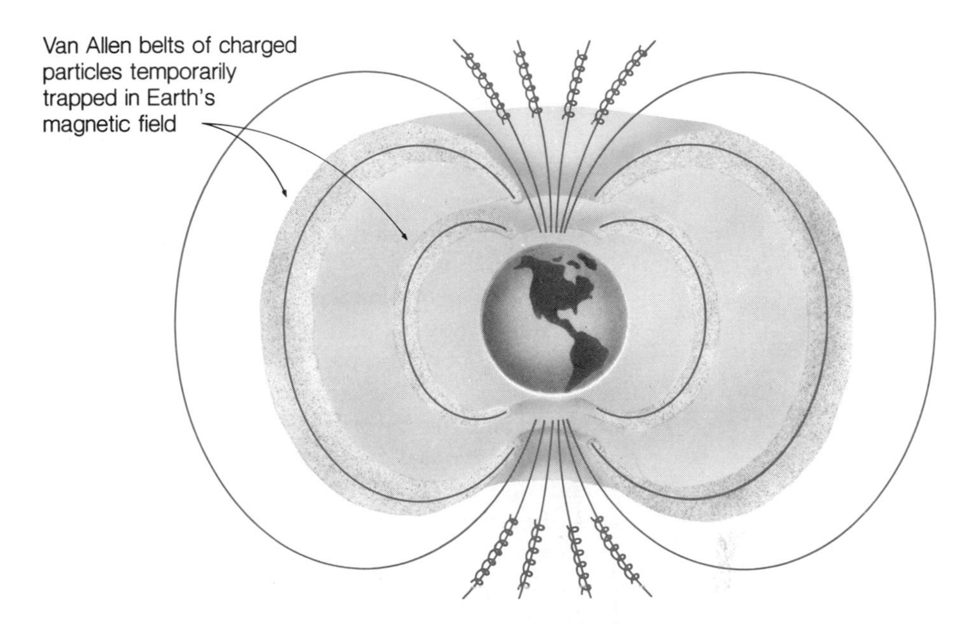

Figure 8-44 When charged particles in the solar wind encounter the Earth's magnetic field, they tend to join the particles trapped in the Van Allen belts, which eventually spiral inward toward the Earth's magnetic poles.

Figure 8-45
Spectacular displays of aurorae, or northern lights, appear preferentially in regions close to the Earth's magnetic poles.

Auroral displays shown in Figure 8-45 represent the most visible results of flares on the sun. More important on Earth are the disruptions in the Earth's ionosphere, the layer of charged particles in our atmosphere that reflects long-wavelength radio waves. Flares can disrupt large parts of our radio communications network, sometimes for days at a time, by affecting the reflection characteristics of the ionosphere. Partly for this reason, the United States government maintains a solar weather bureau to keep track of solar flares as they develop.

In addition to solar flares and the associated solar storms, sunspots are also related to solar **filaments** and **prominences.** These are the same thing seen from different perspectives. Photographed against the photosphere, the filaments appear as dark threads of gas (Figure 8-46). Seen at the edge of the sun, the filaments become prominences, loops and streamers of gas often hundreds of thousands of kilometers long (Figure 8-47). Great amounts of Hα emission, photons from the longest-wavelength member of the Balmer series of emission lines, arise in prominences when the electrons of hydrogen atoms jump from the third-smallest into the second-smallest orbit (see page 262). Some prominences last for weeks or months with little change (the quiescent prominences); others come and go within a few days or hours,

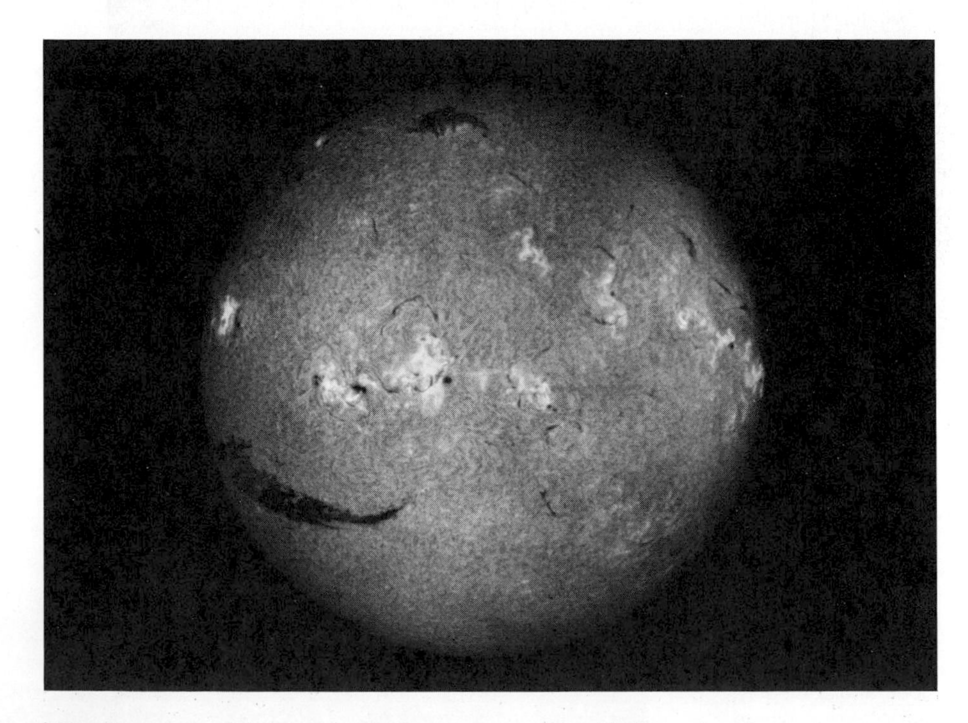

Figure 8-46 This photograph of the sun shows filaments as cooler (darker-colored) threads and patches just above the solar surface. The light-colored areas are called plages.

Figure 8-47 A solar prominence, made of hot gas rising to 200,000 kilometers above the sun's surface, can last for several days.

clearly connected with the sudden release of energy in the photosphere below.

Until we can examine the surfaces of other stars in detail, our sun must serve as the model for all others. Luckily, the sun provides us not only with our daily energy but also with a representative star, right in the middle of its main-sequence lifetime. Much later generations of terrestrial astronomers will have the chance to study a red-giant star at close range, but will have to make sure that the object of their interest does not totally consume them. For now, we rely on the sun's constancy despite its occasional outbursts of flaming gas and high-energy particles, which do not affect the total energy emitted in a significant way.

Summary

Stars apparently form from collapsing clouds of interstellar gas and dust. Although most stars in the Milky Way, and presumably in other galaxies as well, formed billions of years ago, we can still observe this process on a reduced scale. In spiral galaxies, the passage of the denser parts of a density wave pattern serves to trigger the collapse of interstellar clouds and a subsequent burst of star formation. Most stars form in groups, some as large as the million-star globular clusters. But the majority of these groups have too little mass to remain a unit for billions of years.

For all stars, the transition from protostar to energy-liberating star came with the onset of nuclear fusion in the star's center. Stars shine by turning energy of mass into energy of motion (kinetic energy), a process described by Einstein's formula $E = mc^2$. To liberate kinetic energy, most stars use a series of nuclear reactions called the proton-proton cycle. The most massive stars employ a different set of reactions called the carbon cycle. The result is the same: Four protons fuse into a helium nucleus, with the release of some energy of motion made from now-vanished energy of mass.

Deep inside the star, huge numbers of nuclear fusion reactions thus liberate kinetic energy, which appears as extra energy of motion of the particles there. Countless collisions among these particles then spread the newly liberated energy outward through the star. This flow of energy allows the star to resist its tendency to collapse from self-gravitation. At the star's surface, the photons that accompany the high temperature (produced by the liberation of kinetic energy) escape into space, carrying off the same amount of energy each second as the star produces in its central regions.

Most stars can exist for billions of years with a constant rate of energy liberation. During this time, they have surface temperatures and luminosities that depend on the amount of mass they contain. A given star will occupy a location on the main sequence—the region on the temperature-luminosity diagram that characterizes well-developed stars—that reflects the mass of that star.

By far the best-studied and closest star is our sun, whose bright photosphere (visible surface) provides the light and heat on which we depend. This photosphere has a temperature of 5800 K, placing the sun in the middle of the main sequence. Sunspots, cooler regions on the surface, increase and decrease in number over an eleven-year cycle. Above the photosphere lies the chromosphere, only a few thousand kilometers thick, and above that, the gauzy corona, extending millions of kilometers into space and reaching temperatures of 2 million K. The solar wind, a stream of particles blown outward from the sun, extends past the Earth, sometimes affecting our radio communications.

Key Terms

astronomical unit
Balmer series
carbon cycle
chromosphere
color index
corona
coronal holes
deuteron
eclipsing binary stars
filament
forces
H-R diagram
luminosity classes
main sequence
photosphere
positron
prominence
proper motion
proton-proton cycle
protostar
radiation pressure
red-giant star
red supergiant
solar flare
solar wind
spectroscopic binary star
spectroscopic parallax
spicules
Stefan-Boltzmann law
Stefan-Boltzmann constant
sunspot
temperature-luminosity diagram
visual binary star
white-dwarf star
Wien displacement law

Questions

1. What forces hold stars together? What keeps stars from collapsing?
2. Why does nuclear fusion begin within a protostar?
3. What is a visual binary star? What is a spectroscopic binary star?
4. What is a star's proper motion? How can a star's proper motion suggest the existence of unseen companions?
5. What are eclipsing binary stars? Would they appear as eclipsing binaries to any other observer somewhere else in the Milky Way? Why or why not?
6. What raw materials do most stars fuse together to release kinetic energy? Why are high temperatures needed for this process?
7. How does a contracting protostar raise its central temperature to about 10 million K?
8. Why don't stars blow apart as they liberate great amounts of kinetic energy in their interiors?
9. What is the carbon cycle? Why do some stars release kinetic energy through the carbon cycle, although most employ the proton-proton cycle for this purpose?
10. Why do massive stars burn through their supplies of nuclear fuel in less time than less massive stars do?

11. What is the main sequence of the temperature-luminosity diagram? Why do stars of different mass occupy different positions on the main sequence?
12. What is the solar corona? Why does it have a temperature of 2 million K, 350 times hotter than the sun's surface?

Projects

1. *Star Colors.* Look at some of the brightest stars in the sky, such as Betelgeuse (Orion's brighter shoulder), Rigel (Orion's brighter foot), Sirius, Castor and Pollux (the two brightest stars in Gemini), Antares (in the heart of Scorpius), Spica (the brightest star in Virgo), and Deneb, Vega, and Altair (in the Summer Triangle), and try to assign colors to these stars. Since the human eye does not see color well for dim objects, you can make more accurate observations by using color filters (film of the appropriate color) and noting whether a given star seems brighter when seen through a blue or a red filter. (Blue and red stand near the opposite ends of the visible-light spectrum.) Which star appears reddest? Which bluest? What does this tell you about the star's surface temperatures? If a star seems equally bright when seen through red and blue filters, what color is the star?
2. *Binary Stars.* If you have a small telescope or a good pair of binoculars, you can observe some visual binary stars. Particularly good candidates are Albireo (in the head of Cygnus) and Epsilon Lyrae (northwest of Vega); both are well placed for viewing in summer and fall. Albireo shows a marvelous color contrast, and Epsilon Lyrae (a double-double system, in which two pairs of stars orbit each other) can be seen as double without a telescope (if you have good eyesight) but needs a good-sized telescope to show the doubling of each of the two members. In winter, you can look for the members of the Delta Orionis and Sigma Orionis systems in Orion. (See Figure 8-4, where Delta Orionis is called Alnilam.)
3. *The Orion Nebula.* In the evenings of fall and winter, you can see the large constellation Orion, complete with three bright stars in his belt and three somewhat fainter "stars" in the sword that hangs to the south from Orion's belt. The middle "star" is in fact the Orion Nebula, 460 parsecs away, the closest H II region to the sun (see Figure 8-4). With binoculars, you can see some of the Trapezium stars within the haze of the nebula. How many stars seem to be there? How would you judge the apparent brightness of the Orion Nebula in terms of the apparent brightness of the other stars in the constellation? The stars in Orion's belt are almost as far away as the Orion Nebula and rank among the stars with the greatest absolute brightness seen in the night skies. If these stars are 30 times farther away than the star Castor, in Gemini, how does their intrinsic, or true, brightness compare with Castor's? (Note that Castor has an apparent brightness about twice that of the stars in Orion's belt.)

4. *Sunspots.* Arrange to project the sun's image through a telescope, or through well-mounted binoculars, onto a piece of white paper. (Do *not* observe the sun directly!) Can you see any sunspots on the solar disk? Make observations for several days to estimate the length of time it takes a given sunspot to cross the solar disk, and use this estimate to derive the rotation period for the sun. What was the approximate solar latitude of the spot observed for a few days? How does this correlate with what you know about the variation of the sun's rotation period with solar latitude?

Further Reading

Baade, Walter. 1963. *The Evolution of Stars and Galaxies.* Cambridge: Harvard University Press.

Bok, Bart. 1975. "The Birth of Stars." In *New Frontiers in Astronomy.* Edited by Owen Gingerich. San Francisco: W. H. Freeman.

Dickman, Robert. 1977. "Bok Globules." *Scientific American* 236:6, 66.

Eddy, John. 1977. "The Case of the Missing Sunspots." *Scientific American* 236:5, 80.

Gamow, George. 1961. *A Star Called the Sun.* New York: Mentor Books.

Herbig, George. 1975. "The Youngest Stars." In *New Frontiers in Astronomy.* Edited by Owen Gingerich. San Francisco: W. H. Freeman.

Jastrow, Robert. 1969. *Red Giants and White Dwarfs.* New York: New American Library.

Page, Thornton, ed. 1968. *The Evolution of Stars.* New York: Macmillan.

Schatzman, Evry. 1974. *The Structure of the Universe.* London: George Weidenfeld and Nicolson.

Zeilik, Michael. 1978. "The Birth of Massive Stars." *Scientific American* 238:4, 110.

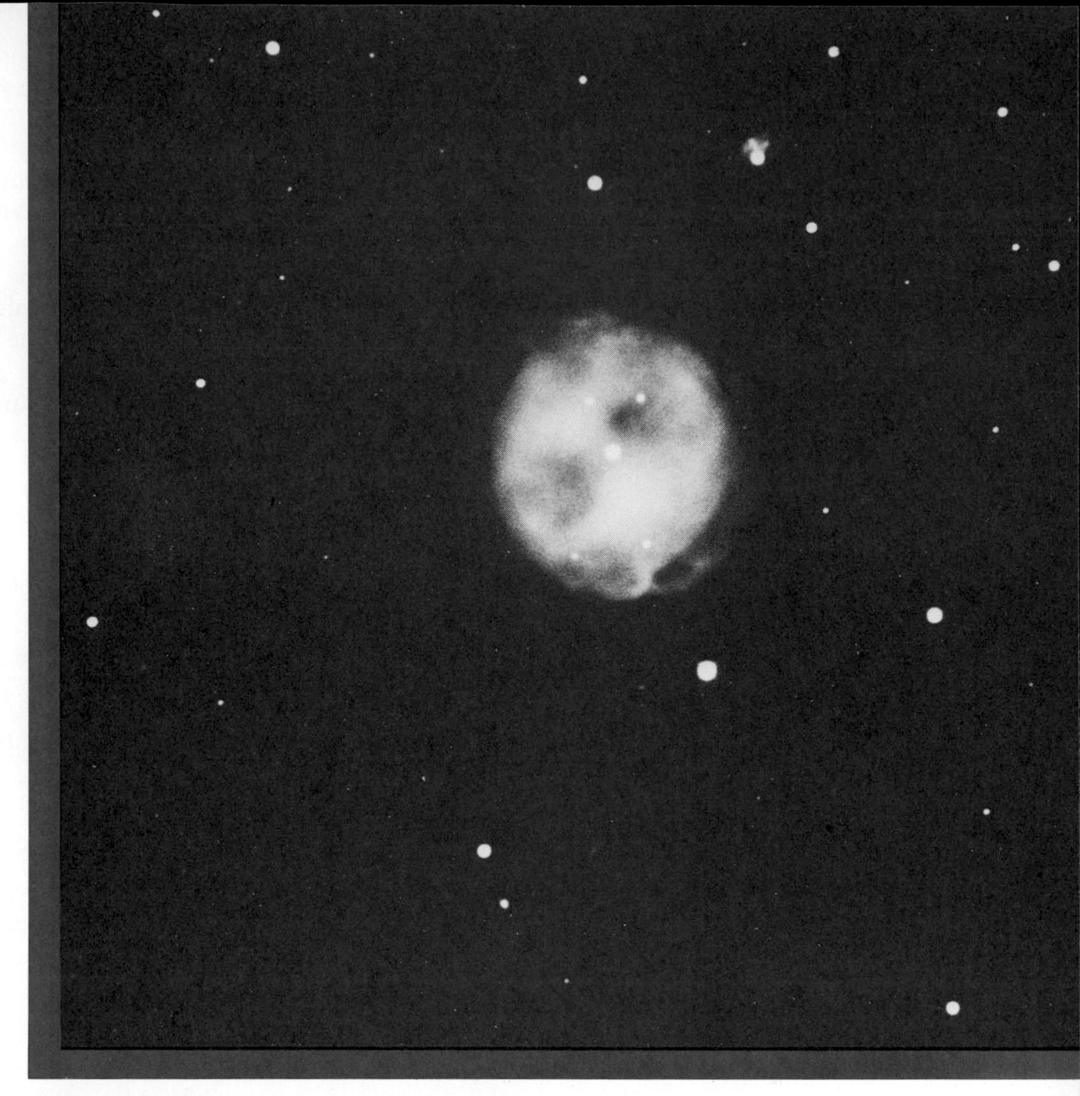

9

The "Owl Nebula," M 97, is a planetary nebula, the ejected shell of gas from an old star.

How Stars Age and Die

Astronomers once thought that stars begin to shine at their hottest, blue-white temperatures and cool down steadily thereafter. Reality turned out to be more complex than this. Although we do not yet know the full story of how a star passes through its life cycle, we do know that the inside and outside of a star may behave quite differently. Astronomers must try to learn how stars evolve by observing the stars' outermost layers (Figure 9-1) and by constructing models of the stellar interiors that explain how stars release energy. They must then apply their skills at coordinating observations and model building to explain the changes that occur as stars of various types grow old. Despite their lack of understanding of certain details, and of the time to watch a star age, astronomers have succeeded in combining calculations and observations to reach a partial comprehension of the stellar aging process.

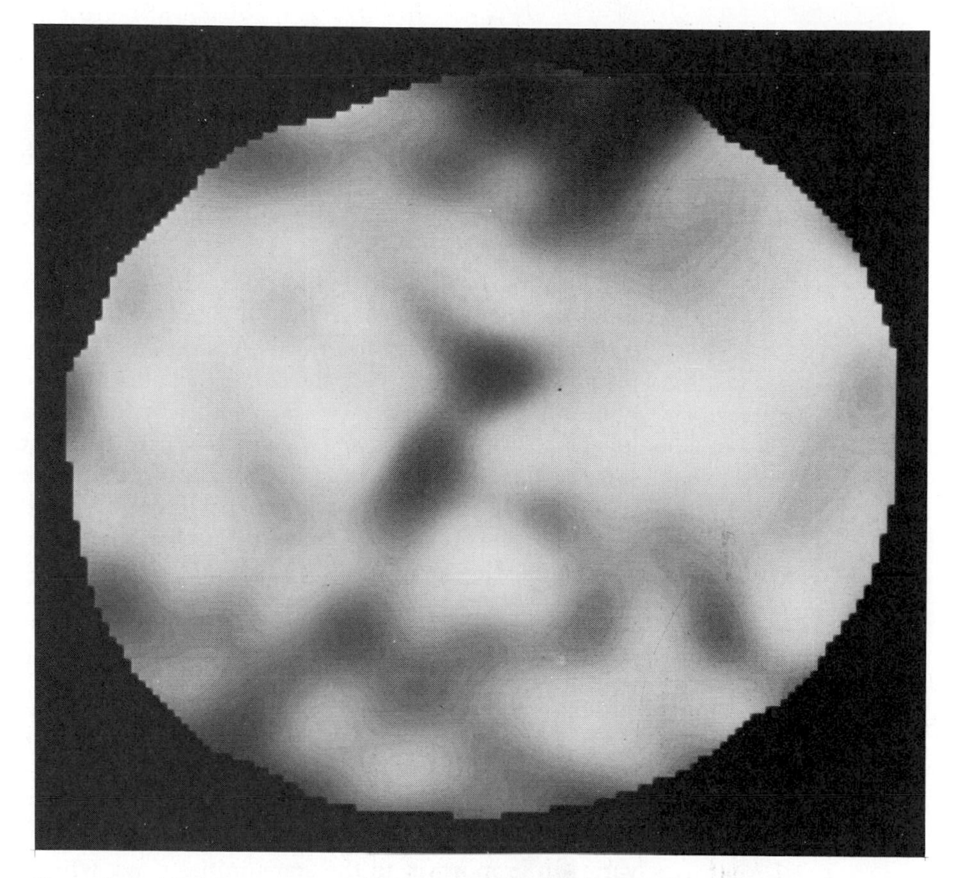

Figure 9-1 The red giant Betelgeuse, second-brightest star in Orion, has a diameter of about 100 million kilometers. This photograph, taken by a special technique called speckle interferometry, seems to show details of Betelgeuse's surface. In fact, we can barely see the disk of Betelgeuse in this picture, and the details are spurious.

9.1 Evolution from the Main Sequence

We have seen that the nuclear reactions of both the proton-proton cycle and the carbon cycle have the same effect: Four protons turn into one helium nucleus, plus positrons, neutrinos, and photons, and this fusion produces new energy of motion. The positrons are quickly annihilated with electrons to produce still more kinetic energy, whereas the neutrinos escape directly into space from the center of the star. The bulk of the energy of motion liberated by nuclear reactions inside the star must fight its way to the surface through millions of collisions among photons and elementary particles. The process of emerging takes about 10 million years in the sun or in similar stars. Hence a change in the rate of energy production at the star's center would produce a corresponding change in the output from the star's surface only 10 million years later. However, as we discussed on page 254, stars regulate their energy production rates extremely well. The sun, for example, has kept its energy output constant (to within a few percent or less) for at least the past 10 or 20 million years, as we can tell from fossil records of life on Earth.

Thus most of a star's lifetime consists of a steady internal liberation of kinetic energy, which flows outward and ultimately escapes into space as radiation. Although this accounts for most of the star's life, the initial and final stages, like the start and end of biological life, command special attention.

To produce kinetic energy, stars constantly turn hydrogen nuclei (protons) into helium nuclei. When a star first begins to liberate energy in its interior, about 90% of the nuclei (comprising 75% of the mass) inside it are protons.[1] As time goes on, the star turns more and more of these protons into helium nuclei, until finally the star's basic fuel supply, protons for the proton-proton or carbon cycle, becomes scarce in the star's interior. When the star has used up the supply of protons in its interior regions by converting them into helium nuclei, the star has passed from the prime of its life, the main-sequence phase, into the glorious senility of the red-giant phase of stellar evolution.

How Long Does a Star Last on the Main Sequence?

How long does it take for a star to exhaust its supply of protons and become a red giant? The answer depends on the star's *mass*. Strangely enough, more massive stars use up their nuclear fuel far more rapidly, and thus pass through their main-sequence prime of life far more quickly, than less massive stars do. The potential of any star for the liberation of kinetic energy varies in proportion to the star's initial mass, which consists mostly of protons. The *rate* at which stars liberate kinetic energy, however, varies approximately as the cube of the mass or even (for the most massive stars) as the mass raised to the fourth power. A star such as Regulus (in Leo) has five times the sun's mass and thus five times the sun's total potential for liberating energy. But Regulus in fact liberates energy at a rate 200 times the sun's. Therefore, Regulus's total lifetime as an energy-liberating star will be 40 times *less* than the sun's. (Five times the sun's energy potential divided by 200 times the sun's energy liberation rate gives $^1/_{40}$ the sun's lifetime.)

Stars stay on the main sequence of the temperature-luminosity diagram so long as they steadily convert protons into helium nuclei in their interiors. The time during which a star performs this fusion is the star's **main-sequence lifetime.** The sun and other stars with the sun's mass will remain on the main sequence for about 10 billion years, according to astronomers' calculations. In the case of the sun, about half this time (4.6 billion years) has

---

[1]Recall that the big bang (page 116) already had fused some of the protons into helium nuclei to give this proportion. Nuclei heavier than helium did not appear in abundances much greater than one-*millionth* of the proton abundance as the result of the big-bang fusion processes.

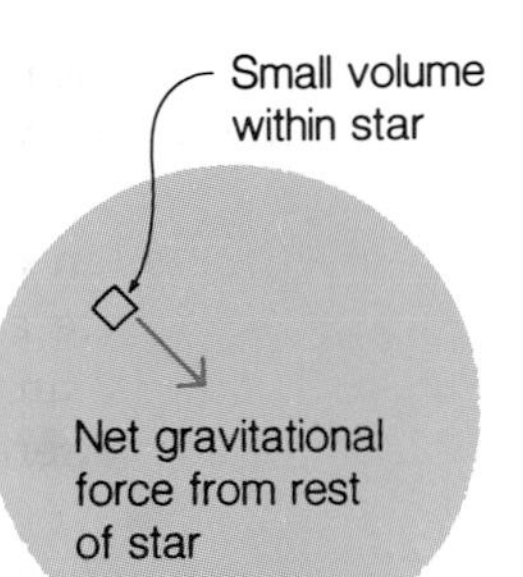

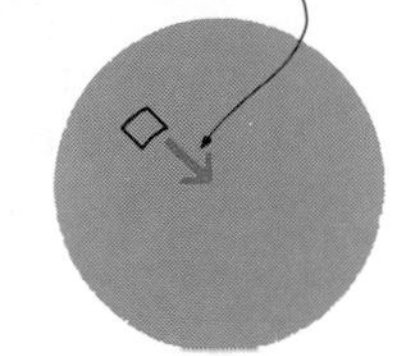

Figure 9-2
As the star's central regions contract during its later stages of evolution, these parts of the star will become closer to each other. As a result, the gravitational force on each small region from the rest of the star will increase.

already passed. A star with 1½ times the sun's mass will spend only 2 billion years on the main sequence; a star with 5 solar masses has a main-sequence lifetime of only 80 million years; and a star with 15 solar masses lasts only 10 million years on the main sequence, a mere instant in cosmic history. Stars with masses smaller than the sun's, which include the vast majority of stars, will remain on the main sequence longer than the sun. For example, a star with 80% of the sun's mass will be a main-sequence star for 20 billion years.

The Exhaustion of Protons: The Red-Giant Phase

What happens to a star as it runs out of protons in its interior? A star running low on nuclear fuel finds itself in a gravitationally induced dilemma. Its self-gravity keeps pulling its outer layers inward, and the energy released at the center must oppose this inward pressure or the star will collapse. As the star's center uses up its supply of protons, the star can avoid collapse only by continuing to liberate kinetic energy. To consume its dwindling supply of nuclear fuel more rapidly, the star contracts, and thus heats, its central regions. The higher temperature (more energy of motion per particle) makes the protons fuse together more rapidly (see page 254). The extra energy released each second tends to keep the core from contracting further, but as still fewer protons remain, only additional contraction can raise the temperature still further to fuse protons still more rapidly. Hence, as its supply of protons diminishes, the star's central regions contract slowly but steadily, always increasing the central temperature to consume the nuclear fuel ever more rapidly. In fact, the total kinetic energy liberated each second *increases,* because the star's center, having contracted, now exerts a greater self-gravitational force (Figure 9-2).

As the central core of the star contracts, however, the star's outer layers actually *expand.* Part of the extra energy liberated in the contracting core pushes the star's outer layers outward, and part of the surplus escapes into space. The star's luminosity increases, because more energy emerges each second from the star's larger surface. The star's surface temperature, however, does not increase, and in fact may decrease, because the expansion of the star's outer layers tends to cool them. In its newly expanded state, each square centimeter of the star's surface will radiate no more energy, and perhaps even less energy, each second than before. But since the total surface now contains many more square centimeters, the *total* amount of energy radiated from the star each second may increase dramatically. The star becomes a **red giant** of low surface temperature and large luminosity. A typical red giant may have a radius 100 times larger than the sun's, with a helium-rich core only a few times larger than the Earth's radius (Figure 9-3).

Figure 9-4 shows how a star's surface temperature and luminosity change as the star's core contracts and its outer layers expand. Stars of different masses, which pass most of their lives at various points on the main sequence, behave somewhat differently as their core regions exhaust their protons. Still, all of the stars expand and grow cooler in their outer parts as their cores contract and grow hotter.

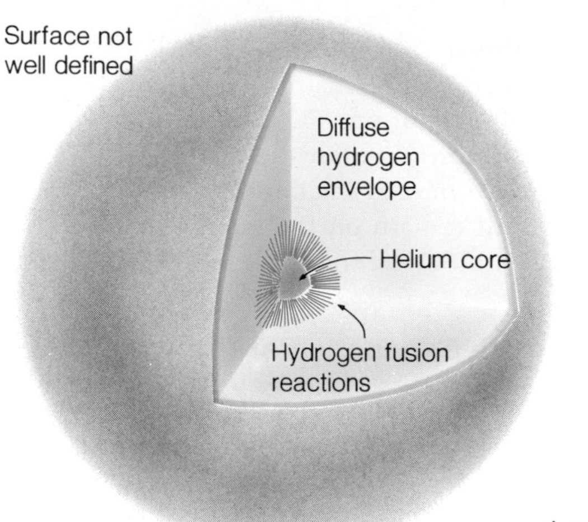

Figure 9-3
Inside the large, fluffy ball of a red giant lies a small, helium-rich core, only a few thousand kilometers across. This core continues to release the kinetic energy that makes the star shine.

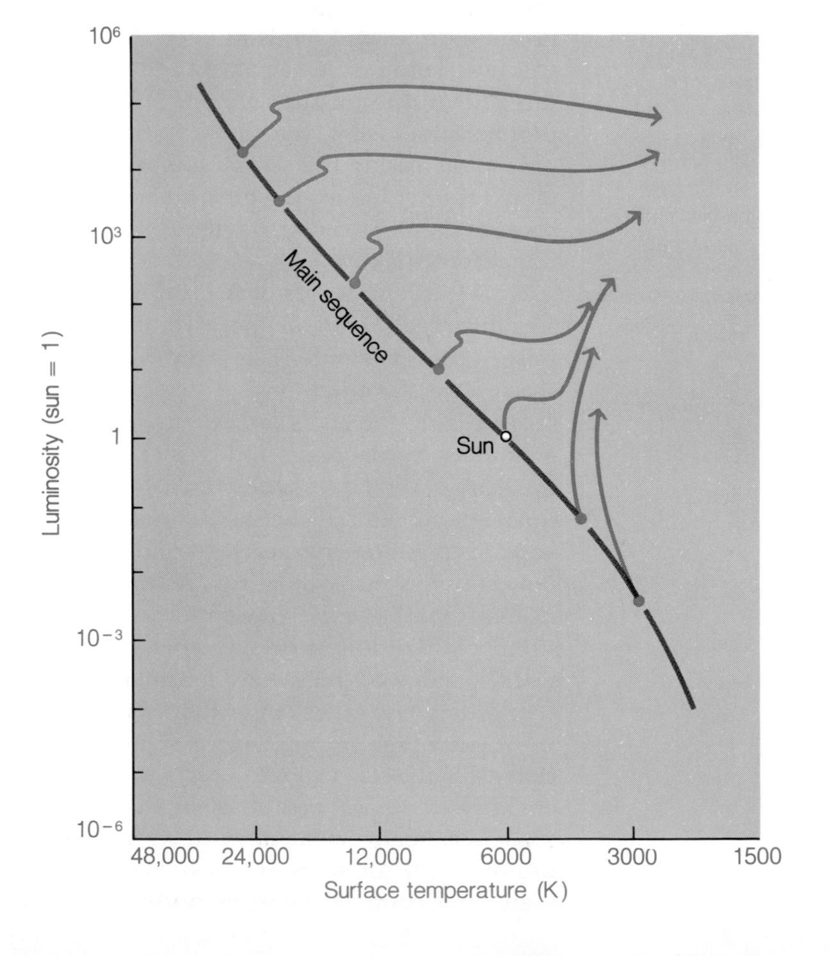

Figure 9-4
As stars exhaust the hydrogen in their central regions, their surface temperatures and luminosities change. The stars leave the main-sequence part of the temperature-luminosity diagram to enter the red-giant domain.

9.2 Estimating the Ages of Star Clusters

Consider the temperature-luminosity diagram of a star cluster (Figure 8-30). All stars that lie above and to the right of the main sequence are called red giants. Not all of these stars are truly red (surface temperatures from 1500 to 3500 K), but they are all redder (lower surface temperature) than main-sequence stars with the same absolute brightness. All of the stars in the red-giant part of the temperature-luminosity diagram have contracting cores, and their supply of protons for nuclear reactions has dwindled.

The Relationship Between Mass and Main-Sequence Lifetime

As we discussed on page 257, stars with greater masses have larger absolute brightnesses during their main-sequence lifetimes. Since the stars of large mass consume their nuclear fuel more rapidly than stars of small mass do, we can see that the large-mass stars will spend less time as main-sequence stars. Such stars will be the first to become red giants within a group of stars that all began to shine at the same time. Less massive stars have smaller luminosities when they are main-sequence stars and take longer to consume their fuel. Because of this inverse relationship between a star's absolute brightness as a main-sequence star and its lifetime on the main sequence, we can estimate the age of a cluster of stars that we assume all formed at about the same time. We do this by measuring how far the cluster's main sequence on the temperature-luminosity diagram extends upward to greater surface temperatures and greater luminosities. That is, we determine the surface temperature and luminosity of the hottest, most luminous stars in that cluster. The location of these stars on the temperature-luminosity diagram marks the **turnoff point** for that cluster.

The Turnoff Point

The youngest star clusters still have stars of extremely great brightness and surface temperature at the top of the main sequence. Perhaps not even the most massive stars in the cluster have had time to become red giants. In a young cluster, the turnoff point from the main sequence to the red-giant region lies at a high surface temperature and a large absolute brightness (Figure 9-5). In an older cluster, the turnoff point occurs at lower surface temperatures and lower luminosities, because less massive stars will have had the time to become red giants. Thus we can find the age of the cluster's stars by measuring the location of the turnoff point from the main sequence in the temperature-luminosity diagram. Since we believe that the original main sequence for all clusters appears in just about the same place on the temperature-luminosity diagram, it is sufficient to measure the surface temperature of the stars at the turnoff point from the main sequence. The lower this temperature is, the older is the star cluster.

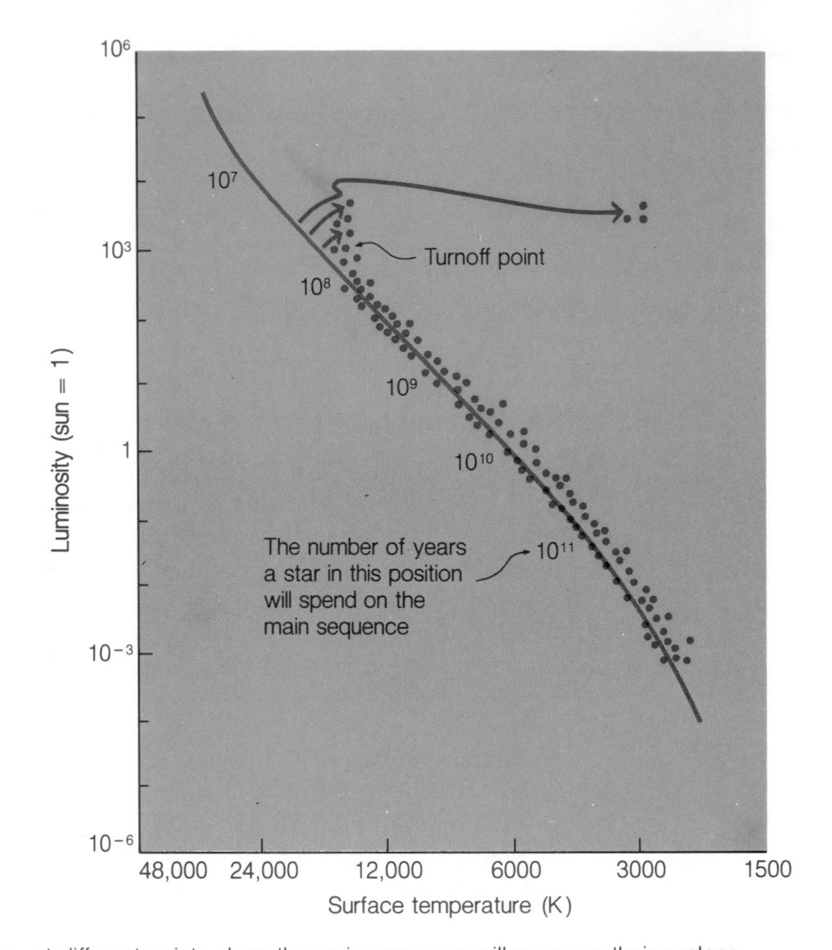

Figure 9-5 Stars at different points along the main sequence will consume their nuclear fuel at different rates and thus will show a turnoff to become red giants at different ages. High-luminosity stars become red giants at much younger ages than low-luminosity stars.

Using this technique, astronomers have found that globular clusters all have ages measured in billions of years, typically 8 or 10 or 14 billion. (Such age assignments depend on computer models of how rapidly stars of different masses age.) In contrast, open clusters have ages that range from 5 *million* years to as much as 10 billion years. A young cluster such as the Double Cluster in Perseus (Figure 6-15) shows a main-sequence turnoff point at a surface temperature of 25,000 K, indicating an age of 10 million years (Figure 9-6). An old open cluster, M 67, shows a turnoff point at surface temperatures of 6500 K, implying an age of 4 billion years for the stars in this cluster. These ages refer to the time since the stars began to liberate energy at a steady rate, that is, since they first appeared on the main sequence. This time, however, was not long (by cosmic standards) after the stars began to condense from protostars—only a few millions or tens of millions of years (see page 239).

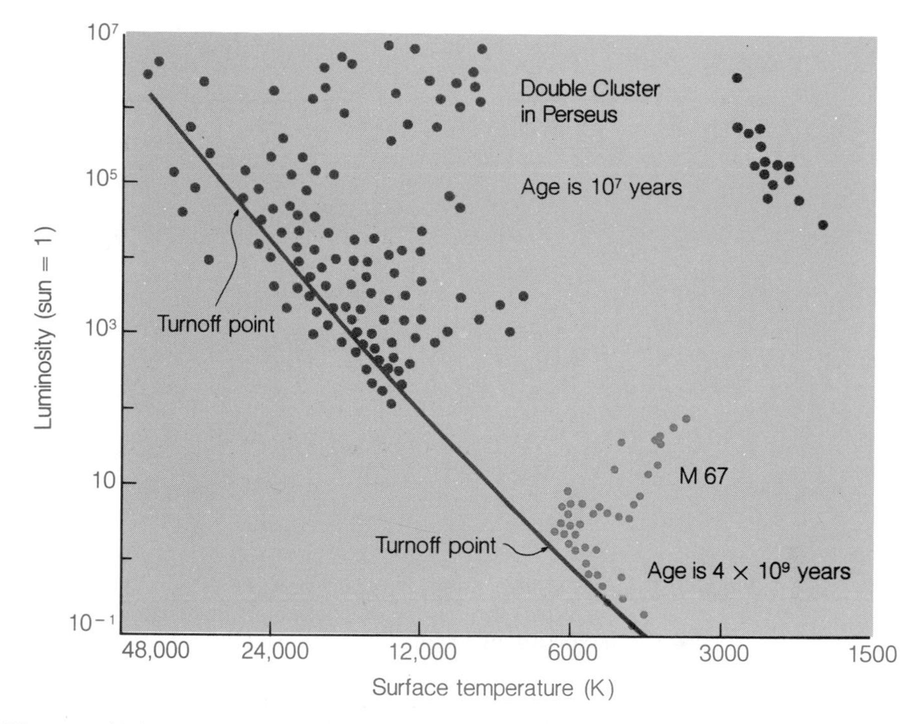

Figure 9-6 The stars in the old open cluster M 67 all have ages of about 4 billion years, based on the location of the turnoff point in their temperature-luminosity diagram. The much younger stars in the Double Cluster in Perseus have ages of only 10 million years.

9.3 Further Evolution of Stars: Continuing Fusion Reactions

What happens to a red-giant star as it continues to process its remaining supply of protons?[2] The star has several new ways to keep liberating energy of motion, each of which lasts a successively shorter time. After a star has fused all of its central hydrogen into helium, the star's interior consists of a nearly pure helium core, surrounded by a shell where protons still fuse into helium nuclei (Figure 9-7). As time passes, the proton-consuming shell embraces more and more of the star's middle regions, but all of these regions continue to contract, raising the temperature within them and thus processing the protons more rapidly. Likewise, the helium core contracts under the pressure from the overlying layers, and its temperature, too, keeps increasing. Eventually, the helium nuclei in the core reach such high temperatures that *they* can fuse together to liberate more kinetic energy.

---

[2]We are specifically discussing the evolution of Population I stars, which astronomers think they understand somewhat better than the evolution of Population II stars. However, in both stellar populations, the general pattern of a star's evolution should be the same.

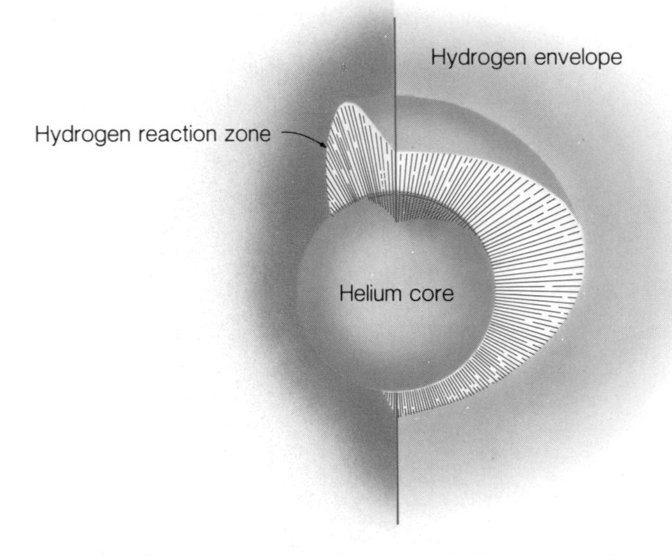

Figure 9-7 As a red giant ages, its core becomes almost pure helium. Around the core, hydrogen fusion reactions continue to liberate energy of motion in a shell of proton-rich material.

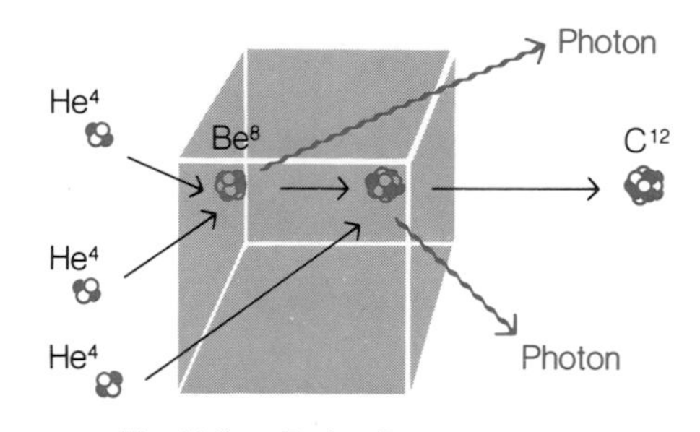

Figure 9-8 At temperatures of about 200 million K, nuclei of helium-4 will fuse together to produce a nucleus of beryllium-8, a photon, and additional kinetic energy. The beryllium-8 nucleus can then fuse with another helium-4 nucleus to produce a nucleus of carbon-12, a photon, and still more kinetic energy.

Helium Fusion

Although protons will fuse together at temperatures of 10 to 20 million K, the fusion of helium-4 nuclei requires a temperature near 200 million K.[3] Inside the helium core of a red giant, the temperature will reach this value

[3]Nuclei with larger numbers of protons repel each other more strongly, so they must have greater velocities upon collision to overcome this mutual repulsion if they are to fuse. An exception to this general rule occurs for helium-3 nuclei (see page 251), which, because of their nuclear structure, fuse easily at temperatures of only 10 million K.

when the contraction of the core has increased the density of particles to 10,000 grams per cubic centimeter, 100 times the present density at the center of the sun. When the temperature grows large enough for the helium nuclei to fuse together, the star's core suddenly liberates a burst of energy called the **helium flash.** The star then continues to fuse helium nuclei into carbon (Figure 9-8).

The decrease in energy of mass during these fusion processes creates new kinetic energy, which deposits itself in the now-dense shell that surrounds the helium core. This energy, chiefly in the form of gamma-ray photons, heats the shell and expands it to a much larger size. The expansion of the proton-fusing shell reduces its density and temperature and thus decreases the rate of nuclear reactions in the shell. Hence the onset of helium fusion has the rather perverse result of lowering the total rate of energy output from the star after the brief helium flash has passed.

Pulsating Variable Stars

Figure 9-9 shows the change in a star's position on the temperature-luminosity diagram as it becomes a red giant and undergoes the helium flash. After the star begins to fuse helium nuclei, it is likely to **pulsate,** that is, to

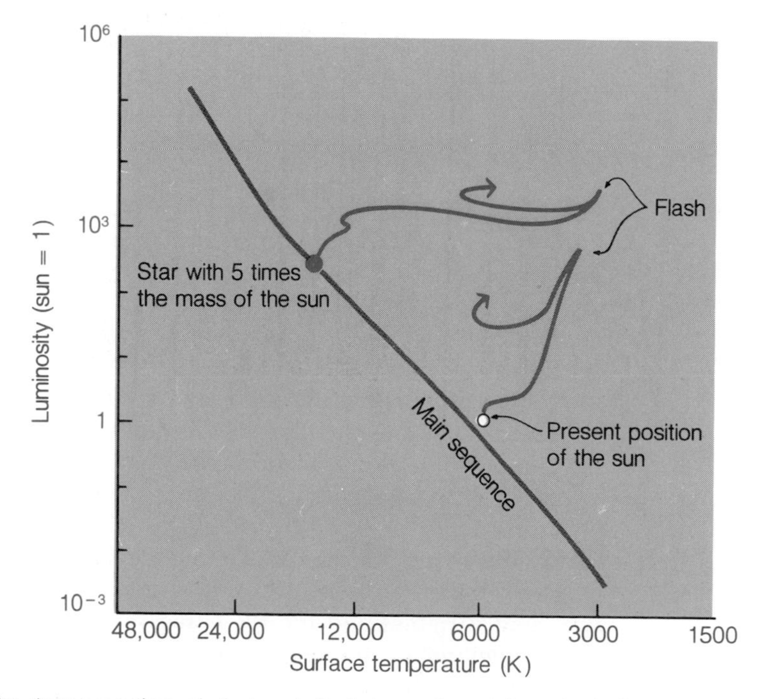

Figure 9-9 As stars pass through their red-giant stage of evolution, the helium flash marks a moment of great luminosity. Afterward, the stars will move toward the left in the temperature-luminosity diagram, achieving a higher surface temperature by shedding some of their outer layers.

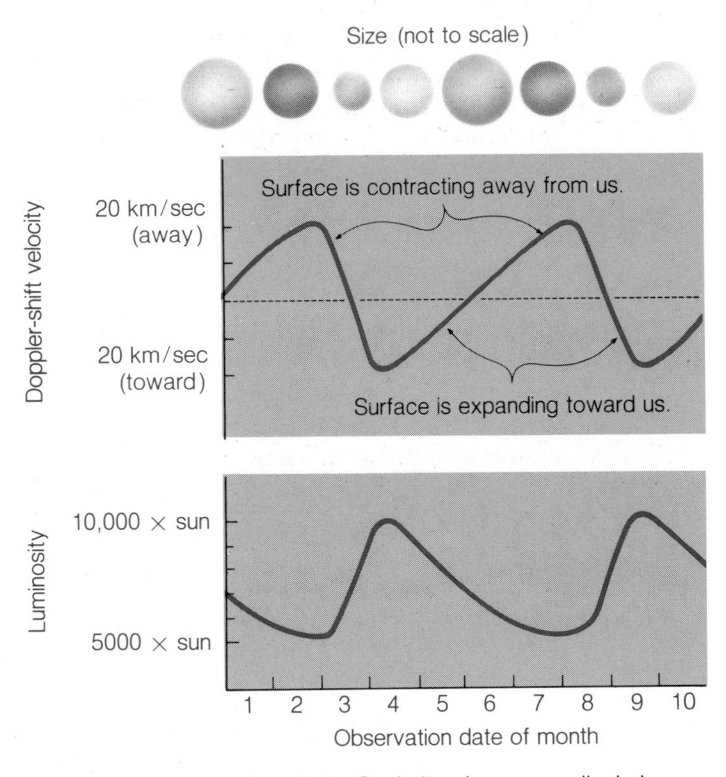

Figure 9-10 A pulsating variable star such as Delta Cephei undergoes cyclical changes in size and luminosity. These stars are brightest in the middle of their contracting phase and dimmest in the middle of their expanding phase, as found from observations of the Doppler shifts in their spectra.

undergo periodic changes in size and luminosity. These periodic changes make the star's surface area progressively larger and smaller (Figure 9-10). Such periodic variations in some red giants have a special importance for astronomers, who can observe these variations as fluctuations in a star's brightness, with a definite period and even a definite pattern in the shape of the graph of brightness plotted against time (Figure 9-10).

Some of these stars are the **Cepheid variable stars** so useful in determining astronomical distances (see page 171). The Cepheid variables have large luminosities, so they can be seen far away, even in some other galaxies. A subclass of Population II Cepheid variables called **RR Lyrae variables** differs somewhat from the Population I Cepheids: The RR Lyrae variables have slightly lower luminosities and shorter periods of light variation. The Cepheid variables and the RR Lyrae variables show up in certain well-defined regions of the temperature-luminosity diagram called **instability strips** (Figure 9-11). *All* of the stars whose surface temperatures and luminosities place them in these regions are "unstable," pulsating in size and energy output during the time when their evolution carries them through this part of the diagram.

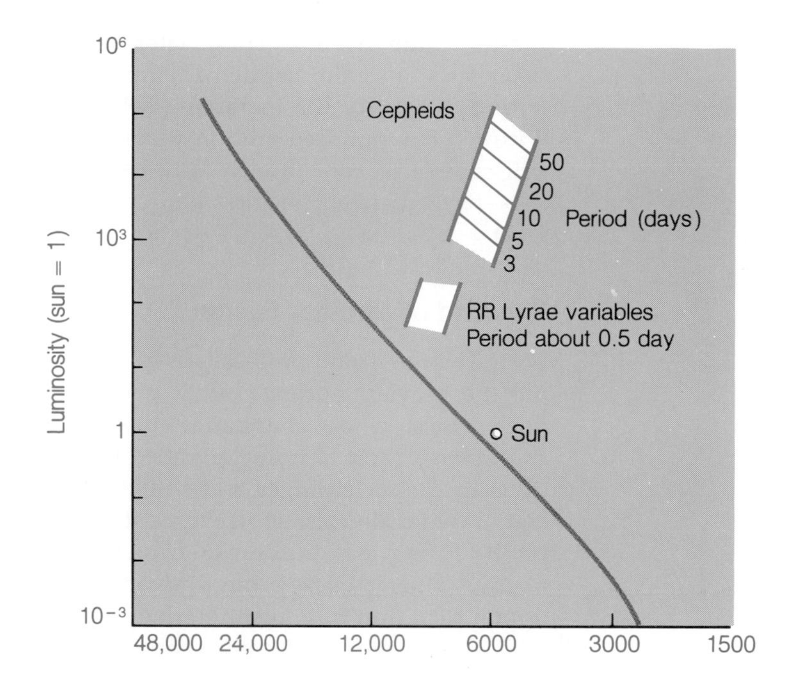

Figure 9-11 Stars whose evolution takes them into certain regions of the temperature-luminosity diagram called instability strips will pulsate in size and luminosity. Such stars are the Cepheid and RR Lyrae variable stars useful in determining distances.

The Rate of Nuclear Fusion

We have seen that stars convert hydrogen into helium, and then helium into carbon. Although both of these processes turn energy of mass into energy of motion, each hydrogen-to-helium fusion releases fifteen times as much energy as each helium-to-carbon fusion. The later stages of a red giant's life therefore unfold far more rapidly than its earlier, main-sequence stage. A star with the sun's mass will spend 10 billion years on the main sequence. The entire red-giant phase, including the stage when the protons fuse in a shell around the helium core and the unstable stage after the helium flash, will take a total of 2 or 3 billion years, less than one-third the star's main-sequence lifetime. As the star's red-giant phase ends, almost all of the protons in the star's core have been changed into helium nuclei, and most of the helium nuclei have themselves fused into carbon nuclei.

What can occur within the star to keep the core from collapsing? More nuclear fusion reactions are indeed available: Carbon plus helium makes oxygen, oxygen plus helium makes neon, neon plus helium makes magnesium, carbon plus carbon also makes magnesium, and so on. Each of these fusion reactions releases some kinetic energy, but the total energy liberated in these reactions remains far less than that released by the original hydrogen-to-helium fusions. These later reactions must occur at progressively faster rates, since the star demands an increasing total energy liberation each second to balance the increasing self-gravitation caused by its contraction. Within a few hundred million years, a star with a few times the sun's mass will change its core from mostly helium all the way through heavier nuclei—oxygen, neon, magnesium, silicon—all the way up to iron (Figure 9-12).

The End of Nuclear Fusion

With iron the fusion process fails to liberate any more kinetic energy. Iron is not the heaviest nucleus, but it is the heaviest of the relatively abundant nuclei (those whose abundance equals at least a millionth of the hydrogen abundance). Most iron nuclei, called iron-56, contain 26 protons and 30 neutrons each. Nuclei larger and heavier than iron, such as lead, silver, gold, and uranium, do exist. However, a star cannot release energy of motion simply by fusing iron into heavier nuclei. The fusion of iron nuclei does not liberate kinetic energy but instead *absorbs* kinetic energy. When nuclei lighter than iron fuse together, the products have *less* mass than the original total: Energy of mass becomes energy of motion. But when, for example, we fuse an iron-56 nucleus and a helium-4 nucleus to make nickel-60, the product has *more* mass than the total mass before the fusion. The increase in mass, and in energy of mass, must come from the kinetic energy of the colliding particles. Thus the fusion process decreases the energy of motion contained in the particles that undergo such a reaction.

Hence iron marks the end of the line for the release of kinetic energy through nuclear fusion. A star with a central core that has steadily decreased

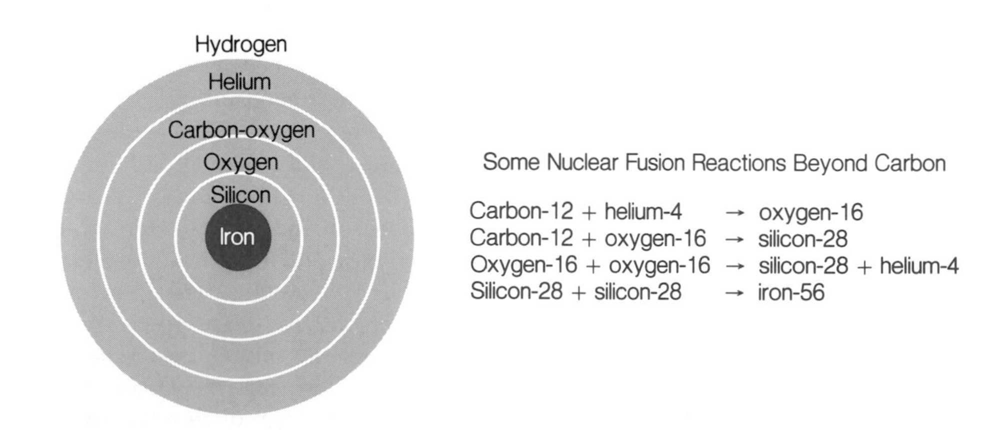

Figure 9-12 Stars can make nuclei heavier than carbon and thus release some additional kinetic energy. But the fusion process stops producing more energy once iron nuclei are created. Thus the star's core will become mostly iron, with surrounding shells of progressively lighter nuclei.

in size and increased in temperature, that has fused helium nuclei into carbon, magnesium, silicon, sulfur, and argon, all the way to iron, has run out of ways to liberate the kinetic energy that will continue to oppose its self-gravity. Such a stellar core has reached the point of catastrophe, and disaster must occur. Before we examine this process, however, we must pause to determine which stars will reach this brink of collapse and which stars will find some path that avoids this apparently inevitable calamity.

9.4 The Exclusion Principle

Stars with masses less than about 1.4 times the sun's mass can avoid collapsing because of a strange resistance of elementary particles to being pushed together. All matter resists compression to some extent. But elementary particles such as electrons, protons, and neutrons have a special resistance to being packed tightly, called the **exclusion principle** by physicists.

What Is the Exclusion Principle?

If we try to pack together a bunch of electrons, the electrons eventually refuse to be squeezed together with far more resistance than would be expected from their mutual electromagnetic repulsion. The exact, mathematical form of this exclusion principle states that if we try to confine a group of electrons within a certain volume, the electrons' positions and velocities cannot all be identical. If we try to move all the electrons closer to the same position in space, their velocities will differ more and more. If we try to get all the electrons to have almost the same velocity, their positions must differ more widely.

In stars, the exclusion principle does not play an important role until the density of matter rises to a million times the density of water, far denser than anything we can make in a laboratory. At the atomic level, however, we live with the effects of the exclusion principle every day, for it is this principle that governs how many electrons can fit into a particular atomic orbit. Thus the exclusion principle determines the rules of chemistry, which deals with the relations among electrons in neighboring atoms.

Powerful as its effects are, the exclusion principle acts only among certain kinds of particles. Electrons, protons, and neutrons obey the exclusion principle. Photons do not, nor do helium-4 nuclei or other nuclei made of an even number of nucleons (protons plus neutrons). Nature moves in mysterious ways, and the exclusion principle allows the electrons inside a star to act as the key determinants of the star's evolution, just when the nuclei show a declining ability to release energy of motion. In our discussion of stellar interiors, we have concentrated on the nuclei (protons, helium-4, carbon-12, and so on) rather than the electrons, since the fusion of nuclei liberates the kinetic energy that makes stars shine. Electrons do not fuse with each other, or with any other particles, at the densities and temperatures inside stars, so

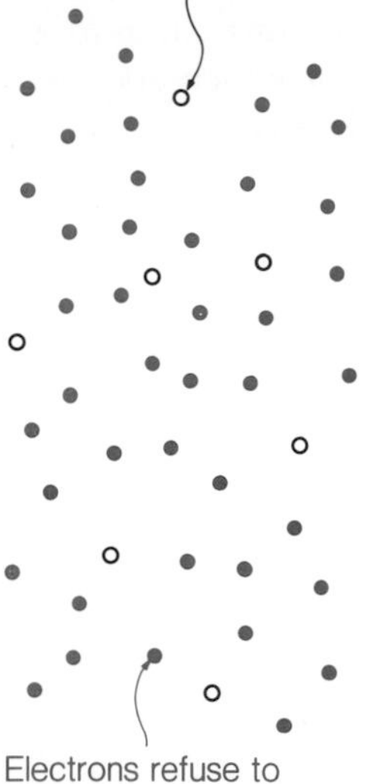

Figure 9-13
The electrons and the nuclei in a star attract each other through electromagnetic forces. Hence the electrons and nuclei must remain thoroughly mixed: If the electrons cannot contract into a smaller volume, neither can the nuclei.

they do not play an important part in the star's liberation of energy.[4] Each star does contain enough electrons, however, to balance the positive electric charges of its nuclei and thus have a total electric charge of zero.

The Effect of the Exclusion Principle in Old Stars

As an aging star's central core contracts, the nuclei and electrons inside it are squeezed closer and closer together. Each cubic centimeter contains more and more electrons and nuclei, and the electromagnetic forces between positive and negative electric charges produce a total charge of zero in each cubic centimeter. The electrons always move more rapidly than the nuclei, because the temperature of the matter represents the average kinetic energy per particle. For velocities much less than the speed of light, each particle's kinetic energy varies as its mass times its velocity squared. Since each electron has a mass thousands of times less than the mass of a single nucleus, the electrons have much larger velocities than the nuclei within the same region and at the same temperature. The electrons are thus inherently more difficult to confine, and the exclusion principle makes this restriction absolute, once the matter reaches a density close to a million times the density of water.

Physicists call matter **degenerate** if the exclusion principle plays an important role in its bulk behavior. Such degenerate matter must always have an immensely high density, 50,000 times the density of gold, for the electrons to be affected strongly by the exclusion principle. Once the electrons are affected, they in turn affect the other particles—the nuclei. The electrons and nuclei exert a mutual electromagnetic attraction, since they have opposite electric charges (Figure 9-13). Thus, if a star becomes degenerate in its interior, the nuclei cannot go on contracting into a smaller volume. Instead, the electrons hold them "up" against gravity through the electromagnetic forces they exert on the nuclei.

In this way, the star's entire core—both electrons and nuclei—can maintain its size against self-gravity indefinitely. This degenerate support occurs only at a density of a million times the density of water, so each cubic centimeter of such degenerate matter would weigh a ton on Earth! But at this density, the exclusion principle intervenes, and no further contraction of the star's core can occur.

The Effects of Mass on Degenerate Cores

Which stars develop degenerate cores? The answer depends on the star's *mass*, which in turn governs the density within the core during the star's evolution. Lower-mass stars develop degenerate cores more easily because

---

[4]A small fraction of the energy released during the proton-proton or carbon cycle does come from the mutual annihilation of positrons (made in the fusion reactions) with electrons (already present inside the star). Such matter-antimatter annihilations do not require high temperatures to occur, since the particle and antiparticle have opposite electric charges and attract each other through electromagnetic forces.

the density in a star's center varies inversely with the star's mass. All main-sequence stars have roughly the same temperature in their interiors, 7 to 35 million K, because this is the temperature at which protons fuse. The more massive stars can achieve such a temperature with a lower central density because the stars have more mass pressing in on the center. The general rule that larger-mass stars have lower central densities remains valid as the stars age and produces an important distinction between large-mass and small-mass stars.

If a star's mass is less than about 1.4 solar masses, its core (which comes to include almost the entire mass of the star as it ages) will become degenerate at about the time all the protons there have fused into helium nuclei. When helium nuclei start to fuse inside the core, the energy released by the helium flash expands the core and removes the degeneracy. The core then starts shrinking again, and after the helium nuclei have fused into carbon nuclei, the core again becomes degenerate, this time for good. After degeneracy sets in for the second time, the star's core contracts no further. Almost no nuclear fusion reactions occur in the core, because the exclusion principle that supports the star also inhibits nuclei from colliding at high velocity. The star's outer layers slowly expand outward from radiation pressure (page 256), forming a **planetary nebula** (Figure 9-14), until only the naked core remains. Eventually, the outer layers will be completely lost. The cores are the **white-dwarf stars** that fill the part of the temperature-luminosity diagram below and to the left of the main sequence, slowly fading into ever-fainter obscurity (Figure 9-15).

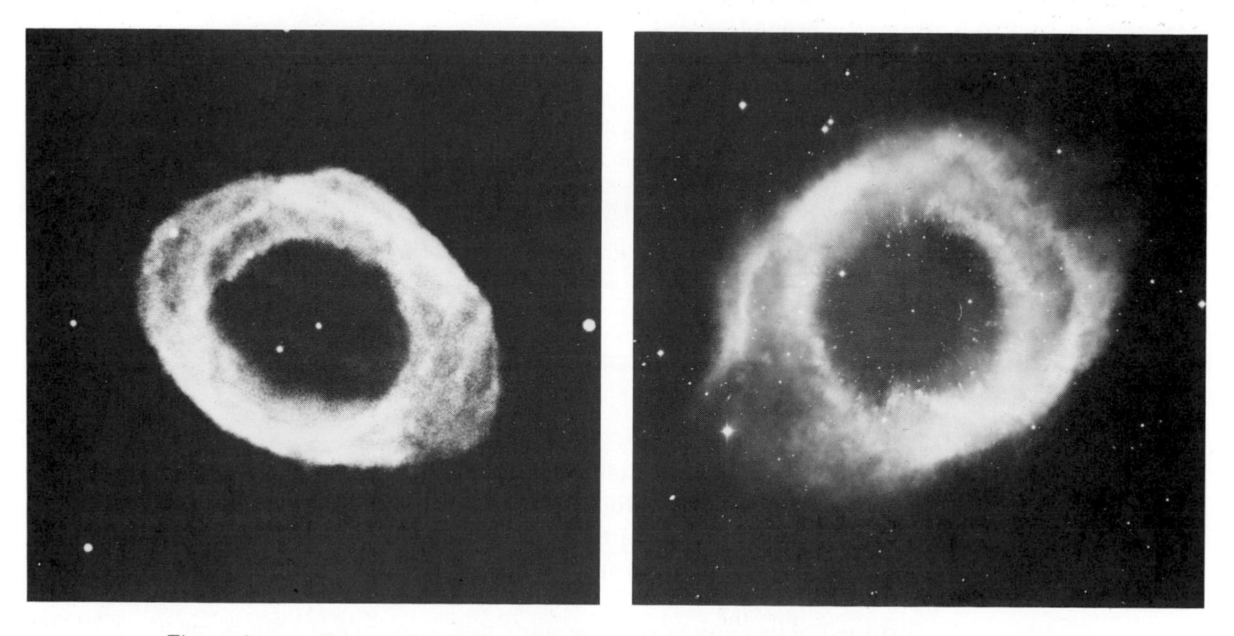

Figure 9-14 Two representative planetary nebulae are the Ring Nebula in Lyra (left) and the planetary nebula in Aquarius (right). The central star that powers each nebula can be seen, glowing only faintly in visible light.

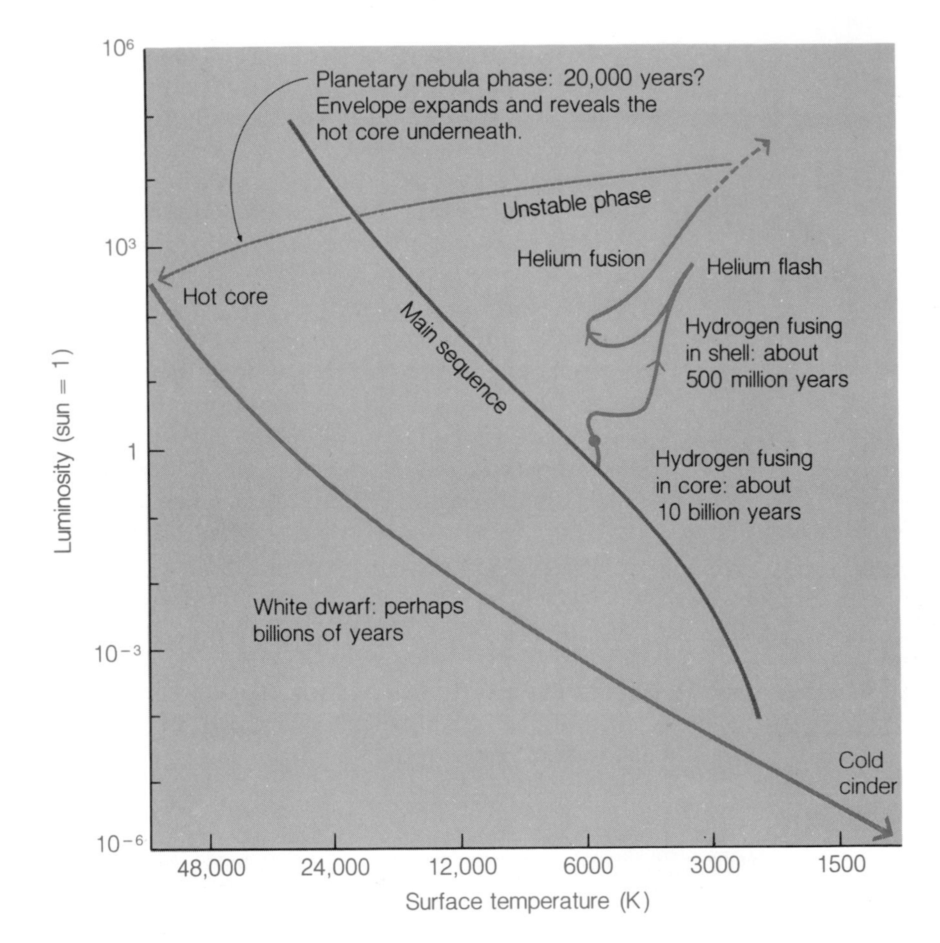

Figure 9-15 A star with the sun's mass will pass through the instability strip of the temperature-luminosity diagram after its red-giant phase. The star will then shed some mass and become a planetary nebula. Eventually, the star's core, now devoid of its outer layers, will shine alone as a slowly cooling white dwarf.

9.5 Further Evolution: Toward the White-Dwarf Stage

Planetary Nebulae

Many aging stars eject their outer layers as a roughly spherical shell that surrounds the star, typically with a diameter of 0.1 or 0.2 parsec, tens of thousands of times larger than the diameter of the Earth's orbit around the sun. Such a shell of gas does not contain much of the star's mass, but its appearance, lit from within by the dying star, can be spectacular (Figure 9-16). Stripped of its outer layers, the star's new surface has a temperature of 80,000 to 100,000 K, more than ten times hotter than the sun's surface. The high-energy photons from the star then ionize many of the atoms in the shell of gas. As electrons and ions recombine to form atoms, they emit photons that make the entire shell glow (Figure 9-17). Nineteenth-century astronomers called these luminous shells of gas **planetary nebulae** because the shells looked like fuzzy (nebular) planetary disks. In fact, planetary nebulae have nothing to do with planets, although the passage of such a shell through the planetary system around an aging star would be a noteworthy event in the history of these planets!

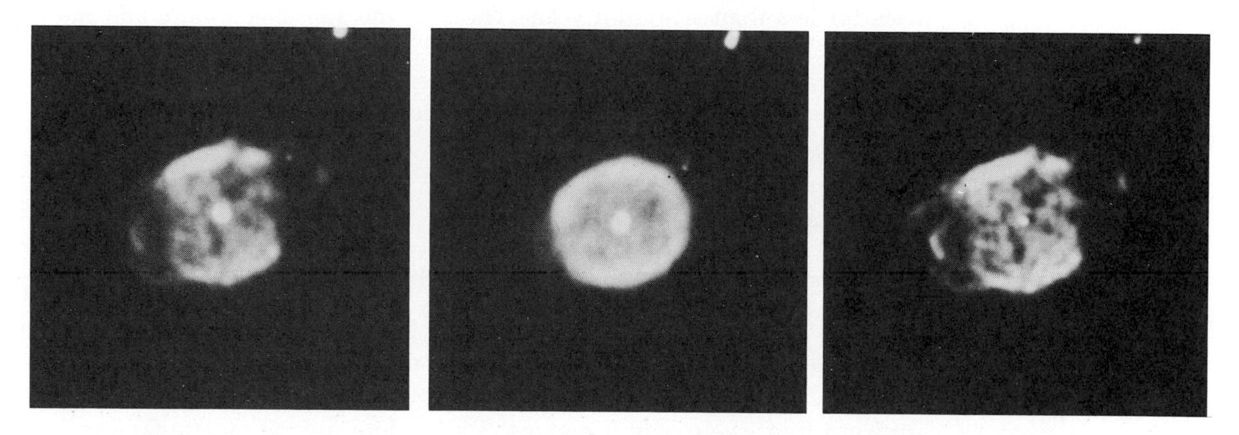

Figure 9-16 The planetary nebula NGC 40, photographed at three different wavelengths, shows emission from hydrogen atoms (left), oxygen ions (center), and nitrogen ions (right). Each kind of atom or ion produces photons of a particular wavelength and frequency upon de-excitation. Photographs at one particular wavelength reveal ions or atoms of one particular type.

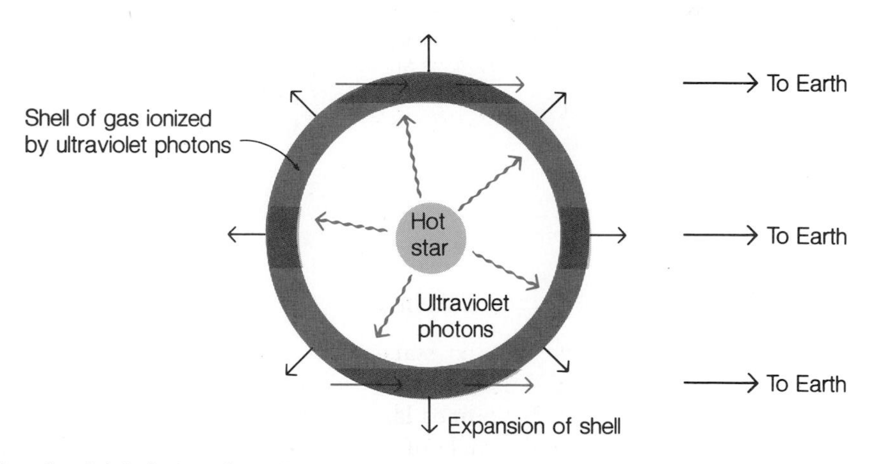

Figure 9-17 The ultraviolet photons from the star at the center of a planetary nebula ionize some of the hydrogen atoms in the gas around it. When these atoms recombine, they can emit visible-light photons. Material in a shell around the star will shine most brightly at the edges, because more of the glowing gas lies along our line of sight in those regions.

Large telescopes reveal the shell structure of typical planetary nebulae, such as the Ring Nebula shown in Figure 9-14. The central star that powers this nebula has such a high surface temperature that most of its photons have *ultraviolet* frequencies and wavelengths. Hence, the star itself appears misleadingly faint in visible-light wavelengths. But *all* the energy emitted by the nebula comes originally from the hot central star in the form of ultraviolet photons.

Stars in their planetary nebula phase pass over the top of the main sequence on the temperature-luminosity diagram (Figure 9-15). This phase lasts for less than a million years. Then, as the star sheds more and more mass, the core stands revealed as a white-dwarf star. The energy radiated by the central star of a planetary nebula comes from its continuously contracting core star, which ceases to liberate kinetic energy when it becomes degenerate.

Mass Limits on White Dwarf Stars

If a star's core becomes so dense that degeneracy sets in, the exclusion principle both supports the star and prevents any further nuclear fusion reactions. The core must then become a white dwarf, possessing a supply of kinetic energy stored as heat, but with no way to liberate more kinetic energy. White dwarfs shine because they slowly radiate kinetic energy conducted outward from their central regions. Since white dwarfs radiate far less energy per second than most main-sequence stars, they can last a long time on the reserves of energy stored from their years of nuclear fusion. The white-dwarf companion of Sirius A, called Sirius B, radiates energy at a rate 300

Figure 9-18 A large telescope can reveal the faint companion of Sirius A, the white-dwarf star Sirius B. The diffraction spikes around Sirius A arise from the support struts in the telescope.

times less than the sun and 10,000 times less than Sirius A (Figure 9-18). This white dwarf's surface temperature almost equals that of Sirius A (10,000 K), but its radius is about 100 times less, about twice the Earth's radius. Yet Sirius B has a mass nearly as large as the sun's, so the average density of matter inside it must be several hundred thousand times the density of water. Sirius B will easily shine for several billion years, cooling and fading so slowly that we cannot yet detect any changes in its brightness.

Calculations made by astrophysicists have demonstrated a remarkable fact: No star can exist as a degenerate white dwarf if its mass exceeds 1.4 solar masses. In line with this calculation, no white dwarf among the dozen or so with well-determined masses has a mass that exceeds this limit. The most massive white dwarf yet found has a mass barely greater than the sun's. But if no star can be a white dwarf if it has more than 1.4 solar masses, what happens to the massive stars? Two possibilities exist. Either a star can lose enough mass during the later stages of its evolution to fall below the **1.4-solar-mass limit**, or the star cannot be supported by the exclusion principle. More massive stars can simply overwhelm the exclusion principle with a sudden rush that we describe on pages 313–15. Before we examine this catastrophic collapse of massive stars, we must consider the first possibilty, the relatively gentle loss of mass that allows some stars to "slim down" enough to become white dwarfs.

Mass Loss From Stars

Observations of aging stars have shown that many of them, especially the more massive red giants and red supergiants, are losing mass from their outer layers. All main-sequence stars eject some matter from their outermost regions, just as the sun does in the solar wind described on page 280. But for red-giant stars, the mass loss proceeds at a much greater rate, amounting to the loss of a sizable part of the star's mass in a few million years. Red giants lose mass more rapidly and easily than other types of stars because their tenuous outer layers are farther from the core of the star and therefore feel less gravitational force (Figure 9-19). It is tempting to think of a red giant's mass loss as the star's attempt to pass below the upper limit for a white-dwarf star. The high rate of mass loss indeed may allow some red-giant stars to avoid catastrophic collapse and thus permit them to end their lives as serene cinders rather than exploding stars.

Consider, for example, the star Sirius B, which presumably was born at the same time as Sirius A. Since Sirius B has evolved further, all the way to the white-dwarf stage, we may easily believe that it began with more mass than Sirius A, which has 2.3 times the sun's mass. Yet Sirius B now contains less than one solar mass, which strongly suggests that it managed to lose more than half of its original mass on its way to becoming a white dwarf with less than the upper limit of 1.4 solar masses.

What happens to the mass expelled by aging stars? Some of it forms planetary nebula shells and some of it disperses through the interstellar gas, ready to be mixed into a new generation of stars. The interstellar dust particles described on page 183 probably came from the outer atmospheres of

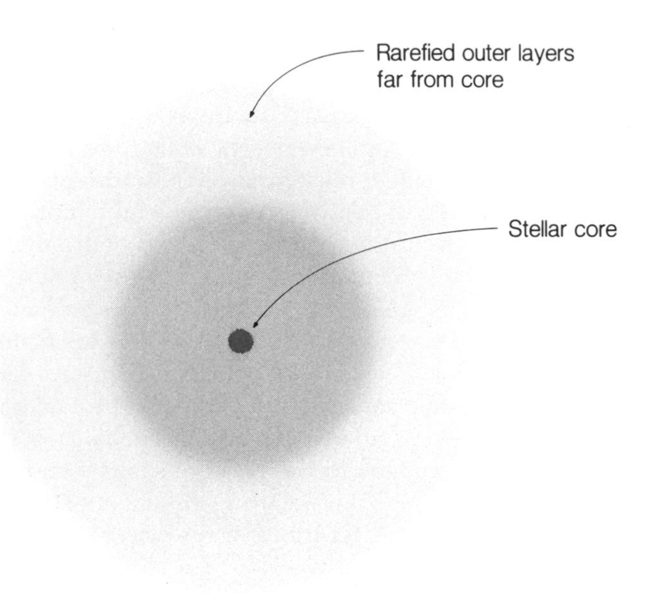

Figure 9-19 Because the outer layers of a red giant are far from the star's center, the star can lose matter from its outer layers rather easily.

red-giant stars, where the low temperature allows atoms to combine into molecules, and the molecules into long chains, without being broken apart by collisions. Some of the gas that red giants expel emerges as individual atoms rather than as large dust grains. We can observe the expansion of the outer layers of some red-giant stars because the atoms and molecules in these layers show a Doppler blue shift in the characteristic frequencies of the light they absorb from below. We can interpret this Doppler shift as the result of the motion of the stars' outer layers toward us (Figure 9-20).

Astronomers expect that the sun will pass through a red-giant phase, shed a small fraction of its mass, and end its life as a white dwarf. Since the sun's mass already lies below the upper limit of 1.4 solar masses for any white dwarf, the sun's future mass loss is not an essential factor. However, we shall probably not be around as a species to see what happens, because as the sun enters its red-giant phase some 5 billion years from now, its luminosity will increase a thousandfold, causing the temperature at the Earth's surface to quadruple from 300 to 1200 K. By this time, humanity must have found a safer haven if it is not to perish along with the oceans and atmosphere of our original home, the Earth.

9.6 Supernovae

As a star approaches the end of its energy generation, its mass determines its fate. If a star has less than 1.4 solar masses, or if it manages to reduce its mass below this limit during its red-giant phase, then the star almost certainly will become a white dwarf, quietly letting its stored kinetic energy seep out-

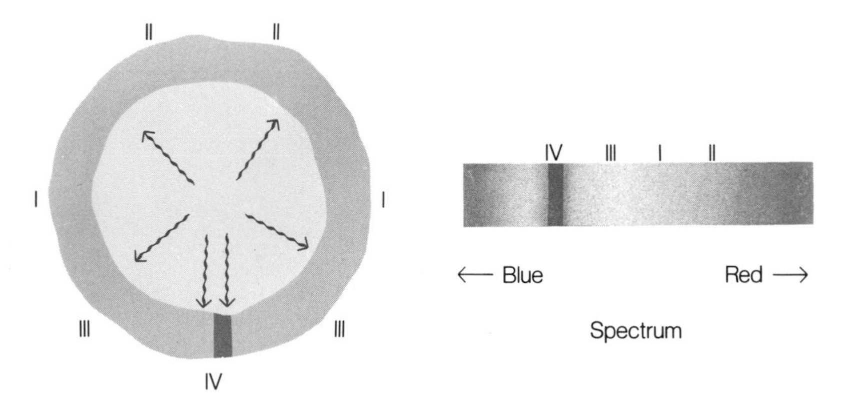

Figure 9-20 If we observe absorption lines produced in the outer layers of a star, we can determine the velocity of the absorbing material along our line of sight by the Doppler effect. This method works if we think we know which atoms or molecules absorb the starlight, so we know the frequency the absorption line would have if no relative motion existed. In this example, we observe the star from direction IV and find a Doppler shift to higher frequencies and shorter wavelengths.

ward bit by bit. As the white dwarf cools, it will leak less and less energy per second into space, giving itself more and more time to disperse its hoarded supply. Only after tens of billions of years will the star cease to shine almost entirely, earning the name **black dwarf**, as opposed to the white dwarfs we can see. Stars of small mass follow this degenerate path toward oblivion, with each stage in the process (main-sequence lifetime and white-dwarf obscurity) taking billions of years.

But what of the stars with large mass, which exhaust their nuclear fuel far more quickly than this, only to confront the possibility of disastrous collapse? We left our description of these more massive stars on the brink of the abyss (page 305), when they had used all possible nuclear fusion reactions to liberate kinetic energy by fusing their nuclei all the way to iron-56. Since the fusion of iron nuclei consumes kinetic energy rather than liberating it, the star no longer can oppose its self-gravitation with energy from fusion reactions and faces collapse. And collapse it will, so long as its center has too little density to become degenerate when the nuclei have fused into iron. Deprived of support from newly released kinetic energy, the inner regions all fall toward the star's center (Figure 9-21). This collapse occurs so quickly—in less than one second—that we may think of it as taking the exclusion principle by surprise.

Collapse of a Stellar Core

Although we can hardly hope to observe the center of a collapsing star directly, recent calculations show that during this collapse, the electrons fuse with the nuclei, and the nuclei themselves fuse together! For an instant, the entire star becomes one gigantic nucleus, with some 10^{57} nucleons (protons

Not to scale

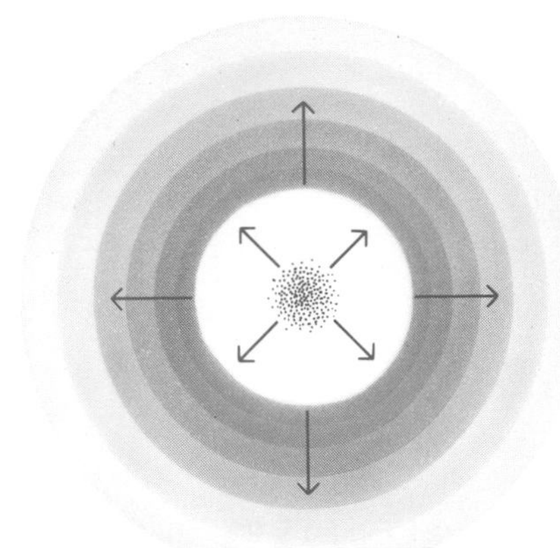

Collapse of core leads to. . .

single dense nucleus which. . .

bounces back to slightly larger size, starting a shock wave that accelerates matter outward.

Figure 9-21 A star's central regions will collapse if it has fused most of the material there into iron nuclei and then has no further way to release kinetic energy. This collapse will produce a single "nucleus" a few kilometers across, which bounces outward to start an accelerating shock wave through the outer layers of the star.

and neutrons) squeezed into a "nucleus" several kilometers in diameter. At the fantastically high density of 10^{15} grams per cubic centimeter—a billion times the density of a white dwarf—the nucleus resists further compression and in fact "bounces" outward like a rubber ball that has been squeezed too hard. This "bounce" starts a shock wave that travels outward through the star, accelerating as it passes into regions of lower density (Figure 9-21).

The Shock Wave

As the shock wave moves outward, it accelerates and heats the material it encounters. This outward blast of energy follows the core's collapse and "bounce" by only a few seconds. Within the matter swept outward by the shock wave, the temperature rises so high that particles collide with enormous kinetic energy. In these collisions, new nuclei can form in a tiny fraction of a second. Thus the shock wave reproduces on a miniature scale some of the features of the primeval fireball, when all kinds of particles were made, destroyed, remade, and destroyed again in high-energy collisions (see page 114).

Result: A Supernova

What do we see from all this? The outward explosion, following just after the unobservable implosion, produces a great wave of photons amid the particle collisions. Such an outburst has the name **supernova** ("super-new" in

Table 9-1 Time Intervals in the Life Stages of Stars

Stage of Stellar Evolution	Length of Time in Each Stage of Evolution	
	1-Solar-Mass Star (Sun)	10-Solar-Mass Star
Before protostar contraction	Billions of years	Billions of years
Protostar contraction to main sequence	Few tens of millions of years	Few tens of millions of years
Main-sequence lifetime	10 billion years	100 million years
Red-giant phase	2 to 3 billion years	20 to 30 million years
Planetary nebula phase	Tens of thousands of years	Tens of thousands of years
White-dwarf phase	More than 10 billion years	[None]
Supernova remnant (if any is left)	[None]	More than 10 billion years

Latin), since we often see a bright "new" star where no individual star had been seen before. Such supernovae, the last outward gasp of collapsing stars, far outshine any other single object within most galaxies. Even within a giant galaxy a supernova stands out as a bright object, capable of shining with 1% of the entire galaxy's output for a few months (Figure 9-22). During the year that follows its explosion, a supernova can radiate as much energy as the star that produced it has radiated during the millions or billions of years of main-sequence life before the explosion. This energy comes from the star's collapse, that is, from the energy stored by the star while supporting itself against its own gravitational force. Table 9-1 shows the length of

Figure 9-22
The supernova that appeared in the galaxy NGC 4303 in 1961 shone for a few weeks with 300 million times the absolute brightness of the sun.

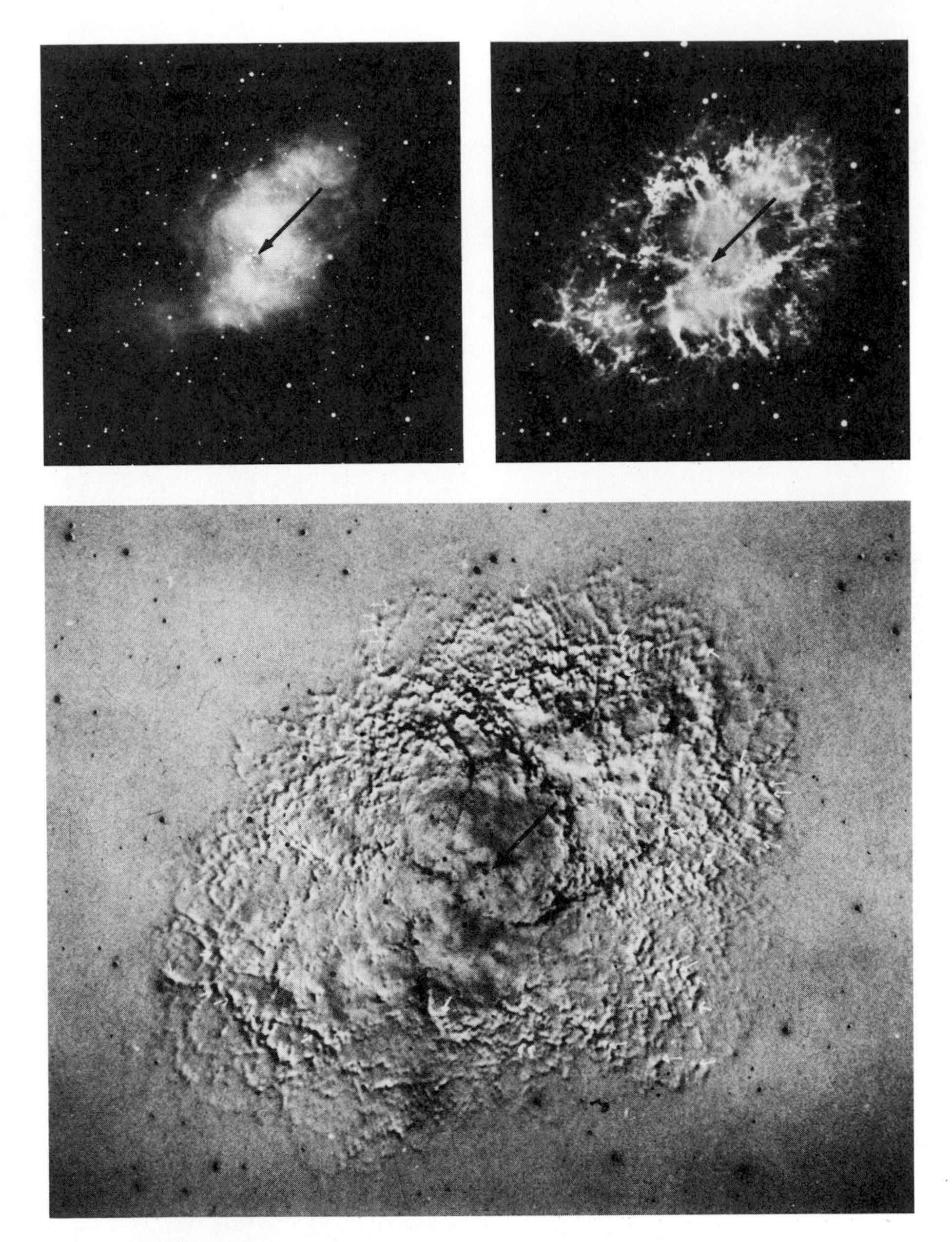

Figure 9-23 The supernova seen in the constellation Taurus in 1054 A.D. left behind the Crab Nebula, which emits visible-light synchrotron radiation (top left). If we photograph the nebula with a special filter to show only the red light emitted by hot hydrogen atoms, we can see that these atoms mostly lie in filaments (top right). The arrow marks the position of the star that exploded. Measurements of the motions of the filaments over a period of 14 years show that the nebula is expanding. The bottom photograph combines a positive photographic print taken in 1950 with a negative print taken in 1964. If no motion had occurred, the combination would show a blank. Instead, we see evidence for the expansion of the nebula.

time stars of 1 and 10 solar masses will spend in the various stages of their lifetimes.

Observations of Supernova Explosions

In a giant galaxy such as our Milky Way, supernova outbursts occur about once a century, as another massive star meets disaster. The last supernova seen in the Milky Way from Earth appeared in 1604 and the one before that in 1572. Astronomers can recognize the remnants of supernova explosions through the residue of expanding gas, which usually glows with synchrotron radiation in both the visible and radio domains of the spectrum (Figure 9-23). As the supernova remnant ages, it expands less rapidly, and the synchrotron emission occurs primarily at radio wavelengths (see page 208). The visible light from the remnant then arises primarily from electron jumps to smaller orbits in atoms that have been involved in collisions that kicked the electrons into larger orbits (Figure 9-24).

The Crab Nebula, best-studied of supernova remnants, arose from the supernova explosion seen on Earth in A.D. 1054 (Figure 9-23). European scientists apparently failed to notice a new star in Taurus, but Chinese scientists

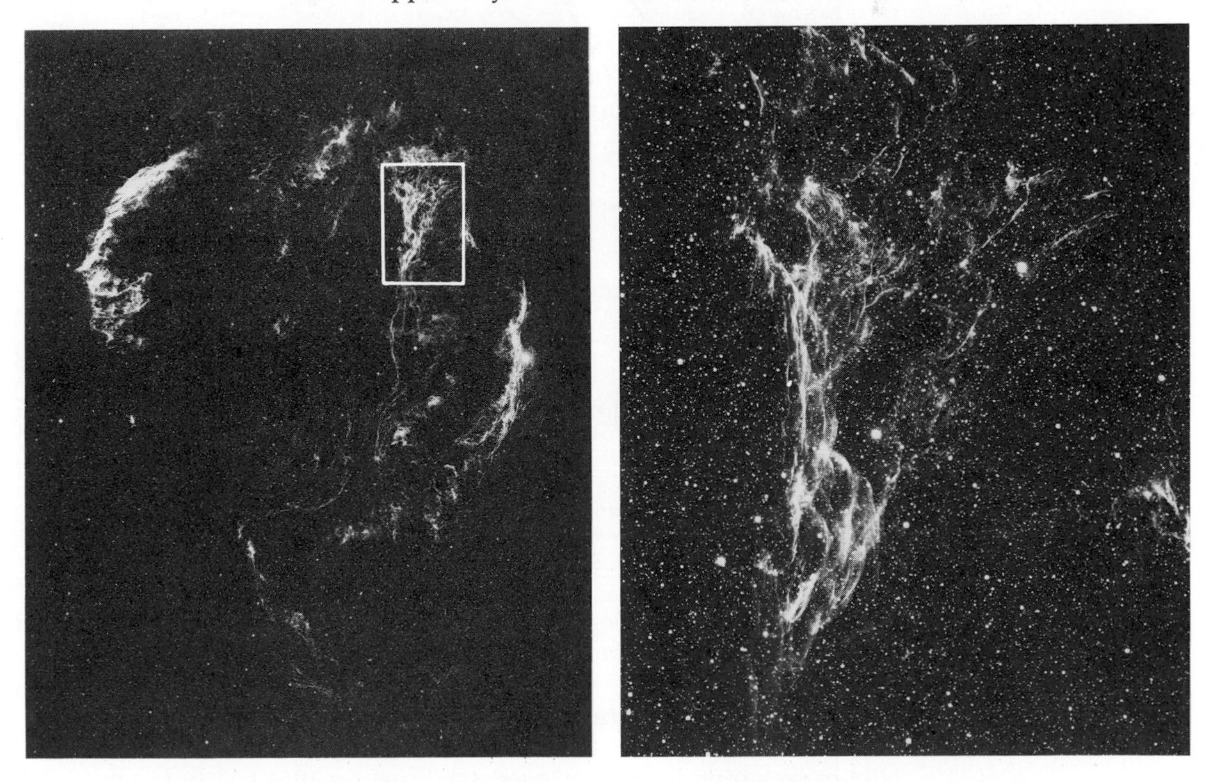

Figure 9-24 The Veil Nebula in Cygnus appears to be the remnant of a supernova that exploded about 30,000 years ago. Most of the visible light comes from electron jumps to smaller orbits in the atoms within the filaments of gas that form a roughly spherical structure.

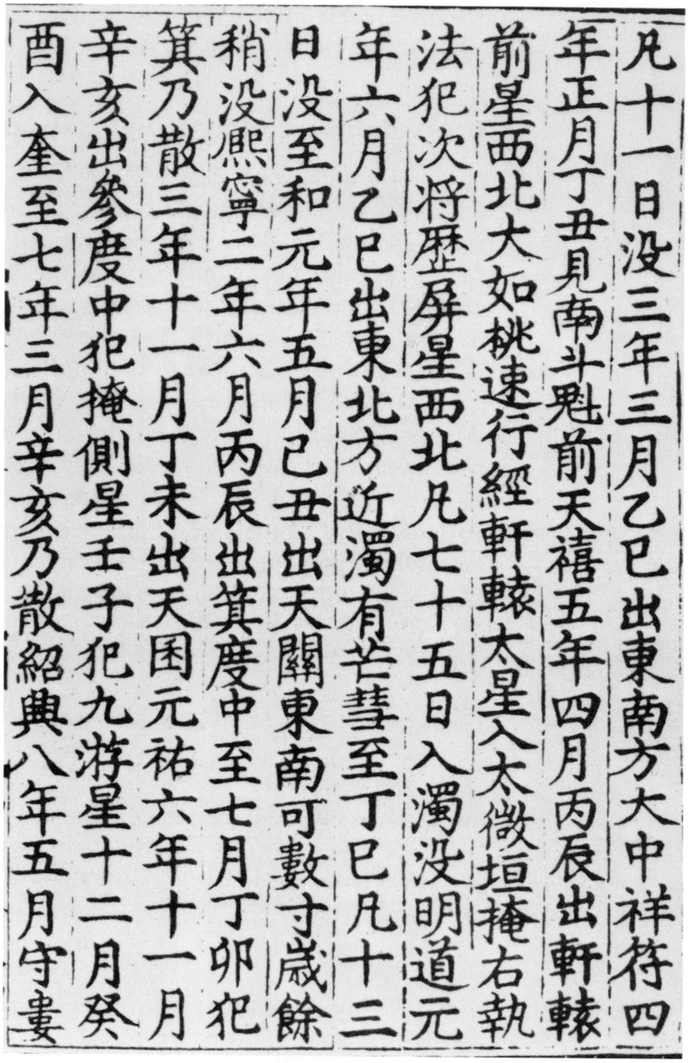

凡十一日没三年三月乙巳出東南方大中祥符四
年正月丁丑見南斗魁前天禧五年四月丙辰出軒轅
前星西北大如桃速行經軒轅太星入太微垣掩右執
法犯次將歷屏星西北凡七十五日入濁没明道元
年六月乙巳出東北方近濁有芒彗至丁巳凡十三
日没至和元年五月己丑出天關東南可數寸歲餘
稍没熙寧二年六月丙辰出箕度中至七月丁卯犯
箕乃散三年十一月丁未出天囷元祐六年十一月
辛亥出參度中犯掩側星壬子犯九游星十二月癸
酉入奎至七年三月辛亥乃散紹興八年五月守婁

Figure 9-25 This Chinese record describes the appearance of the "guest star" in the year 1054 A.D. Note the recurrence of the character 月, which means both "month" and "moon."

made extensive observations of this "guest star," whose brightness exceeded that of any other star for a few weeks (Figure 9-25). The intriguing rock painting shown in Figure 9-26 may represent a native American observation of the Crab Nebula supernova, but we can't be sure.

We should note that since the Crab Nebula supernova occurred about 1700 parsecs from Earth, the "guest star" seen in 1054 had in fact exploded about 5500 years earlier (Figure 9-27). Similarly, the supernova that appeared in the Andromeda galaxy in 1885 must have exploded about 2 million years previously. If it were only 30 parsecs farther away, astronomers would be

Figure 9-26 This pictograph from northern Arizona may represent the crescent moon and the supernova of 1054, since the moon did pass close by the supernova one morning when the supernova was close to its brightest.

just now discovering it and could study it with much better instruments than they had in 1885.

If supernovae appear in most large galaxies every hundred years or so, we must be overdue for another. In fact, the next few supernovae in the Milky Way that will be seen on Earth have almost certainly exploded already, and the news has been traveling toward us at the speed of light ever since.

Furthermore, news of supernova explosions may have reached the Earth during the nearly four centuries since 1604 without the supernovae being seen in visible light. The best example of such a possibility occurs in the source of radio emission called Cassiopeia A. This radio source, the most intense region of radio emission observed from Earth, shows a distribution of synchrotron radiation characteristic of the expanding gas from a supernova remnant. In addition, Cassiopeia A produces x rays and shows some wisps reminiscent of supernova remnants in visible light (Figure 9-28). The spectra of these wisps or filaments show emission lines whose Doppler shifts indicate huge velocities, up to 8000 kilometers per second along our line of sight. Photographs of the wisps taken several years apart reveal outward motions across our line of sight, as well as changes in the appearance of the filaments.

This all suggests that Cassiopeia A represents a young supernova remnant. We can extrapolate the observed velocities of the filaments back in time to find the interval that has passed since they all occupied the same position: about 300 years. Hence astronomers believe that Cassiopeia A was a super-

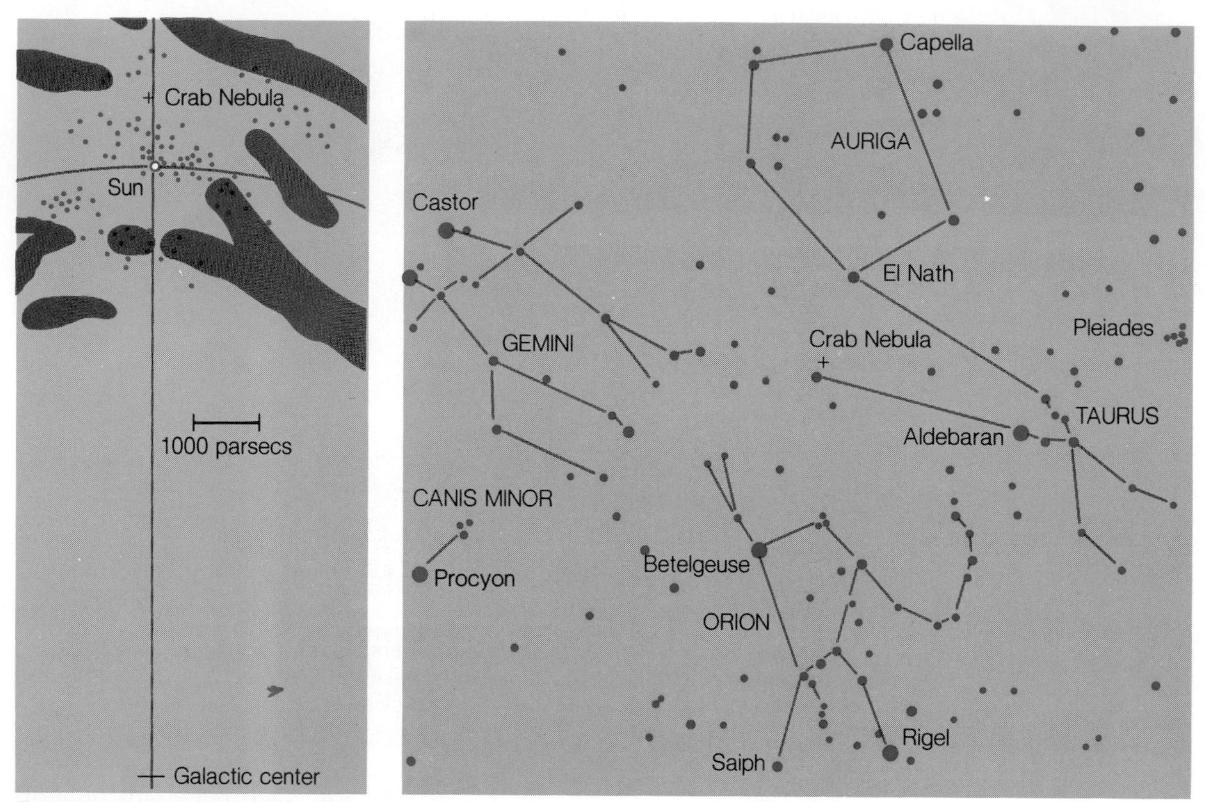

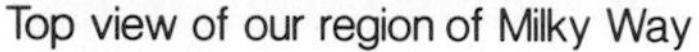
Top view of our region of Milky Way

As seen from Earth

Figure 9-27
The Crab Nebula lies in almost the opposite direction from the center of our galaxy, at a distance of about 1700 parsecs (left). Thus we can see the nebula (with a small telescope) when we look at the region where Taurus joins Orion (right).

nova that would have been seen to explode in about 1670.[5] But no one saw such an explosion! The failure to observe a supernova at a time when astronomy was well advanced in Europe might rest with large amounts of absorption of starlight by interstellar dust in the direction of Cassiopeia A. But it takes huge amounts of absorption to obscure a supernova. The Soviet astrophysicist Josef Shklovskii has suggested that the supernova itself may have obscured our view of it by expelling huge amounts of dust. If this explanation proves true, and if Cassiopeia A is not an isolated example, then other supernovae may also have passed unnoticed, at least in their visible light.

"Seeding" of Nuclei by Supernova Explosions

Supernovae apparently play a crucial role in the overall evolution of galaxies like our own by seeding their galaxies with nuclei heavier than helium. These heavier nuclei can be included in later generations of stars and planets

[5]Since the distance to Cassiopeia A is about 3000 parsecs, the actual explosion occurred some 10,000 years earlier.

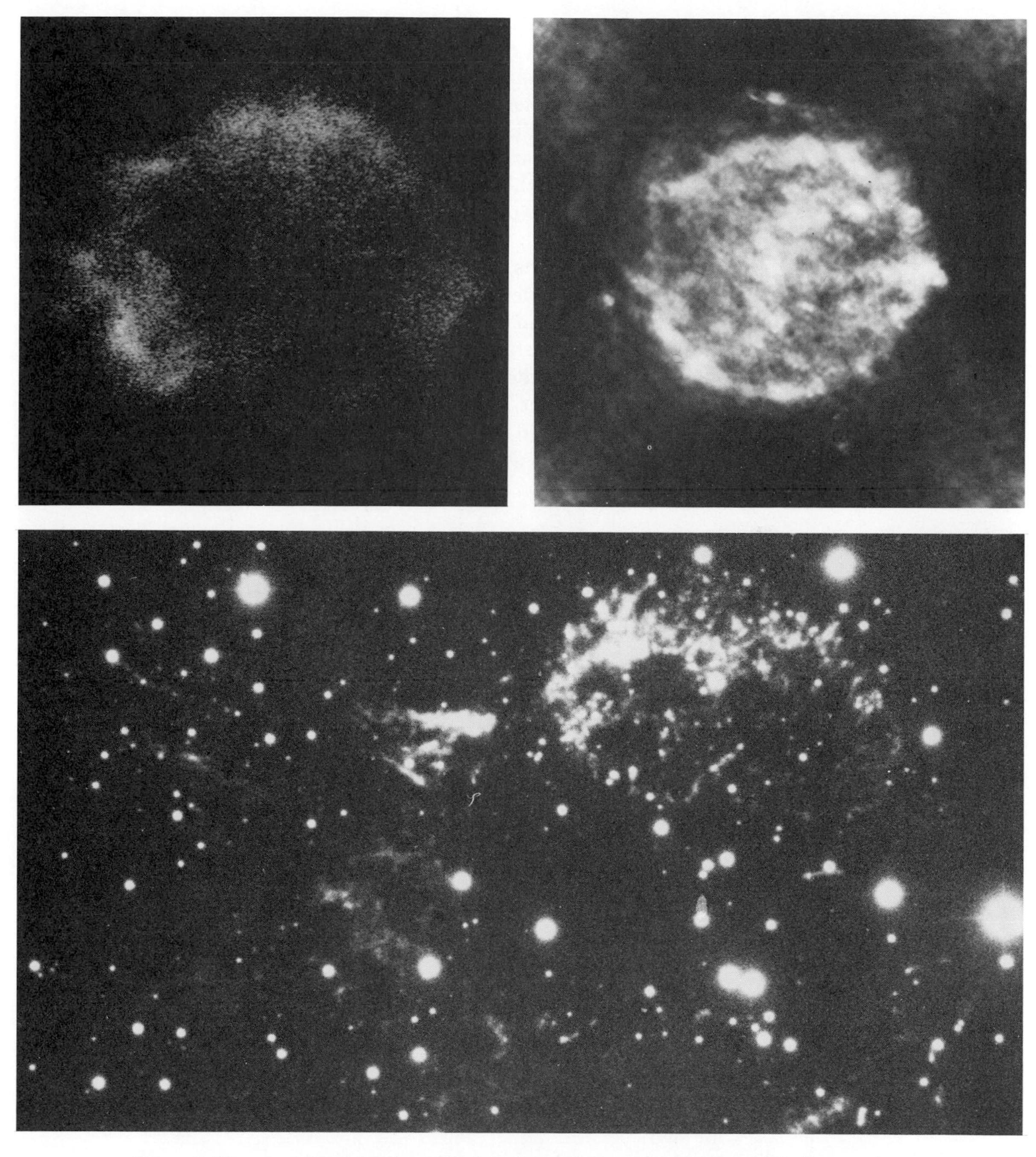

Figure 9-28 A visible-light picture of the supernova remnant Cassiopeia A shows only faint wisps of gas (bottom). The expanding shell of gas appears in x-ray emission (top left) and in radio emission (top right). The radio emission arises primarily by the synchrotron process, and the x-ray emission probably also consists of synchrotron radiation.

that condense in interstellar clouds. We may recall that the big bang produced most of the helium nuclei now found in stars, but did not form appreciable amounts of carbon, nitrogen, oxygen, or any heavier nuclei (see page 117). All of these elements, which are essential to living organisms on Earth and indeed form the bulk of the Earth itself, came from fusion processes inside stars. Although stars expel some heavier elements during their red-giant phases (page 311), the amount of this expelled matter falls far short of the observed abundances of heavy elements in stars like our sun (see Table 9-2).

Supernova explosions provide the best theoretical explanation of how stars got their heavy elements (where "heavy" means anything heavier than helium). Astronomers find a reasonable match between the abundances they calculate in theories of stellar evolution and the observed abundances of many of the elements—those from carbon through iron—in the sun and in similar stars. If the outer layers of stars that become supernovae produce these elements, we would expect to find the elements in the same proportions as they were made during the presupernova stage. Finally, the super-

Table 9-2 Average Abundance of Certain Elements in the Sun and the Solar System

Atomic Number (Number of Protons in Atomic Nucleus)	Element	Number of Nuclei for Every 10^{12} Hydrogen Nuclei
1	Hydrogen	1,000,000,000,000
2	Helium	80,000,000,000
3	Lithium	1,500
4	Beryllium	25
5	Boron	10,000
6	Carbon	370,000,000
7	Nitrogen	115,000,000
8	Oxygen	670,000,000
9	Fluorine	75,000
10	Neon	110,000,000
11	Sodium	2,000,000
12	Magnesium	30,000,000
13	Aluminum	2,500,000
14	Silicon	30,000,000
15	Phosphorus	300,000
16	Sulfur	15,000,000
17	Chlorine	180,000
18	Argon	3,500,000
26	Iron	25,000,000
29	Copper	70,000
47	Silver	15
79	Gold	6
82	Lead	125
92	Uranium	0.8

nova explosions themselves may well have manufactured the small amounts of elements heavier than iron, such as lead, gold, and uranium. The much lower abundance of these elements reflects the small amount of time in which they could be made, compared to the millions of years before the supernova explosion.

We thus lean toward the conclusion that just about every nucleus heavier than hydrogen and helium formed inside a massive star, shot into space during a supernova explosion, perhaps to be incorporated billions of years later in another star, a planet, or an interstellar cloud. With this scenario we can explain the difference in abundance of heavy elements between Population I and Population II stars. Population I stars typically have 1% or 2% of their mass in the form of elements heavier than helium. This may seem a small amount, but it exceeds by 10 or 100 times the abundance of heavy elements in Population II stars. We conclude that when galaxies began to form, an initial burst of star formation produced many massive stars that soon became supernovae. These supernovae seeded the galaxies with enough heavy elements to explain the heavy-element abundance in the Population I stars, which formed from the "enriched" matter. The less massive Population II stars remain visible today, unlike their more massive coevals that exploded long ago. Since even Population II stars contain some heavy elements, we must imagine a still more primeval population of stars that gave them the small fraction of heavy elements we observe in them.

Cosmic Rays

We have seen that the elements heavier than helium appear as star stuff, processed in stellar interiors and flung into interstellar space. Supernovae thus do the work of "cosmic evolution," spreading "evolved" matter through the universe and even making new elements from old. As if this were not enough, supernovae may also produce **cosmic rays,** elementary particles that move through interstellar space at almost the speed of light (Figure 9-29). These particles include protons, electrons, positrons, helium nuclei, and a small amount of heavier nuclei. Each cosmic-ray particle has a huge kinetic energy (for an elementary particle), the result of its enormous velocity. They may well arise in the outermost layers of supernovae, which acquire the largest velocities as the star explodes. An alternative theory, which still makes supernovae responsible for cosmic rays, suggests that the cosmic-ray particles are stripped from interstellar dust grains by the blast of protons from a supernova.

From continual supernova outbursts, a galaxy such as our own would acquire a relatively uniform density of cosmic-ray particles. Although they contribute only a tiny amount to the galaxy's mass, cosmic rays are important for another reason. Cosmic rays may be the cause of mutations in living organisms (see page 486). Since mutations play a key role in the evolution of life, we may owe not only the cosmic evolution of elements, but also some of the biological evolution on Earth, to supernovae.

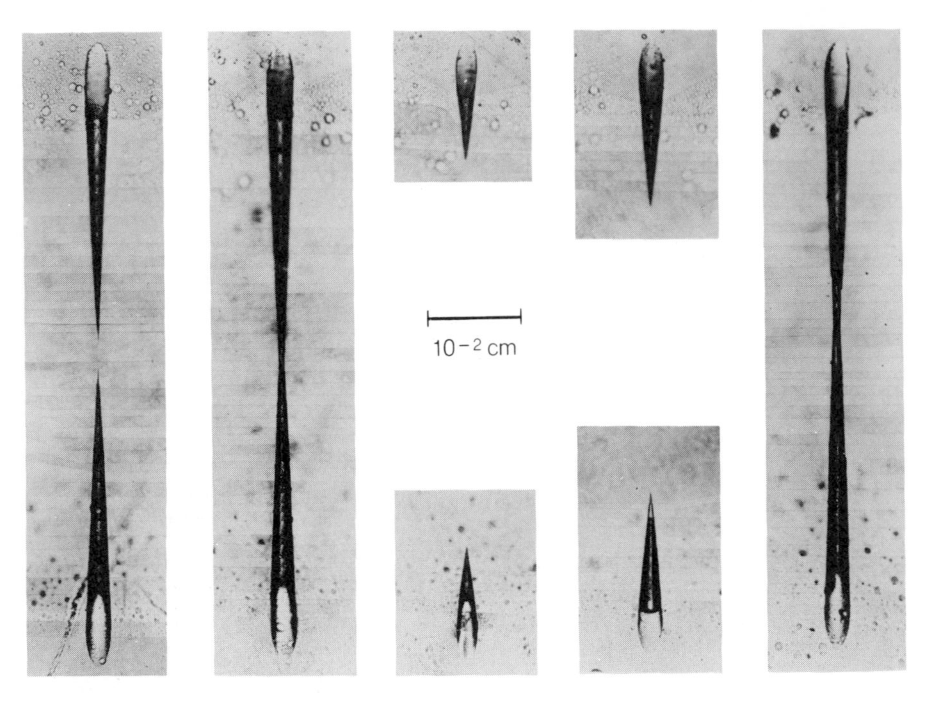

Figure 9-29 Cosmic-ray particles passing through lexan plastic left behind these tracks a few millimeters long, testimony to the large kinetic energy in a single electron or nucleus moving at almost the speed of light.

In short, supernovae do almost everything. Adrift on our lump of cosmic debris, itself the accreted remnant of repeated supernova explosions, some of the heavier nuclei have formed into organic molecules. The continuing flux of cosmic rays may change the way some of these molecules replicate, to bring forth what we call intelligent life. The hand that wrote this book, the paper on which it appears, the eye that reads it, and the mind that interprets it all may owe their existence to the stars that exploded during the past 5 to 10 billion years. We are living on the product, as the product, and by the product of supernovae. Cosmic disaster? For us, the demise of stars may have spelled life itself. William Blake spoke of the chance "to see a world in a grain of sand." The grain of sand and our ability to see the world come from the stars whose death led to our life.

Summary

As a star grows older, it steadily converts protons into helium nuclei at its center. As the supply of protons there runs low, the star loses its ability to maintain a constant rate of energy liberation. For a time, the star can compensate for its diminishing supply of protons by contracting its central regions, thereby raising the temperature to fuse the remaining protons at an ever-increasing rate. This contraction actually increases the rate of energy liberation; part of the extra energy of motion expands the star's outer layers, cooling them slightly and producing a red-giant star, with a large outer surface but a small, fast-burning core. More massive stars will exhaust their supplies of protons far more rapidly than less massive stars. Hence in a star cluster (all of whose stars began to shine at almost the same time), the more massive stars will leave the main sequence of the temperature-luminosity diagram before the less massive stars do.

A red-giant star will eventually contract its core to the point that helium nuclei begin to fuse into carbon. This helium flash temporarily expands the core and sets the stage for a period of instability, during which the star is likely to pulsate in size and brightness. The Cepheid variable stars, so important in determining the distances to nearby galaxies, have reached this stage in their evolution.

When helium nuclei have fused into carbon nuclei, most stars will become degenerate in their centers. At this point, the central regions have such high density that the exclusion principle prevents the electrons from packing any tighter. The electrons hold the nuclei by electromagnetic forces, so the exclusion principle supports the entire star against its self-gravitation. After the star's outer layers evaporate, the core remains as a white dwarf, capable of slowly fading into dimness.

No white dwarf can exist, however, if its mass exceeds 1.4 times the sun's mass. Stars with more than this mass are likely to fuse carbon into heavier and heavier nuclei, liberating a bit more energy, until their central regions collapse upon the exhaustion of all possible nuclear fuel. In such stars, the central density does not rise so high that the exclusion principle can support the star. With all energy-liberating processes exhausted, the star's core collapses in a time of about one second, momentarily becoming a single nucleus of starlike mass. This nucleus rebounds a bit in size to initiate a supernova explosion, which blasts the star's outer layers into space at enormous speeds. Such supernova explosions seed nuclei heavier than helium throughout the galaxy and probably produce most of the cosmic rays, which are nuclei and electrons traveling at almost the speed of light.

Key Terms

black-dwarf star
Cepheid variable star
cosmic rays
degenerate matter
exclusion principle
helium flash
instability strips
1.4-solar-mass limit
planetary nebula
pulsating star
red-giant star
RR Lyrae variables
supernova
turnoff point
white-dwarf star

Questions

1. How long has the sun been converting protons into helium nuclei? How long will it take for the sun to run out of protons in its interior?
2. As the sun begins to exhaust its supply of protons, why will its central regions contract? What effect will this contraction have on the sun's outer layers?
3. Do we expect all stars to pass through a red-giant phase? What happens after the red-giant phase?
4. In a cluster of stars that all begin to liberate kinetic energy at the same time, which stars will become red giants first? How can we use this fact to find the ages of the stars in a cluster?
5. The Pleiades and the Hyades are both open clusters with about 150 stars each. In the Pleiades the hottest main-sequence stars have spectral type B5. In the Hyades the hottest main-sequence stars are type A1. Which cluster is likely to be the older?
6. A star with 10 solar masses liberates energy of motion about 2000 times more rapidly than the sun does. If the sun can liberate kinetic energy for 10 billion years, how long will the more massive star be able to liberate kinetic energy?
7. What happens to a star when the helium flash occurs? Does this make the star's interior hotter or cooler?
8. At what stage in a star's lifetime is it likely to become a Cepheid variable star? What characteristics of such a star produce a variation in light output?
9. Why does the fusion of helium nuclei into carbon nuclei fail to provide as much kinetic energy as the fusion of protons into helium nuclei?
10. What effect does the exclusion principle have on helium nuclei? On carbon nuclei? On electrons? How does the exclusion principle support an entire white-dwarf star against collapse?
11. Which kind of stars are likely to eject matter into space over a time span of several million years? Which kind of stars are likely to eject much of their mass into space in a period of a few minutes?
12. Is the sun likely to produce a planetary nebula as it ages? Will the sun become a supernova? Why or why not?
13. Why are supernova explosions important in the evolution of other stars?
14. What are cosmic rays? Why do we think they may come from supernova explosions?

Projects

1. *Density.* Large stars do not necessarily have larger masses than small stars. Although this violates our intuition, it is in fact true. In particular, if a given star swells to a larger size without gaining any mass, it must spread the same amount of mass (or less mass, if it loses some) through a larger volume. Calculate, for the case of the sun, how many times the

average density of matter will decrease if the sun's radius increased by 50 times during the sun's red-giant phase. Now suppose that half the sun's mass remained in its contracting core, while the other half spread into this larger volume. In that case, how many times would the average density of matter decrease? Why do you get the same answer? Now insert some numbers: Average density in sun's core now = 50 grams/cm^3; average density in rest of sun = 1 gram/cm^3. (By the "core" we mean here simply the inner half of the sun's mass.) What will the average density in the noncore parts of the sun become during the sun's red-giant phase if the sun's radius increases 50 times? How does this compare with the atmosphere of the Earth at sea level (density = 0.0005 gram/cm^3)?

2. *Temperature and Density: The Diesel Engine.* Consider what happens inside the cylinders of a diesel engine, where oxygen combines with the hydrocarbon molecules in diesel fuel to create a chemical explosion that drives the pistons. What ignites the air-fuel mixture, since a diesel engine has no spark plugs? Why does this ignition process recur in a steady cycle, rather than going on at all moments? What lowers the temperature inside the cylinders? If we seek to make an analogy between a star's core and the cylinders of a diesel engine, what force in the star would we compare to the compression part of the diesel cycle? What would we compare to the explosive part of the cycle? Why don't nuclear fusion reactions occur among the protons inside the cylinders of a diesel engine?

3. *Star Clusters.* The most obvious star cluster, as seen from northern latitudes, is the Pleiades, a compact group just west of the bright star Aldebaran in Taurus. Aldebaran itself stands at the point of the V formed by the inner parts of the Hyades cluster. Observe the Pleiades and the Hyades and compare their angular extent. If the two open clusters had the same true diameter in space, what could we learn about their relative distances from us by comparing their apparent diameters on the sky? Compare the colors of the brightest members of the Pleiades and Hyades. Which cluster has the bluer stars? What does this tell us about the ages of the clusters? Why does it suffice to observe the colors of the *brightest* members of a star cluster to find the approximate age of the stars in the entire cluster?

4. *Supernovae.* The rate of occurrence of supernova explosions in a large galaxy such as our own is about one per century. Suppose these supernovae occur within the galactic disk, about 100,000 light years in diameter and 2000 light years thick. At any given time, about how many supernovae have in fact exploded without our having had time to receive the news? Consider the supernova of A.D. 1054 (the year it was first seen on Earth), which left behind the Crab Nebula and its pulsar. Why is it possible that scientists in Europe left no record of the supernova, despite its appearance as the brightest star in the sky for several weeks? (A knowledge of European history will help here.) In what sort

of records should we search, if we hope to uncover a hitherto-overlooked sighting of the Crab Nebula pulsar in Europe?

Further Reading

Greenstein, Jesse. 1970. "Dying Stars." In *New Frontiers in Astronomy.* Edited by Owen Gingerich. San Francisco: W. H. Freeman.

Jastrow, Robert. 1969. *Red Giants and White Dwarfs.* New York: New American Library.

Kirshner, Robert. 1976. "Supernovas in Other Galaxies." *Scientific American* 235:6, 88.

Page, Thornton, ed. 1968. *The Evolution of Stars.* New York: Macmillan.

Stephenson, F. Richard, and Clark, David. 1976. "Historical Supernovas." *Scientific American* 234:6, 100.

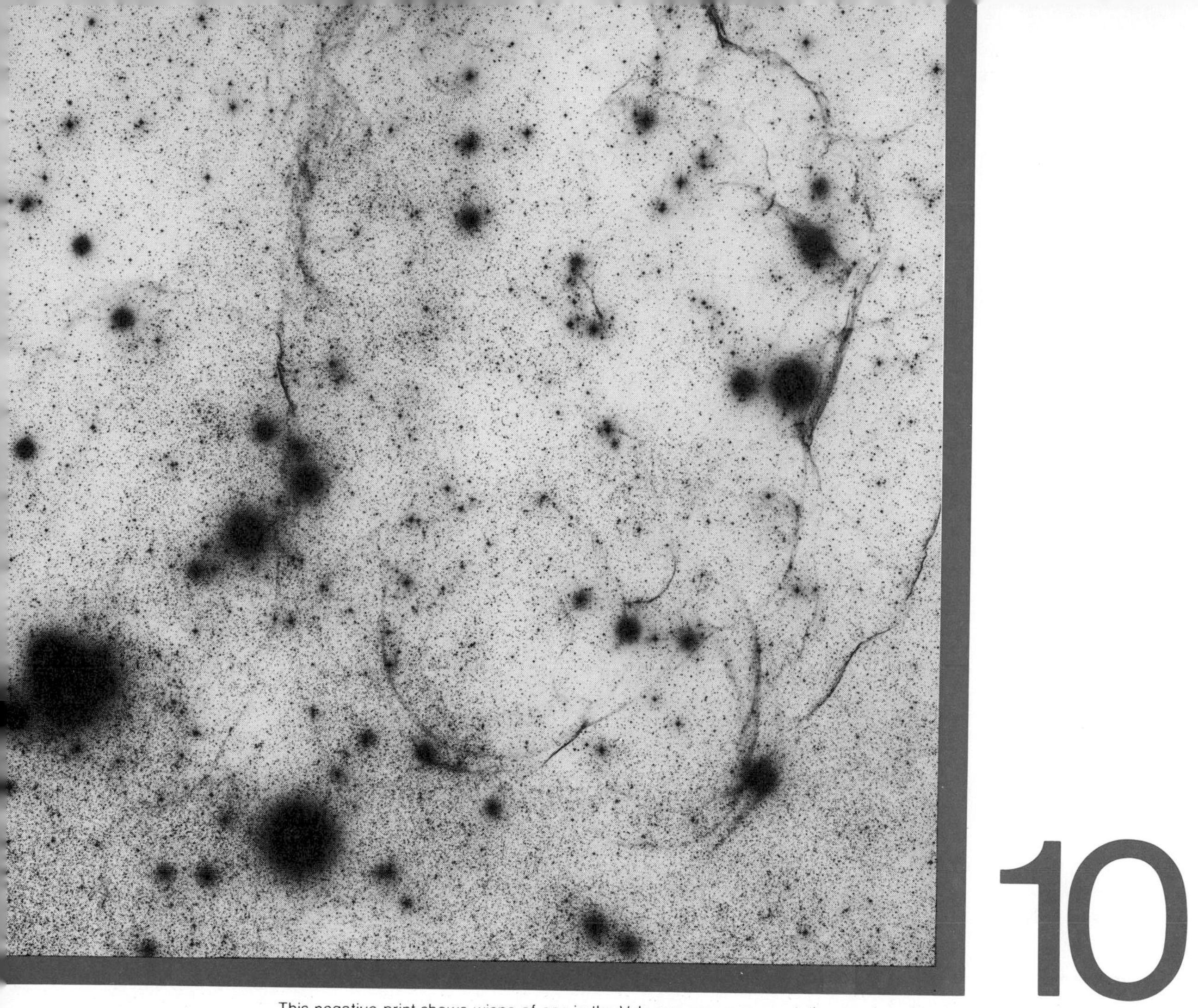

10

This negative print shows wisps of gas in the Vela supernova remnant, the remains of a star that exploded tens of thousands of years ago.

Neutron Stars, Pulsars, and Black Holes

Stars that do not have small enough masses to become white dwarfs, slowly cooling for billions of years, undergo catastrophic collapse in their central regions. Such collapse, which arises from the star's inability to support itself against its own gravitation through the liberation of kinetic energy, can lead to one of two final states. The star's interior will become either a neutron star, incredibly dense and compact, or a black hole, still smaller and denser, evidence of the triumph of gravity (Figure 10-1). We must examine these two possibilities to see which stars may produce them and to appreciate the ways in which neutron stars and black holes can affect the rest of the universe.

20 km

Neutron star

Black hole

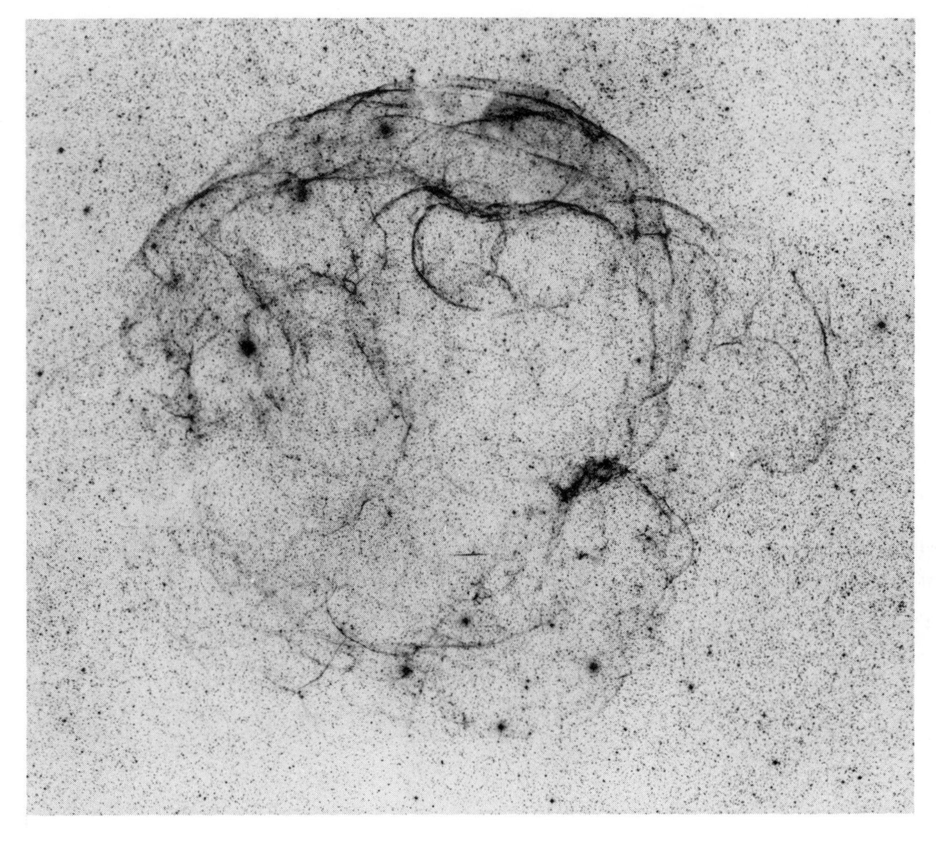

Figure 10-1
Supernova explosions blow the outer layers of a star into space, as shown by the expanding filaments of gas in this photograph of the supernova remnant S147. The inner parts of the star may become a neutron star, only 20 kilometers in diameter, or an even smaller and denser black hole.

10.1 Neutron Stars

We have seen that stars with less than 1.4 times the sun's mass will become white dwarfs as they age. Stars with more than 1.4 solar masses become ready to collapse, and they do indeed collapse after their central regions have become mostly iron nuclei and electrons (page 313). This collapse proceeds so swiftly and so violently that the exclusion principle among the electrons cannot stop it. Instead, the electrons are squeezed into the other particles, which all become a single gigantic nucleus (page 314), a billion times denser than the matter in a white dwarf. This single nucleus is called a **neutron star.**

In white dwarfs, the exclusion principle keeps the electrons apart. The electrons in turn prevent the nuclei from all rushing together by exerting electromagnetic forces on them. This works well when many electrons exist. Most stars, such as white dwarfs, have plenty of electrons. Neutron stars do not. As part of the collapse that initiates a supernova explosion, the electrons merge with the protons broken from the nuclei (Figure 10-2). These reactions are called **weak reactions,** in contrast to the strong reactions of nuclear fusion, when two nuclei merge. Electrons are extremely reluctant to fuse

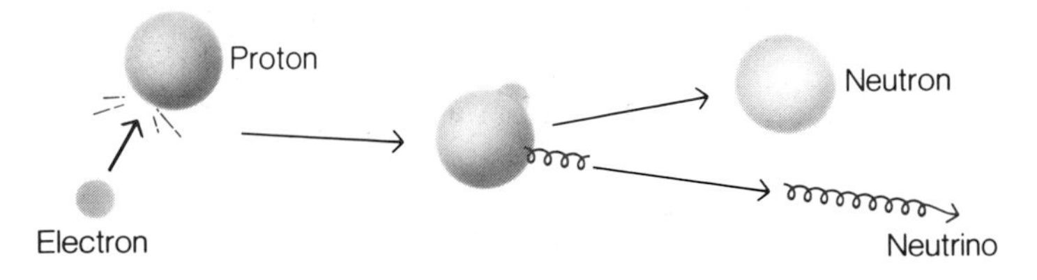

Figure 10-2 The tremendous densities involved in the collapse of a star's central regions make the electrons merge with the nuclei in weak reactions. A typical weak reaction, such as the proton-electron fusion shown here, produces a neutrino along with the new nuclear particle (in this case a neutron).

with nuclei, and only the immense density and temperature, created by the star's sudden collapse, force them to do so.

The Neutron Star as a Single Giant Nucleus

Once the electrons have merged with the nuclei, the entire star may be thought of as a single nucleus. As we described on page 314, the collapse stops at a density close to 10^{15} grams per cubic centimeter, and matter falling in from the outer layers of the collapsing star "bounces" outward to form the supernova explosion. The collapse ceases because the exclusion principle among the nucleons (protons and neutrons) in the single nucleus prevents them from being squeezed any closer together. Protons and neutrons, like electrons, feel the effect of the exclusion principle and become "degenerate," but only at densities much greater than the 10^6 or so grams per cubic centimeter that brings about electron degeneracy.

In the twinkling of an eye, then, the collapsed star becomes a giant nucleus. After the "bounce," the matter that remains will expand slightly to become the neutron star. A neutron star may be thought of as a single atomic nucleus that consists mainly of neutrons. At any given time, some of these neutrons decay into protons and electrons and antineutrinos (see page 115), but the tremendous densities inside the neutron star soon make an equal number of protons merge with electrons to re-form neutrons. The entire neutron star supports itself against its (enormous!) self-gravitation through the exclusion principle, which refuses to allow neutrons to be squeezed together beyond a density of just under 10^{15} grams per cubic centimeter.

Properties of Neutron Stars

Calculations show that no neutron star can exist with a mass greater than three to five times the sun's mass. (The inexactness of this mass limit testifies to the difficulty of the calculations.) In objects with masses above this limit, even the exclusion principle among neutrons does not work, for the collapse occurs before the exclusion principle can make itself felt. Thus the most massive stellar cores collapse completely to form black holes (page 337). We may

Table 10-1 Sizes and Densities of Various Types of Stars and Objects

Type of Star	Typical Mass (Solar Masses)	Radius (Kilometers)	Average Density (Grams per Cubic Centimeter)	Fraction of Photon's Original Energy Lost in Escaping from Star's Surface
Main sequence star	1	700,000	1	0.000005
Red giant star	1	7,000,000	10^{-3}	0.0000005
Red supergiant star	10	150,000,000	10^{-6}	0.00000023
White dwarf star	1	7,000	10^{6}	0.0005
Neutron star	2	10	10^{15}	0.60
Black hole	6	Less than 18	More than 5×10^{14}	1.0

therefore expect neutron stars to have masses ranging from about 1.4 solar masses up to 3 to 5 solar masses, with white dwarfs below this range and black holes above it.

Table 10-1 gives the sizes and densities of various types of stars. The density of a white dwarf, although enormous, falls far below the density of a neutron star. A typical neutron star with a diameter of 20 kilometers may have twice the sun's mass, with a density a billion times that of a white dwarf and a million billion (10^{15}) times the average density of the sun. A star with twice the sun's mass would become a black hole if it collapsed to a radius of less than 6 kilometers. If it becomes a neutron star instead, with a radius of perhaps 12 kilometers, it will have a density eight times less than it would at 6 kilometers radius. This difference may seem small when we talk of densities 10^{15} times the density of water, but it can be crucial. Astronomers believe that a neutron star can last indefinitely, supported against gravity by the exclusion principle, with twice the radius that would lead to the star's complete disappearance as a black hole.

10.2 Pulsars from Neutron Stars

During the late 1930s, when physicists first calculated the properties of neutron stars and black holes, few astronomers thought that such strange objects really existed or would soon be detected. A generation later, astronomers may have found black holes, and they certainly have observed neutron stars. Neutron stars produce **pulsars,** radio sources that emit photons in regularly spaced pulses. These pulses of photons apparently arise near the surface of rapidly rotating neutron stars, and their rhythmical emergence from the pulsars reflects the rotation cycles of the neutron stars.

The Rotation of Neutron Stars

Neutron stars should all be rotating quite rapidly. If a star retains about the same mass as it collapses, then it must spin more quickly to conserve its angular momentum, just as a high diver spins more rapidly when she contracts her body (see page 148). The star's rate of spin will increase in proportion to 1 over the square of its radius, if it loses a negligible fraction of its total mass. Thus if the star becomes 1000 times smaller, it will spin a million times more rapidly. A neutron star that forms from the collapse of a stellar interior might begin its collapse with a radius of 120,000 kilometers. If it forms a neutron star with a radius of 12 kilometers, the neutron star will be 10,000 times smaller than the original collapsing star. Hence if the star's initial rate of spin was one rotation each month, the neutron star will rotate 100 million times more rapidly, or 40 times per second.

The Magnetic Field of a Neutron Star

Along with this great increase in the star's rate of spin goes a similar increase in the strength of the star's magnetic field. Most stars have at least a modest magnetic field, and as the star rotates, so does its magnetic field. Since electrically charged particles feel the electromagnetic forces produced by the star's rotating magnetic field, these forces tend to push the particles around as the star and its magnetic field rotate. For weak magnetic fields, this produces only moderate effects. But if the magnetic field acquires enormous strength, its electromagnetic forces can control the motions of nearby charged particles in the same way that magnets in a particle accelerator do, making the charged particles trace a curved path instead of the straight line that their momentum calls for (Figure 10-3). The strength of the magnetic field at a star's surface, like the star's rotation rate, increases as 1 over the square of the star's radius. The magnetic field in a star that collapses to one ten-thousandth of its original size will increase in strength by 100 million

Figure 10-3 The Fermi National Accelerator Laboratory uses superconducting magnets to make charged particles move in a circle 2 kilometers in diameter at almost the speed of light. The particles are then directed down the tracks to the left of the main circle.

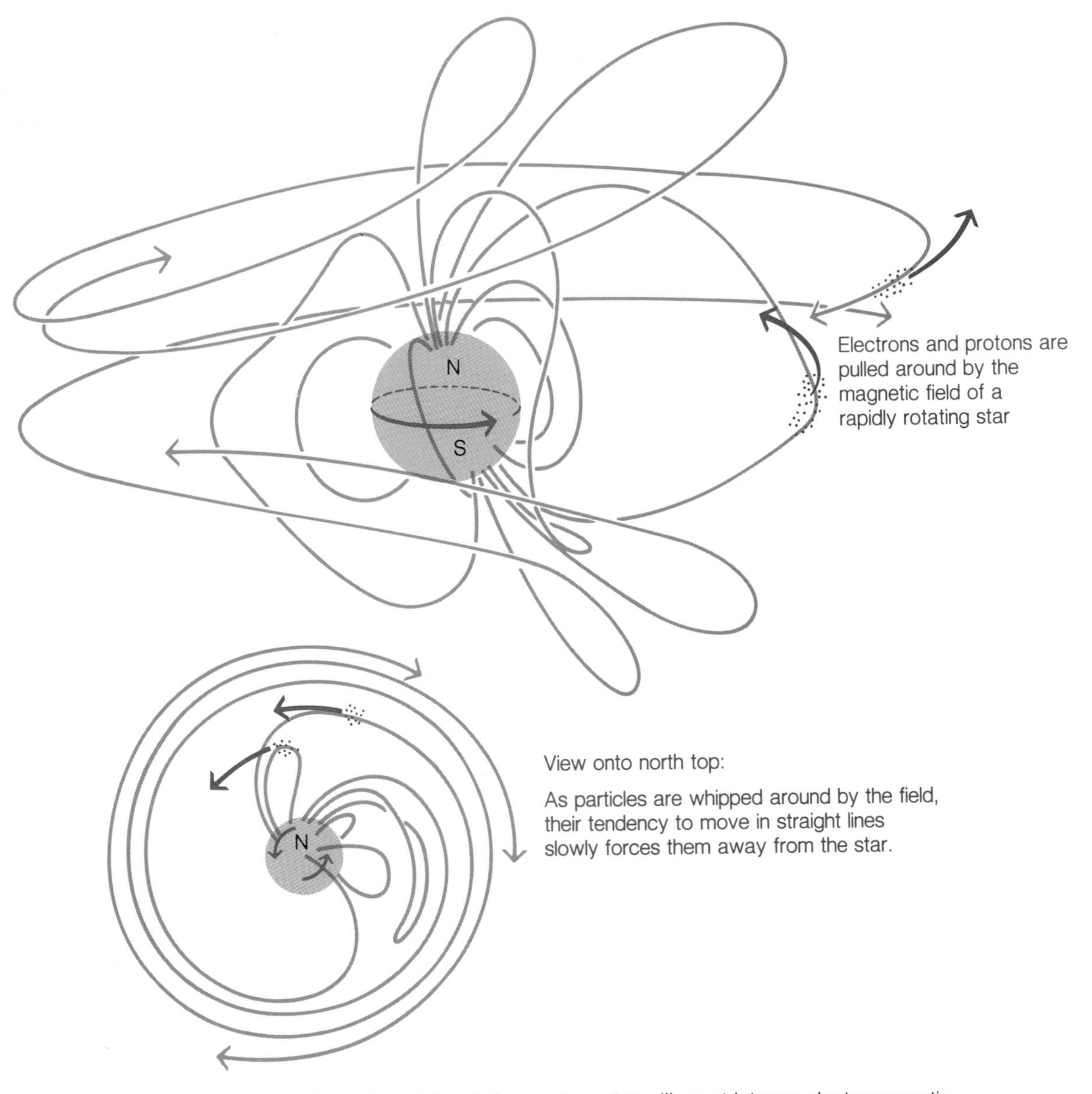

Figure 10-4 A collapsed, rapidly rotating neutron star will exert intense electromagnetic forces on nearby charged particles. Hence the charged particles will be forced into rapid rotation along with the star, slowly spiraling away from it. As the particles move, they produce synchrotron radiation.

times. This enormous increase makes the neutron star a giant spinning magnet, which sweeps along any charged particles nearby (Figure 10-4). Though the particles can spiral outward along the lines of the magnetic field, they produce photons through the synchrotron emission process as they move.

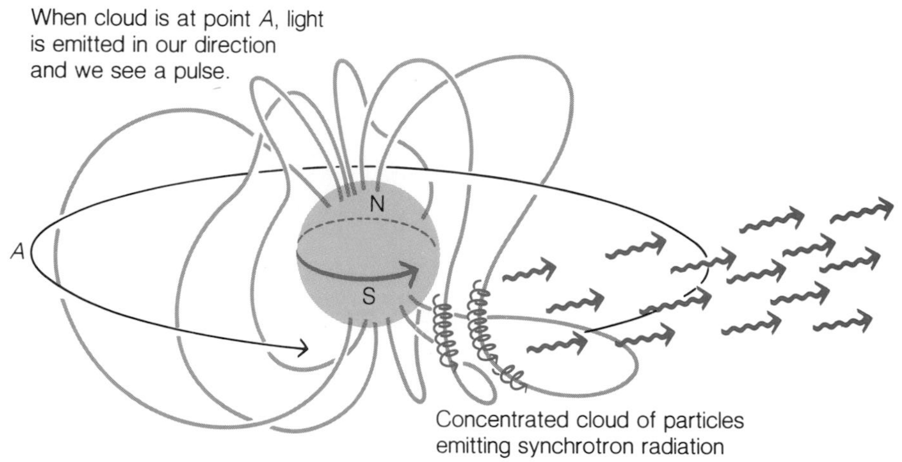

Figure 10-5
Pulsars apparently contain some regions that emit more photons per second than other regions. As the regions of more intense emission sweep by us, we see a flash of photons.

How Pulsars Pulse

Such rapidly rotating, strongly magnetic neutron stars produce pulsars. The term *pulsar* comes from the original designation of *pulsating radio source*, and indeed all known pulsars emit bursts of radio photons, repeated in periodic fashion many millions of times.[1] Some pulsars also emit pulses of visible-light and x-ray photons, though most do not, so far as we can tell.

The time interval between pulses almost certainly corresponds to the rotation period of the neutron star that produces them. It seems likely that some part of the pulsar emits more photons than other parts. As the more highly emitting region sweeps past us, we observe a pulse like the flash from a rotating lighthouse beacon (Figure 10-5).

Why Pulsars Slow Down

As the neutron star's rotating magnetic field sweeps nearby charged particles around the star, the particles spiral outward, emitting photons through the synchrotron process (see page 208). The charged particles that acquire the most kinetic energy emit gamma-ray and x-ray photons; particles with less kinetic energy emit visible-light photons; and the particles with the least energy of motion (but still enough to be moving at nearly the speed of light) emit radio-wave photons. As these charged particles produce photons by the synchrotron process, they lose some of their energy of motion which now appears as the photons' energy. The star's rotating magnetic field can continue to supply kinetic energy for a time, but not forever. As a direct result of producing the synchrotron emission, the neutron star, and thus the pulsar,

[1]Pulsars emit *some* radio waves at all times, but they emit much more radio power during the "on" part of their pulse cycle than during the longer "off" part of the cycle.

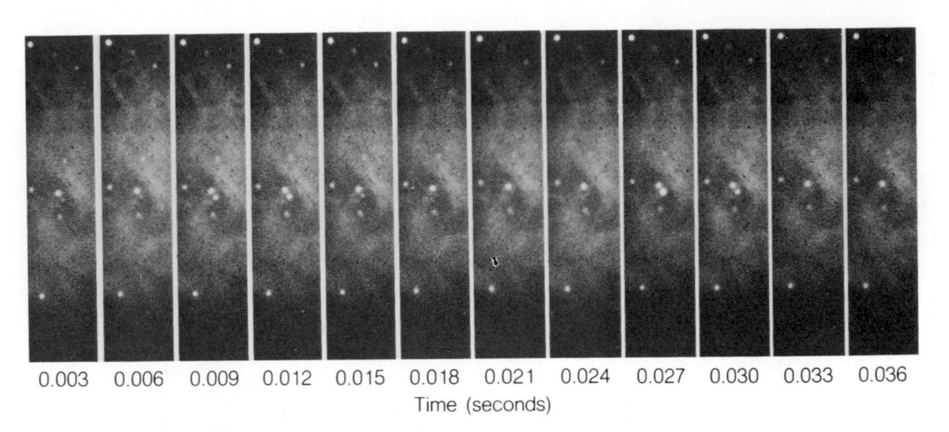

Figure 10-6
The pulsar at the center of the Crab Nebula (Figure 9-23) shows a regular cycle of light variation that repeats 30 times per second. This cycle includes a subpulse (0.009 second) as well as the main pulse (0.027 second).

must gradually slow down and emit pulses at increasingly greater time intervals.

Now that astronomers have observed pulsars for more than a decade, they have established that pulsars indeed slowly increase their pulse intervals. Of the 200 pulsars so far detected, the best studied is the Crab Nebula pulsar. As we discussed in Chapter 9, this web of hot gaseous filaments represents the remnant of the supernova seen in explosion in A.D. 1054. At the center of the nebula, a radio source pulsating 30 times per second was discovered in 1968. This source also pulses, and at the same rate, in visible light, x rays, and gamma rays (Figure 10-6). Precise measurements have established that the interval between successive pulses lengthens each year by thirteen-millionths of a second. In other words, the pulsar is slowing down by about 0.33% each year because it radiates away some of its kinetic energy (page 335). This slowing does not occur with complete regularity, however, because astronomers have measured sudden deviations from the overall slowing trend. Some of these deviations indicate small, temporary *decreases* in the interval between pulses. These decreases may arise from "starquakes" in the outer crust of the rotating neutron star that produces the pulses.

The Crab Nebula pulsar has the fastest pulsar spin rate known today. The third-fastest pulsar, the Vela pulsar, appears in the Gum Nebula (Figure 10-7) and flashes on and off in radio and x-ray photons 11 times per second. Other pulsars have periods that range up to almost 4 seconds, and quite probably longer-period pulsars exist but have not yet been discovered. We expect to find that the youngest pulsars pulse most rapidly, since all pulsars gradually slow down. The Crab Nebula pulsar will halve its spin rate in a few hundred years, the Vela pulsar in 5000 years. Other pulsars will take from a few thousand to a few million years to double the intervals between successive pulses. We may reasonably conclude that a few thousand to a few million years have passed since these pulsars first began to flash on and off in our skies.

As we have seen, pulsars arise from rapidly rotating neutron stars, which in turn originate from some supernovae. But other supernovae produce still more spectacular results of stellar collapse—the most compact objects of all, the black holes.

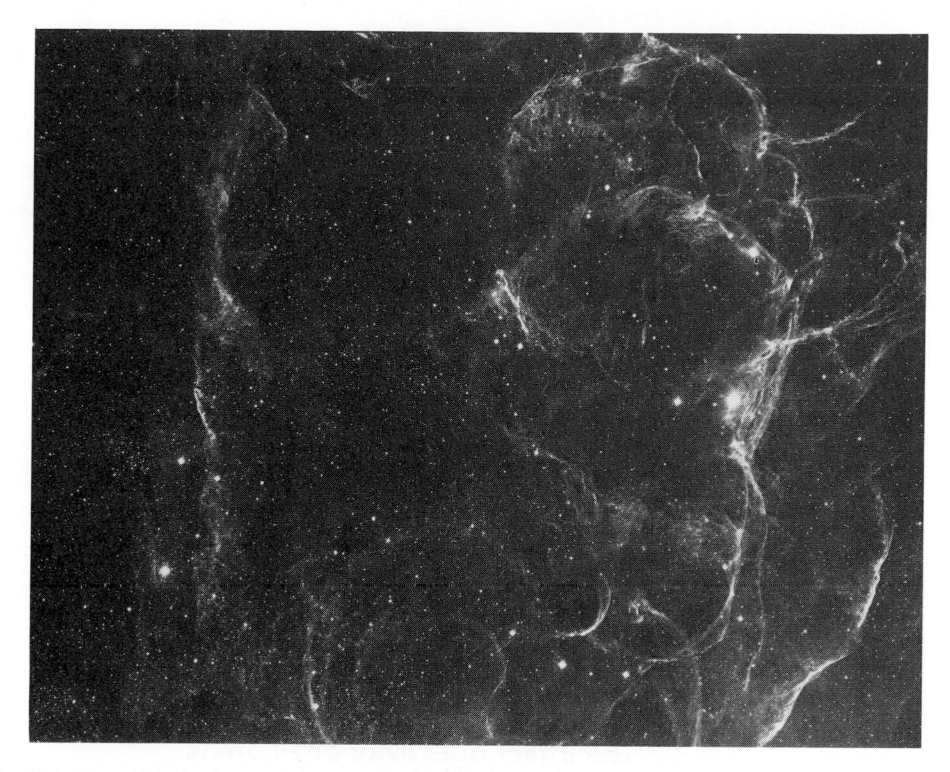

Figure 10-7 The Gum Nebula (named after its discoverer, Colin Gum) consists of a web of gaseous filaments about 500 parsecs from the sun. This photograph shows only the central parts of the Gum Nebula.

10.3 Black Holes

Black holes are the point of no return in the universe, the ultimate collapse. A **black hole,** as astronomers use the term, is an object whose gravitational forces prevent any photon or any other kind of particle from leaving the object. Hence a black hole must be invisible, with only its gravitational force continuing to interact with the rest of the universe.[2] Since neither light, nor radio waves, nor particles can escape from a black hole, we can never have two-way communication with such an object. Our knowledge of black holes must therefore remain circumstantial, based on the physics that we know and on the observations that may reveal the immediate surroundings of a black hole (see page 352).

[2]To be precise, a black hole can interact with the rest of the universe in three ways: through its gravitational force, its total electromagnetic force, and its total spin. We shall see later (page 346) that a spinning black hole will tend to drag in-falling particles into a spiraling disk outside the black hole.

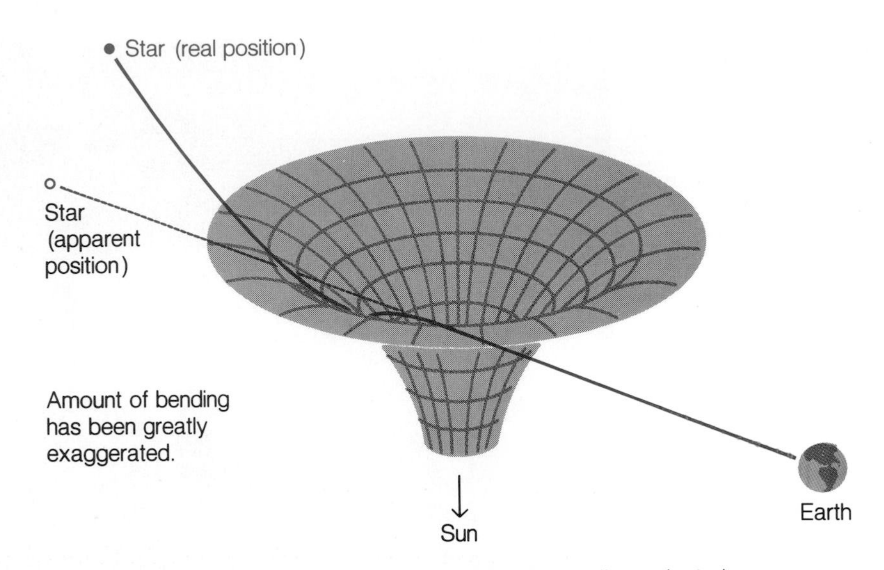

Figure 10-8 Photons passing close by the sun will deviate from straight-line trajectories. We can think of this deviation as the result of the sun's bending of space.

The Effect of Gravity on Photons

What does it mean to say that gravity "prevents" any light or other particles from leaving a black hole? Gravity affects *everything*, even particles with no mass, such as photons. For particles with mass, the mutual gravitational attraction varies as the product of any two particles' masses, divided by the square of the distance between their centers. But particles with no mass still obey the gravitational force from a massive object such as a star. Photons, the most important massless particles, will deviate from straight-line trajectories as they pass close to the sun (Figure 10-8). The observational proof of this fact in 1919 brought Albert Einstein fame throughout the world, for Einstein (Figure 10-9) had predicted just such a deviation in his general theory of relativity three years before (see page 226).

Since everything feels the effect of gravitational forces, physicists often find it convenient to think of space as curving, or bending, in response to gravity. In this view, gravitational forces are equivalent to "low spots" in curved space, toward which objects tend to fall as the result of the curvature. Light rays passing close to a massive object such as the sun will continue to travel in a straight line, but since the sun bends space itself, the "straight line" looks curved to us. The gravitational effect remains identical to that calculated on the supposition that the sun bends the light rays, rather than bending space.)

Thus photons passing close to the sun will feel the effect of the sun's gravity. But what about photons escaping *from* the sun? They too feel the effect of gravitational force from the sun. This force does not decrease the

Figure 10-9
Albert Einstein (1879–1955) developed the general theory of relativity. This theory predicts that gravitational forces will make light rays deviate from the straight-line trajectories they would have in the absence of gravitation.

photons' speed (still the speed of light), but it does reduce the photons' energy. Every photon that escapes from the sun, or from another star, loses a tiny fraction of its energy in escaping. In attempting to retain the photons, the sun's gravitational force robs each of them of about 0.0005% of its kinetic energy. In Chapter 7 we discussed, but rejected, the possibility that such a **gravitational red shift** (decrease in photon energy caused by gravity) could explain the large red shifts observed in quasars' spectra.

The fraction of energy lost by an escaping photon will increase as the gravitational force of the star increases. The force of gravity will increase either if the star contracts while keeping the same mass, or if the star acquires more mass while maintaining the same size (Figure 10-10). The first possibility is likely for some stars; the second highly unlikely. If we halve a star's radius while keeping its mass constant, the gravitational force at its

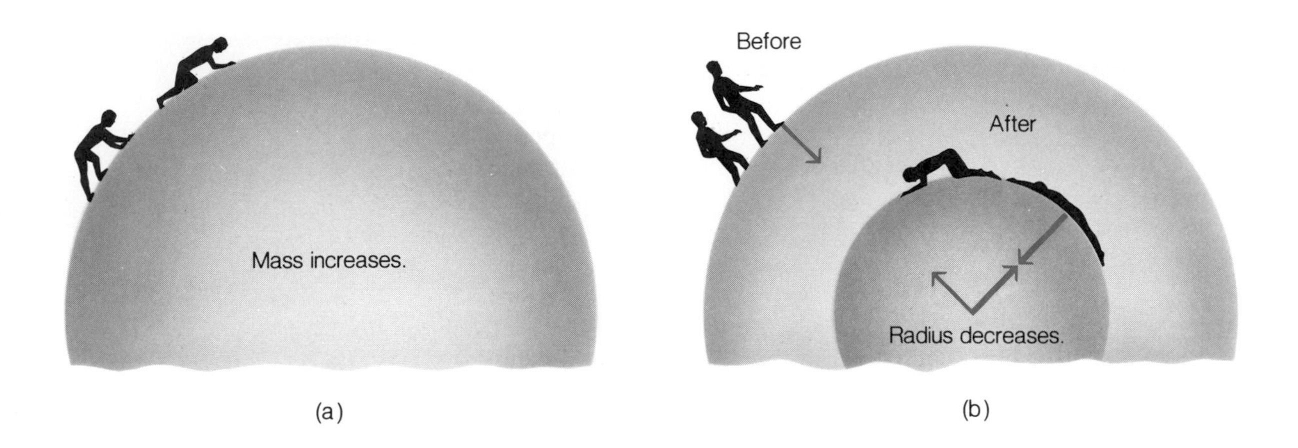

Figure 10-10 The force of gravity that a star exerts at its surface will increase in either of two cases: (a) if the star gains mass while remaining the same size, or (b) if the star shrinks while keeping the same mass.

surface will increase 4 times. If we make the radius 10 times smaller, the gravitational force will grow 100 times stronger. Einstein's relativity theory predicts that an escaping photon will lose a fraction of its original energy that increases in inverse proportion to the star's decreasing radius, provided that the star's mass remains constant.

From what we have just said, we can see that if the sun's radius decreased by 2000 times, escaping photons would lose not 0.0005% but a full 1% of their original energy. We would then receive 1% less energy from the sun, assuming it somehow managed to produce the same amount of photon energy at its surface as before. If the sun's radius shrank not 2000 times but 200,000 times, to a radius of 3 kilometers, escaping photons would lose *100%* of their original energy. In other words, they would not escape at all! The sun would then be a black hole: If photons cannot escape, then neither can any other kind of particle, for photons are the most mobile of particles, having no mass whatsoever.

The Black Hole Radius

Any object can become a black hole if it shrinks far enough. The sun is unlikely to do so, nor the moon, nor the Earth, nor ourselves. But every object has its critical **black hole radius** (also called the Schwarzschild radius), at which it must become a black hole. For the sun, or for any object with the sun's mass, the black hole radius equals 3 kilometers. The critical radius varies in direct proportion to the object's mass. Thus a 10-solar-mass star has a black hole radius of 30 kilometers. The Earth, with 1/300,000 the sun's mass, has a black hole radius 300,000 times smaller than 3 kilometers: 1 centimeter! If something could shrink the Earth within such a radius, it would become a black hole—bad news for us, but hardly likely to occur. A person with a mass of 70 kilograms would have to shrink to a radius of 10^{-23} centimeter (far smaller than a single atomic nucleus) to become a black hole!

We might pause to reflect that if the sun *did* become a black hole (tremendously unlikely, but let's suppose), the sun's gravitational force on the Earth would not change. The gravitational force that the sun exerts on the Earth depends only on the sun's mass, the Earth's mass, and the distance between their centers (Figure 10-11). None of these would change if the sun became a black hole. What changes in such a situation is the gravitational force at the sun's *surface,* since the surface moves much closer to the sun's center as the sun shrinks (Figure 10-10). We could therefore continue in our usual orbit around the sun, deprived only of its light and heat.

What objects are likely to become black holes? The most favored candidates are stars that collapse and have too much mass to become neutron stars (see page 331). The upper limit on the possible mass of a neutron star is three to five times the sun's mass. If an object with a greater mass collapses, the collapse can probably overwhelm the exclusion principle in a fraction of a second, in which case the object will shrink within its critical black hole radius. From this state it cannot emerge.

Conditions Inside a Black Hole

A black hole with, for example, ten times the sun's mass must have an enormous density. If we pack ten times the sun's mass inside a radius of 30 kilometers, we must fit this mass into a volume that is 10^{13} times less than the sun's volume. Hence the density of matter within this volume must be at least 10^{14} times greater than the density of matter within the sun. We say "at least" because 30 kilometers represents the critical radius for the black hole. The actual black hole may have a smaller radius than this. No one knows just what form matter would assume inside a black hole—and any volunteer who went to find out would be unable to report back to us!

Mini–Black Holes and Supermassive Black Holes

What about black holes that do not form from stars? Here astronomers think of two classes of black holes: those with small masses and those with enor-

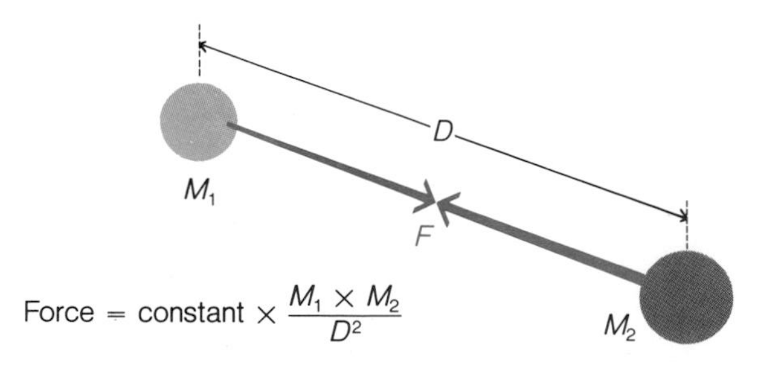

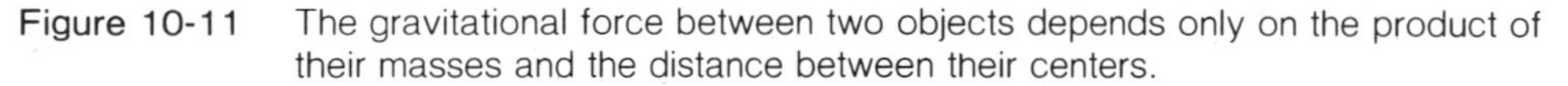
Figure 10-11 The gravitational force between two objects depends only on the product of their masses and the distance between their centers.

mous masses. Numerous low-mass black holes might have formed during the early years of the universe. Such **mini-black holes** could have served as gravitational "seeds" in the formation of clumps of matter after the time of decoupling (see page 146). An amazing sidelight on black holes shows that any black hole with less than a mass of 10^{16} grams (about the mass of a mountain) would have "evaporated" during the 10 or 15 billion years since the big bang. What physicists call *quantum mechanical effects,* and which amount to the fact that we can never be entirely sure of just where a particle is, apply to black holes as well as to the rest of the universe. Hence at any given time, a number of particles we think of as "inside" a black hole may momentarily be outside it. If enough particles happen to find themselves outside the black hole, it will no longer have enough mass to remain a black hole and will in effect evaporate, showering its particles back into space. The length of time needed for a given black hole to evaporate varies with the black hole's mass. A black hole with 10 solar masses (2×10^{34} grams) will evaporate after 2×10^{28} years—1 or 2 billion billion times the age of the universe!

On the other end of the scale from mini-black holes are **supermassive black holes,** those with millions or billions of times the sun's mass. Such huge black holes would have a critical radius millions or billions of times 3 kilometers, the black hole radius for a 1-solar-mass object. Even so, their sizes would barely exceed the size of the solar system, despite their enormous masses. Such supermassive black holes may exist at the center of every large galaxy. Matter falling into such a black hole in our own galaxy might be the cause of the emission we detect from the centers of the Milky Way and of Seyfert galaxies, and from quasars (see pages 187 and 214).

What would the density be like inside a black hole with a billion times the sun's mass? The answer: extremely low! An object with a billion times the sun's mass has a black hole radius of 3 billion kilometers. Its volume, which increases in proportion to the *cube* of its radius, exceeds the volume inside a 1-solar-mass black hole by the cube of 1 billion, or 10^{27} (Figure 10-12). Thus, even though the object has a billion solar masses, its density (mass divided by volume) equals only one part in 10^{18} of the density inside a 1-solar-mass black hole. This density is 1/50 the density of water, or of the average density of matter inside the sun!

Astronomers have evidence for the existence of such a massive black hole in the giant elliptical galaxy M 87 (Figures 5-12 and 7-10). Spectroscopic observations of the galaxy's central regions imply the existence of an enormous mass that is making the stars there move much more rapidly than they would if no such mass were present. The mass deduced from the stars' motions equals just about a billion solar masses! Although this does not prove that a black hole of this mass exists at the core of the galaxy, the observations do show that whatever form this mass takes, it occupies no more than a few parsecs of diameter. At the present time, astronomers see no way to pack a billion solar masses into such a volume without having it contract to form a black hole less than 3 billion kilometers (0.0001 parsec) in radius. With a diameter of a few parsecs, the object's self-gravitation would be too great to avoid such a fate.

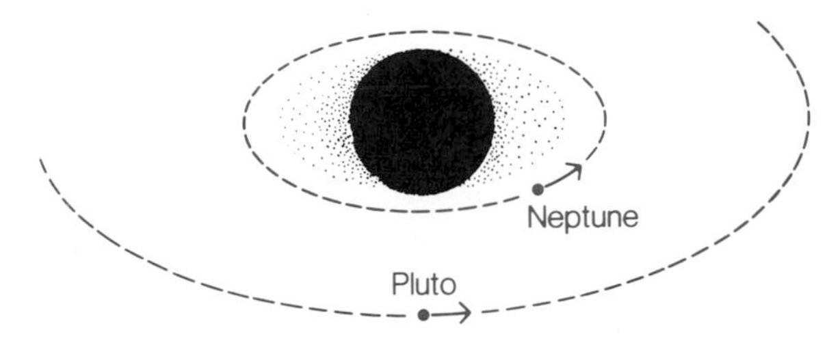

Figure 10-12 An object with a billion times the sun's mass will become a black hole if its radius decreases to 3 billion kilometers or less. At a radius of 3 billion kilometers, the black hole would be as large as the orbit of Uranus around the sun.

Supermassive black holes may thus prove to be the key to giant galaxies, explaining how the galaxies formed and how their centers emit so much energy. (But how the *black holes* formed would require additional explanation.) To learn more about black holes, we would like to be able to study them in our own galaxy. But how many black holes with starlike masses are there in the Milky Way? A few? Thousands? Millions? The question has importance in understanding the motions of stars within our galaxy, since the existence of great numbers of black holes would provide an unseen, but highly significant, source of gravitational force.

The best way to find black holes of starlike mass is to look for their effects on nearby stars. To understand this effort, we must return to our study of binary stars, trying to find which of them may contain a black hole as one member of a double-star system.

10.4 Close Binary Stars

In Chapter 8, we discussed double-star systems that consist of two normal stars in orbit around their mutual center of mass. We must now consider double-star systems in which the two stars have only a small separation, and in particular those systems in which at least one of the two stars has become a compact object—a white dwarf star, a neutron star, or a black hole. Aside from offering our best chance to detect black holes in our own galaxy, these systems also reveal an intriguing aspect of stellar evolution, in which one star can perturb the development of another.

The Roche Lobe

Stars in close proximity to one another combine their gravitational forces to act upon particles in the stars' immediate vicinity. When we calculate the forces that act upon these particles, we can see that a particle extremely close to one of the two stars will feel an attractive force toward that star, despite a weaker gravitational pull from the other star. If we take into account not only gravity but also the effect of "centrifugal force"—the momentum that comes from the motion in orbit around the center of mass—we can calculate

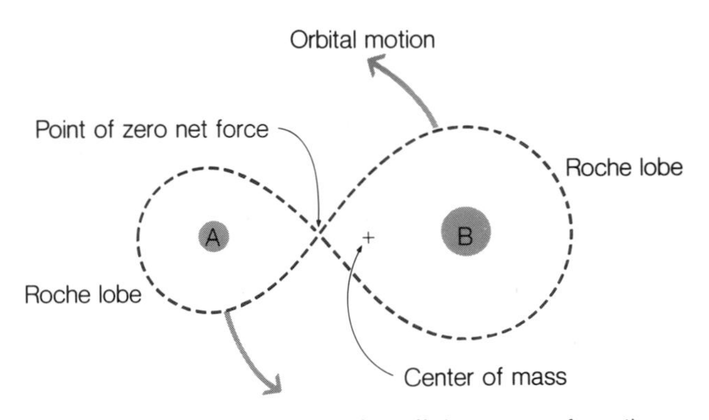

Figure 10-13 When we take into account the "centrifugal force" that comes from the orbital motion of the binary pair, we find that each star possesses a sphere of influence—a tear-drop-shaped region around the star called the star's Roche lobe. Particles within each star's Roche lobe will feel a net force that pulls them toward that star.

how particles will move in the vicinity of the two stars. Such calculations predict an important effect for the transfer of mass between the two stars.

As Figure 10-13 shows, if we include the effects of "centrifugal force," then the point between the stars at which a particle feels no net force (gravitational force plus centrifugal force) no longer corresponds to the center of mass. Other things being equal, a particle placed at the point of zero net force is as likely to move to the left as to the right. Points to the left of this zero marker fall within the sphere of influence of star *A*; those to the right come within the sphere of influence of star *B*.

The **sphere of influence** we are referring to is defined by an elongated tear-drop of space around each star called the **Roche lobe.** Particles within the Roche lobes of stars *A* and *B* tend to stay there, but an exception occurs at the point where the two Roche lobes meet. At this point, mass can pass from star *A* to star *B*, or vice versa (Figure 10-13).

The Evolution of Close Binary Systems

During the later phases of a star's evolution, when it becomes a red giant, it swells its outer layers enormously. A red giant that belongs to a closely spaced binary may expand to fill its entire Roche lobe. This creates a **semi-detached binary**—one lobe full—from which the red giant will then lose matter to its companion and to space. The famous variable star Algol (page 17) represents such a binary. The originally more massive of the two stars in the system has transferred so much mass to its companion that it now has a much smaller mass than it once had, and than its companion now has.

If both stars in a double system fill their respective Roche lobes, their material will spill over into the space around the Roche lobes, producing what astronomers call a **contact binary** system (Figure 10-14). These contact

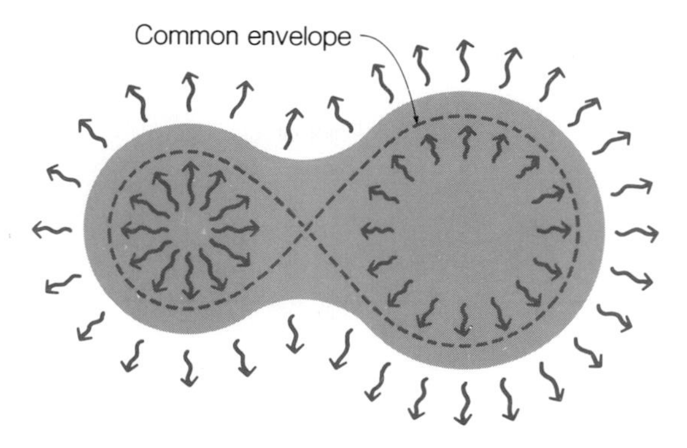

Figure 10-14 In contact binaries or W Ursae Majoris stars, two stars orbit at such close range that they share a common envelope of gas. This occurs because both stars overflow their Roche lobes and put gas outside them.

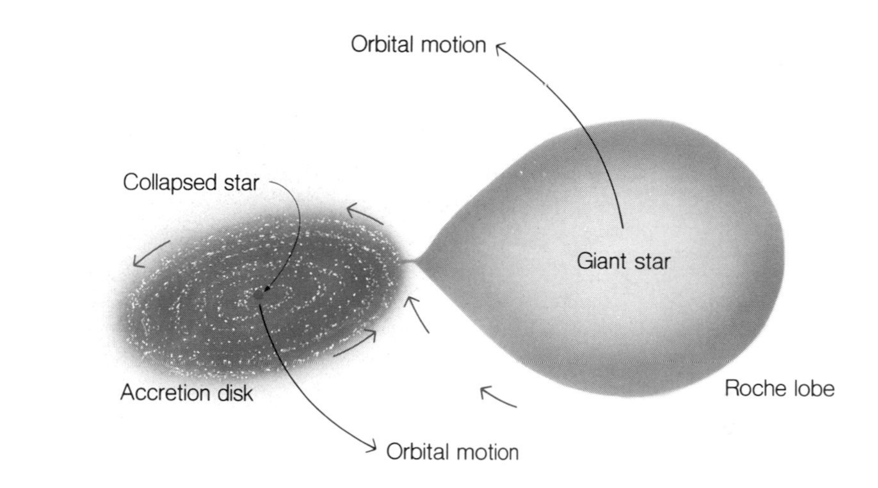

Figure 10-15
When a giant star and a compact object (a neutron star or black hole) move in orbit around their common center of mass, material from the Roche lobe of the giant can fall onto the compact object. As the material falls inward, it forms an accretion disk because of the orbital motion it acquires.

binaries are often called W Ursae Majoris stars after the best-studied member of the class. Typical W Ursae Majoris stars have orbital periods of less than 8 hours.

Binary Systems with Compact Companions

Consider next what happens in a close binary pair if one of the stars has become a white dwarf, a neutron star, or a black hole. The other star may continue its evolution, no doubt somewhat affected by the explosion that left behind its companion neutron star or black hole. As it does so, it is likely to pass through a stage in which it swells so large that it fills its Roche lobe. When this happens, the material that passes through the tip of the Roche lobe will have a fairly large amount of space to cover before falling onto the compact object (Figure 10-15).

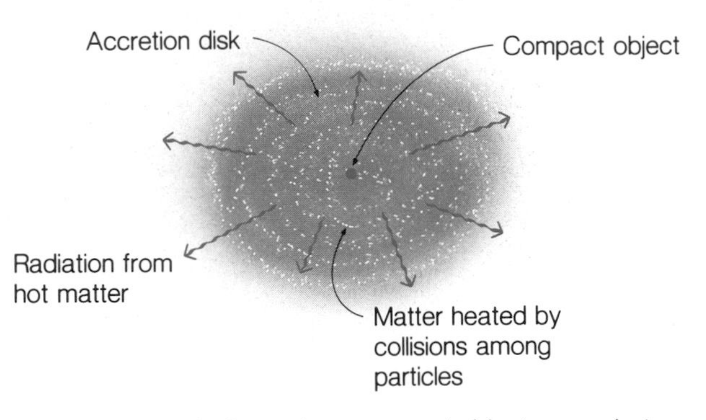

Figure 10-16 The accretion disk of matter spiraling onto a compact object grows hot, especially in its inner parts, because of the large kinetic energies the particles acquire by falling inward.

The infalling material will not strike the compact object (white dwarf, neutron star, or black hole) until it has made at least several orbits around the object. As a result, the stream of infalling gas from the enlarged companion forms an **accretion disk,** a flattened distribution of matter spiraling inward toward the compact object. The mutual orbits of the two objects give to the gas that crosses from one Roche lobe to the other an overall direction, and hence creates a spiral maelstrom rather than a simple cloud of gas falling inward (Figure 10-15).

Within the accretion disk of infalling matter, collisions among the particles heat them to high temperatures. In effect, the accretion disk turns part of the total kinetic energy of infall into the individual kinetic energies of the particles that form the accretion disk at any given time (Figure 10-16). In the inner regions, where the motion is most violent, the temperature within such a disk can rise to millions of degrees. Matter at such temperatures will produce x rays. Thus the model of an accretion disk around a compact object may explain what astronomers observe in sources of x-ray emission.

10.5 X-Ray Sources

During the 1970s, astronomers detected about 100 relatively strong x-ray sources within our galaxy (Figure 10-17). X-ray detection had to wait until that time because x rays do not penetrate our atmosphere: We must send a rocket or satellite to altitudes of several hundred kilometers to observe them.

The Case of Hercules X-1

For many variable x-ray sources, careful measurement has revealed a periodicity that suggests the existence of a neutron star. A good example of such a source is Hercules X-1, whose x rays pulse on and off with a period of 1.24 seconds (Figure 10-18). Hercules X-1 is not a conventional pulsar, because it does not emit detectable radio emission, certainly not at regular intervals.

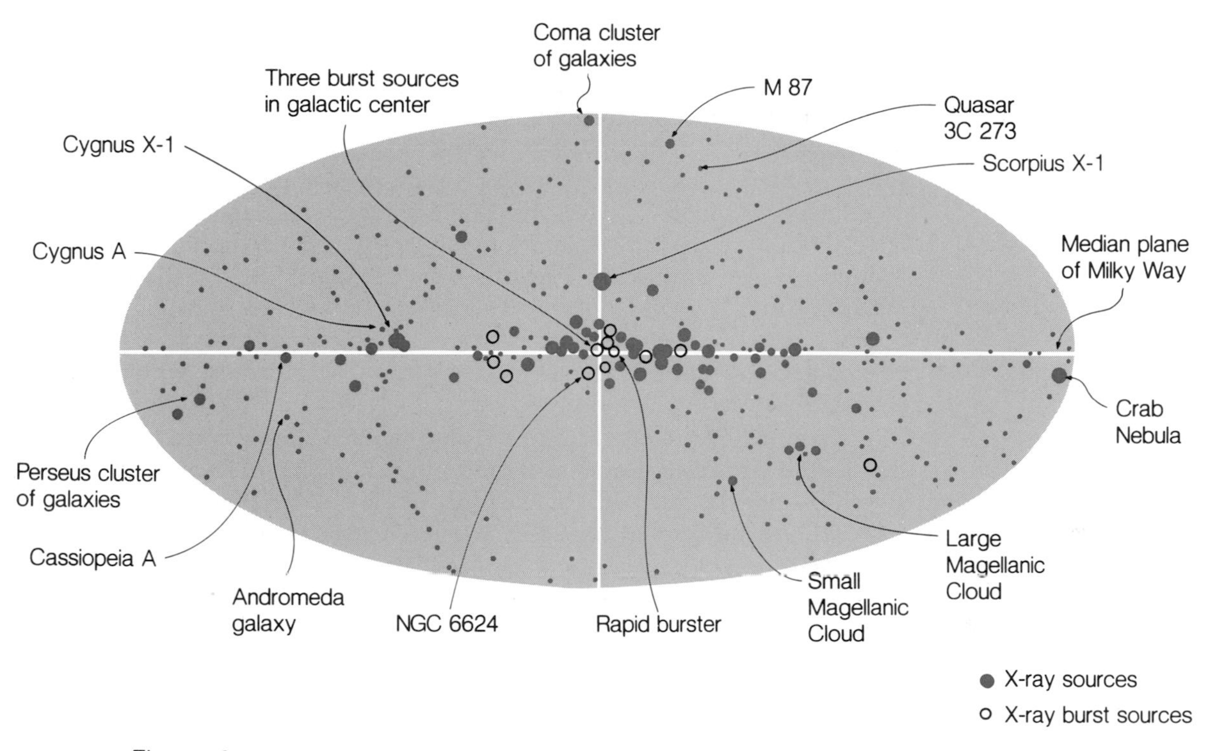

Figure 10-17 A map of the locations of the x-ray sources observed by the Uhuru satellite shows a strong concentration toward the plane of the Milky Way (central line in map). This indicates that most x-ray sources lie in the disk of our galaxy, though some nearby galaxies and at least one quasar have been detected in x-ray emission.

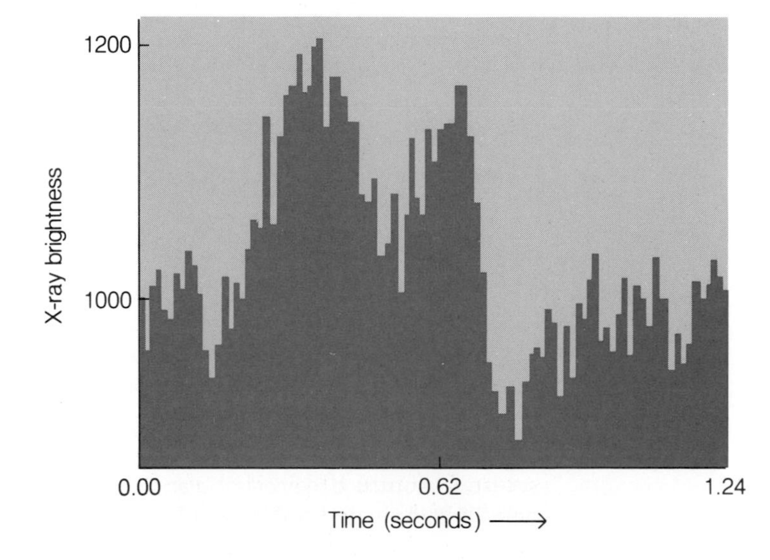

Figure 10-18 The x-ray source Hercules X-1 emits regular pulses at intervals of 1.24 seconds. The pulses show a complex substructure that basically repeats from pulse to pulse, with some exceptions.

But Hercules X-1 certainly is a member of a binary system, because careful timing of its x-ray pulsations shows a regular variation in the time interval between pulses. (In the case of a pulsar that belongs to a binary system, these variations should arise from the motion of the source of the pulses in orbit around the center of mass.) When astronomers looked (in visible light) at the spot from which the x rays come, they found the companion to Hercules X-1: a variable star called HZ Herculis. The light from HZ Herculis varies regularly with a period of 1.7 days—precisely the orbital period implied by changes in the x-ray pulse intervals from Hercules X-1! Finally, the x-ray emission from Hercules X-1 shows recurring eclipses, each lasting about six hours, which occur every 1.7 days. These eclipses would be expected whenever the x-ray source orbits behind the star HZ Herculis, which has expanded to fill its Roche lobe.

If we believe in the accretion-disk model of x-ray sources, we can explain the Hercules X-1 system with the diagram shown in Figure 10-19. A relatively normal star, HZ Herculis, fills its Roche lobe and transfers about one-billionth of a solar mass of matter each year to the neutron star. This matter falls toward the neutron star faster and faster, swirling into itself as it falls and reaching a temperature of 100 million K. Since the neutron star has a strong magnetic field, the infalling gas may be led only to certain regions of the neutron star, those close to its magnetic poles. The neutron star spins with a period of 1.24 seconds, and as it spins, we see x rays from the "hot spots" on the neutron star sweep by us with this period. The emerging x rays heat the side of HZ Herculis that happens to face the neutron star, making a hot, bright "blister" on the star. As the two stars orbit their center of mass, the changing orientation of the "blister" with respect to us produces the 1.7-day period we observe in the light variation of HZ Herculis.

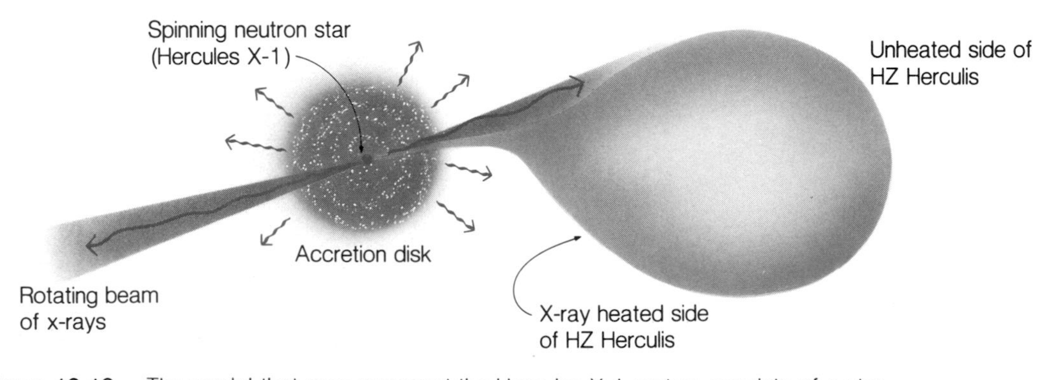

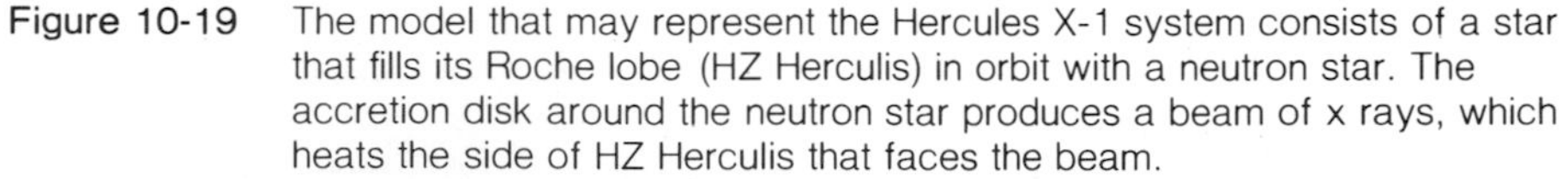
Figure 10-19 The model that may represent the Hercules X-1 system consists of a star that fills its Roche lobe (HZ Herculis) in orbit with a neutron star. The accretion disk around the neutron star produces a beam of x rays, which heats the side of HZ Herculis that faces the beam.

The Case of a White-Dwarf Companion

A number of other binary x-ray sources besides HZ Herculis/Hercules X-1 have been detected (see Table 10-2). Some of these are presumably similar situations, where material from a relatively normal star falls onto a rotating neutron star. But what if the material falls not onto a neutron star but onto a white-dwarf star? What then? Astronomers think that this configuration usually produces a **nova**, a temporary flare-up of the white dwarf, in which nuclear fusion reactions on the white dwarf's surface temporarily release large amounts of energy (Figure 10-20). This process begins as hydrogen-rich material from the normal star collects on the white dwarf. Nuclear fusion can occur because the material has a high temperature and the white dwarf has an abundant supply of carbon nuclei to serve as the catalyst for the carbon cycle of fusion (see page 254). With a single violent flash, the new material may be consumed within a period of several hours or days. Then the nova subsides until it can accumulate enough material from its companion for a new eruption.

Thus close binaries with at least one compact object can, perhaps, explain most of the x-ray sources found in the Milky Way, as well as most of the nova outbursts we observe. Incidentally, the current record for the shortest orbital period in a binary system, 17 minutes, belongs to AM Canum Venaticorum, apparently a pair of white dwarfs less than 50,000 kilometers apart. To complete our story about the possibilities of mass transfer in binary systems, we must consider the remaining configuration, in which at least one member of the pair has become a black hole.

Cygnus X-1

Cygnus X-1, the first x-ray source discovered in the constellation Cygnus, provides the best evidence so far for a black hole in the Milky Way galaxy.

Table 10-2 Characteristics of Some Binary X-Ray Sources

Source Name	Binary Period (Days)	Characteristics of X-Ray Emission	Characteristics of Visible Star
Cygnus X-1	5.6	Variable on time scales of 1/1000 to 1 second	Blue supergiant, about 30 solar masses
Centaurus X-3	2.09	X-ray eclipses, duration 0.49 day	Blue giant, about 16 solar masses
Hercules X-1	1.7	X-ray eclipses, duration 0.24 day	HZ Herculis, about 2 solar masses
Cygnus X-3	4.8	Varies on time scale of 4.8 hours	None visible; source of infrared with 4.8-hour variation
Circinus X-1	More than 15	X-ray eclipses, duration about 1 day	Not found
Vela X-1	8.95	X-ray eclipses, duration 1.7 days; x-ray flares lasting a few hours	Blue supergiant, about 25 solar masses
Small Magellanic Cloud X-1	3.9	X-ray eclipses, duration 0.6 day	Blue supergiant, about 25 solar masses

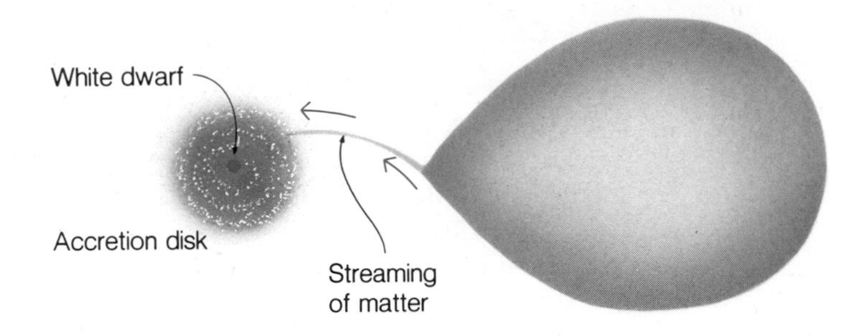

Figure 10-20 A model for nova explosions suggests that material from one member of a binary pair (perhaps a cool star that fills its Roche lobe) falls onto the other member, a white-dwarf star. When enough matter rich in hydrogen has collected on the white dwarf's surface, an explosion resembling a super hydrogen bomb can occur.

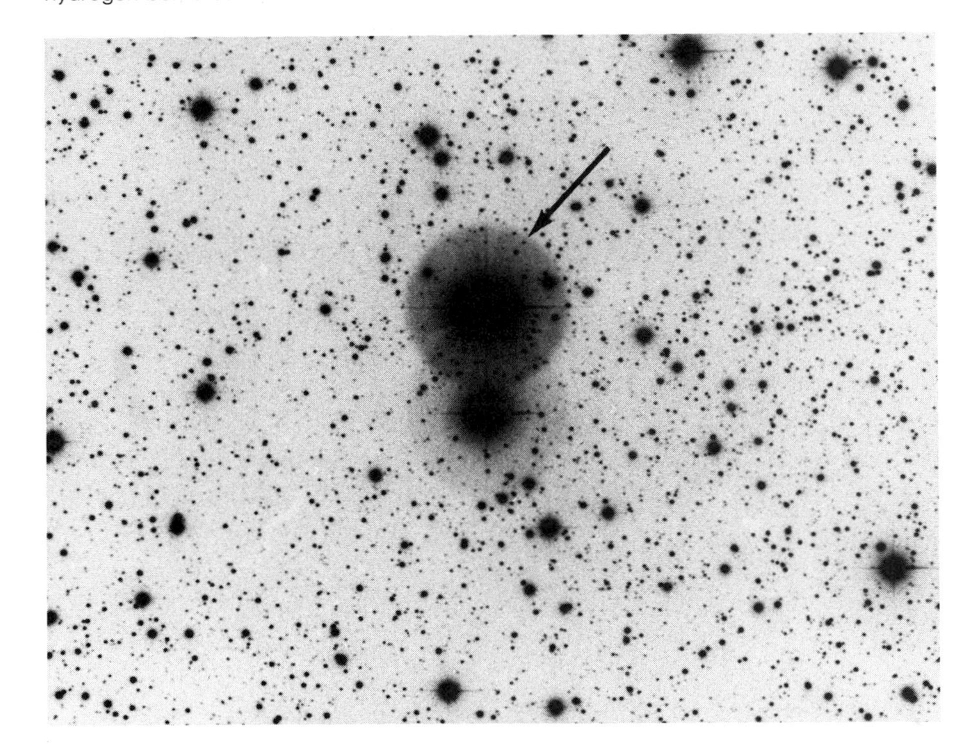

Figure 10-21 At the precise location in the sky from which the x-ray emission of Cygnus X-1 comes, we can see a B0 supergiant star called HDE 226868.

The x-ray emission from Cygnus X-1 varies irregularly, unlike the regular x-ray pulsations of Hercules X-1 and similar sources. These variations occur on time scales as short as a few hundredths of a second. In this amount of time, light travels only a few thousand kilometers. Hence the source of x-ray emission can be only a few thousand kilometers across. We used a similar argument to set an upper limit to the sizes of quasars whose light output varies in less than one day (see page 220).

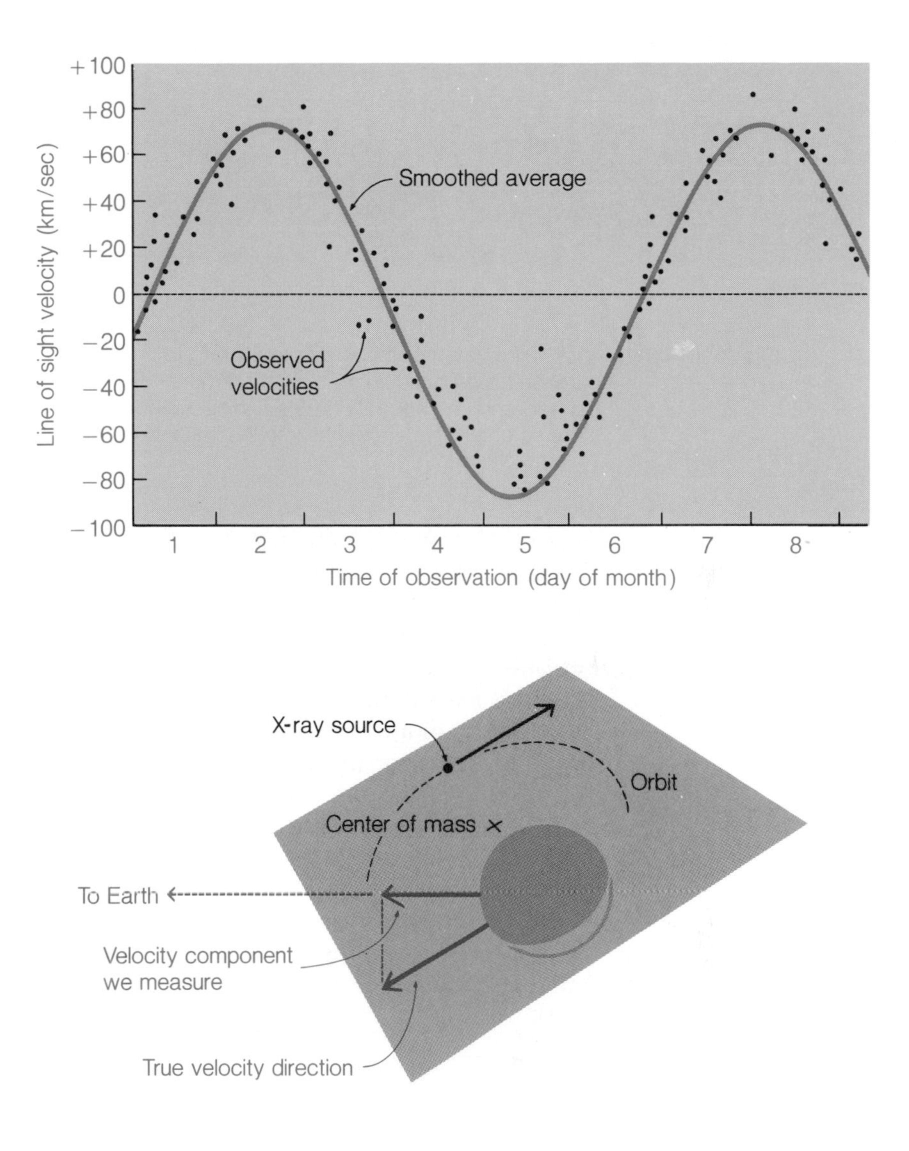

Figure 10-22
The observed Doppler shifts in the spectrum of HDE 226868 imply an orbital motion with a period of 5.6 days (top). The changes in radial velocity amount to 70 kilometers per second away from the average radial velocity. The true orbital velocity may be greater, since our line of sight probably does not coincide with the plane of the star's orbit (bottom).

When we look at the exact location on the sky from which Cygnus X-1's x-ray emission comes, we find a B0 star called HDE 226868 (Figure 10-21). This star's spectrum shows periodic shifts, presumably arising from the Doppler effect, with a period of 5.6 days (Figure 10-22). We may assume that the B0 supergiant moves in orbit with another object. If this object is Cygnus X-1, we can obtain some information on the mass of the x-ray source from our knowledge of how massive objects move in orbit around their common center of mass.

We know that a blue supergiant star like HDE 226868 has about 30 times the sun's mass. To make HDE 226868 show Doppler shifts of the observed amount, its companion must have a mass of at least 6 times the sun's mass.

Figure 10-23 A possible model for the x-ray emission from Cygnus X-1 shows an accretion disk around an invisible black hole. Matter flowing out of the Roche lobe of the star HDE 226868 joins the accretion disk and produces x rays as it falls in spiral paths into the black hole.

We cannot know the companion's mass with complete accuracy, because we cannot determine its motion in orbit[3] (in fact, we cannot see the companion at all—only the x rays from its immediate vicinity). But the 6-solar-mass figure appears to be a rather good *lower limit* for the mass of HDE 226868's companion.

If this mass limit is valid, then the companion of HDE 226868 cannot be a neutron star or a white dwarf, since its mass lies above the calculated possible upper limits for those stars. The simplest conclusion is that the x-rays from Cygnus X-1 come from the immediate surroundings of a black hole with at least 6 times the sun's mass. The x rays calculated to emerge from the accretion disk around a black hole resemble in amount and spectrum the x rays from the disk around a neutron star. We would not, however, expect the black hole's accretion disk to produce regularly pulsed x-ray emission, since we expect nothing that corresponds to the rigidly fixed "hot spots" from matter falling onto a neutron star, guided by its magnetic field (Figure 10-23).

The Detection of Black Holes

Cygnus X-1 does not provide the only example of a possible black hole inside an x-ray source, though it does furnish the most likely case. Of course, we can find a black hole in this way only if the black hole is part of a binary system, and only if the companion star is visible, so that we can estimate its mass and thus the mass of the possible black hole. Thus we might easily miss detecting the vast majority of black holes that exist in our galaxy. On the other hand, we might have already found a significant fraction of the Milky Way's black holes just with Cygnus X-1! Astronomers look forward to a new generation of x-ray satellites that will help them make better observations to resolve this uncertainty. But if we ever make a "certain" identification of a

[3]The x-ray pulses from the companion are not regular enough to determine the companion's orbit by careful timing (page 354).

black hole, it must rely on our observations of the black hole's surroundings (along with a correponding deduction of the black hole's existence), rather than on seeing the black hole itself.

What Good Are Black Holes?

One fascinating aspect about the x-ray emission from the vicinity of the black hole concerns the currently fanciful idea that we could throw our waste matter into such a black hole and obtain useful energy in the form of x rays. If we could somehow find a black hole with one-millionth of the Earth's mass (and thus with a black hole radius of one-millionth of a centimeter) and set it into orbit around the Earth, we could easily dispose of a million tons of waste products each day. As the waste fell into the black hole, it would produce about 10^{20} ergs of x-ray energy each day—more than the daily consumption of energy by human beings!

10.6 The Binary Pulsar

The second-fastest known pulsar has a particular claim to our attention. It is the first pulsar found to be a member of a double-star system. This **binary pulsar,** called by the unassuming name PSR 1913+16, has an interval of 1/17 second between pulses. But closer measurements reveal that the interval between pulses shows a tiny increase for a time, then a tiny decrease, then an increase again, and so on. The most straightforward interpretation of these changes seems to be that they derive from the motion of the pulsar away from us, then toward us, then away from us again, and so on (Figure 10-24). If we adopt this hypothesis, we can translate the changes in pulse intervals into the velocity of the pulsar toward us and away from us.

Determining the Orbit of the Pulsar

The changes in the time of arrival of the pulses suggest that the pulsar is moving in orbit with some other object. We can calculate many features of this orbit from accurate timings of the pulse intervals, and astronomers have done just that since the pulsar was first discovered in 1974. They have found that the pulsar's orbit has a semi-major axis of just 70,000 kilometers, less than six times the diameter of the Earth! The object in mutual orbit with the neutron star must therefore be another neutron star (one that does not emit radio waves) or perhaps a compact white dwarf. The orbital period of the pulsar and its companion amounts to only 7¾ hours, and the two objects each have a mass close to 1.4 solar masses. (It is intriguing that the objects have masses at just about the critical upper limit for white-dwarf stars.)

Astronomers take great interest in all of these parameters. Their ability to obtain such accurate data—for example, that the pulsar has a period (as of 1981) equal to 0.05902999697 second—testifies to the progress we have made in measuring time intervals. But the most significant result to emerge from the study of binary pulsars is the discovery that the binary system emits gravity waves.

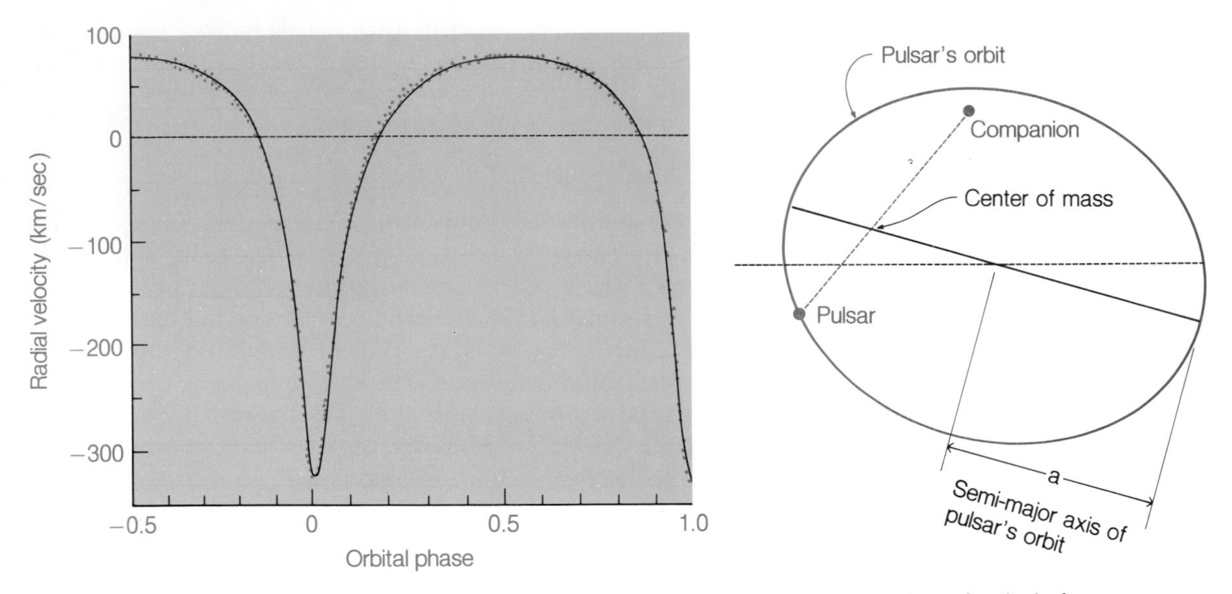

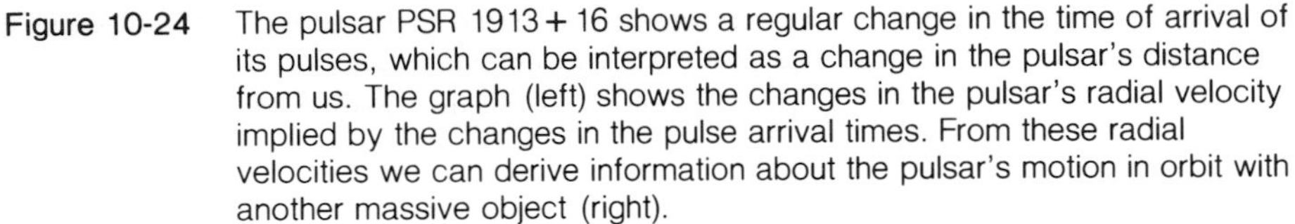

Figure 10-24 The pulsar PSR 1913 + 16 shows a regular change in the time of arrival of its pulses, which can be interpreted as a change in the pulsar's distance from us. The graph (left) shows the changes in the pulsar's radial velocity implied by the changes in the pulse arrival times. From these radial velocities we can derive information about the pulsar's motion in orbit with another massive object (right).

Gravity Waves

Gravity waves are an essential feature of Einstein's general theory of relativity. Part of Einstein's theory predicts that if an object moves in an elliptical orbit under the influence of gravity, its orbit will rotate in space. As a result, the point of closest approach will move as time passes. For the planet Mercury in orbit around the sun, this motion amounts to 43 seconds of arc per century. For the objects that form the binary pulsar, much closer together and hence exerting much greater gravitational forces on each other than the sun and Mercury do, this *apsidal motion* reaches 4°.2 per year—35,000 times the effect for Mercury's motion! But Einstein's theory also predicts that objects moving in mutual orbits must emit **gravitational radiation,** also called **gravity waves.** We can make an analogy between gravity waves and electromagnetic radiation (photons), which can be produced by the motion of charged particles relative to one another. In a similar way, the motion of particles with mass relative to one another produces radiation of a totally different nature, consisting of **gravitons** instead of photons.[4]

We would hear more about gravitational radiation if we could detect it more easily. But physicists have found that such gravity waves are *weak*. For example, the gravitational radiation produced by Mercury's noncircular mo-

[4]However, in the case of two objects of equal mass moving in perfectly circular orbits around their center of mass, no gravitational radiation will be emitted.

tion around the sun amounts to less than one part in 10^{24} of the photon energy from the sun. Gravitational forces are in fact the weakest of the four kinds of forces known to science (see box, page 250). Only because gravitational forces always attract, and because astronomical objects always consist of huge numbers of particles, can gravity dominate astronomical situations. On a smaller scale, electromagnetic forces rule, as in the binding of the atoms and molecules that form our bodies. On a still smaller scale—that of atomic nuclei—strong and weak forces dominate the scene, as in the fusion of protons (see page 249).

In the binary pulsar, gravity waves count a great deal. Because two objects, each with more than the sun's mass, orbit one another at extremely close range, gravitational forces become enormous. Because these objects move in highly noncircular orbits (the orbital eccentricity equals 0.62, compared to an eccentricity of 0.21 for Mercury's orbit), the distance between them changes by large amounts over a period of an hour or so. Hence the binary pulsar emits significant amounts of gravity waves—not enough to be detected directly, but enough to produce a noticeable effect.

The effect consists of a reduction in size of the orbits of the two objects. The act of emitting gravity waves constantly robs the moving objects of some energy. To make up for this loss, the objects come a bit closer. Oddly enough, by approaching more closely, the objects must move still faster in orbit, and thus will radiate more gravity waves each second. Hence, the effect must increase with time.

The binary pulsar's orbital period is decreasing at just the rate Einstein's theory predicts. For astronomers, the discovery that Einstein's theory of gravity waves is correct—that the orbits indeed shrink—has been impressive enough. Since astronomers only recently (1979) observed that the orbital period had decreased slightly, they have not yet been able to see the *acceleration* of the decrease that we predicted above. For the time being, though, we may rest content with the first direct confirmation of Einstein's prediction that massive objects, in changing their distance from each other, must emit a new form of radiation, gravity waves.

Attempts to observe gravity waves directly have involved extremely careful monitoring of objects with large masses in laboratories (Figure 10-25). When gravity waves encounter such an object, they should make it vibrate, but only gently, since the waves are inherently weak. To date, all attempts to detect gravity waves by their impact on laboratory masses have failed to produce clear-cut evidence for gravitational radiation. Of the three binary pulsars that have so far been detected, PSR 1913+16 presents the best demonstration that gravity waves exist, even though we observe the effects of gravitational radiation rather than the gravity waves themselves.

10.7 Strangest of All: SS 433

By 1979, astronomers had grown fairly well accustomed to finding strange objects. Quasars, pulsars, neutron stars, binary x-ray sources, and even black holes had apparently done their best to surprise the astronomical commu-

Figure 10-25
A gravity-wave detector consists of an aluminum cylinder hung with extreme care and monitored with incredible precision. No gravity waves have been definitely detected until now.

nity. But astronomers received yet another surprise when they came to study in detail the 433rd object in a list of "stars" with emission lines. This object, called SS 433 after the compilers of the catalog of emission-line stars (Bruce Stephenson and Nicholas Sanduleak), had a spectrum that appeared to defy explanation.

As in the case of quasars' spectra two decades earlier, the emission lines in SS 433 appeared at unfamiliar wavelengths. But unlike the quasars' spectra, the emission lines in the spectrum of SS 433 changed their frequencies and wavelengths by large amounts over a period of several weeks (Figure 10-26). Further study showed that although two sets of emission lines undergo such changes, some emission lines in the spectrum maintain a fairly constant frequency and wavelength. A year of observation sufficed to show that the changes in emission-line frequency repeat with a period of 164 days. The amount of change, however, verges on the incredible. If the frequency changes arise from the Doppler effect, they imply velocities of one-quarter of the speed of light, both toward us and away from us!

When astronomers try to construct a model of SS 433 that can explain all the changes we observe, they come up with a binary system whose two members are a relatively normal star and a compact object, either a neutron star or a black hole. Somehow the compact object produces jets of matter, streaming in opposite directions at one-quarter of the speed of light (Figure 10-27). A precession, or wobble, of the axis along which the jets emerge must have a 164-day period to explain the changes we observe in the spectral emission lines of SS 433. This precession resembles the wobble of the Earth's axis in space over a period of 26,000 years (see page 389).

SS 433 almost certainly is part of the Milky Way, since it lies almost precisely in the plane of our galaxy. When we map the region around SS 433 in radio waves, we find that it appears to lie within a supernova remnant. SS 433 itself emits radio waves and x rays in varying amounts, with no pattern to the variation. The source remains an enigma, holder of the record Doppler blue shift for the beam pointed toward us at one-quarter of the speed of light. What makes SS 433 emit such beams—if indeed it does—and why we have found only one source like SS 433 remain unanswered questions. With sufficient research, astronomers hope to resolve them.

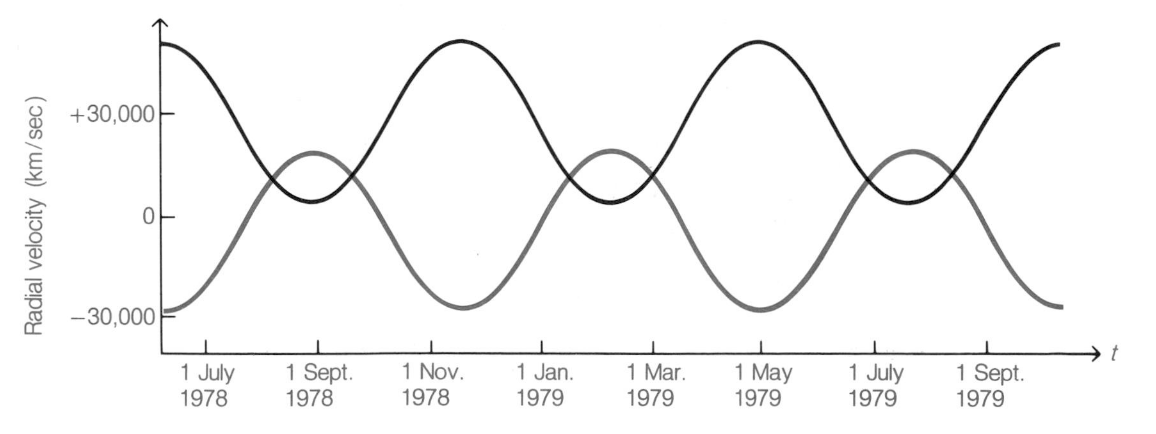

Figure 10-26 SS 433 shows three sets of emission lines. One of these sets shows no Doppler shifts, while the other two, shown here, exhibit large Doppler shifts that change in alternating phase, with a period of 164 days.

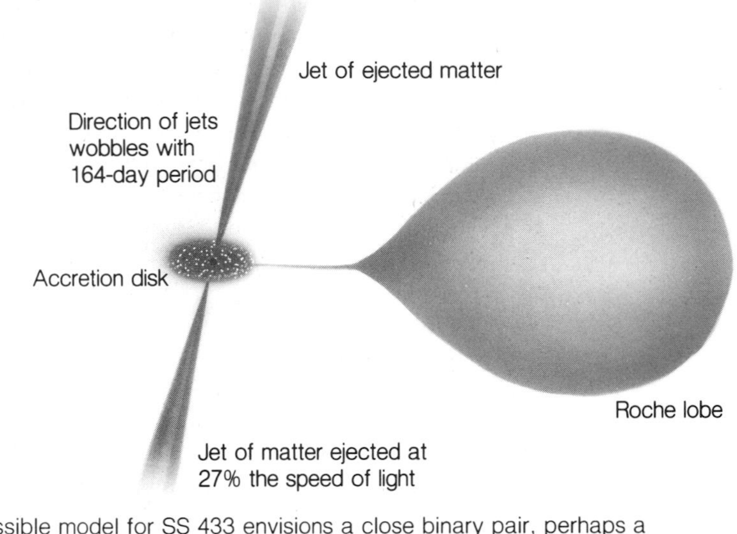

Figure 10-27 A possible model for SS 433 envisions a close binary pair, perhaps a neutron star and a star that fills its Roche lobe, with the compact object somehow squirting out two jets of matter in opposite directions, each at 27% of the speed of light. The 164-day period could arise from a "wobble" in the direction of the jets under the perturbing influence of the companion star.

Summary

Some stars collapse after exhausting all possible nuclear fusion reactions that can liberate kinetic energy to oppose the star's self-gravitation. From these collapses sometimes come neutron stars, incredibly dense collections of neutrons, 10 kilometers or so in radius, held against total collapse by the exclusion principle among the neutrons. Such collapsed objects should be rotating quite rapidly. Apparently this rotation drags the greatly increased magnetic field around with the star, providing the mechanism for pulsars. Pulsars emit photon pulses at regular time intervals, typically in the form of radio waves, sometimes also as visible light, x rays, and gamma rays.

No neutron star can exist with a mass greater than three to five times the sun's mass, according to astronomers' calculations. If a star with more than this mass in its interior collapses, it is likely to become a black hole, an object whose gravitational force prevents any particles—including the photons that form light waves and radio waves—from escaping. Any object will become a black hole if it shrinks so far that its radius becomes less than 3 kilometers times its mass in solar-mass units. Although collapsed stars are the most likely to become black holes, we now have evidence that the galaxy M 87 may have a black hole with a billion times the sun's mass at its center.

If a compact object such as a neutron star or a black hole forms one member of a binary system, we have a good chance to deduce its existence and its properties from its effects on the other member of the system. If that other member is an "ordinary" star, then as the star swells into its red-giant phase, it will fill its Roche lobe, the sphere of influence that marks the region within which particles "belong" to the star. Matter that overflows the Roche lobe will pass into the Roche lobe around the compact object, to form an accretion disk of matter spiraling onto the object's surface (if it is a neutron star) or into oblivion (if the object is a black hole). As the matter falls into the accretion disk and spirals inward, it can reach high temperatures and emit photons of visible light, x rays, and even gamma rays. Matter that falls onto the surface of a white-dwarf star from the Roche lobe overflow may trigger a nova outburst, apparently the result of providing new hydrogen-rich material to the hot surface of the white dwarf.

Several of the x-ray sources that astronomers have recently detected seem to be binary systems in which the x rays arise from the accretion disk of infalling matter. Hercules X-1 provides the best example of matter falling onto a neutron star, while Cygnus X-1 furnishes the most likely case of matter spiraling into a black hole. In the case of Cygnus X-1, the best evidence for a black hole consists of the mass that has been deduced for the compact companion of a relatively normal blue star. This estimated mass, at least six times the sun's mass, would be too large for a neutron star, and hence points to the object's being a black hole.

Key Terms

accretion disk
binary pulsar
black hole
black hole radius
contact binary star
gravitational radiation
gravitational red shift
gravitons
gravity waves
mini–black hole
neutron star
nova
pulsar
Roche lobe
semidetached binary star
sphere of influence
supermassive black hole
weak reaction

Questions

1. What produces neutron stars? What is the connection between neutron stars and pulsars?
2. Can a neutron star expand to become a white-dwarf star? Can a white-dwarf star shrink so much that it becomes a neutron star? Will neutron stars eventually become black holes?
3. How does a neutron star compare in size and density with a white-dwarf star? What holds neutron stars against collapse induced by self-gravitation? Why do neutron stars rotate rapidly?
4. What characteristic of the mechanism producing a pulsar's pulses accounts for the almost constant interval between pulses?
5. Why does the interval between a pulsar's pulses slowly change? Which is likely to be younger, a pulsar that pulses 10 times per second or one that pulses once every 2 seconds?

6. What are gravity waves? How does the pulsar in a binary system help prove the existence of gravity waves?
7. What is a black hole? What makes a black hole capable of preventing all particles, including photons, from leaving?
8. If the sun decreased its radius about 200,000 times, it would become a black hole. How many times stronger would the gravitational force at the sun's surface become as a result of such a contraction?
9. Is the sun likely to become a black hole? What types of stars are most likely to contract so far that they become black holes?
10. Where might we find a black hole with a billion times the sun's mass? What is the maximum radius such a black hole could have? How dense would the matter inside such a black hole be?
11. What is a contact binary star system? How does it differ from a semi-detached binary? What role does the Roche lobe play in these star systems?
12. What is the accretion disk around the compact member of a binary star system? What is its importance in the production of x rays?
13. How can we hope to discover whether Cygnus X-1 is or is not a black hole? On what chain of reasoning does such a conclusion depend?
14. What is unusual about SS 433? What sort of objects might be involved in producing its observed features?

Projects

1. *The Force of Gravity*. Gravitation ranks as the weakest of the four types of forces known to physicists, yet on astronomical scales of distance, gravitation typically is the most important force. Part of the reason for this is that gravitational forces always attract. The Earth contains about 10^{52} elementary particles (protons, electrons, and neutrons in atomic nuclei), while your body contains about 10^{28} elementary particles. Think of a situation in which your body temporarily overcomes the force of gravity from the Earth. Something stronger than gravity is at work: electromagnetic forces among the molecules in your body. Per elementary particle, at least how many times stronger must electromagnetic forces be than gravitational forces for you to overcome the Earth's gravity? (The actual ratio is about 10^{39}.) How much easier would it be to overcome the force of gravity at the Earth's surface if the Earth had twice its present radius with its present mass? What if the Earth had 1/10 its radius and the same mass? Given that the Earth would have to shrink to a radius of 1 centimeter to become a black hole, how small would your radius have to become before you would be a black hole?
2. *The Slowing of Pulsars*. Pulsars emit flashes of radio waves at intervals that steadily grow longer. The increase in pulse period with time can be used to assign ages, at least roughly, to various pulsars, assuming

that they began pulsing with about the same period and that they slow down in about the same way. Studies of the Crab Nebula pulsar suggest that the pulse period is proportional to the ⅔ power of the time since the pulsar was formed. If this is true, how many times older would a pulsar with 8 times the Crab pulsar's period (0.033 second) be? How many times older than the Crab pulsar would a pulsar with a period of 33 seconds be?

3. *Measuring Orbital Motion through Time Delay.* The motion of the pulsar that belongs to a binary system can be measured by accurate timing of the pulses that reach the Earth. These pulses have successive delays as the pulsar moves away from us, followed by successive speed-ups in arrival time as the pulsar moves toward us. Discuss whether this would still happen if the speed of light received a "boost" or a decrease from the motion of a source of photons toward us or away from us. What *does* change as the result of the source's motion? If we think we know the time when an event should occur, show how we can use our timing of the event as we see it to determine the speed of light. This method was used for the first accurate determination of the speed of light by Ole Roemer in 1675, when he observed eclipses of Jupiter's satellites by the giant plant. Once Roemer knew when the eclipses ought to occur (from theoretical calculations), his observations of the actual eclipses from two different places along the Earth's orbit (at two different times of the year) allowed him to compute the speed of light in terms of the distance across the Earth's orbit. Draw a diagram illustrating Roemer's procedure.

Further Reading

Asimov, Isaac. 1977. *The Collapsing Universe.* New York: Doubleday.

Clark, George. 1977. "X-Ray Stars in Globular Clusters." *Scientific American* 237:4,42.

Gursky, Herbert, and van den Heuvel, Edward. 1975. "X-Ray Emitting Double Stars." In *New Frontiers in Astronomy.* Edited by Owen Gingerich. San Francisco: W. H. Freeman.

Hawking, Stephen. 1977. "The Quantum Mechanics of Black Holes." *Scientific American* 236:1,34.

Ostriker, Jeremiah. 1975. "Pulsars." In *New Frontiers in Astronomy.* Edited by Owen Gingerich. San Francisco: W. H. Freeman.

Sagan, Carl. 1973. *The Cosmic Connection.* New York: Doubleday.

Shipman, Henry. 1976. *Black Holes, Quasars, and the Universe.* Boston: Houghton Mifflin.

Sullivan, Walter. 1979. *Black Holes: The Edge of Space, the End of Time.* New York: Anchor Press/Doubleday.

Taylor, John. 1974. *Black Holes.* New York: Avon.

Thorne, Kip. 1975. "The Search for Black Holes." In *New Frontiers in Astronomy.* Edited by Owen Gingerich. San Francisco: W. H. Freeman.

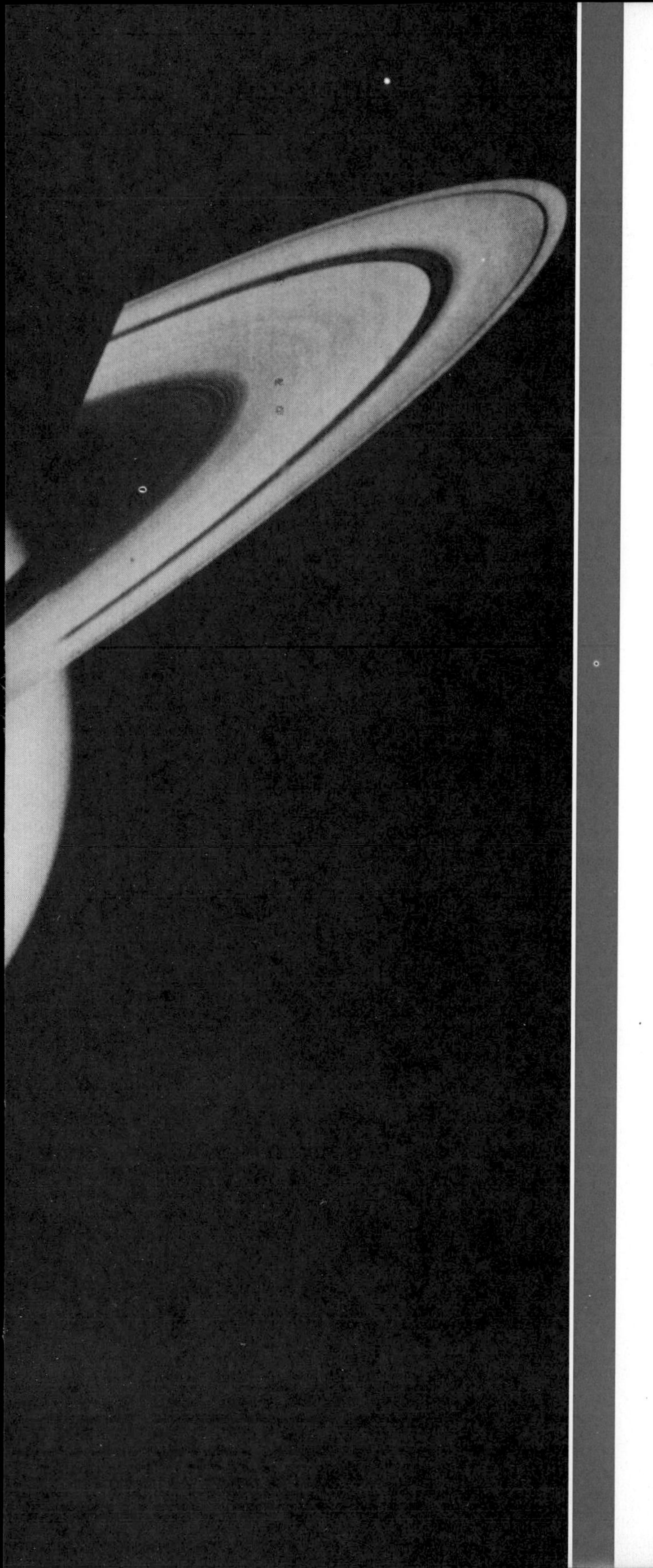

PART 4

The Solar System

Planets are minuscule objects compared to stars, and yet we take a special interest in them, living as we do on the planet Earth. We therefore study the planets in our own solar system to gain insight into what other planets, in orbit around other stars, might be like.

In our brief lifetime, we can reach the closest planets but not the stars. We have already crossed the vastness of space to our closest neighbor worlds, millions of times nearer than the stars of the night skies. Our exploration of Mercury, Venus, Mars, Jupiter, and Saturn has brought us new knowledge of the structure and evolution of the sun's family. There we have found satellites previously unknown, along with a wealth of detail about the planets and their moons. The Voyager studies of Jupiter and Saturn represent a peak achievement in human attempts to understand the cosmos.

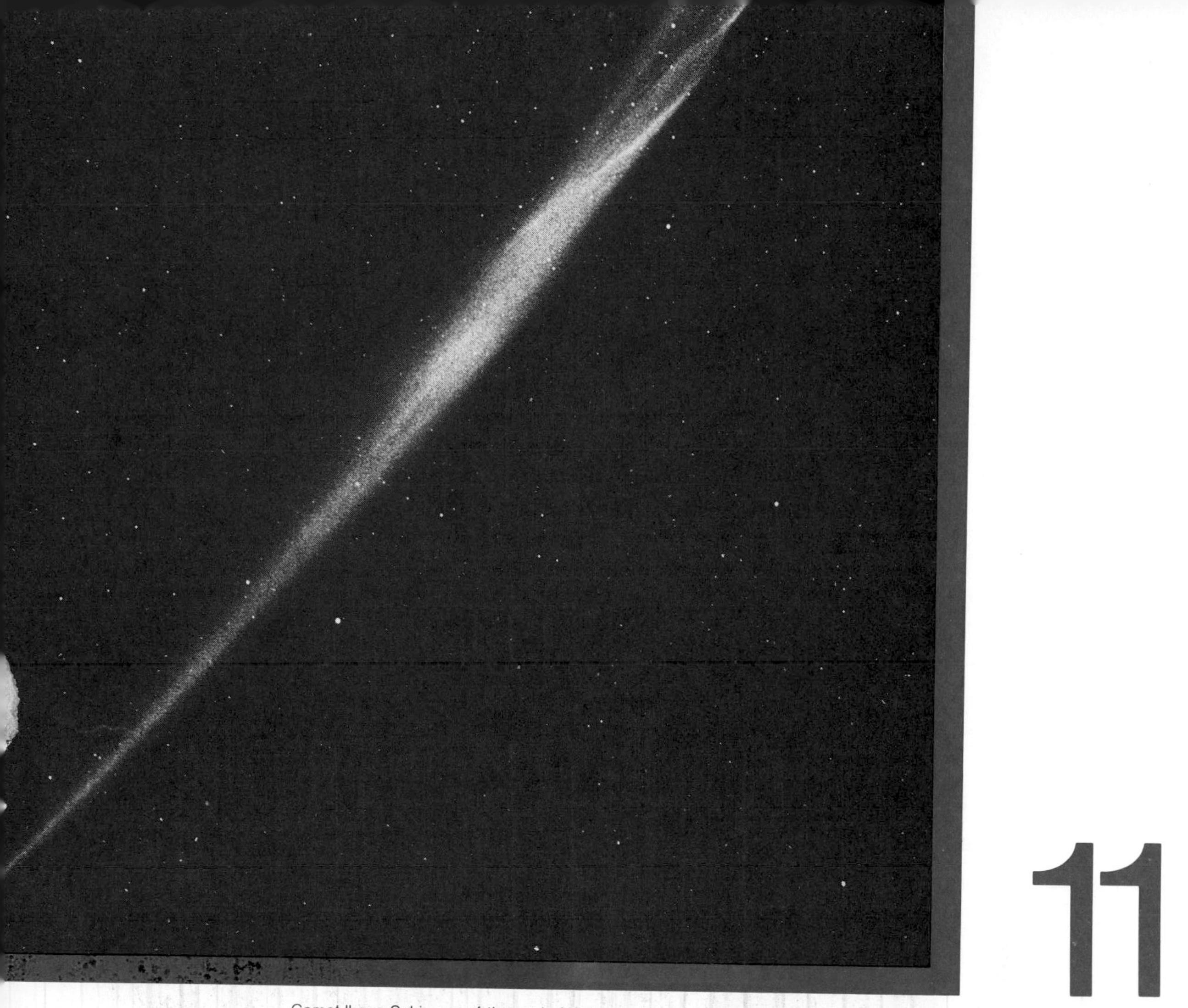

Comet Ikeya-Seki, one of the sun's family of millions of comets, formed along with the sun and its planets 4.6 billion years ago.

11

The Solar System

Among the 400 billion stars of the Milky Way galaxy, all in orbit around the galactic center, there exists a representative orange-yellow ball of gas, with an age of 4.6 billion years, a surface temperature of 5800 K, and a mass of 1.99×10^{33} grams, 333,000 times the Earth's mass. Caught by the gravitational pull of this star, and whirling in orbits around it, are nine cooler objects whose total mass barely equals one-thousandth of the central star's mass. These nine planets, together with at least 39 still smaller satellites, several thousand rocky asteroids, millions of minuscule meteoroids, and an equally great number of small cometary nuclei, form what we grandly call the **solar system**.

If we judge the importance of objects on the basis of mass, then the solar system consists of the sun and nothing else. If we judge on the basis of numbers, then the comets and meteoroids comprise the solar system. Our affection for our own existence and our hopes for finding life similar to our own elsewhere in the universe make us regard planets as objects worthy of attention and exploration. For many thousands of years, human interest in the solar system has focused on the sun, the moon, and the planets, those bright objects that appear to wander among the fixed stars (see page 45). The story of human understanding of the planets has taken some dramatic turns during the past two decades, as immensely capable, automated probes have visited five of our planetary neighbors and have landed on two of them (see Chapters 12 and 13). The knowledge we have gained about Mercury, Venus, Mars, Jupiter, and Saturn, together with our increased understanding of the Earth and its moon, has enabled us to begin to unravel the mysteries of how the solar system formed, why its planets are so similar in some respects and so different in others, and whether life may yet be found on any planet or satellite outside our familiar, hospitable Earth.

11.1 The Formation of the Solar System

Astronomers now have good understanding of the ages of the sun, the Earth, and the moon, based on calculations of how stars age (in the case of the sun) and on radioactive dating of rocks from the Earth and the moon. All three of these ages fall between 4 and 5 billion years. The sun began to shine 4.6 billion years ago; the oldest known moon rock is 4.2 billion years old; and the oldest rock found on Earth has an age of 4 billion years. Astronomers have arrived at an interpretation of this evidence that includes their calculations of how a contracting cloud of gas and dust could form the sun and its planets.

Scientists believe that the entire solar system—sun, planets, satellites, and comets—has the same age, just about 4.6 billion years. They base this conclusion on the fact that a cloud of gas that began to contract through its own gravitational forces would contract quickly, in astronomical terms, once it had decreased to the present size of the solar system (Figure 11-1). No more than 100 million years, and quite possibly only a few million years, should separate the ages of the various planets, satellites, and the sun. In other words, we can consider that the entire solar system formed almost overnight, so far as the history of our galaxy is concerned, just as we may imagine a redwood tree to have sprung up overnight in terms of the history of biological evolution.

The Size of the Solar System

When we speak of the rapid formation of the solar system, we are really considering only the final stages of this process, those that occurred when the solar system had shrunk to about its present size. The sun's family of

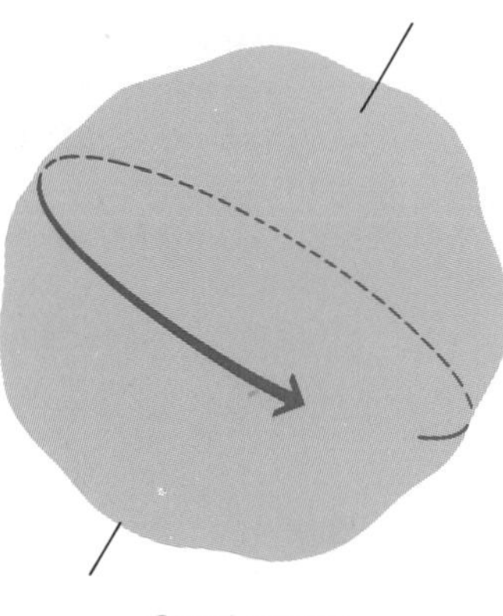

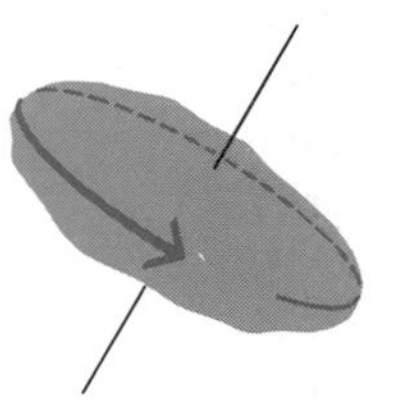

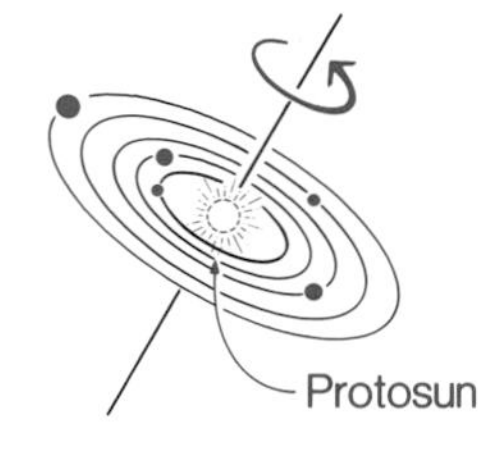

Figure 11-1
As the solar system contracted, the later stages of the contraction occurred more rapidly than the initial stages. These later stages saw the flattening of the system into a disklike shape and the formation of protoplanets within the disk.

nine planets nestles at distances from the parent star that range from Mercury's 58 million kilometers to Pluto's 5.9 billion kilometers. Large though these distances are in human terms, they are minute in comparison to the distances between stars. Alpha Centauri, the sun's closest neighbor, has a distance of 41 *trillion* kilometers, 275,000 times the Earth's distance to the sun, and 7000 times the distance from the sun to Pluto. If we imagine a map of the galaxy in which the stars are represented by tiny light bulbs, each a centimeter in diameter, then the average distance between stars in this map should be about 300 kilometers, the distance from New York to Boston (Figure 11-2). In this map, the sun's planets would orbit a 1-centimeter bulb at distances that range from 40 centimeters (for Mercury) to 40 meters (for Pluto). Our solar system, with the exception of the comets, would thus easily fit inside a football stadium, and the closest stars, along with their planets (if they have planets), would be in fairly distant cities. The Milky Way galaxy, in this model, would span 10 million kilometers, 25 times the distance from the Earth to the moon!

Thus when we consider how the solar system formed from a cloud of gas and dust, the largest question is not how the sun and planets condensed once this cloud had shrunk to the size of the solar system now, but rather how and why the cloud *began* to contract in the first place. Similarly, the key question about the age of the solar system is not how much time has passed since the planets formed and the sun began to shine, but rather how long before that did the contraction that produced the sun and planets actually begin.

How Did the Solar System Begin to Form?

We do not know why part of an interstellar cloud of gas and dust ever began to contract, or whether this contraction began 4.8 or 5.2 or even 6 billion years ago. The reason for this lack of knowledge is simple: The evidence has

Figure 11-2 If we model the stars in our galaxy with light bulbs, then the average distance between lights should equal 300 kilometers.

just about disappeared. What we see is the solar system in its present configuration, which has existed for the past 4.6 billion years. Of the earlier form of the solar system, only the comets, which we have yet to examine close up, may contain material unaltered for longer periods of time. The earliest history of the Earth and the moon, no longer available to us from a study of surface rocks, is the hardest to reconstruct.

During the late 1970s, however, scientists found some evidence suggesting that a supernova explosion occurred within about 10 parsecs of the solar system just before it began to form. This evidence consists of careful measurements of the abundance of magnesium isotopes in the oldest meteorites (see page 399). A larger-than-expected abundance of magnesium-26 (see Table 3-1) suggests that aluminum-26 was present in the material that formed the meteorite, because aluminum-26 turns into magnesium-26 after a few tens of millions of years, and no other good explanation exists for the over-abundance of magnesium-26 relative to the other magnesium isotopes.

But to produce aluminum-26 is not an easy job. Supernova explosions are the most likely source of this isotope. Detailed examination of the meteoritic material pinpoints the time of formation of the aluminum-26 at just 10 or 20 million years before the meteorite itself formed. Since astronomers believe that the oldest meteorites are among the most primitive material in the solar system, we are led to the conclusion that a supernova exploded relatively nearby only a brief time (astronomically speaking) before the sun and its planets started to condense. If this is true, it is hard to avoid the conclusion that the supernova explosion somehow triggered the formation of the solar system.

11.2 Planetary Orbits

The nine larger objects that orbit the sun show a similarity in their trajectories that immediately suggests that they did not form independently. All nine planets orbit the sun in the same direction and in nearly the same plane (Figure 11-3). Thus it is reasonable to assume that the planets all formed from a single, rotating, pancakelike cloud of gas and dust, whose central regions became the sun.

The sun itself rotates once a month in the same direction that the planets orbit around the sun. This too can be explained. As the cloud that formed the solar system shrank to the present size of the planetary orbits, it flattened and spun faster, like a miniature version of the Milky Way in formation (see page 150). We simply do not know where the proto-solar system acquired its initial spin. We do know, from our understanding of physics, that an initial slow spin would become a more rapid spin as the cloud contracted under its own gravitational forces.

The contracting cloud that formed the solar system would not have been spinning at the same rate in all its parts. Instead, the outer parts would always be rotating more slowly than the inner parts. When the individual planets condensed within the disklike, rotating cloud, their motions would reflect the cloud's spin rate at the place and time of their formation.

The generally accepted scenario for the formation of the solar system fits the final stages into a few million years of time, 4.6 billion years ago. During that interval, the planets grew out of clumps called **protoplanets** within the pancake-shaped proto-solar system. These clumps themselves had

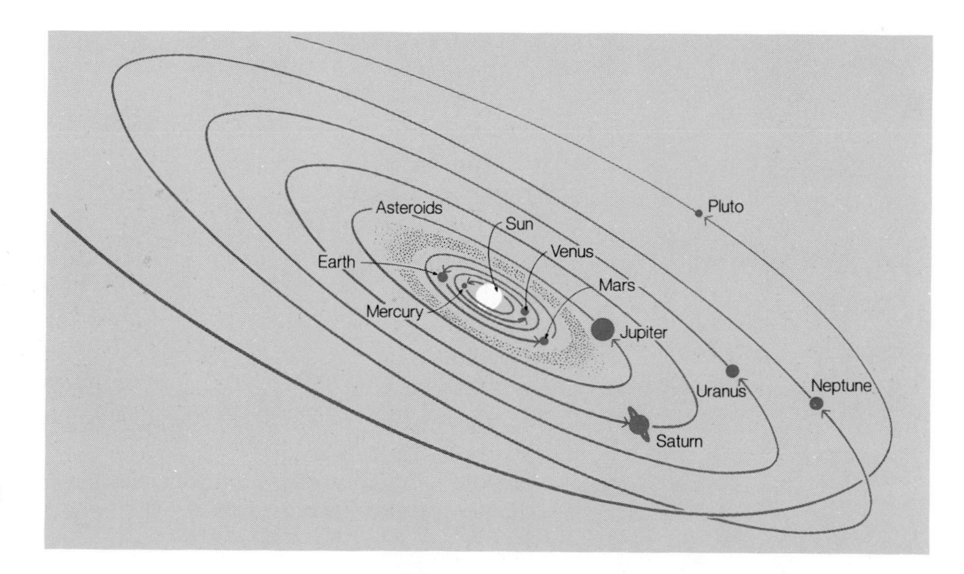

Figure 11-3
The sun's nine planets orbit the sun in the same direction and in almost the same plane.

formed from random swirls and eddies within the rotating disk. Dust particles in collision tended to produce an enlarging nucleus for each protoplanet, as well as for the smaller clumps that became asteroids and meteoroids. As a clump grew to significant size (several hundred kilometers), its own gravitational force could help to attract more gas and dust from within the rotating cloud. In this way, the material in the proto-solar system, previously processed by countless supernova explosions, became the sun, moons, planets, and interplanetary debris that we see today.

Table 11-1 shows the distances of the planets from the sun, the time it takes each planet to complete an orbit, the eccentricity of each planet's orbit, and each planet's average velocity in orbit. As we discussed on page 59, eccentricities larger than zero measure the deviation of an orbit from a perfect

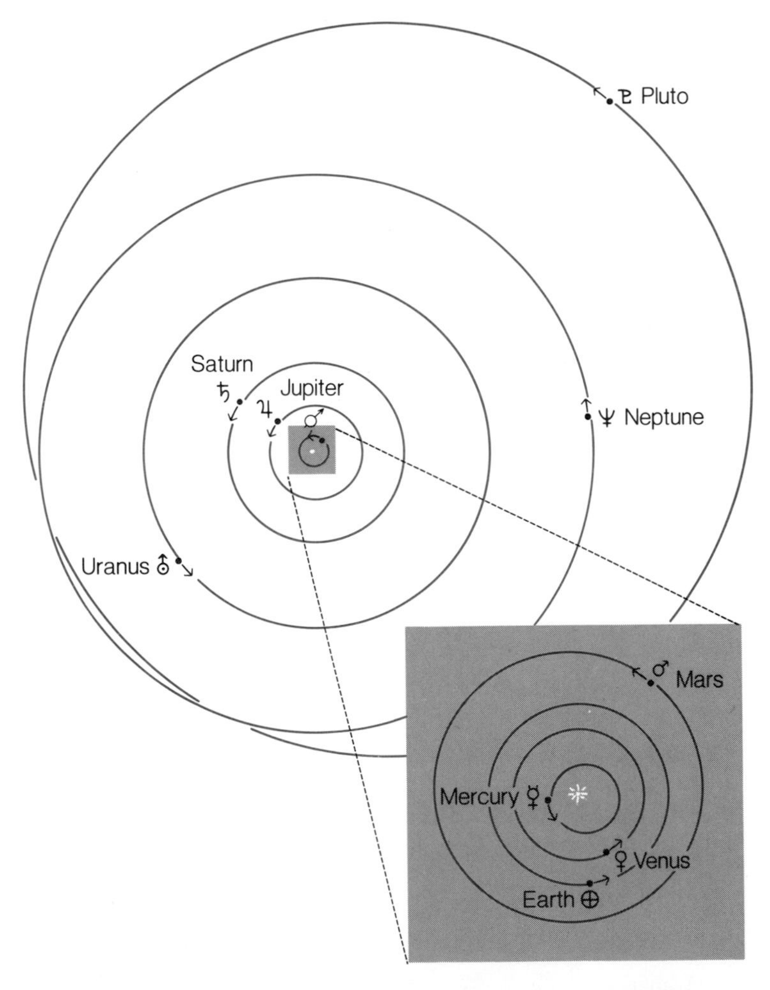

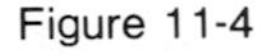
Figure 11-4 The orbits of Mercury, Mars, and Pluto deviate noticeably from circularity, while the other six planets have orbits that are almost, but not quite, perfect circles.

Table 11-1 Characteristics of Planetary Orbits within the Solar System

Planet	Semi-Major Axis of Orbit (km)	Semi-Major Axis of Orbit (A.U.)	Time to Complete One Orbit (Years)	Orbital Eccentricity	Average Velocity of Planet in Orbit (km/sec)
Mercury	57,900,000	0.39	0.241	0.206	47.9
Venus	108,200,000	0.72	0.615	0.007	35.0
Earth	149,600,000	1.00	1.00	0.017	29.8
Mars	227,900,000	1.52	1.88	0.093	24.1
[Asteroids]	414,000,000	2.77	4.60		
Jupiter	778,300,000	5.20	11.86	0.048	13.1
Saturn	1,427,000,000	9.54	29.46	0.056	9.6
Uranus	2,869,600,000	19.2	84.01	0.047	6.8
Neptune	4,496,600,000	30.1	164.8	0.009	5.4
Pluto	5,900,000,000	39.4	248	0.250	4.7

circle. Only Pluto, Mercury, and Mars have significantly noncircular orbits, and even their orbits have only modestly elongated shapes (Figure 11-4).

11.3 Gravitation and the Laws of Motion

Three centuries ago, perhaps the greatest physicist of all time showed why planetary orbits are elliptical. Isaac Newton did so with a few assumptions about the ways in which objects move and interact—and an exceedingly powerful brain.

Newton's Law of Universal Gravitation

In 1665 and 1666, Cambridge University, in England, was threatened by an outbreak of the plague. As the scholars retired to less threatened, more isolated habitation, Isaac Newton went home to his family's farm. There he conceived the **law of universal gravitation,** supposedly (so his favorite niece said 60 years later) after having watched an apple fall from a tree. Newton's mind linked the fall of objects to Earth with the motion of the moon in its orbit. He saw that the moon also falls, but it falls *around* the Earth (Figure 11-5). As the moon moves in its orbit, the Earth attracts the moon, but the moon's **momentum**—its tendency to keep moving in the same direction at the same speed—allows the moon to keep falling around the Earth.

By 1685, Newton was ready to compose his masterpiece, the *Philosophiae Naturalis Principia Mathematica (Mathematical Principles of Natural Philosophy).* This book, published in 1687, explained *why* the Copernican model works, *why* Kepler's laws are valid, and *why* the moon orbits around the Earth, and not the Earth around the moon. With Newton's work humanity continued the quest begun in Greece two millenia before: to find out why the cosmos behaves as it does.

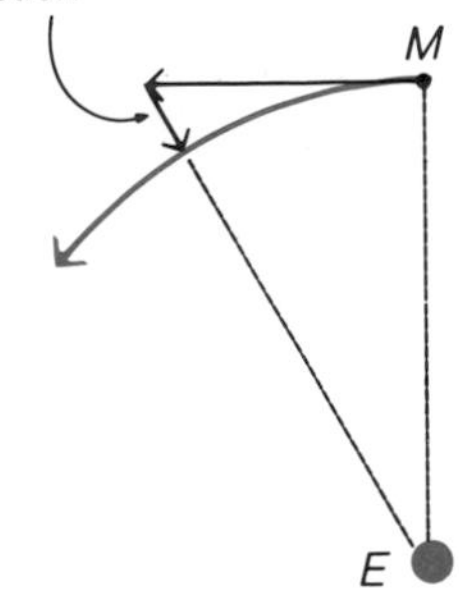

Figure 11-5
Newton saw that the moon, like an apple from a tree, is falling, but it is falling *around* the Earth. The moon's momentum would carry it off into space in a straight line if the Earth's force of gravity did not at every instant produce an acceleration, or change in direction of motion, upon the moon.

Newton's law of gravitation has a majestic simplicity, verified by experiment but still a great leap in deduction. It states that every object in the universe attracts every other object with a force that varies in proportion to 1 over the square of the distance between their centers. Until Newton could prove that the Earth attracts other objects as if all its mass were concentrated at its center, he hesitated to publish his work. Once he achieved this proof, the power of his law became clearer (Figure 11-6). But if all objects attract one another with forces that vary in inverse proportion to the square of the distances between them, why don't they all rush together in one place? The answer lies, at least in part, in Newton's three laws of motion.

Newton's Three Laws of Motion

The first of Newton's laws says that in the absence of an external force, an object will continue to move in the same direction, and at the same speed, that it did a moment before. We know that if we drive a car down a highway and lift our feet off the accelerator, the car will coast on its own momentum. To change either our speed or our direction (for example, to turn a sharp corner), we must accelerate the car by applying a force to it, typically with the brakes, the motor and drive train, or the steering mechanism.

Newton's Law of Universal Gravitation and Three Laws of Motion

	Statement in Words	Algebraic Expression[a]
Law of universal gravitation	Every object with mass attracts every other such object, with an amount of force that varies in proportion to the product of the two masses, and in inverse proportion to the square of the distance between their centers.	$F_{1,2} = G\frac{M_1 \times M_2}{D^2}$
First law of motion	An object not acted upon by a net force will continue to move in the same direction and at the same speed.	If $F = 0$, $a = 0$.
Second law of motion	A net force acting on an object will produce an acceleration (change in speed, direction, or both) that varies in proportion to the amount of force, and in inverse proportion to the mass of the object. The acceleration is produced in the same direction as the force.	$\vec{a} = \frac{\vec{F}}{M}$
Third law of motion	When two objects exert forces upon each other, the two forces have the same magnitude but act in opposite directions.	$\vec{F}_{1,2} = -\vec{F}_{2,1}$

[a]We use vector symbols, such as $\vec{a}$ for acceleration, to denote quantities with a magnitude (a with no arrow) plus a direction.

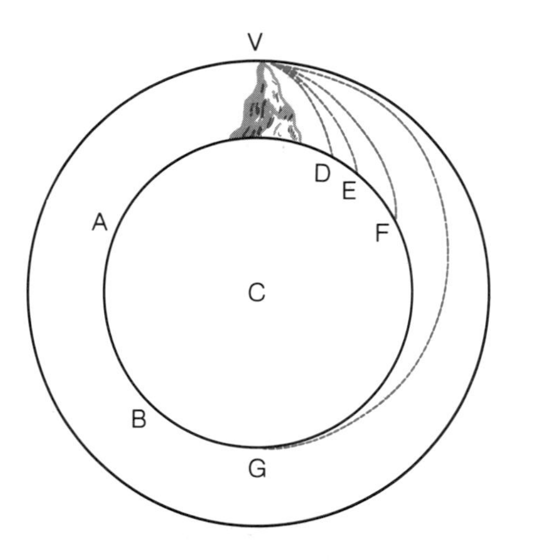

Figure 11-6 Once Newton showed mathematically that the Earth attracts other objects as if all its mass were concentrated at its center, he could show the exact way in which the moon's motion around the Earth relates to the fall of objects near the Earth's surface.

Aristotle had declared that any motion demands a continuous applied force. Galileo had shown this to be untrue, and Newton provided the basic principle of momentum. In his second law, Newton expressed the effect of a given force. Any net force that acts on an object will produce an acceleration in the same direction that the force acts. The amount of the object's acceleration will vary in proportion to the strength of the force, and in inverse proportion to the object's mass. Thus, a golf club can accelerate a golf ball to 50 meters per second in one-tenth of a second (thus giving the ball an acceleration of 500 meters per second per second), but the same force will accelerate a football to only 2 meters per second.

Newton's third law states that when two objects exert forces upon one another, the first object's force on the second exactly equals in amount, and opposes in direction, the second object's force upon the first. Thus, for example, the moon attracts the Earth with exactly the same amount of force as the Earth exerts upon the moon, though the force acts in the opposite direction. But by Newton's second law of motion, the moon will accelerate 81 times more readily than the Earth to a given force, since the moon has 1/81 of the Earth's mass. Hence the moon orbits the Earth, not the Earth the moon. More correctly, both the moon and the Earth orbit the center of mass of the Earth-moon system, which lies 81 times closer to the Earth's center than to the moon's center (Figures 11-7 and 11-8).[1]

[1]If we consider a system made of any two objects, the system's center of mass lies on the line joining the objects' centers. If one object has x times the mass of the other, the center of mass lies x times closer to the center of the more massive object.

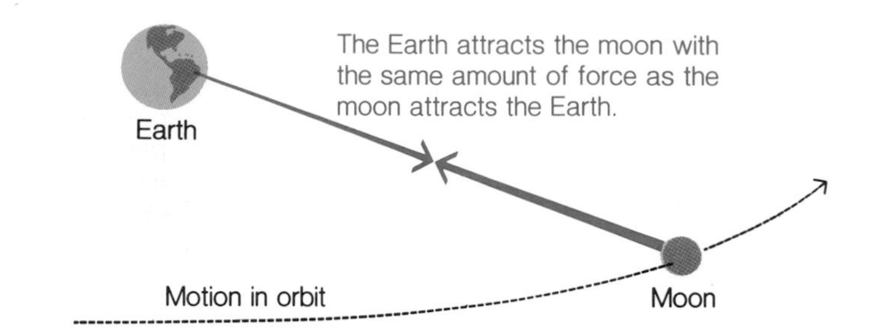

Figure 11-7
The moon and the Earth—or any two objects with mass—attract each other with gravitational forces. These forces act along the line joining the centers of the two objects and tend to produce accelerations in the same direction as the forces. However, since the Earth has 81 times the moon's mass, the moon accelerates 81 times more readily to the Earth's force on the moon than the Earth accelerates to the moon's force on the Earth. The two forces are equal in magnitude, opposite in direction, and highly unequal in the accelerations they produce.

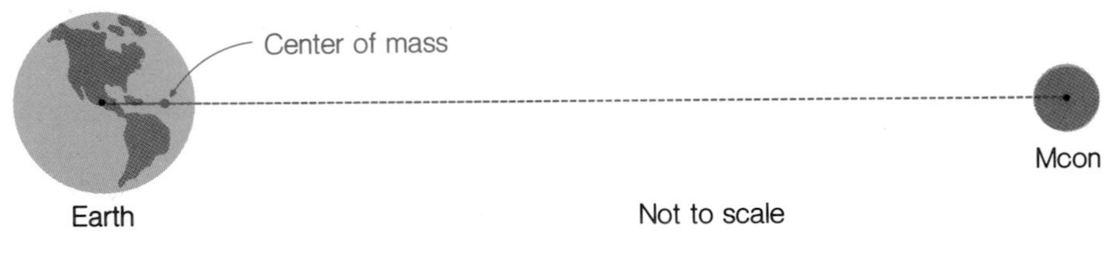

Figure 11-8
The center of mass of the Earth-moon system lies 81 times closer to the Earth's center than to the moon's center, because the Earth has 81 times the moon's mass. This puts the center of mass inside the Earth, though closer to its surface than its center.

Orbital Motion under the Law of Gravitation

Using his three laws of motion and the law of gravitation, Newton produced mathematical proof of a startling series of assertions. *Any* two objects that attract one another by gravitational forces will tend to move in elliptical orbits around their common center of mass (Figure 11-9). Orbits in the form of parabolas and hyperbolas, which do not close back on themselves, are also possible, but clearly these are one-time affairs, while elliptical orbits are permanent.

Furthermore, the motion about the center of mass must obey Kepler's second law, so that the line joining either object to the center of mass will sweep over equal areas in equal amounts of time (Figure 11-10). Finally, Kepler's third law can likewise be shown to be a mathematical and physical necessity, provided we modify the law slightly. If, for example, we consider the

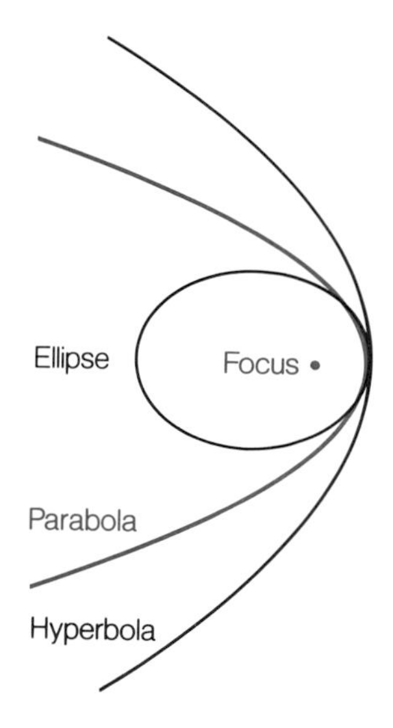

Figure 11-9
Newton showed that two objects that exert gravitational forces on each other will each orbit their common center of mass. This orbit can be elliptical (eccentricity, *e*, less than one), parabolic ($e = 1$), or hyperbolic (*e* greater than one). Parabolic orbits mark the point of transition from elliptical orbits, which close on themselves, to hyperbolic orbits, which extend to infinity and thus take an infinite amount of time to complete.

planets in orbit around the sun, then the square of each planet's orbital period will be proportional to the cube of the orbital semi-major axis divided by the sum of the planet's and sun's masses.[2] But since the sun's mass so much exceeds any planet's mass, the sum of the planet's and sun's masses may be taken as nearly the same for any planet, which restores the original form of Kepler's third law as a good approximation: The square of the orbital period varies in proportion to the cube of the semi-major axis.

Newton's monumental work produced an impact on philosophy second only to Copernicus's reworking of the cosmos more than a century earlier. Scientists began to see how all the motions observed in the heavens might be explained through a small number of sweeping generalizations, the laws of Newton. For the next two centuries, the science of **celestial mechanics** (the motions of objects in space) seemed to consist basically of improving calculations based on Newton's laws. Eventually, a genius named Albert Einstein would show how more complex rules applied to motions approaching the speed of light and to situations of immensely strong gravity. But while Newton's laws reigned unquestioned, their successes multiplied, seemingly without end.

In 1759, the return of Halley's comet, whose orbit had been predicted by Newton's laws, seemed a stunning triumph of the Newtonian ordering of the universe. During the 1760s, Joseph Louis Lagrange used Newton's mathematics to explain details of the motions of the moon and Jupiter's satellites. In the 1780s, Pierre Simon Laplace had a still greater triumph, as he resolved the question of how the planets Jupiter and Saturn perturb one another, and their neighboring asteroids, from the orbits they would have in the simpler case of just two objects in orbit around their mutual center of mass (Figure 11-11). Laplace also explained further details of the moon's motion around the Earth and even showed why the eccentricity of the Earth's orbit slowly varies, first toward slightly larger eccentricities and then toward slightly smaller ones.

---

[2]In algebraic terms, we may express the relationship between the orbital period, P, and the orbital semi-major axis, a, as

$$P^2 \sim \frac{a^3}{M_1 + M_2}$$

where $\sim$ denotes proportionality and M_1 and M_2 are the masses of the sun and planet.

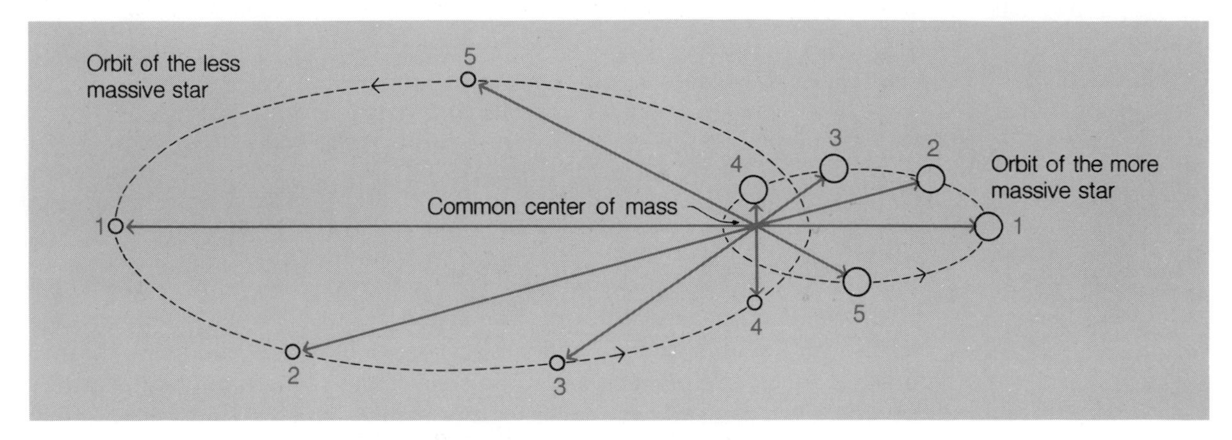

Figure 11-10 Newton also showed that Kepler's third law of orbital motion refers to motion around the center of mass, which forms one of the two focal points of the elliptical orbits. In the solar system, the sun has such an overwhelming majority of the mass that it lies quite close to the center of mass. In systems composed of two stars, however, the center of mass may be noticeably far from either star.

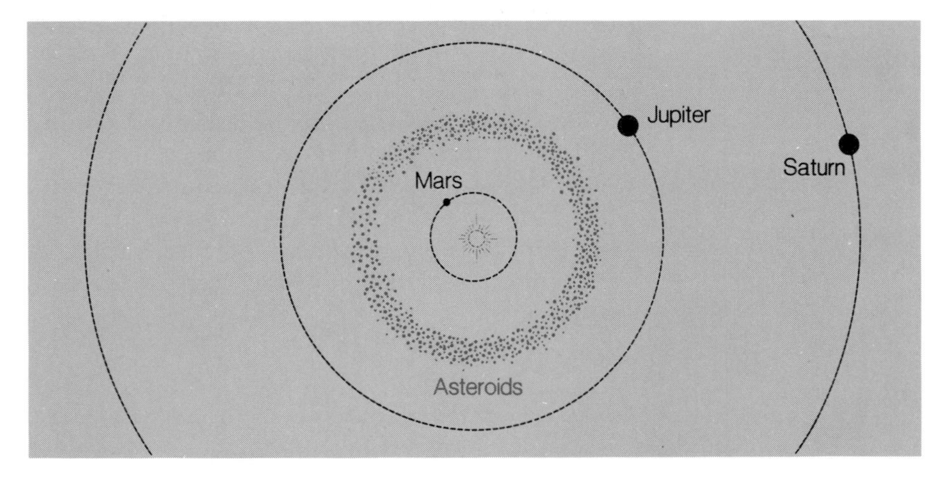

Figure 11-11 Jupiter and Saturn each have more than 1/3000 of the sun's mass, enough to produce noticeable perturbations on each other's orbits and on the motions of the asteroids that orbit the sun. These perturbations give the planets and asteroids orbits slightly different from those they would have if only a single object moved around the sun.

11.4 The Two Types of Planets

The solar system contains four **inner planets,** which are all relatively small, rocky, and dense. Outside the orbits of these four planets, and beyond the belt of asteroids, orbit the four **giant planets,** which are large, gaseous, and rarefied. Pluto, the outermost planet, resembles an inner planet in size, but it appears to be a great ball of ice, not rock, perhaps an escaped satellite of Neptune (see page 467).

The contrast between the four inner planets—Mercury, Venus, Earth, and Mars—and the four giant planets—Jupiter, Saturn, Uranus, and Neptune—remains significant and obvious. Earth, the largest and most massive of the four inner planets, has just 1/333,000 of the sun's mass. In contrast, Jupiter has 318 times the Earth's mass, or almost one-thousandth of the sun's mass. Saturn's mass is 95 times Earth's, and even Uranus and Neptune, the smaller of the giant planets, have 14.6 and 17 times the Earth's mass, respectively. If we compare the giant planets to the other three inner planets, the contrast in masses becomes still more striking. As Table 11-2 shows, Venus's mass equals just 82% of the Earth's; Mars's mass is 11% of our planet's; and Mercury has just 5.5% of the Earth's mass. Table 11-2 also gives the volume of each planet and reveals an even greater contrast between the inner planets and the giant planets. Jupiter's volume exceeds the Earth's by more than a thousand times, and the other giant planets each have between 57 and 836 times the volume of the Earth (Figure 11-12)!

With such large volumes, the giant planets are far less dense than the inner planets. Mercury, Venus, Earth, and Mars all have average densities of matter that range from 3.9 to 5.5 times the density of water. In contrast, Jupiter, Saturn, Uranus, and Neptune have average densities that are several times less. Neptune, the densest of the four giant planets, has only 1.7 times the density of water. Saturn, the most rarefied, has just 0.63 times water's density, and would float if we could find an ocean large enough to contain it! Pluto's density, 0.8 times the density of water, is close to the density of Saturn.

Table 11-2 Physical Characteristics of the Planets

Planet	Equatorial Diameter (km)	Equatorial Diameter (Earth = 1)	Volume (Earth = 1)	Mass (Earth = 1)	Average Density (gm/cm^3)	Surface Gravitational Force (Earth = 1)
Mercury	4,873	0.382	0.056	0.0553	5.44	0.38
Venus	12,120	0.950	0.86	0.815	5.24	0.90
Earth	12,756	1.0	1.0	1.0	5.52	1.0
Mars	6,790	0.532	0.151	0.1075	3.93	0.38
Jupiter	142,600	11.18	1,397	317.83	1.26	2.54
Saturn	120,000	9.42	836	95.15	0.63	1.07
Uranus	52,000	4.08	67	14.6	1.2	0.88
Neptune	49,000	3.85	57	17.2	1.7	1.16
Pluto	3,000	0.24	0.013	0.002	0.8	0.03

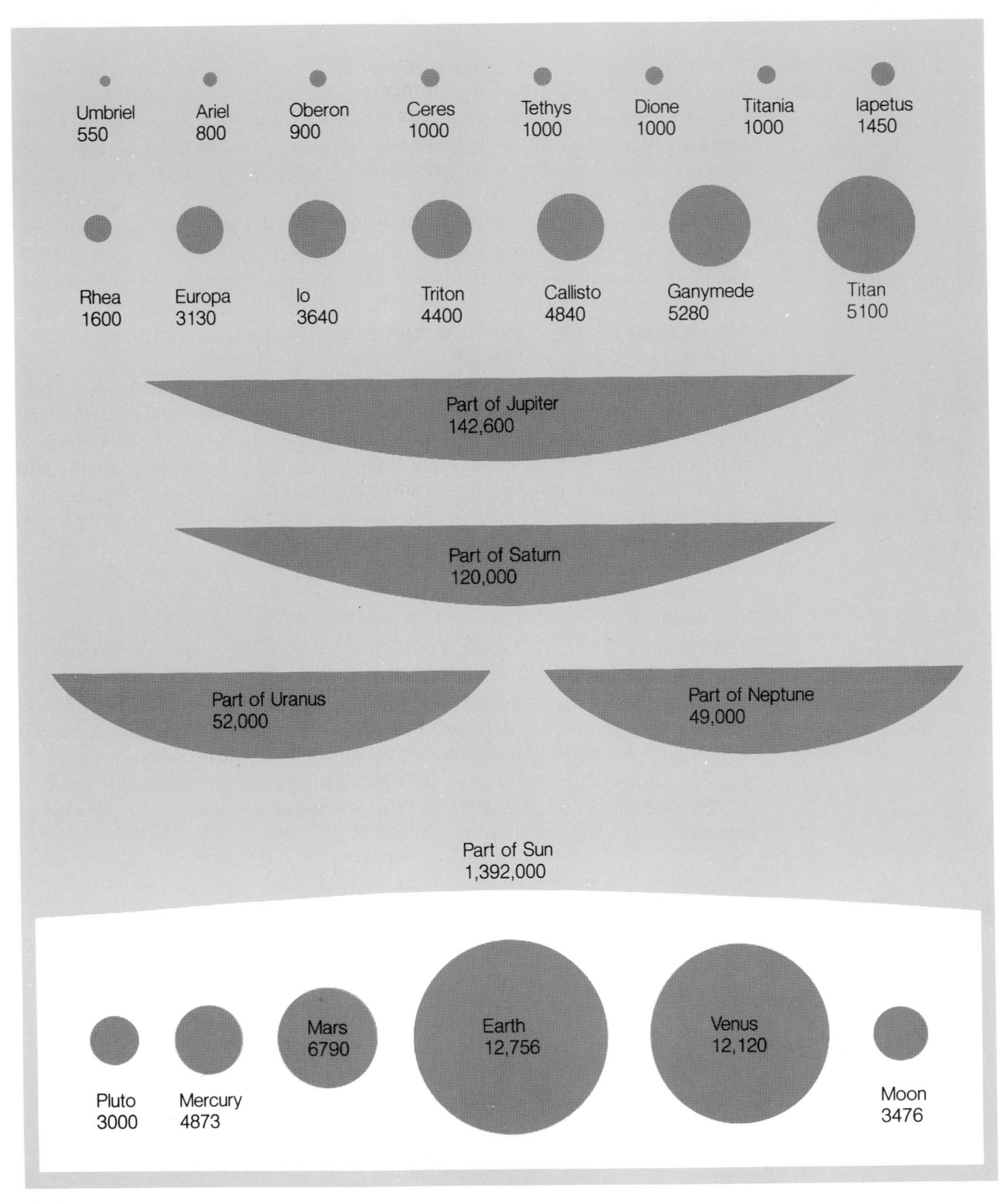

Figure 11-12
A scale drawing shows the 26 largest objects in the solar system (diameters in kilometers). Pluto's satellite, Charon, has a diameter between 1200 and 1800 kilometers.

11.5 The Inner Planets

Mercury, Venus, Earth, and Mars have much in common, despite great differences in their outward appearance. Most significantly, these four inner planets consist primarily of elements heavier than hydrogen and helium, the two most abundant elements in the sun and indeed in the universe (see page 322). In fact, the inner planets are startling exceptions to the universal rule, since they are mostly made of such heavy elements as silicon, oxygen, magnesium, and iron, with almost no hydrogen or helium.

The Escape of the Light Elements

Why don't the inner planets possess the lightest and most abundant of elements? The answer seems to lie with the fact that these are the innermost of the sun's planets. Once the sun began to shine, the lightest atoms, moving the most rapidly, escaped from the vicinity of the four closest protoplanets (planets in formation). The four giant planets, on the other hand, managed to retain large quantities, if not all, of their original hydrogen and helium, and so ended up far more rarefied, and far more massive, than the inner planets.

The critical factors in the evolution of each of the protoplanets were the *mass* of the object and its *distance from the sun*. These two factors together determined the ability of each protoplanet to retain the lightest gases, hydrogen and helium. Smaller protoplanets could not retain these two gases, especially if they orbited the sun at relatively close distances and thus acquired greater temperatures than the planets farther out. As the planets formed within the rotating disk of gas and dust, the crucial mass of each protoplanet was the mass that had built up *before* the sun began to shine. Once the sun had turned on, hydrogen and helium tended to escape from *all* the protoplanets. The question was then whether most of these gases would escape, as happened for the inner planets, or whether most would be retained, as occurred for the giant planets (Figure 11-13).

All of the factors favorable to the retention of hydrogen and helium lay with the giant protoplanets: greater distance from the sun and greater masses within the protoplanets (because the outer parts of the disk of material had more mass than the inner parts). The giant planets won the struggle to keep their hydrogen and helium; the inner planets lost. If we *now* moved Jupiter to the Earth's distance from the sun, Jupiter could still retain its hydrogen and helium, because it has 318 times the Earth's mass. The great mass of Jupiter gives it enough gravitational force to hold on to even the lightest gases, even if it were five times closer to the sun than it is now, but it could never have grown to this mass if it had formed at the Earth's distance from the sun. We might note that if we added hydrogen and helium to the Earth's mixture of elements in the same proportion that these elements appear in the sun, the Earth's mass would grow about 100 times, to equal approximately the mass of Saturn.

Sun

Not to scale

Inner Planets

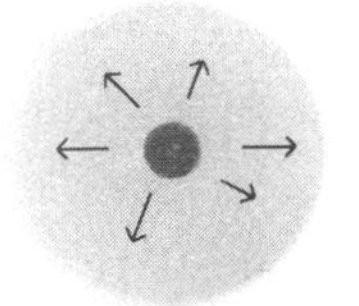

Planet closer to sun cannot retain hydrogen and helium as it forms.

Outer Planets

Planet farther from sun can retain hydrogen and helium.

Figure 11-13 When the sun began to shine, the inner regions of the solar system were heated more than the outer regions. This made the escape of the lightest gases, hydrogen and helium, easier in the inner regions of the solar system.

Internal Structure of the Inner Planets

Their smaller protoplanet masses and nearness to the sun left the inner planets almost devoid of hydrogen and helium. Instead of the great gas spheroids of the outer solar system, the inner planets are made of rock and metal. Furthermore, the internal structure of each of the inner planets is divided into a rocky, less dense mantle surrounding an iron-rich, denser core (Figure 11-14). Although this fact has been proven only for the Earth, everything we know about geology suggests that Mercury, Venus, and Mars have similar internal structures, though Mercury has a significantly different composition.

Consider what must have happened as the protoplanet Earth passed through the final stages of its formation. The Earth's **primitive atmosphere**—the hydrogen and helium gases that once formed most of the local agglomeration—escaped as the sun began to shine. The protoplanet also

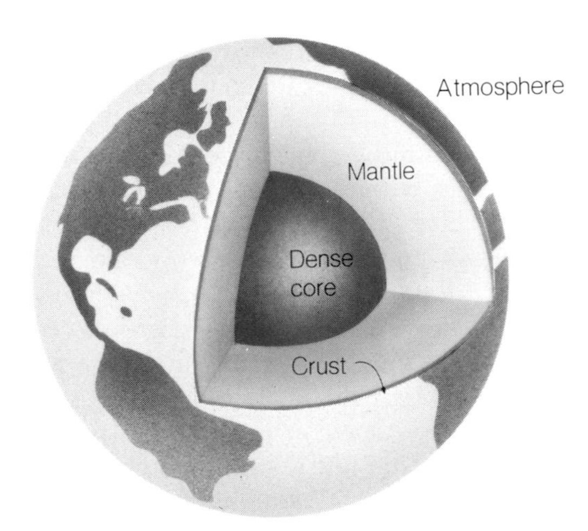

Figure 11-14
The four inner planets have an internal structure that consists of a dense, iron-rich core surrounded by a less dense, silicon-oxygen mantle.

contained a small amount of radioactive elements (mostly potassium and uranium), significant because they produce heat as they decay into other elements. This heat, released most rapidly during the Earth's early history, melted most of the iron in the protoplanet, allowing this denser element to concentrate in the planet's central core. The average density of the rocks near the Earth's surface equals 2.6 times the density of water. The mantle has a density of 3.5 times the density of water. But the Earth's core, the innermost eighth of the Earth, has a density 12 times the density of water. A similar **differentiation** into a dense core and a less dense mantle should characterize the other inner planets.

Mercury represents the extreme case of a planet close to the sun. It is so close, in fact, that elements such as silicon and oxygen must have been lost, in part, after the sun began to shine. Mercury is almost as dense as the Earth, yet it does not have enough gravitational force to produce a core as dense as the Earth's. If we compare Mercury with the moon, which has a diameter 71% of Mercury's, we find a surprising contrast. The moon's average density is 3.34 times the density of water, typical of the density in the Earth's mantle and indicative of the absence of a large dense core, as we would expect for such a small object. On the other hand, Mercury's density, 5.4 times the density of water, exceeds that of Mars, which is twice as massive, and almost equals the Earth's. The explanation for Mercury's large density can be found in its proximity to the sun: Greater solar heating caused the escape not only of hydrogen and helium, but also of some of the lighter "heavy" elements, some 4.6 billion years ago. Thus Mercury retained a superabundance of the heavier and denser elements.

Motions of the Earth's Crust

Radioactive elements, which melted the iron in the inner protoplanets, have an important geological role even today. Deep in the Earth, radioactive potassium, uranium, and thorium continue to produce heat. The motions of the Earth's crust, the famous **continental drift**, occur precisely because of the heat released below the Earth's surface.

The heat that builds up within the Earth's crust makes its subsurface layers almost ready to melt. Pieces of the crust called **plates** can therefore slide over the subsurface layers. The mechanism that drives **plate tectonics**, the motion of the plates of the crust, is the upwelling of new material, typically along ridges beneath the oceans, where the Earth's crust is thinnest (Figure 11-15). As a result of this "sea-floor spreading," some plates, such as the ones that carry Africa and the Americas, are being forced apart with a speed of a few centimeters per year, while other plates are colliding with one another at about the same speed. Collisions of the crustal plates typically make one plate ride over another, creating mountains where none were before (thus the Himalayas were born 30 to 40 million years ago, when the Indian plate encountered the Eurasian plate) and often producing volcanoes and earthquakes. The best-known region of plate encounter lies at the edge

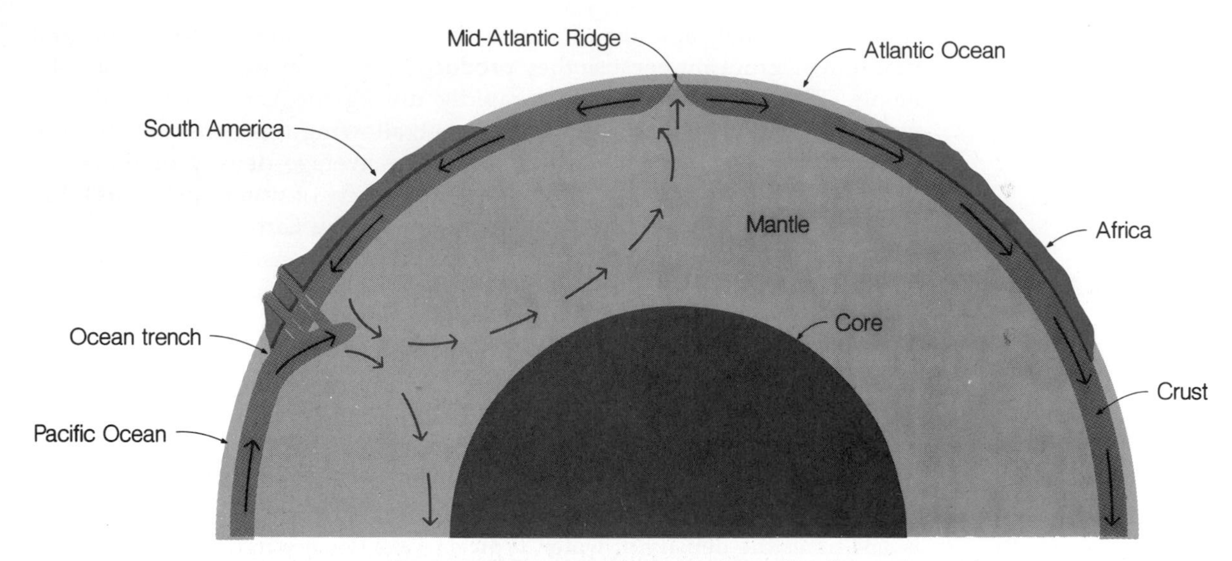

Figure 11-15 Heat released within the Earth's mantle by radioactive rocks produces convection currents that provide the driving force for the motion of the crustal plates. Such plate-tectonic motion occurs as new crust wells up in the middle of sea floors, where the crust is thinnest, and moves the plates apart.

of California, where the San Andreas fault divides the Pacific plate from the North American plate for a distance of 1000 kilometers (Figure 11-16).

Mars and Mercury are smaller than the Earth and Venus. Since smaller objects have a larger ratio of surface area to total volume, the two smaller planets can more readily cool themselves by conducting the heat of their subsurface layers to their surfaces and then radiating it into space. Thus we would expect to find less plate-tectonic activity on Mars and Mercury than we would on Venus and Earth, and this conjecture has received partial verification from the Viking studies of Mars.

11.6 The Earth and the Moon

Although we shall later discuss the exploration of the moon in detail, along with the results from other space probes, we ought to pause now to admire the uniqueness of the Earth-moon system. With the likely exception of Pluto's satellite, no other satellite in the solar system has as much as one-thousandth of the mass of its planet; the moon has 1/81 of the Earth's mass. No other satellite save Pluto's has even 1/20 of its planet's diameter; the moon's diameter exceeds one-quarter of the Earth's. Thus, although five satellites in the solar system exceed the moon in size, and four exceed the moon in mass, the moon remains outstanding among solar system satellites in the effect it has on its parent planet, our Earth.

Figure 11-16 This view of the San Andreas fault near Taft, California, shows 5 kilometers of slippage of the Pacific plate against the North American plate.

The Tides

Chief among the effects caused by the moon are the *tides*, in importance, and *eclipses*, for show. The **tides** raised by the moon, and to a lesser extent by the sun, reflect the difference in gravitational force that these objects produce on different parts of the Earth. The moon's gravitational force attracts the entire Earth (just as the Earth's force attracts the moon), but the pull on the near side of the Earth exceeds the pull on the center, and the force on the Earth's center exceeds that on the far side of the Earth (Figure 11-17).

Because of the difference in gravitational attraction from one place to another, the Earth tends to bulge *toward* the moon (because the force on the near side is greater than the force on the center) and also *away from* the moon (because the force on the center exceeds the force on the far side). All of the Earth responds to these differences in amount of gravitational force, but the oceans, being fluid, can respond far more easily than the land. What we see as the ocean tides represents the greater ability of water to bulge toward and away from the moon, though the land itself has tides of a good fraction of a meter that go almost unnoticed.

The bulge of water toward the moon and away from it creates two high tides and two low tides per day, as the Earth rotates underneath the moon

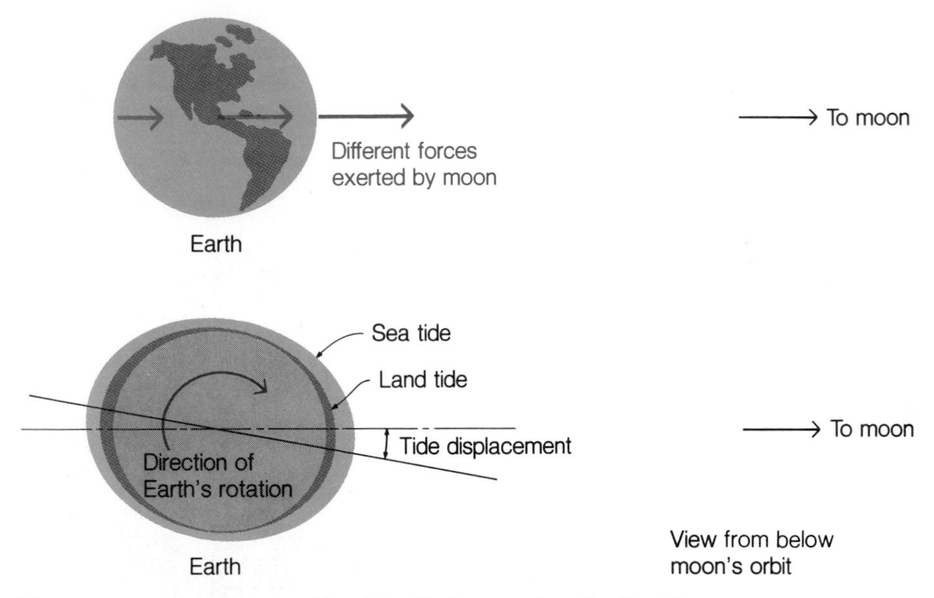

Figure 11-17 The moon attracts the near side of the Earth more than the Earth's center, and the center more than the far side. These differences in gravitational attraction make the Earth tend to bulge toward the moon and away from the moon. Water can respond more readily than land to this tendency to bulge, and so the bulge produces two high and two low ocean tides each day, as a given spot rotates through the bulges.

(Figure 11-17). The general tendency is for high tide to occur when the moon is overhead or underfoot, plus an hour or two for the lag time needed for the water and land to respond to the moon's attractive force. However, the irregular shape of the coastline makes this only a rough approximation. The most spectacular high and low tides come when a mass of water must funnel into, and out of, a V-shaped opening, such as the Bay of Fundy in Nova Scotia. The water level can then change by as much as 10 meters between high and low tides (Figure 11-18).

The Sun's Tide-Raising Ability. The sun also helps raise tides on Earth, but the sun's tide-raising ability is less than the moon's. This may seem odd, since the sun exerts far more gravitational force on the Earth than the moon does. But we must recall that tide-raising power depends on the difference in gravitational force from one place to another. This *difference* varies as 1 over the *cube* of the distance from the object exerting the force, while the force itself varies as 1 over the square of the distance. Thus although the sun exerts 180 times more gravitational force on the Earth than the moon does, the moon has about twice the sun's tide-raising ability on Earth.[3]

[3]The sun's mass is 27,061,000 times the moon's mass, and its distance from Earth is, on the average, 390 times the moon's distance. 27,061,000 divided by the square of 390 equals 177.9, while division by the cube of 390 gives 0.456.

Figure 11-18 Two views of the harbor at Cutler, Maine, at high tide and low tide show the great tidal variations that arise in the Bay of Fundy.

Spring Tides and Neap Tides. Consider, then, what happens to the tides as the moon orbits around the Earth. Each orbit takes 27 1/3 days, but since the Earth moves part way around the sun in that time, we must wait a bit longer, 29 1/2 days in all, for the moon to regain the same alignment with respect to the Earth and the sun (Figure 2-10). Hence, the interval between two full moons (when the moon is on the far side of the Earth from the sun) or between two new moons (when the moon is on the same side of the Earth as the sun) equals 29 1/2 days.

At full moon and new moon, the tide-raising powers of the sun and moon combine, and the result is especially high tides and especially low tides. This condition of **spring tides** recurs at intervals of 14 3/4 days. Halfway in between, when the moon is at first quarter or last quarter, the sun's tide-raising force counteracts some of the moon's (Figure 11-19). The resultant **neap tides** are less pronounced than the high and low tides of full and new moon. Tides that occur between spring tides and neap tides are, as might be expected, intermediate in their heights.

Eclipses

Once each month, at new moon, the potential exists for the moon's shadow to fall on the Earth, thus producing a **solar eclipse,** or **eclipse of the sun.** Once each month, at full moon, the Earth's shadow may cover the moon, making a **lunar eclipse,** or **eclipse of the moon.** We might expect such eclipses to occur every time the moon traveled around the Earth, but in fact

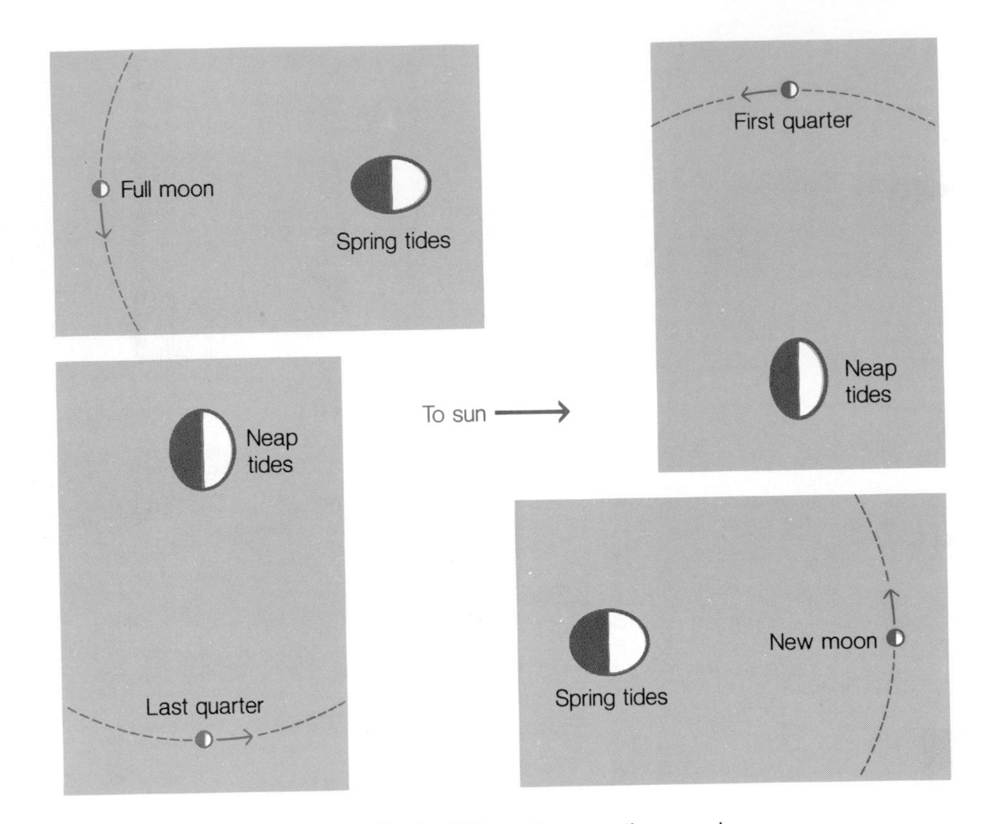

Figure 11-19 The sun's tide-raising force on Earth, 45% as strong as the moon's, can either add to the moon's tide-raising ability or oppose it. The addition of forces, which produces spring tides, occurs at full moon and at new moon. The opposition of forces, leading to smaller neap tides, occurs near first quarter and last quarter of the 29½-day lunar cycle of phases.

eclipses are much rarer than this. The reason that most months fail to bring eclipses is that the moon's orbit is tilted with respect to the Earth's orbit around the sun (Figure 11-20). Thus at most full moons and most new moons, the moon misses being in a direct line with the Earth and the sun by a significant amount. Only if the moon happens to be passing through the plane of the Earth's orbit at the time of full moon or new moon can an eclipse occur. Conditions prove right for an eclipse of the moon at about one full moon in every seven, and for an eclipse of the sun at about one new moon in every six.

Lunar Eclipses. Eclipses of the moon can be seen from an entire hemisphere on Earth, the one that faces the moon at the time of the eclipse. Eclipses of the sun, on the other hand, can be seen from only a tiny fraction of the Earth's surface, the part on which the moon's shadow falls. The reason for this difference is that the moon's diameter barely exceeds one-quarter of

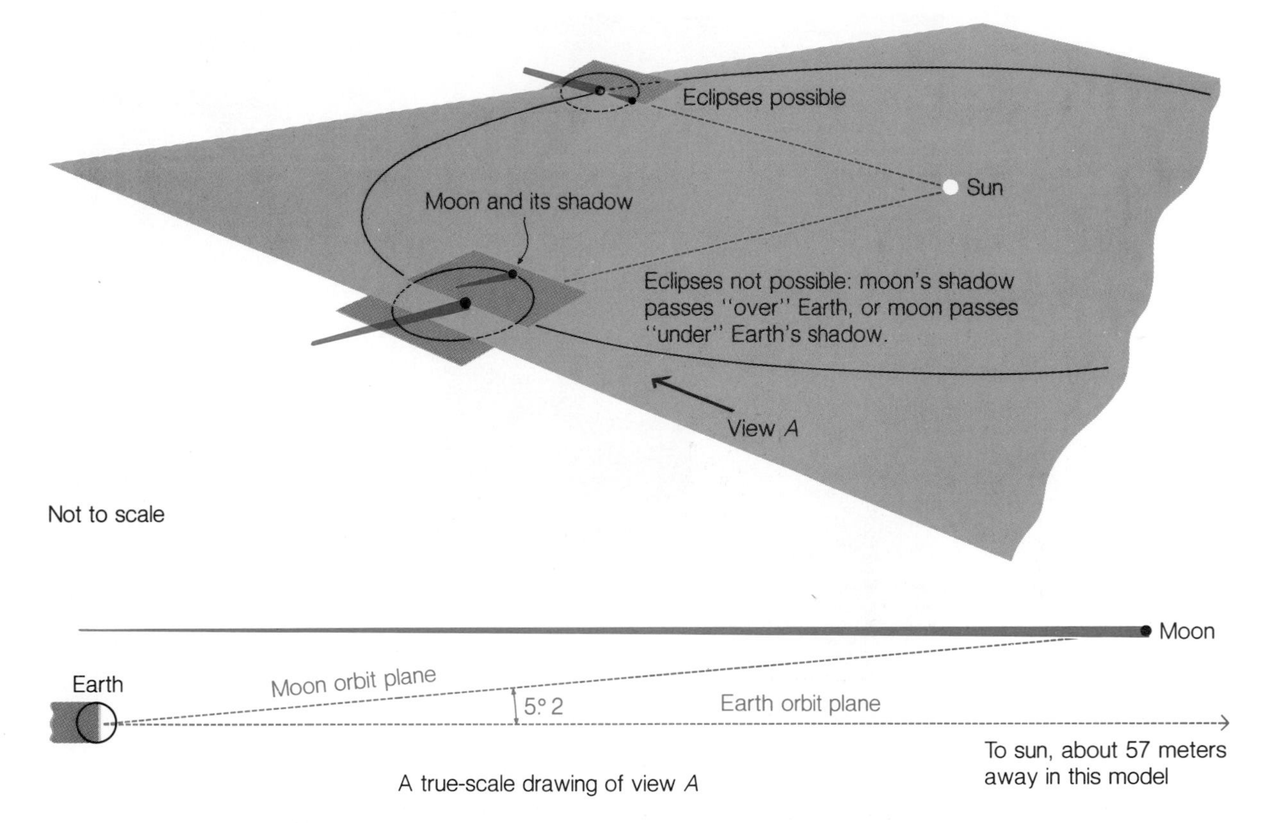

Figure 11-20 Because the moon's orbital plane is tilted by 5°.2 with respect to the Earth's orbital plane, the moon's shadow misses the Earth at most new moons and the Earth's shadow misses the moon at most full moons. Only when the moon happens to be crossing the Earth's orbital plane can an eclipse of the moon occur at full moon or an eclipse of the sun at new moon.

the Earth's, so the Earth's shadow in space is proportionately wider and longer than the moon's.

During an eclipse of the moon, the Earth's atmosphere acts as a lens, bending some light from the sun onto the moon's surface. Since the Earth's atmosphere scatters blue light more easily than red light, the light rays that reach the moon without being scattered are mostly red. The moon therefore usually assumes a dull copper color during the hour or more it spends inside the Earth's shadow as it moves along its orbit. Atmospheric conditions on Earth make some lunar eclipses come closer to the near-total disappearance of the moon than do others.

Solar Eclipses. Eclipses of the sun, perhaps the most majestic of celestial spectacles on Earth, owe their splendor to the remarkable coincidence that the sun, with a diameter just 400 times the moon's, is about 400 times farther

Figure 11-21 During a total eclipse of the sun, the moon completely covers the solar disk, and we can see the sun's outer atmosphere, called the corona, which is usually too faint to be visible in the sun's glare.

from Earth. Therefore the moon and sun have almost the same apparent diameter on the sky. If the moon comes directly between the sun and Earth at the time of new moon, it can just cover the sun's disk, revealing the pearly-white **corona** of blazing gas that surrounds the much more luminous solar disk (Figure 11-21 and Color Plate L).

Annular and Total Solar Eclipses. The moon's orbit around the Earth is elliptical, and its distance from Earth varies from 1/369 to 1/412 of the average Earth-sun distance (which itself varies by just plus or minus 1.7%). At the farther points in its orbits, the moon cannot cover the sun's disk completely, and a small ring of light remains even when the moon is directly in front of the sun. Such **annular eclipses,** or ring eclipses, are almost as frequent as **total eclipses** of the sun, when the moon *does* cover the sun completely, revealing the otherwise invisible corona.

When the moon has reached the closest point in its orbit during the time of a solar eclipse, its shadow on the Earth's surface can be 300 kilometers wide. As the Earth rotates beneath the shadow, a band several thousand kilometers long may fall beneath the shadow, but the maximum fraction of the Earth's surface that can see a given total eclipse of the sun never exceeds 0.5%. Thus a given spot must wait a long time—360 years, on the average—for a total solar eclipse to occur there. The longest duration of totality at any spot barely reaches 8 minutes.

But what minutes they are! No one who has ever witnessed a total eclipse of the sun will forget the feelings of awe that arise as the familiar, bright sun loses first a slice of its disk, then progressively more and more during the hour it takes for the moon to move directly between the sun and the watcher. In the last few seconds before totality, you can see the "Baily's beads," rays of sunlight filtering between the mountains of the moon, followed by the "diamond ring effect" (Color Plate K). Then totality itself reveals the whiteness of the corona, no brighter than the full moon, to replace the glare of day. Birds and insects, having made their songs of dusk, fall silent. The darkness over the land can be perceived to be localized, with sunlight still streaming down many kilometers away. At the sun's rim, the pink and red prominences are visible, perhaps with binoculars. Then the reappearance of the diamond ring and Baily's beads at the opposite rim from that which first covered the sun marks the end of totality and the second half of the partial phases of the eclipse. Figure 11-22 shows where the eclipses of the sun will occur between 1981 and 1995, for those who want to plan their travels in advance to be sure of seeing one of nature's great dramas.

As long as we are looking ahead, we might note that the time will eventually come when no total solar eclipses can occur. The moon's shadow extends 375,000 kilometers behind the moon, but the moon's average distance from Earth equals 384,390 kilometers. Thus already we can have total solar eclipses only when the moon is at one of the closer points in its orbit at the time of eclipse, but the moon is steadily moving farther away from the Earth! Why? The friction of water against the ocean floors, especially in shallow oceans such as the Bering and Irish Seas, gradually causes both a slowing down of the Earth's rotation and an increase in the Earth-moon distance. Since this friction occurs because of tidal motion, the moon itself makes the Earth-moon distance increase, at a rate of a few centimeters per year. In a few hundred million years, the moon will orbit the Earth at too great a distance for any total solar eclipses to occur, so we must enjoy our remarkable coincidence while we have the chance to do so!

Precession

In addition to producing tides and eclipses, the Earth's relatively large satellite has another interesting effect on our planet, called **precession.** Precession, or luni-solar precession to be exact, is a slow wobble of the Earth's rotation axis. Although we can say with accuracy that the rotation

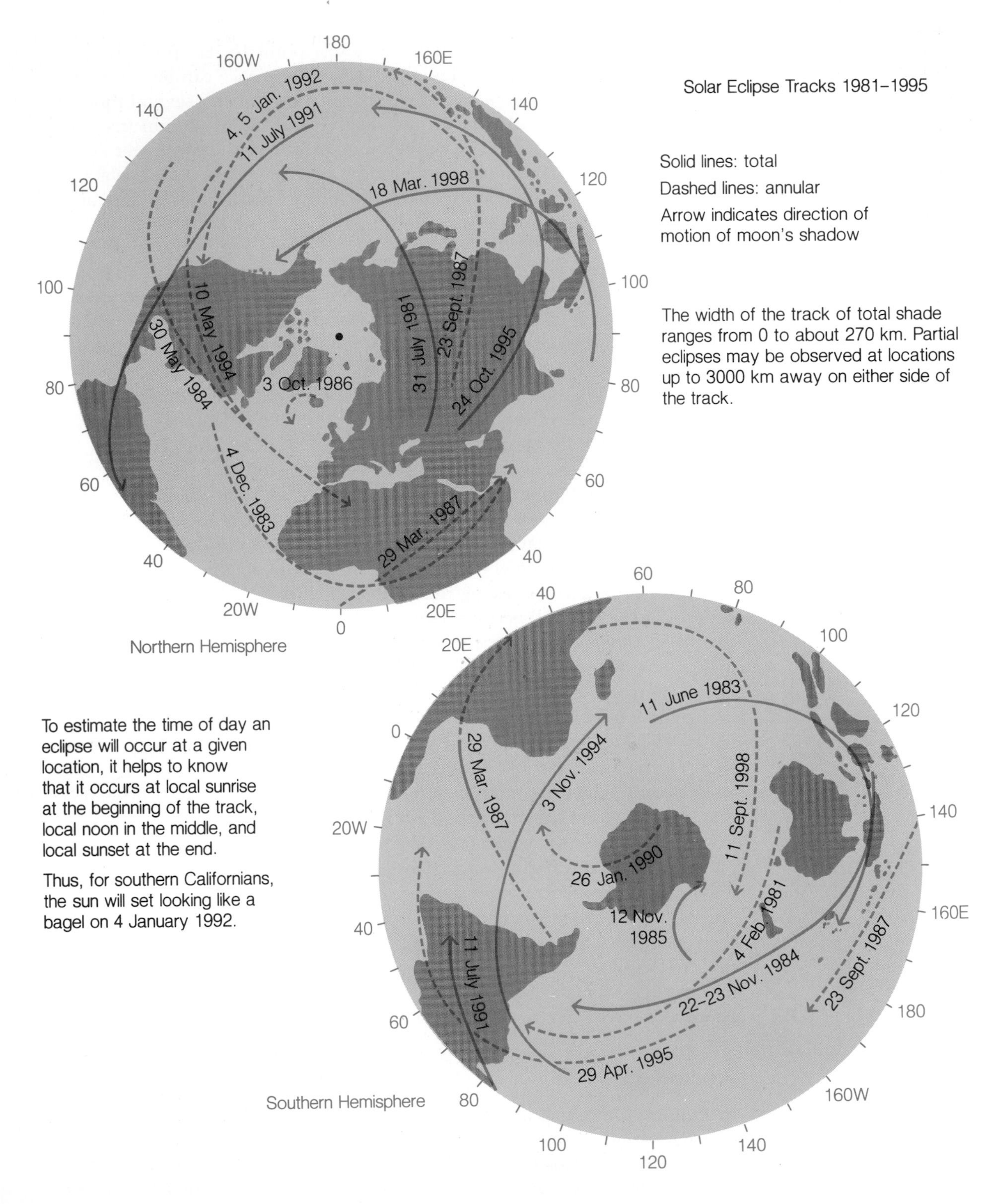

To estimate the time of day an eclipse will occur at a given location, it helps to know that it occurs at local sunrise at the beginning of the track, local noon in the middle, and local sunset at the end.

Thus, for southern Californians, the sun will set looking like a bagel on 4 January 1992.

←
Figure 11-22
Solar eclipses for the years 1981 through 1995 will occur throughout the globe, often at highly inaccessible points. On either side of the total eclipse tracks, for several thousand kilometers, observers can see a partial solar eclipse, when the moon covers only part of the sun's disk.

axis points to the same place night after night, and even year after year, the situation changes when we look at longer time spans. The Earth's axis wobbles in a circle of 23 1/2° radius over a period of 26,000 years (Figure 11-23). This precessional motion changes the astronomical coordinates of the stars on the sky, since the coordinates refer to the points directly above the Earth's north and south poles.

Because of precession of the Earth's axis, our present north star, Polaris, has not always been the north star, nor will it always be in the future. Polaris now lies within 1° of the north pole of the sky, and for a few hundred years, the Earth's precessional motion will move the celestial north pole even closer to Polaris, making Polaris an even better north star. But after that, the celestial pole will drift farther and farther from Polaris. In about 14,000 years, the bright star Vega will lie almost directly above our north pole, not because Vega will have moved, but because of the precession of the Earth's rotation axis. About 12,000 years after that, Polaris will once again serve well as the north star.

The ancient Egyptians, including those who built the famous pyramids almost 4000 years ago, did not enjoy the luxury of having a north star as bright as Polaris. Instead, they had to make do with Thuban, a relatively faint star in the constellation Draco. Thuban was apparently used in the

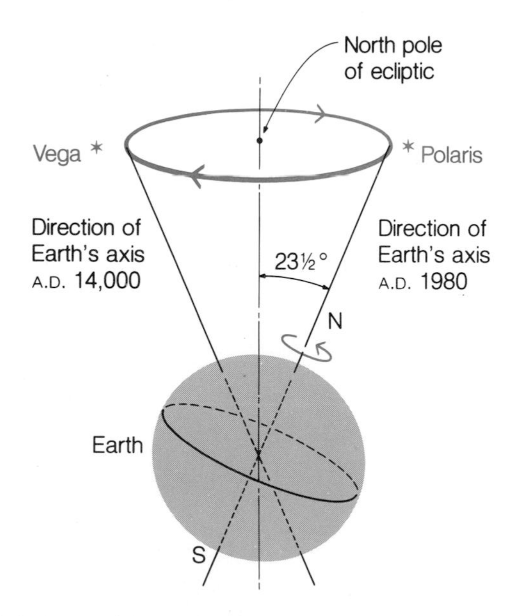

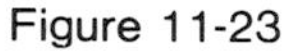
Figure 11-23 Because the Earth bulges at its equator (exaggerated in this diagram), the moon and the sun exert gravitational forces that try to drag the bulge into line with the directions to the sun and to the moon, which lie close to the Earth's obital plane. Instead of this occurring, however, the Earth's rotation axis performs a slow circle in the direction it points in space. The radius of this circle is 23½°, the angle by which the rotation axis tips from being perpendicular to the Earth's orbital plane.

alignment of the Great Pyramid of Khufu. Although Egyptian civilization endured long enough to allow for the discovery of precession, this discovery evidently did not occur until the second century B.C., when the Greek astronomer Hipparchus (page 50) realized that the sun's motion around the ecliptic brought it back to the spring equinox a few minutes earlier each century (Figure 11-24).

The reason the sun returns to the spring equinox a bit earlier than it would without precession is that the wobble of the Earth's rotation axis makes the celestial equator move in space. As the axis points to different places, the points on the sky directly above the equator mark out a different circle on the sky (Figure 11-23). The ecliptic remains the same, since the Earth's orbit around the sun does not really change, but the equator's points of intersection with the ecliptic, the spring and autumn equinoxes, slide backward along the ecliptic by 50 seconds of arc each year. Thus if we use the traditional twelve constellations of the zodiac to delineate the ecliptic (see page 39), the spring equinox eventually moves from one constellation to another. In Hipparchus's time, the spring equinox fell in Aries; now it falls in Pisces and will soon enter the constellation Aquarius.

Since traditional astrology dates from Hipparchus's era, an interesting sidelight of precession is that almost every birth sign given in this century must be "wrong." Most people born when the sun was in the constellation of Pisces think they are "Aries"; most born with the sun in Taurus think they are "Pisces"; and so forth. This egregious error in assigning birth signs would have tremendous importance if astrology had any validity (see page 47).

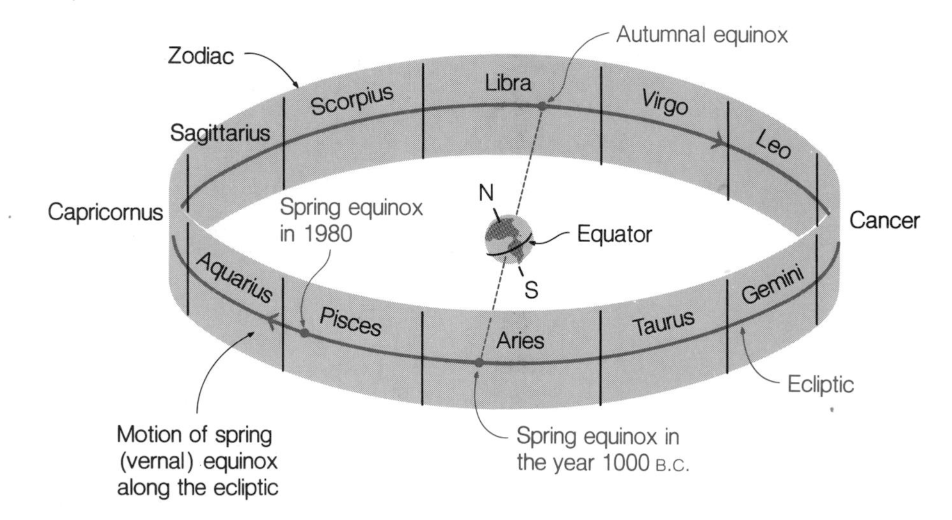

Figure 11-24 As precession occurs, the points above the Earth's equator, which mark out the celestial equator, will not be the same as before. The ecliptic will remain the same, however, since it marks the sun's position in the sky as the Earth orbits, and this orbit does not change. Since the spring and autumn equinoxes mark where the celestial equator intersects the ecliptic, the location of the equinoxes on the ecliptic will slowly change with time, making a complete circle every 26,000 years.

Precession is caused by the sun and the moon, but mostly the moon, just as in the case of the tides. The Earth's rotation makes the Earth bulge slightly at the equator. The gravitational forces from the sun and the moon constantly try to make the Earth stand "upright," with its equatorial bulge in the plane of the sun and the moon. (We can ignore the 5° inclination of the moon's orbit to the ecliptic plane.) Instead of assuming an "upright" position for its rotation axis, the Earth instead wobbles, much as a spinning top wobbles instead of falling straight over. So long as the Earth keeps rotating, it will continue the precession of its rotation axis, and the north celestial pole will continue to trace out a circle on the sky.

Astronomers who set their telescopes by catalogued star positions must allow for precession. Even a few months can change the stars' positions by a noticeable amount (if you have a large telescope). Thus every star catalog must have a designated "epoch," such as January 1, 1950, from which time the precessional changes in position can be reckoned.

11.7 The Giant Planets

In sharp distinction to the hard, rocky spheroids of the inner solar system, the four giant planets are gaseous, diffuse, and rich in the hydrogen and helium that characterize the composition of stars. These four giants formed with enough mass, and at a great enough distance from the sun, that they managed to retain most of their primordial gas and dust through the 4.6 billion years that separate their formation from the present. We may think of these giant planets as miniature, never-to-be stars that failed to begin thermonuclear fusion because their masses did not suffice to raise their central temperatures to the necessary level (Figure 11-25). The four inner planets, by contrast, represent the denuded cores of material left behind when the vast bulk of hydrogen and helium that once formed their protoplanets escaped as the sun began to shine. The most direct evidence that the giant planets consist mainly of hydrogen and helium comes from their low densities. Further confirmation comes from spectroscopic examination of the planets' outer layers, which provides evidence of common hydrogen compounds such as methane (CH_4) and ammonia (NH_3).

Chapter 13 presents a more detailed survey of human investigation of the giant planets. For now, however, we should remind ourselves that to an unprejudiced observer, the giant planets would *be* the sun's family, with the inner planets and Pluto, as well as the comets, asteroids, and meteoroids, noticeable only upon much closer inspection. To put things another way, the five most massive bodies in the solar system—the sun, Jupiter, Saturn, Uranus, and Neptune—all have about the same composition, the relic of their original process of condensation through self-gravitation. Less massive objects, such as Mercury, Venus, Earth, and Mars, have a more complicated history that includes the loss of most of their original material, resulting in their consequent minuscule masses in comparison with the giant planets.

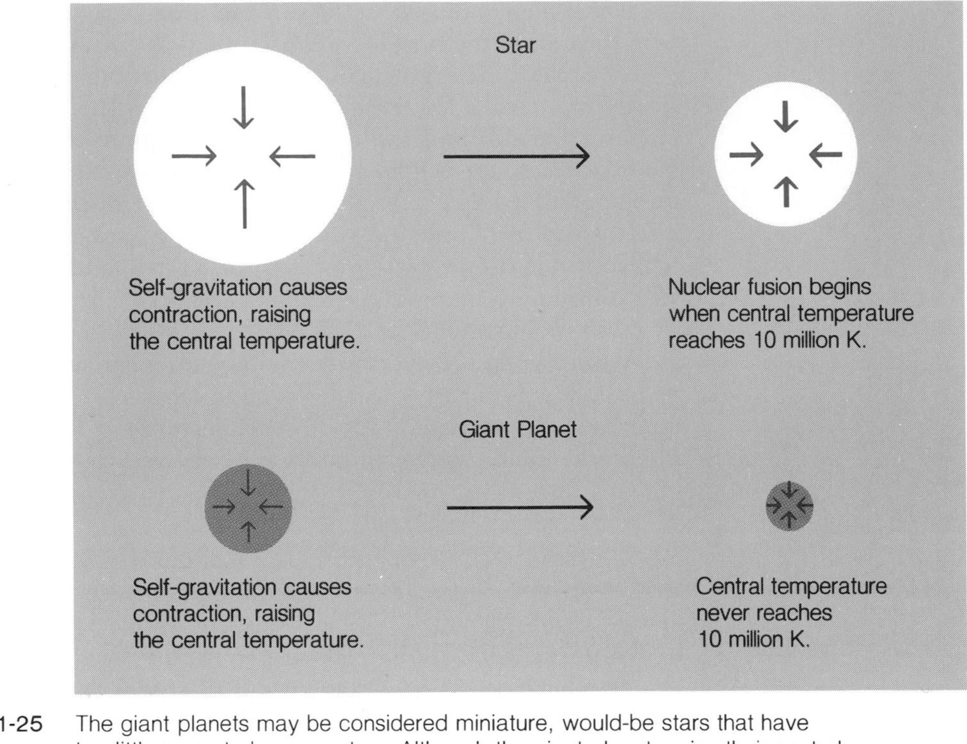

Figure 11-25 The giant planets may be considered miniature, would-be stars that have too little mass to become stars. Although the giant planets raise their central temperatures by contracting their interiors, they never achieve a temperature high enough for nuclear fusion to begin.

11.8 Comets: Frozen History

The most primitive, least altered objects in the solar system are the **comets,** small lumps of material that may have condensed directly from interstellar gas and dust 4.6 billion years ago. The sun has a family of many millions, perhaps billions, of comets, orbiting the sun at great distances, far beyond the realm of the planets (Figure 11-26). Once in a while, mutual interaction among the comets deflects one of these cosmic relics into an orbit that comes close to the sun. The fact that these highly elongated orbits show no particular concentration toward the disk of the solar system implies that comets formed before the proto-solar system had assumed a disklike shape.

The Composition of Comets

Current theories of comets, based on limited though increasingly detailed spectroscopic observations, suggest that a comet may be thought of as a huge, dirty snowball, with the mass of a large mountain and a diameter of a few kilometers. The "snow" consists of ordinary water ice, plus frozen methane, ammonia, carbon dioxide, and other as yet undetermined molecules,

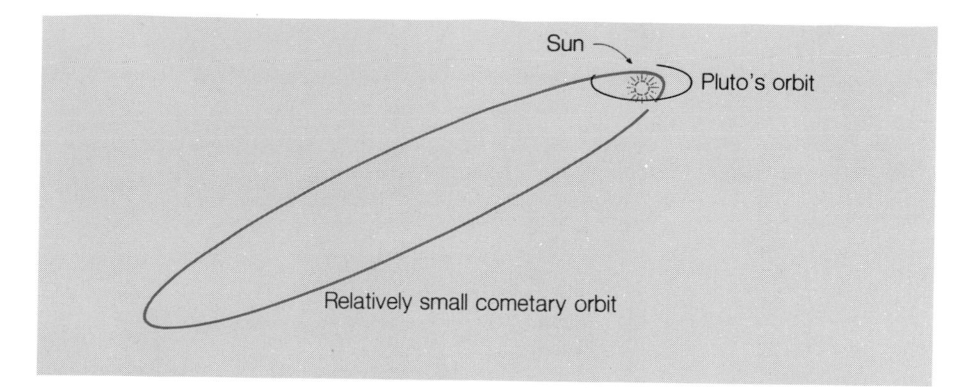

Figure 11-26 The sun's millions, or even billions, of comets typically have orbits that carry the comets to hundreds or thousands of times the distance of Pluto from the sun. These orbits are inclined in all directions to the plane of the planets' orbits.

which may include more complex compounds such as formaldehyde and cyanoacetylene. The "dirt" particles in the comet are grains of rocky material of various sizes. These grains have apparently not been subjected to melting or to any other transformation since the comet formed, for they are still imbedded in the cometary ices. Because of this lack of chemical processing, comets probably represent almost pristine samples of the original material from which the solar system formed. It is therefore not surprising that astronomers would dearly love to get their hands on such a comet.

Close Approach of Comets to the Sun

Since most comets spend most of their lives at immense distances from the sun, they remain in a cosmic deep freeze, only a few degrees above absolute zero. If a comet is deflected into an orbit that comes close to the sun, nothing much happens to the snowball until it reaches a distance to the sun comparable to Jupiter's. At this point, some of the ice in the comet begins to vaporize, and the gas and dust released from the *nucleus*, or original snowball, spread out to form a fuzzy envelope, or **coma,** around the nucleus (Figure 11-27).

As the comet comes still closer to the sun, more and more gas and dust separate from the nucleus and trail away from it. The comet acquires a spectacular *tail*, often millions of kilometers long, always pointing away from the sun (Figure 11-27). Some comets have two tails, one made of gas and one made of dust, and occasionally comets have several tails of each kind. Astronomers have seen some comets break apart as they pass close to the sun, destroyed by the tidal forces exerted by the sun's gravitation. On occasion, the gravitational pull from one of the giant planets can deflect the comet into a new orbit much smaller than the one along which it was originally traveling toward the sun. The resultant **short-period comets,** dozens of which are known, take only a few years to orbit the sun, in comparison with the millions of years for the original long-period comets.

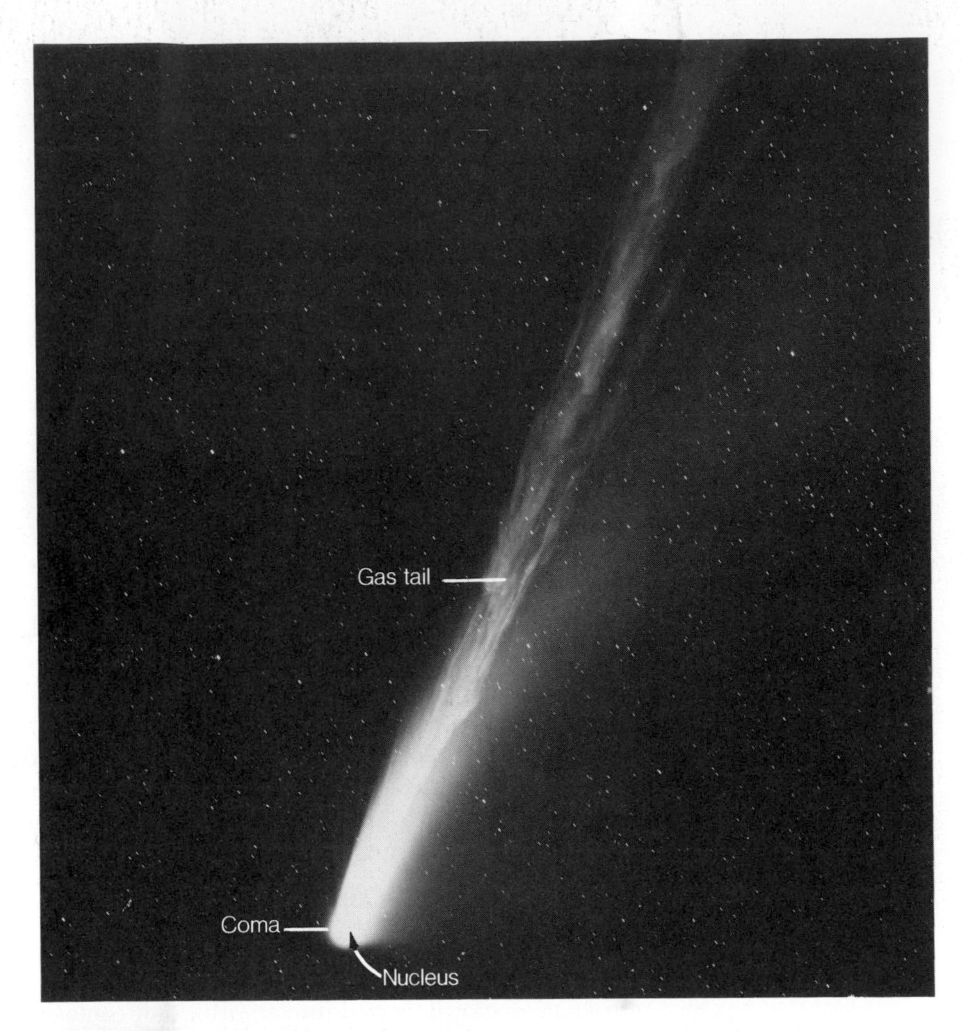

Figure 11-27 When a comet, such as Comet Mrkos shown here, comes close to the sun, heating of the nucleus produces a fuzzy coma and a far more diffuse tail. Particles in the solar wind push on the gas in the tail, and sunlight radiation pressure pushes on the dust, often producing two or more tails for a single comet, which stretch for millions of kilometers away from the sun.

Halley's Comet

The best-known of all comets, Halley's comet, has an orbit larger than most of the short-period comets but far smaller than those of the long-period comets. This orbit carries Halley's comet out past the orbit of Neptune and brings it close by the sun and the Earth every 76 years (Figure 11-28). The repeated visits of this comet, which were first recognized to be evidence of the *same* comet by Edmund Halley in 1682, have often been taken as forerunners of tragedy. For example, the Bayeux tapestry shows the return of Halley's comet in 1066 as an omen of King Harold's defeat by William the Conqueror that year (Figure 11-29). Similarly, the return of the comet in 1453 was seen as a prediction of the downfall of Constantinople.

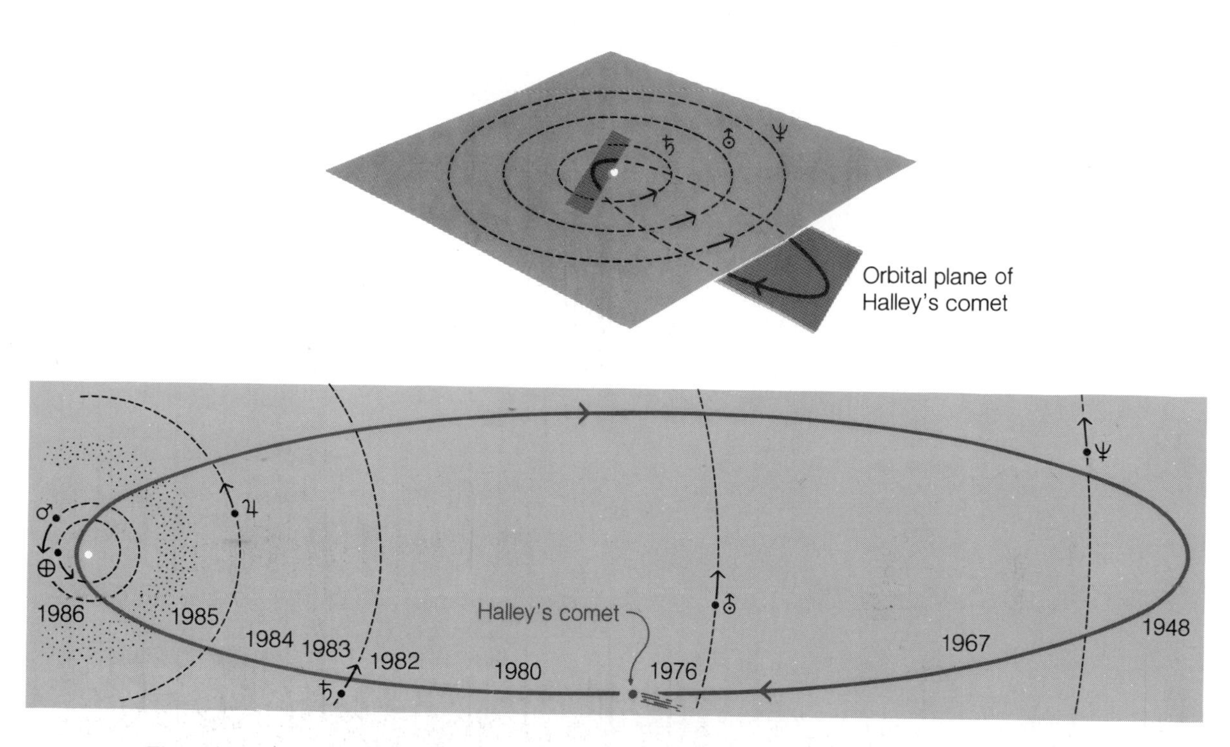

Figure 11-28 The orbit of Halley's comet, which is highly inclined to the planets' orbital planes, carries the comet past the orbit of Neptune and inside the orbit of Venus. In a few years, Halley's comet will again be in our part of the solar system, but not so spectacularly visible as it was during its last close encounter with the Earth, in 1910.

Figure 11-29 The Bayeux tapestry, which commemorates the Norman invasion of England, shows the English king and his subjects in terror of Halley's comet, supposedly an omen of the coming disaster.

Table 11-3 Molecules Detected in Comets

Coma			Tail	
HCN[a]	CN	NH_2	H_2O^+	N_2^+
CH_3CN	CH	C_3	CO_2^+	CO^+
H_2O[a]	OH			CH^+
CO_2	CO			OH^+
NH_3	NH			
	C_2			
	CS			

[a]These identifications require further verification.

[b]Suggested parent molecules of less complex molecules such as CO and NH, but not yet directly detected.

When Halley's comet came near the Earth in 1910, the Earth actually passed through the tail of the comet. Though the tail of a comet is far more rarefied than the best vacuum we can make on Earth, and shines so impressively only because of its length and the reflecting power of the particles within it, many humans feared what might happen within the tail of Halley's comet. The wiser among the populace went unperturbed, and the canniest made money by selling "anti-comet pills."

Do Comets "Seed" Planets?

From spectroscopic analyses of Halley's comet, of short-period comets, and of long-term comets that venture by the sun, astronomers have found a surprisingly large number of relatively complex molecules in the gases released by these primitive snowballs. Table 11-3 lists the molecules found so far. There exists a noticeable, though hardly unexpected, correlation with the molecules found in the interstellar medium. The discovery that comets contain water, carbon monoxide, hydrogen cyanide, and methyl cyanide strengthens the theory that they condensed from interstellar gas and dust.

If still more complex molecules exist in comets, then we might expect that these cosmic snowballs contain amino acids and similar organic compounds. Comets then might have played a significant role in bringing organic molecules to the protoplanets as they formed, or even to planetary surfaces. In that case, we would expect that other planetary systems would have undergone a similar "seeding," since astronomers consider the sun's family of comets to be entirely representative of the (as yet unknown) comets that circle other stars.[4]

[4]Two British astronomers, Fred Hoyle and Chandra Wickramasinghe, have suggested not only that comets carry organic molecules, but also that life itself may well exist in comets, and that repeated visits by comets have caused plagues and other catastrophes on Earth. Though this theory provides a natural explanation of the superstition that comets are harbingers of evil, the weight of scientific opinion holds that life would have an extremely difficult time developing in comets.

11.9 Asteroids and Meteoroids

The comets represent the oldest, most wide-ranging debris of the original solar system. Next oldest and next least altered since the time of formation come the **asteroids** and **meteoroids,** lumps of rock, sometimes containing iron and nickel, that have circled the sun basically unchanged during the past 4.6 billion years.

Asteroids

Asteroids represent the remains of a would-be planet that apparently could never form because Jupiter's gravitational pull disrupted its attempt to collect in one place. The orbits of the asteroids, usually much more elliptical than the planets' orbits, lie mostly between the orbits of Mars and Jupiter. The continuing effects of Jupiter's gravitational force have made certain regions of the **asteroid belt** almost empty of asteroids. There is an absence of asteroids, for example, whose orbits have semi-major axes equal to simple fractions of Jupiter's (Figure 11-30).

The largest asteroid, Ceres, has a diameter of 750 kilometers, just one-fifth of the moon's diameter. The next largest, Pallas and Vesta, have diameters of 500 kilometers. Thousands of asteroids, most with diameters of only a few kilometers, have been discovered so far. Millions more, with diameters measured in meters or tens of meters, are assumed to exist.

A special group of asteroids, from our point of view, consists of those whose orbits cross the Earth's orbit. These rare asteroids have highly elliptical orbits, so they approach closer to the sun than the Earth does at their perihelion points. A few have orbital semi-major axes less than the Earth's. None of these asteroids has a diameter of more than 1.4 kilometers. The largest of them, Icarus, has been observed to approach within a million kilometers of the Earth and could conceivably someday collide with our planet. The Earth would still survive, since it has a trillion times the mass of Icarus, but the impact could be devastating to large portions of the planet's surface.

Meteoroids

Similar to asteroids, and distinguished only because many of them *do* reach the Earth, are the meteoroids, interplanetary wanderers that range in size from a few hundred meters across to a tiny fraction of a millimeter. Meteoroids have significantly elliptical orbits around the sun, whereas the Earth's orbit is nearly a perfect circle. Therefore a meteoroid can have a large velocity with respect to the Earth, as much as 40 kilometers per second, at the point where the orbits intersect (Figure 11-31). This large velocity creates a tremendous frictional heating in the Earth's atmosphere that consumes any small meteoroid, turning it into a **meteor,** or **shooting star** (Figure 11-32). If the meteoroid begins with the mass of a small house, it can survive its frictional passage through the atmosphere, and its remnant will reach the Earth's surface as a **meteorite.**

Every day, hundreds of tons of meteoritic debris reach the top of the Earth's atmosphere, to be consumed almost without exception by atmospheric friction. On any clear night, you can see several meteors per hour, which rank among the largest of the millions of meteoroids to reach the

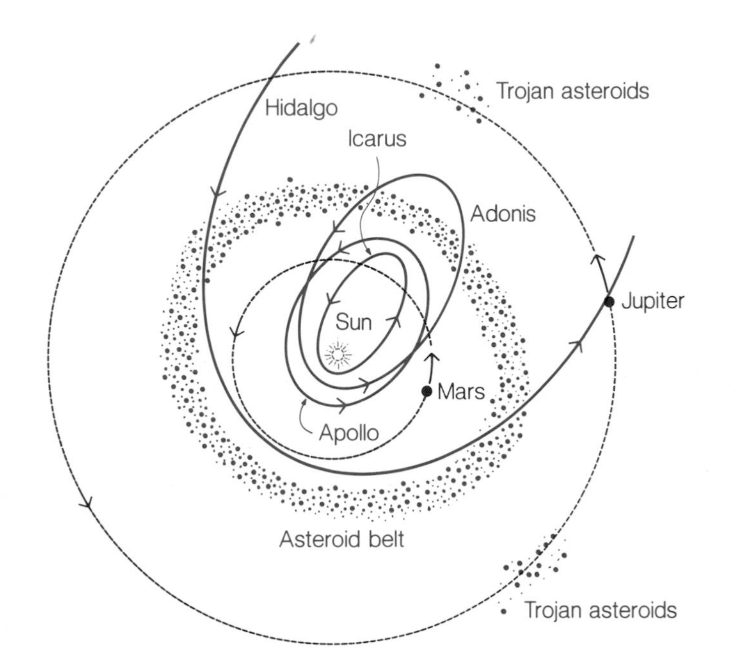

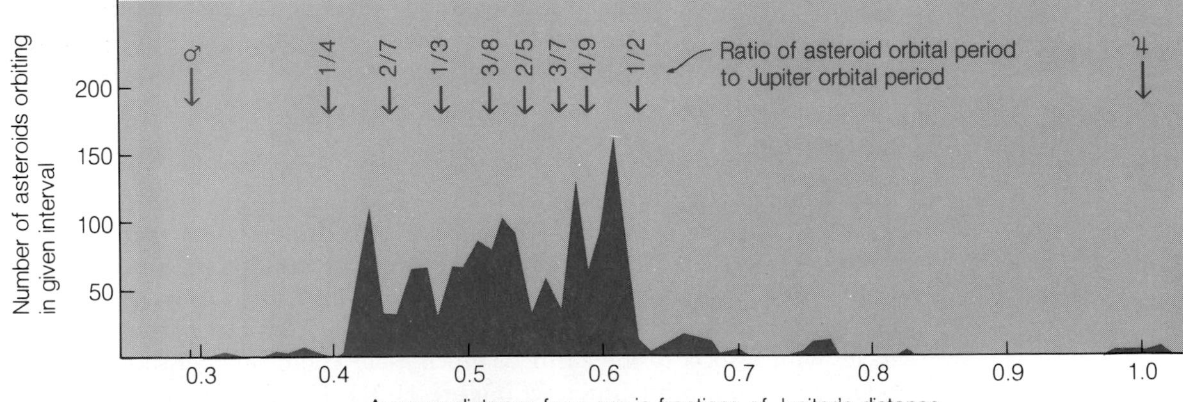

Figure 11-30 Most of the asteroids have somewhat elongated orbits that keep them in the asteroid belt between Mars and Jupiter. A few have orbits that carry them unusually close to the sun or unusually far from it. Jupiter's gravitational forces have produced an absence of asteroids with orbital periods equal to a simple fraction of Jupiter's orbital period.

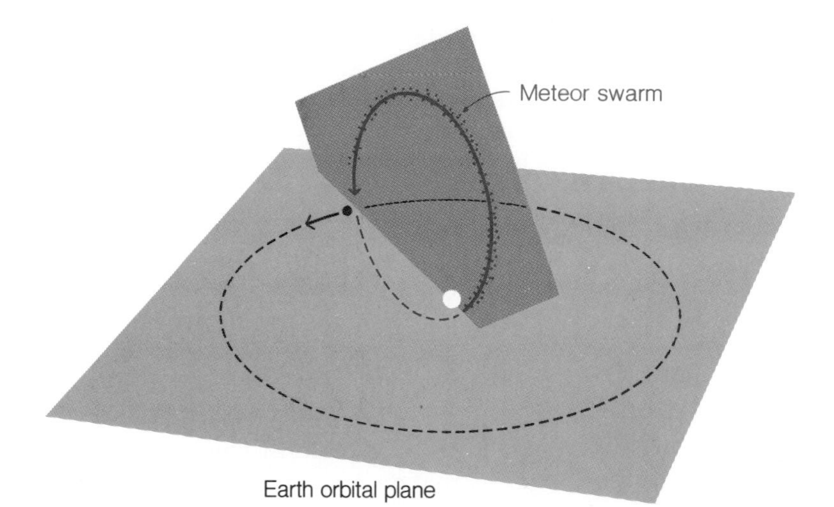

Figure 11-31 Meteoroids move in elongated orbits, often highly inclined with respect to the Earth's orbital plane. Because their orbits differ from the Earth's, the meteoroids can have a large velocity relative to the Earth when they intersect the Earth's orbit. This relative velocity, which is increased by the Earth's gravitational force on the meteoroids, produces great frictional heating on the meteoroids when they enter the Earth's atmosphere.

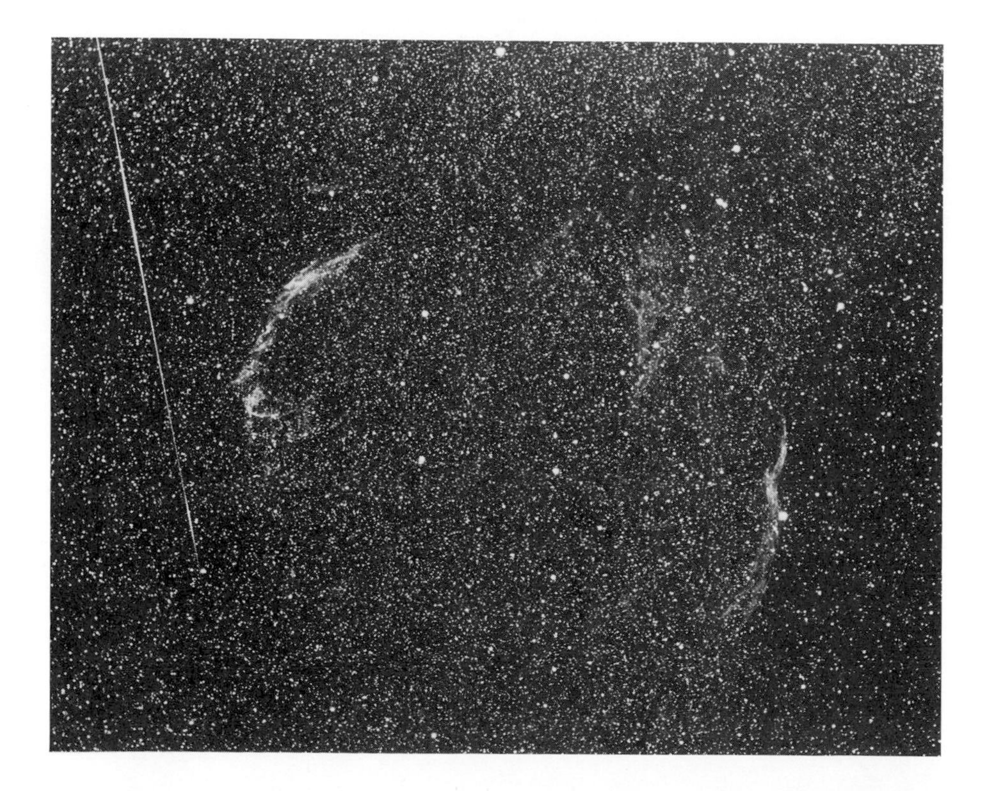

Figure 11-32
Meteoroids heated by friction in the Earth's atmosphere produce meteors, or shooting stars, as they burn.

Earth that night. Occasionally, an extremely large meteoroid will flash out as a bright "fireball" that leaves a trail for several seconds before dissipating in the atmosphere, tens of kilometers above the Earth's surface. Meteors often look as though they are headed just over the next hill (if there are hills on the horizon) and have fooled many observers who were positive that a meteorite had landed "just half a mile down the road."

Meteorites and Meteor Craters. Of course, some meteoroids *do* reach the ground. The heaviest meteorite yet recovered and brought back to civilization weighs 35 tons and is on display in New York. A woman in Alabama was once struck by a meteorite that came crashing through the roof of her mobile home, but fortunately she recovered. Still more impressive impacts left behind the Meteor Crater in Arizona, more than a kilometer wide (Figure 11-33), and a basin in Quebec 65 kilometers wide. A spectacular near miss occurred on August 10, 1972, when a meteoroid burned its way through the skies of Wyoming, missing the surface of the Earth by just 58 kilometers (Figure 11-34). Although this object had a diameter of only 4 meters, it had a mass of 1000 tons and would have made an impressive crater had it hit the Earth. Perhaps a similar object, or the nucleus of a comet, did hit the Earth in Siberia in 1908. The impact felled trees for miles around and has often been cited as evidence of the explosion of the atomic drive of extraterrestrial visitors to Earth, on grounds more economic than scientific.

All of the meteor craters on Earth are insignificant when compared with the surfaces of Mercury, Venus, Mars, and the moon, which are dotted with hundreds of thousands of craters (Figure 12-9). Most of these craters were

Figure 11-33 The crater near Winslow, Arizona, was made by a meteoroid that largely vaporized upon impact about 20,000 years ago. Only small meteoritic fragments have been found by boring in and around the crater.

Figure 11-34 The meteoroid that almost hit the Earth near Wyoming made a trail bright enough to be seen in daytime in Grand Teton National Park.

made within a few hundred million years after the solar system formed. During this epoch, meteoroids by the millions must have rained down on all the inner planets. Although our own planet has only a few visible impact basins, the Earth did not escape this bombardment. The first few hundred million years of our geological record, which would reveal the effects of the bombardment, have vanished because of erosion and the movement of the crustal plates.

Classification of Meteorites. From an examination of all the meteorites found on Earth, scientists have created a general classification scheme based on the mineralogical and chemical composition of these objects. The majority of meteorites are stony, basically lumps of rock; a minority are stony-iron, with metal-rich inclusions; and a few are made mostly of iron, nickel, and other metals. Radioactive dating of meteorites gives a maximum age of 4.6 billion years, the age of the solar system itself. Most interesting of all the meteorites are those that form a subclass of stony meteorites called **chondrites.** These contain rounded inclusions, called *chondrules,* which are noticeably distinct from the rest of the material. Of the chondrites, the most unchanged, and hence the most significant for reconstructing the early years of the solar system, are the **carbonaceous chondrites,** in which as much as 5% of the meteorite's mass consists of various types of carbon compounds. Of all the meteorites, these objects show the least amount of modification by heating, and are the least altered by the passage of time.

Within the class of carbonaceous chondrites, the most primitive examples, the Type I, contain the highest percentage of carbon, nitrogen, and water of all meteorites. Some astronomers believe that the Type I carbonaceous chondrites may be fragments of old comets, rather than debris from within the asteroid belt. But the most exciting thing about this type of meteorite is that at least two representatives, the Murchison meteorite and the Murray meteorite, have been found to contain amino acids, the basic building blocks of protein molecules. This intriguing discovery suggests that at least certain smallish molecules quite important to life on Earth may have formed in interstellar clouds.

Observation of Meteors. If you would like to observe for yourself some of these interplanetary wanderers as they flame out in one last moment of glory, the best time to do so is after midnight if the moon is down. The meteoroids that reach your vicinity at that time tend to have head-on relative velocities, rather than the come-from-behind motions needed to appear before midnight (Figure 11-35). Then the rotation of the Earth tends to give an additional relative velocity, an additional frictional heating, and additional brightness to the meteor, in comparison with the same meteoroid's passage before midnight.

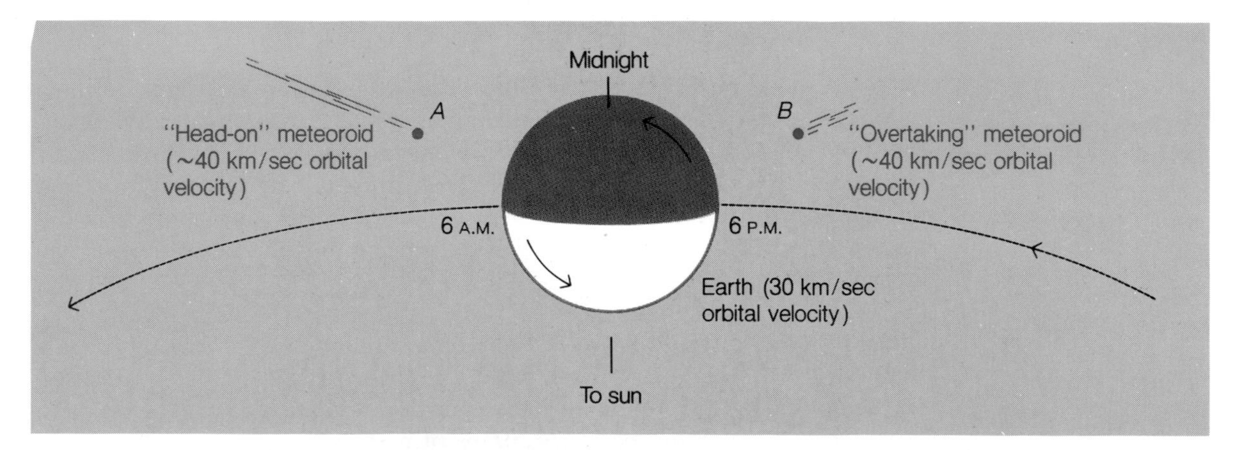

Figure 11-35 Meteoroids that reach the atmosphere above an observer before midnight tend to have to *overtake* the Earth in its orbit. Those that are seen after midnight usually undergo head-on collisions with the Earth. The head-on collisions have greater relative velocities than the overtaking collisions, so a given meteoroid will produce a brighter meteor trail if it appears after midnight than before.

Table 11-4 Principal Meteor Showers

Meteor Swarm	Normal Period of Visibility	Night of Maximum Visibility	Average Number of Meteors Visible per Hour
Quadrantids	Jan. 2–4	Jan. 3	30
Lyrids	Apr. 20–22	Apr. 22	8
Eta Aquarids	May 2–7	May 4	10
Delta Aquarids	July 20–Aug. 14	July 30	15
Perseids	July 29–Aug. 18	Aug. 12	40
Draconids	Oct. 10	Oct. 10	
Orionids	Oct. 17–24	Oct. 21	15
Taurids	Oct. 20–Nov. 25	Nov. 4	8
Leonids	Nov. 14–19	Nov. 16	6
Andromedids	Nov. 15–Dec. 6	Nov. 20	
Geminids	Dec. 8–15	Dec. 13	50
Ursids	Dec. 19–23	Dec. 22	12

The best nights of all for observing meteors are the times of **meteor showers,** when especially large numbers of meteoroids reach the Earth. Meteoroids orbit the sun preferentially in bunches, or "swarms," that intersect the Earth's orbit. On the night or nights when the Earth reaches these intersection points, 10 or 100 times the usual number of meteors may be observed. Table 11-4 lists the best-observed meteor showers, which may vary from year to year in impressiveness. But the Perseids are always fine—especially if you go far from city lights. Late at night in the mountains, the skies may seem to teem with meteors, as the Perseids fall almost more rapidly than you can count them.

Summary

The nine known planets of our solar system orbit the sun in nearly circular paths, all in the same direction and in almost the same plane. The orbits apparently owe their regularities to the way that the planets formed from the contracting, rotating cloud of gas and dust that became the protosun and the protoplanets 4.6 billion years ago.

The four inner planets, Mercury, Venus, Earth, and Mars, are much smaller and denser than the four giant planets, Jupiter, Saturn, Uranus, and Neptune. The giant planets consist primarily of hydrogen and helium, while the inner planets have little of these two lightest elements and are formed mainly from oxygen, silicon, magnesium, and iron. These differences in composition arose when the sun began to shine, making the inner protoplanets incapable of retaining hydrogen and helium. The inner planets possess a differentiated structure: Their iron-rich cores consist of preferentially heavier elements than their outer, silicon-oxygen-rich mantles. Pluto, the farthest planet, is basically a modest lump of ice.

Radioactive minerals in the Earth, and presumably in Venus and Mars as well, release enough heat to drive the sea-floor spreading that causes the crustal plates to move and collide. Such plate-tectonic activity on Earth has constantly re-formed the continents and has dragged the first billion years of the geological record down into the Earth's mantle.

Our Earth has a satellite that possesses a significant fraction (though only 1/81) of its planet's mass. The moon, together with the sun, produces tides in the oceans and leads to the spectacle of lunar and solar eclipses. Furthermore, the luni-solar precession, or slow wobbling of the Earth's rotation axis, has shifted the location of the spring and autumn equinoxes on the sky, so that the method of recording birthdays by "sun signs" has become incorrect in nomenclature as well as in presumed effects.

Comets, meteoroids, and asteroids are small lumps of material left over after the formation of the sun and planets. The far-ranging comets contain the most primitive material from which the solar system formed: methane, ammonia, and water ice, frozen around lumps of rock and smaller dust particles. Close passage by the sun vaporizes some of the cometary nucleus, making a gauzy coma and a long, rarefied tail that streams behind the nucleus for millions of kilometers. Meteoroids, which seem to be small asteroids, orbit the sun in trajectories that cause some to intersect the Earth's orbit. Frictional heating in our atmosphere then vaporizes most or all of the object to make a meteor or "shooting star." Large meteoroids survive this heating to reach the Earth's surface as meteorites. An important though rare class of meteorites, the carbonaceous chondrites, have part of their mass in the form of carbon compounds, quite unlike other meteorites, which are made of rock, iron, or a combination of rock and metal.

Key Terms

annular eclipse
asteroid belt
asteroids
carbonaceous chondrite
coma
comets
chondrites
continental drift
differentiation
giant planets
inner planets
law of universal gravitation
lunar eclipse (eclipse of the moon)
meteor
meteorite
meteoroid
meteor shower
momentum
neap tides
plate tectonics
precession
primitive atmosphere
protoplanet
shooting star
short-period comet
solar eclipse (eclipse of the sun)
solar system
spring tides
tides
total eclipse

Questions

1. Why do we think that comets represent the most primitive objects in the solar system?
2. We know that all the planets orbit the sun in the same direction and in almost the same plane. What does this imply about the formation of the sun's planets?
3. Why do the giant planets consist mostly of hydrogen and helium, whereas the inner planets consist mostly of oxygen, iron, silicon, and magnesium?
4. What is a shooting star? Why do most of them never reach the Earth after entering our atmosphere?
5. Are the asteroids remnants of a planet that exploded? What rival theory to explain the asteroids seems more probable?
6. What causes plate-tectonic activity on Earth? What are the visible results of the slow motions of the continental plates?
7. What causes the tides? Why does high tide occur at different times of day? Why does high tide rise higher on some days than on others?
8. Why are total eclipses of the sun visible from only a small fraction of the Earth's surface?
9. What is precession? Why has precession made most astrological pronouncements "out of date"?
10. What happens to a comet when it comes close to the sun? What do we see as a result?
11. How old are the oldest meteorites found on Earth? How does this compare with the age of the solar system and with the age of the oldest rocks found on Earth?
12. What causes a meteor shower? Why do such showers recur on nearly the same day of the year, year after year?

Projects

1. *A Model of the Solar System.* Consider the feasibility of making a scale model of the solar system. The chief difficulty is the enormous range of sizes and distances involved. If a tennis ball (with a diameter of 7 centimeters) represents the sun, then Jupiter will have a diameter of 7 millimeters, and Earth only 0.7 millimeter. The distance from the sun to the Earth should be 7.5 meters, and the distance from the sun to Jupiter should be 39 meters. See Tables 11-1 and 11-2 for other relevant data.
2. *Densities.* Determine the densities of a common rock and of a piece of metal by weighing each object and comparing this weight with the weight of an equal volume of water. The latter can be obtained by submerging the object in a pan of water that has been carefully filled to the brim, and which sits in a larger pan to catch the overflow when the object is submerged. How do the densities of the rock and metal compare with the average density of the Earth?

3. *The Time of High and Low Tides.* If you live near the ocean, determine the time of high tide and low tide over the course of several days, either by direct observation or by reading these times from the tables published in daily newspapers. Compare these times with the times when the moon is most directly overhead or most nearly underfoot. The latter can be found by observation (and interpolation) or by consulting the *American Ephemeris and Nautical Almanac.* The first method is easier.
4. *Newton's Laws of Motion.* Balance yourself on a skateboard and consider Newton's first law, which says that an object at rest will remain at rest so long as no net force acts upon it. What force counteracts the Earth's gravitational pull downward upon you? Now try to accelerate yourself in a horizontal direction by throwing an object, such as a tennis ball, in a given direction. (If you feel unable to do this on a skateboard, find a volunteer and study that person.) In what direction does the skateboard move? How can you explain this, using Newton's second and third laws of motion? How does the acceleration of the tennis ball compare to the acceleration of the person on the skateboard? Why?
5. *Astrology.* Find someone who thinks that astrology has some merit as a general indicator of personalities, and ask this person whether the discovery that his or her birth sign was incorrect would have any effect on that person's belief in astrology. Does the discovery that your own birth sign may be incorrect have any effect on your own attitude? Since the spring equinox has now moved 95% of the way through Pisces toward Aquarius, you can determine which days still have the same birth sign as they did when current astrological beliefs became standardized. Which dates are they?

Further Reading

Dole, Stephen. 1974. *Habitable Planets for Man.* 2d ed. New York: Elsevier.

Press, F., and Siever, R., eds. 1974. *Planet Earth.* San Francisco: W. H. Freeman.

Schramm, David, and Clayton, Robert. 1978. "Did a Supernova Trigger the Formation of the Solar System?" *Scientific American* 239:4,124.

Scientific American. 1975. *The Solar System.* San Francisco: W. H. Freeman.

Whipple, F. L. 1968. *Earth, Moon, and Planets.* Cambridge: Harvard University Press.

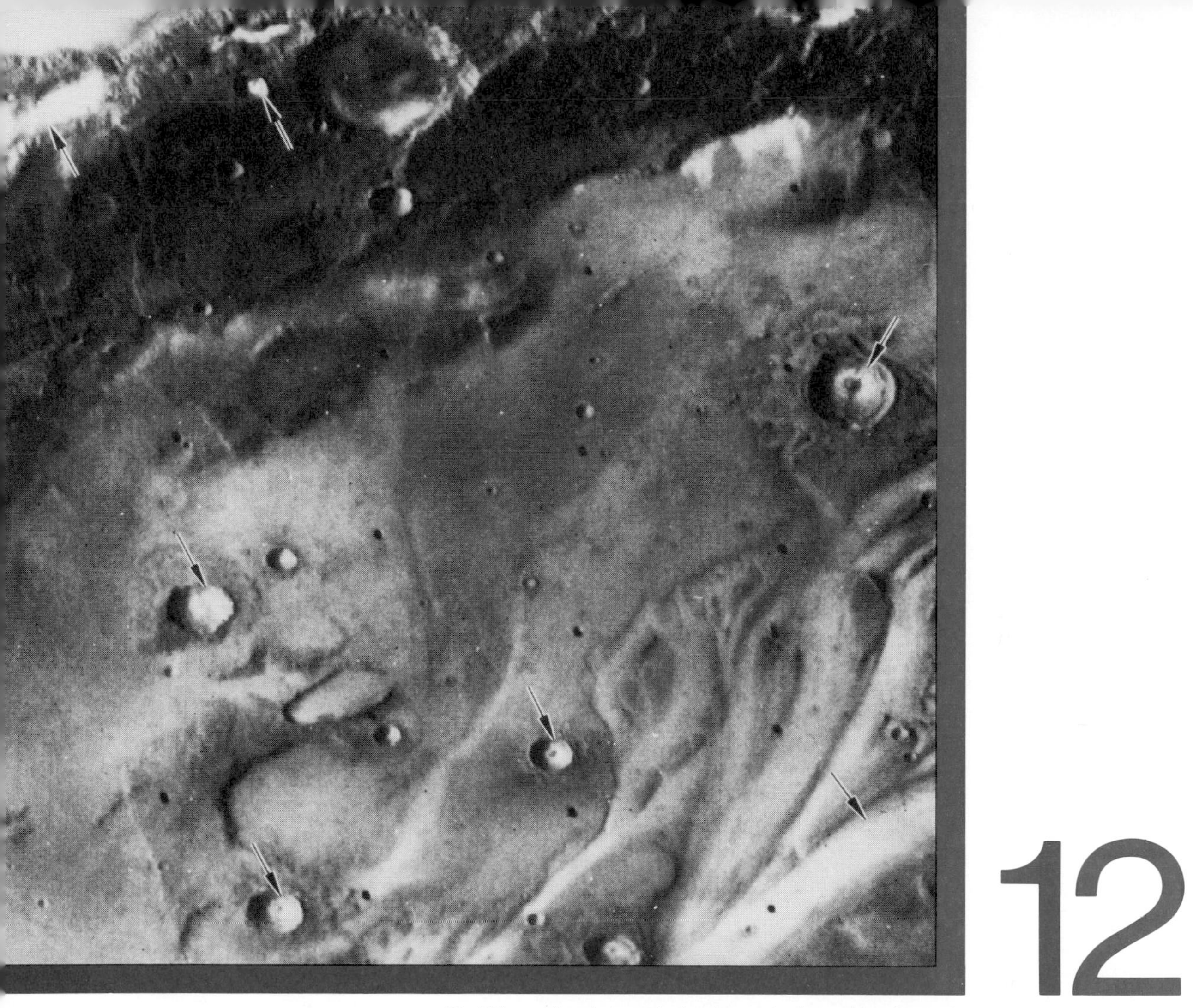

12

The Viking orbiter photographed early-morning fog on Mars (arrows).

The Age of Exploration: The Inner Solar System

Since time immemorial, humans have looked at the skies and have noticed that certain points of light change position with respect to the stars. The Greeks named these points *planets*, meaning "wanderers," and astrologers claimed that their motions influenced the rise and fall of kingdoms and even the fate of individual lives. Generations of astronomers charted the planets' movements and invented and refined the Ptolemaic, Tychonic, and Copernican models of the solar system to explain the planets' paths in the sky (Figure 12-1). Although we have known the reason for the planets' *motions*

Figure 12-1 This engraving from Doppelmayer's *Atlas Novus Coelestis* (*New Atlas of the Heavens*), published in 1742, shows Urania, the muse of astronomy, with maps of the Ptolemaic model of the solar system (left), the Tychonic model (center), and the (correct) Copernican model (right).

for 400 years (see page 55), our knowledge of what the planets' *surfaces* are like has, until recently, been shrouded in mystery. The best telescopic views of the closest planets, Venus and Mars, tell us little about their atmospheres and surface features (Figure 12-2). Yet astronomers had only these and similar views of the planets until the early 1960s.

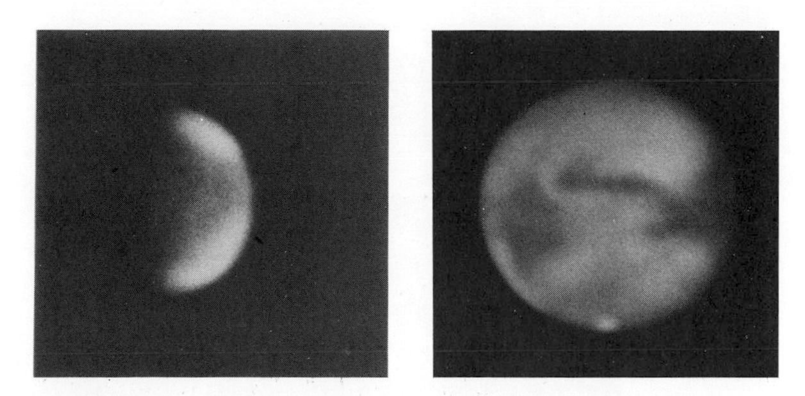

Figure 12-2 Some of the best photographs of Venus (left) and Mars (right) taken by Earth-based telescopes barely reveal cloud patterns in Venus's upper atmosphere and surface features on Mars.

12.1 Human Exploration of the Solar System

Today, only two dozen years after the space age began with the first satellite launching, humans have walked on the moon and have landed automated spacecraft on Venus and Mars. In addition, we have made a photographic and spectroscopic reconnaissance of Mercury, Jupiter, and Saturn and have a spacecraft on the way to Uranus and Neptune (see page 445). These spacecraft missions to the five closest planets, together with our detailed study of the lunar surface, have opened a new era in our understanding of what the solar system is like and how it came to be that way. The discoveries from the Mariner, Pioneer, Viking, and Voyager missions, as well as from Soviet space probes, have tremendously broadened our study of the planets. Moreover, the results from Project Galileo, now scheduled for the late 1980s, should tell us far more about Jupiter than even the highly successful Pioneer and Voyager probes did.

It is with some regret that we note how little exploration of the inner solar system by automated spacecraft is planned for the early 1980s. Perhaps the success of the Mariner and Viking missions to Mercury, Venus, and Mars has left people feeling we have done enough to learn about these planets. Astronomers hope that the United States, as well as other nations, will once again decide that knowledge of our closest celestial neighbors merits a modest investment of our resources. But for the next few years at least, the marvelous achievements described here stand as the peak of our efforts to understand the inner solar system.

12.2 Mercury

Closest planet to the sun, and the only planet known to lack a substantial atmosphere, Mercury possesses an almost unaltered surface recording 4.6 billion years of meteoritic bombardment (Figure 12-3). Only the Earth's moon

Figure 12-3 This photograph of Mercury, taken by the Mariner 10 spacecraft in 1974 at 200,000 kilometers from the planet, shows craters up to 200 kilometers in diameter. The large Caloris impact basin can be seen at the bottom center.

can rival Mercury as the most primitive major object in the inner solar system, for the moon, too, has no atmosphere to produce weathering effects on its surface.

Mercury's surface shows the craters left by countless meteoroids, most of which hit during the first few hundred million years after the planet formed. Since Mercury reflects light in just about the same way as the moon's surface, we may assume that the rocks that form Mercury's surface contain much the same sort of material as lunar rocks.

The Internal Composition of Mercury

Mercury differs from the moon in its interior: The *density* of matter in Mercury far exceeds that in the moon. The mass of Mercury, just 5.5% of the mass of the Earth, turns out to be large enough to give the planet an overall density of 5.44 grams per cubic centimeter. In fact, Mercury ranks second only to the Earth as the densest planet in the solar system (Table 11-2). In contrast, the moon has a density of only 3.34 grams per cubic centimeter. Since Mercury's diameter of 4900 kilometers puts it closer to the moon's 3475 kilometers than to the Earth's 12,756-kilometer diameter, we cannot explain Mercury's high density as the result of the gravitational compression of the planet's core. This explanation makes sense for the Earth, with 19 times the mass of Mercury. But Mercury, like the moon, simply does not have enough mass for its self-gravitation to produce significant compression of the core.

We must therefore conclude that the material that forms the innermost planet has a higher proportion of heavier elements, such as iron and nickel, at the expense of lighter elements, such as carbon, oxygen, and silicon. (The lightest of all elements, hydrogen and helium, have almost vanished from the inner solar system, as we described on page 379.) Mercury's enrichment in iron and nickel and the corresponding depletion of carbon, oxygen, silicon, and magnesium probably occurred as a result of the planet's formation close to the sun. As Mercury grew in size, the local temperature in that part of the protosolar system must have been so high that the lighter elements could not condense as well as the heavier elements. Thus Mercury lost its "fair share" of the silicate rocks, made mostly of silicon and oxygen, that form much of the Earth's crust and mantle, and most of the moon as well. Iron, the chief gainer in relative abundance as the lighter elements evaporated, has a high melting point, making it less susceptible to boiling and evaporation.

The Rotation of Mercury

Today, iron-rich Mercury alternately bakes and freezes its surface during the harsh day and night cycles of the slowly rotating planet. Mercury's rotation has become locked to its orbital cycle around the sun, but in a delicate resonance in which Mercury rotates exactly three times for every two orbits around the sun (Figure 12-4). Thus, as Mercury rotates once every 59 days, each point on the planet's surface undergoes an 88-day night and an equally long day. The surface temperature, which falls to 125 K at night, rises to 700 K during the day, hot enough to melt lead! Any explorers of Mercury's surface would have to deal with a combination of sub-Siberian cold and boiling-oil heat if they paid an extended visit to the planet.

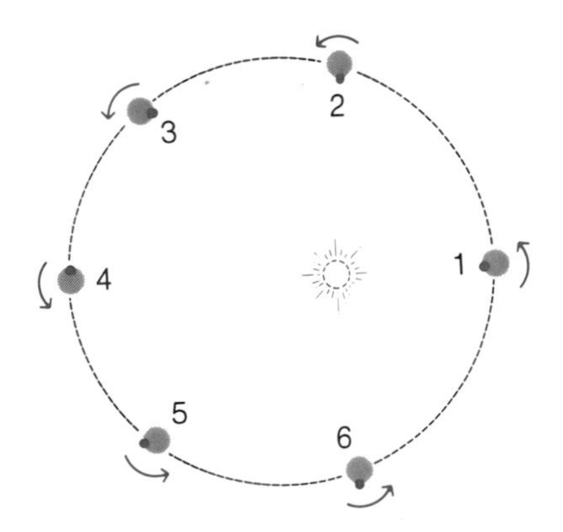

Figure 12-4 Mercury rotates once in 59 days and orbits the sun once in 88 days. Thus the planet spins three times in every two orbits around the sun. At alternate points of closest approach to the sun the same side of Mercury faces the sun.

Does Mercury Have an Atmosphere?

The enormous daytime temperature on Mercury prevents even the heavier atoms, such as krypton and argon, from remaining long in the planet's atmosphere. In fact, the only atmosphere Mercury can claim consists of particles from the solar wind (page 280) that are temporarily captured by the planet's gravity, only to escape after a few years. Thus at any particular moment Mercury possesses a gauzy envelope, less than one-billionth the density of the Earth's atmosphere.

The Absence of Plate Tectonics on Mercury

A final noticeable deficiency of Mercury, attributable to its small size, is an absence of plate-tectonic activity. Unlike the Earth and Venus, Mercury should have no continental drift and no formation of volcanoes along fracture zones. Such plate motions run on the heat released by radioactive rocks, and only the larger inner planets release enough heat to produce significant plate-tectonic activity (see page 381).

Airless, dense Mercury thus has made almost 20 billion trips around the sun without much change, especially during the past 15 billion orbits, after the time of the early bombardment. Without plate-tectonic activity to drag the surface down into the planet's mantle, Mercury preserves its almost pristine record of history in its rocks, ready for interpretation by the first explorers who can brave the blistering conditions and intense ultraviolet radiation from the sun that prevail on the planet's surface. Until then, we have the photographs and other measurements from Mariner 10 and an appreciation of the rigors of exploring the innermost parts of the solar system.

12.3 Venus

The planet Venus has been associated in myth and folklore with love and beauty. Nearly a twin of the Earth in size and mass, Venus is the brightest planet in our skies and often the closest planet to the Earth. But the cause of Venus's brilliant beauty—the atmosphere that reflects sunlight so well—hides the planet's surface beneath an envelope of perpetual cloud. For centuries, humans knew almost nothing of the conditions on Venus and could only speculate about the possibilities for life there. But our explorations have pierced the veil of Venus, revealing not paradise but a close approximation of a planetary inferno, stifling below a smothering, thick blanket of carbon dioxide.

Radio and Spectroscopic Studies of Venus

Through modern observing techniques, we have penetrated the mysteries of our sister planet. Since radio waves can penetrate the atmosphere of Venus, we can bounce them off the planet's surface and study the radar echoes. From such radar mapping of Venus, we have learned that craters cover part

Figure 12-5 The Pioneer-Venus spacecraft photographed the upper level of clouds on the planet (left). Radar observations of Venus permit an artist's reconstruction of the planet's surface (right), showing craters, plateaus, and a mountain called Maxwell, higher than Mount Everest.

of the surface, just as they do on Mercury, Mars, and the moon (Figure 12-5). In addition, we can study the composition of Venus's atmosphere by spectroscopic techniques, and can apply our knowledge of physics to draw conclusions about the ways in which this atmosphere alters the planet's environment. As a check on our conclusions, we can send a spacecraft down through the atmosphere to land on the surface, as the Soviet Union did in 1977, or we can orbit spacecraft around the planet and make close-up studies of the atmosphere, as the United States did in 1978.

The Surface Temperature of Venus

The spectroscopic studies of Venus have led to the most startling conclusions, later verified by the Soviet landings: Venus is hot! Here on Earth we consider 310 K a heat wave, call 320 K unbearable, and marvel at the 700 K of the day side of Mercury. But Venus has a surface that bakes at 750 K all over, all the time! The reason for this tremendous temperature, hottest of any planet's surface, lies within the fantastically thick carbon dioxide atmosphere, which traps solar heat with impressive efficiency.

We might expect that at least the "night" side of Venus would cool below 750 K, but the evidence proves otherwise. Venus's atmosphere is so thick that it efficiently spreads the heat from the day side to the night side, from pole to pole, so that the planet has just about the same temperature all over. This occurs despite the fact that Venus rotates more slowly than any other planet, taking 243 days to spin once.

Because Venus has a perpetual cloud cover, its slow rotation became evident only through radar techniques, which allowed us to analyze the Doppler shift of radio waves bounced from the planet's surface. The 243-day rotation of Venus occurs in the direction opposite to the planet's motion around the sun. But the mystery of Venus's rotation lies in its slowness rather than in its direction.

The Greenhouse Effect on Venus

Spectroscopic observations of the sunlight reflected by Venus show a characteristic set of absorption lines caused by the presence of carbon dioxide gas. As astronomers made more accurate measurements during the 1960s to determine how *much* carbon dioxide exists on Venus, they reached a startling conclusion: Venus has an atmosphere almost 100 times thicker than Earth's! More detailed studies revealed that the atmosphere is about 96% carbon dioxide and 4% nitrogen, with only a tiny amount of oxygen, and that the surface pressure on Venus equals 90 atmospheres—90 times the surface pressure on Earth, equal to the pressure at an ocean depth of almost 1 kilometer.

Later we shall examine the question of where so much carbon dioxide came from (see page 419). For now, the important questions are, what does all this carbon dioxide do, and how does it do it? The way in which the carbon dioxide makes the atmosphere and surface of Venus so hot deserves study both for its own sake and for what it can tell us about our own planet and its future.

The sun emits far more energy in visible-light photons than in infrared radiation (see page 275). Therefore the major input of energy into Venus' atmosphere consists of visible sunlight, just as on Earth. Some of the visible-light photons pass through the clouds and manage to reach the planet's surface, which absorbs them and their energy. These sunlight photons heat the surface to several hundred degrees above absolute zero, and the surface therefore radiates infrared photons (Figure 12-6). But now the importance of a thick carbon dioxide atmosphere appears, for the infrared photons cannot penetrate the blanket of carbon dioxide molecules with any success. Instead, they are absorbed by the carbon dioxide, and only after many successive reradiations and reabsorptions can infrared photons eventually escape from the top of the atmosphere. During these cycles of photon production and photon absorption, the photons emitted by hot carbon dioxide molecules are radiated in all directions, so the entire atmosphere, as well as the surface, grows hot.

The combination of the absorption of visible-light photons, the emission of infrared photons, and the partial trapping of the infrared photons by Venus's atmosphere produces a tremendous heating of the planet's surface and lower atmosphere (Figure 12-6). Only at an altitude of 35 kilometers does the temperature fall to a bearable (for humans) 300 K. The heating of the surface and atmosphere by the trapping of infrared photons has the name **greenhouse effect** because it resembles what happens in a terrestrial greenhouse. Sunlight photons of visible-light frequencies stream through

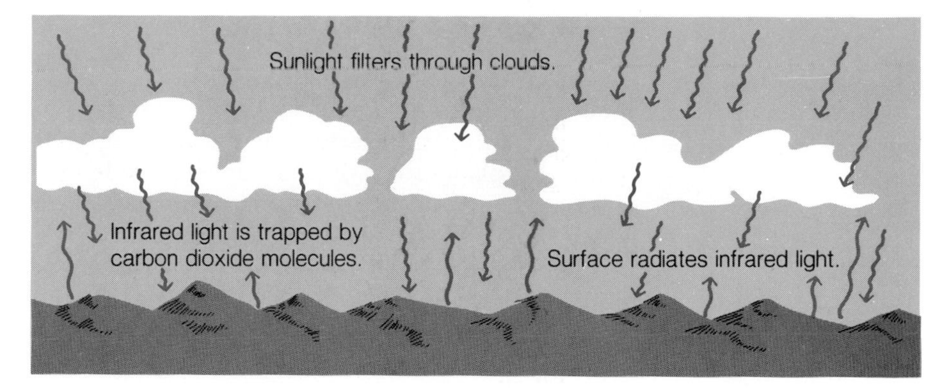

Figure 12-6 Sunlight filtering through the clouds on Venus heats the surface, which then radiates infrared photons. Since the infrared photons cannot easily escape through the carbon dioxide atmosphere, they heat the atmosphere, and the surface as well, to temperatures of many hundred degrees above absolute zero.

greenhouse windows and heat the soil. But the infrared photons emitted by the hot soil and plants cannot easily escape, because the window glass, like the atmosphere of Venus, traps infrared photons more efficiently than visible-light photons. A still more familiar example of the greenhouse effect occurs inside an automobile on a sunny day, when visible-light photons heat the car's interior, but the infrared photons radiated by the interior cannot easily escape through the glass windows.[1]

The Greenhouse Effect on Earth

On Earth, the atmospheric greenhouse effect traps some of the infrared radiation from the ground, and thus maintains the Earth's surface and lower atmosphere at a higher temperature than they would have if no such trapping occurred. Denuded of an atmosphere, the Earth's surface would have an average temperature close to 265 K instead of the actual average of 290 K. The crucial 25 K difference, which allows life to exist almost all over the Earth, arises from the trapping of infrared radiation by water vapor and carbon dioxide molecules in our atmosphere. These two molecular types are particularly efficient at absorbing infrared light, while nitrogen and oxygen, the major constituents of our atmosphere, are not. We have already significantly increased the amount of carbon dioxide in the Earth's atmosphere through the combustion of fossil fuels, and have thus increased the greenhouse effect and made our planet warmer. This trend may cause serious problems during the next few decades if we continue to use fossil fuels at our present profligate rate.

[1]The automobile analogy would be exact only in a car made entirely of glass, since some heat enters and leaves the car through the metal roof and doors.

Venus: A Planet Hostile to Life

On Venus, the greenhouse effect operates at full blast, since the atmosphere, almost 100 times thicker than Earth's, consists mainly of carbon dioxide. Even though only a fraction of the arriving sunlight filters through the clouds, the planet's massive atmosphere, made mostly of efficient absorbers of infrared photons, raises the surface temperature by 400 K over the value it would have if Venus had no atmosphere.

Thus Venus has an unbearably hot surface, though one on which recognizable rocks can be seen (Figure 12-7). The intensity of the sunlight we find there equals the intensity of sunlight on Earth when the sky has a complete cover of thick clouds. The clouds in Venus's upper atmosphere, which make the planet so brilliant in the skies of Earth, consist not of water or ice crystals, but of droplets of sulfuric acid (H_2SO_4), and seem to be the source of intense lightning discharges.

Why is Venus So Different from Earth?

The differences between Venus and Earth make us wonder how two planets so similar in size and mass could develop such divergent surface conditions. Such questions of comparative planetology interest us not only in analyzing our own solar system, but also in considering planets that may orbit other stars. The answer to the question posed above lies close to our hearts: Life and the presence of liquid water on Earth have made our planet markedly different from the Venerean inferno.

The atmospheres of Venus and Earth differ most noticeably in that Venus's atmosphere has a tremendous amount of carbon dioxide, the Earth's very little. Most of the carbon dioxide near the Earth's surface has been locked up in limestone rocks, mostly calcium carbonate ($CaCO_3$). A typical calcium carbonate rock consists of an agglomeration of millions of tiny sea shells, formed by sea creatures from the carbon dioxide dissolved in sea water and from calcium. As the animals use up some of the gas dissolved in the water, more carbon dioxide can enter the water, so we have a natural

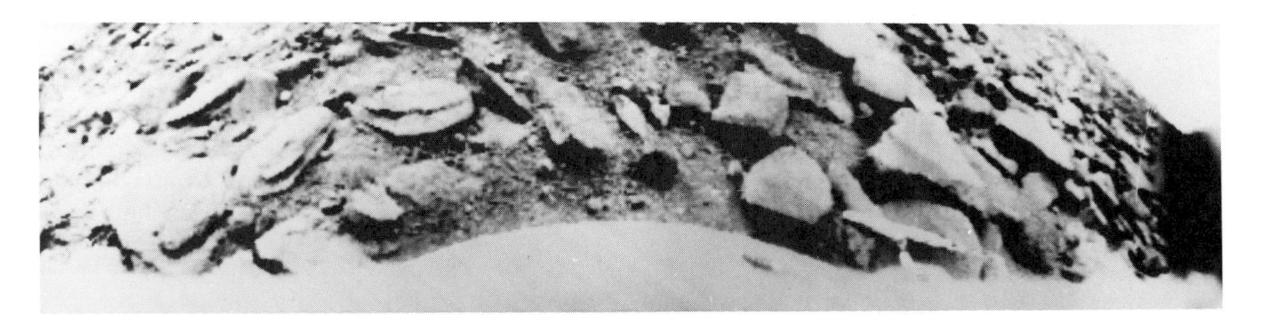

Figure 12-7 The first photograph of Venus, taken in 1974 by the Soviet Venera 9 spacecraft, shows that rocks exist on the surface despite the temperature of 750 K.

sink for atmospheric carbon dioxide (but one that humans are overwhelming with their production of carbon dioxide). If we ground up all of the carbonate rocks in the Earth's crust and heated the rubble, we would release into the atmosphere an amount of carbon dioxide equal to *70 times* its present amount of matter. Then the Earth would have an atmosphere similar to that of our sister planet Venus!

Suppose that the Earth's atmosphere grew much richer in carbon dioxide, so that it had 10 times its present amount (but still less than a thousandth of the amount on Venus). Since carbon dioxide gas absorbs infrared radiation so efficiently, the average temperature on Earth would increase by 10 K, and this increase would completely alter the face of our planet. Or suppose instead that we suddenly moved the Earth closer to the sun, so that it followed the same orbit as Venus. The higher temperature produced by our greater proximity to the sun would make the oceans warmer and evaporate some of the water they contain. The additional water vapor in the atmosphere would trap more of the infrared radiation from the surface, thus raising the temperature still further, evaporating more water, and producing more infrared trapping, until finally the oceans would entirely evaporate into the atmosphere! This **runaway greenhouse effect** would produce an extremely hot Earth. Eventually, the water molecules would be broken apart by ultraviolet sunlight into hydrogen and oxygen atoms. The hydrogen would escape from the Earth's gravity, while the oxygen would remain to combine with other elements.

Just 28% closer to the sun than Earth, Venus could never maintain seas of liquid water on its surface. Since liquid water can dissolve carbon dioxide and can support the living creatures that keep turning carbon dioxide into carbonate rocks so that more carbon dioxide can be dissolved, a planet with no liquid water and no life will maintain its carbon dioxide molecules in its atmosphere, ready to absorb infrared photons radiated by the planet's surface. The result, as we see, is a hellishly hot planet, something like what the Earth might become if we release enough carbon dioxide into our own atmosphere.

12.4 The Moon

Brightest object in our night skies, closest neighbor to the Earth, our moon has exerted an influence on human events far beyond that of any celestial object save the sun. To the moon we owe not only tides and eclipses, but also the dim beginnings of astronomical thought, as humans pondered the changing phases of the moon and tried to design their calendars accordingly. Yet until 1965, we had no sure knowledge of the moon's composition, of whether it formed close to the Earth or came out of the Pacific Ocean, or of the ages of the various features on the lunar surface familiar to us as the "man in the moon" (Figure 12-8).

We made great improvements in our knowledge by sending several dozen automated spacecraft and eight manned rockets to and around the

Table 12-1 Manned Apollo Explorations of the Moon

Mission	Date	Landing Site	Results	Average Age of Rocks Returned (Billions of Years)[a]
Orbital Missions				
Apollo 8	Dec. 24, 1968	—	Orbital mapping from 115-kilometer minimum altitude	—
Apollo 10	May 21, 1969	—	Approach to moon to within 17-kilometer minimum altitude	—
Landing Missions				
Apollo 11	July 20, 1969	Mare Tranquillitatis	Samples of mare material	3.6
Apollo 12	Nov. 18, 1969	Oceanus Procellarum	Samples of mare material	3.3
Apollo 14[b]	Feb. 5, 1971	Fra Mauro (close to Mare Imbrium)	Samples of ejecta from Imbrium basin	3.9
Apollo 15	July 30, 1971	Edge of Mare Imbrium (foot of Appenine Mountains)	Samples of material from Appenine Mountains	3.4
Apollo 16	April 20, 1972	Lunar uplands	Samples of upland material	3.9
Apollo 17	Dec. 11, 1972	Taurus Mountains	Samples from region thought to have recent volcanism	4.1

[a]For comparison, typical rocks found on Earth are a few tens of millions, or hundreds of millions, of years old. Only a few specimens are as old as 3.9 billion years.
[b]Owing to a malfunction, the Apollo 13 spacecraft did not land on the moon.

moon (see Table 12-1). The Apollo program of lunar exploration cost approximately 40 billion dollars. For the price, we enormously increased what we know about the history of the moon and, indirectly, what we know about the origin and development of the solar system.

Lunar Craters

Telescopic observations of the moon began in 1609, when the great Italian scientist Galileo discovered the thousands of craters on the lunar surface. Ever since, astronomers have debated how the craters might have formed. Are they the result of lunar volcanoes? Are they produced by gas bubbles that rose through molten rock and burst on the surface? Or are they the scars left by the impact of rocky projectiles? Later astronomers saw that the moon's brighter areas, called the **uplands**, have many more craters than the dark areas, called **maria** (the Latin word for "seas"), as we can see in Figure 12-8. This difference in the number of craters between the uplands and the maria provided the key to the craters' origin. We can explain the difference in cratering if craters arise from impact and the more heavily cratered uplands are older than the maria, which might have been covered with new lava some time after the moon formed. Such immense lava flows, stretching for hundreds of kilometers, could have arisen from collisions with huge meteoroids that eradicated the cratered terrain and replaced it with a newly made "sea" of molten rock (Figure 12-9). Analysis of the rocks brought back from the moon proved that for the samples tested, impacts *did* produce the craters.

Figure 12-8 The features of the "man in the moon" turn out to be the great lava basins on the side of the moon that always faces the Earth.

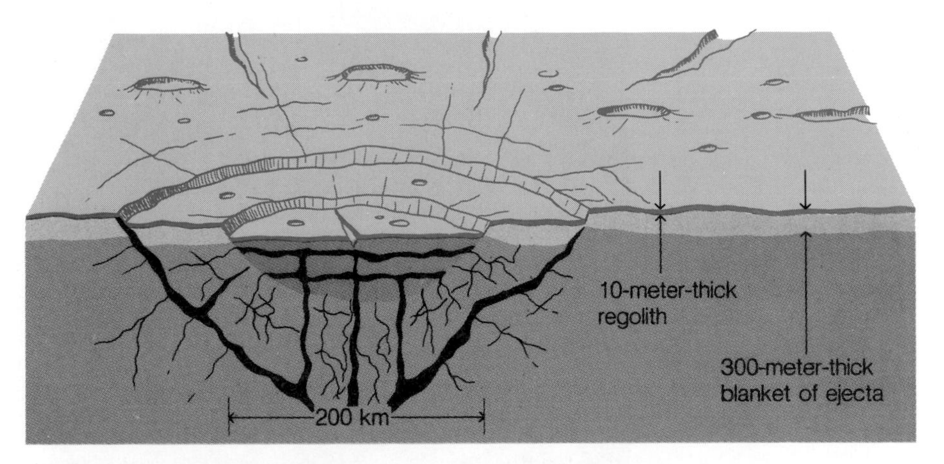

Figure 12-9 The great impact basins arose from titanic meteoroid encounters, which melted the lunar surface for several hundred kilometers. The vertical scale in this drawing has been exaggerated by a factor of about 20.

Lunar Rocks

Human exploration of the moon, starting with the automated Ranger VII probe in 1964 and continuing through the Apollo program until 1972, has shown that all of the moon's surface has a cover of fine dust and scattered rocks. This covering, called the lunar **regolith** ("rocky layer" in Greek), appears to be the residue of material blasted from craters upon meteoroid impact. Long streaks of dust point away from some of the most prominent craters (Figure 12-8). By analyzing rocks brought back from the moon, scientists determined that the great majority of them are **igneous** (fire-formed) rocks that were made from once-molten material. These rocks closely resemble the basaltic rocks found near volcanoes on Earth, but most of them are 3.9 billion years old, far older than the average terrestrial basalt.

Although most lunar rocks have an average age in excess of the age of any rocks found on Earth, we have discovered no moon rocks older than 4.2 billion years, though we have plenty with ages between 3.9 and 4.2 billion years. The fact that the first 400 million years of the moon's history appear to be missing from the lunar rocks suggests that these years saw an intense bombardment, which altered the rocks over the entire half of the moon that we see. Some scientists think that still older rocks may be lying on the far side of the moon, where fewer large meteoroid impacts have occurred (Figure 12-10).

Like Mercury, the moon has no atmosphere and so has no weathering to change the condition of the igneous rocks formed after meteoritic impact.

Figure 12-10
The far side of the moon can never be seen from Earth, because the moon always keeps the same side facing the Earth as it orbits around us. Spacecraft have photographed the far side and have found it consists of rugged terrain much like the uplands on the side we see, with few impact basins or maria.

Figure 12-11 Geologist (later Senator) Harrison Schmidt, well protected against the harsh lunar environment, examines a large boulder during the Apollo 17 expedition in 1972.

With no atmosphere to mitigate the temperature variations, the moon's surface has a temperature that rises to 400 K during the two-week lunar daytime and falls to 150 K during the equally long night that follows. The twelve men who took rock samples from the moon, examined the structure of the moon's surface, and left behind seismographic and other instruments had to be well protected against the heat, the ultraviolet solar radiation, and the lack of air that they found on the moon (Figure 12-11). Their success in carrying out the tasks of the Apollo missions depended on human understanding of the lunar surface conditions and on human ingenuity in constructing the apparatus to deal with these conditions.

Did the Moon Come Out of the Earth?

Detailed studies of the composition of the lunar rocks have provided scientists with several conclusions about our oversized satellite. First of all, the lunar rocks contain much smaller amounts of volatile elements (those that boil at low temperatures) than are found in terrestrial basalts. The most abundant volatile elements, carbon, nitrogen, and oxygen, are *hundreds* of times less abundant in lunar rocks than in rocks found here on Earth. Since the Earth is itself deficient in volatile elements in comparison to the sun, to other stars, and to the giant planets, this implies an extremely low abundance (on a cosmic scale) for volatiles on the moon. Hence we could hardly expect to find life on the moon, since life as we know it consists mostly of these volatile elements (see Chapter 14).

Moreover, we can prove that the moon never formed part of the Earth, because small but significant differences in composition of the *non*volatile elements, presumably unaffected by boiling and evaporation, testify to a difference in the basic material from which the Earth and the moon formed.

But even though the theory that the moon came from the Pacific Ocean basin no longer holds water, no one knows for sure just where the moon did come from, whether it accreted in about its present location or formed much farther from the Earth, to be later captured by the Earth's gravitation. An emerging consensus of lunar specialists favors the coformed hypothesis that the moon and Earth formed as a sort of double planet, with the moon having 1/81 of the Earth's mass. Still, no complete explanation of the composition differences between the Earth and the moon has proven valid, so prominent advocates of alternative models of the moon's origin continue to debate their hypotheses. If we could find lunar rocks that took us farther back through the 4.6-billion-year history of the moon, we might be able to choose among these hypotheses with certainty.

12.5 Mars

Because of the red color of the planet Mars, many cultures on Earth came to associate Mars with the gods of war. Tycho Brahe's observations of the red planet allowed Johannes Kepler to formulate his laws of planetary motion (see page 58), and telescopic observations of the planet revealed suggestions of life on Mars. During the early part of this century, observational evidence for life on the fourth planet from the sun consisted of seasonal changes in the planet's surface contrast. As part of these changes, a yearly **wave of darkening** begins in late spring at the polar cap and then spreads toward the equator as summer advances (Figure 12-12). Although the direction of the wave of darkening, from pole to equator, just reverses the direction that spring travels on Earth, astronomers conjectured that the changes on Mars might somehow reflect the growth of Martian vegetation as water from the polar caps arrives. Today, astronomers think the wave of darkening arises from the seasonal removal and redeposition of surface dust particles.

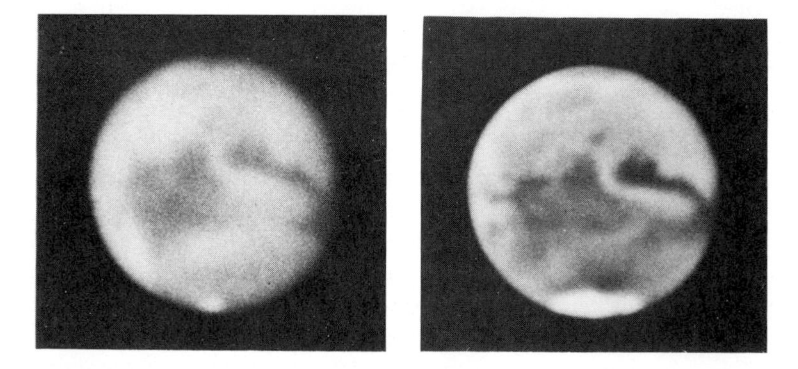

Figure 12-12 As spring turns to summer in a given Martian hemisphere, we can observe the dark areas become darker, starting near the poles and moving toward the equator. This wave of darkening is more easily observed with the human eye than photographed through a telescope, because atmospheric blurring deprives a long-exposure photograph of the chance to use the moments of best "seeing," when our atmosphere is least turbulent.

The Martian Canals

A far more exciting, though highly controversial, set of observations pointed to life on Mars: the Martian canals. These apparently linear markings were first discovered in 1869 by the Italian astronomer Angelo Secchi, who named then *canali*, meaning "channels" or "canals." These canals drew the life-long attention of Percival Lowell, who built his own observatory in Flagstaff, Arizona, just to study Mars. Lowell drew maps of the planet that showed a complete network of canals, and he claimed that this network proved the existence of intelligent Martians, capable of constructing a planet-wide network of irrigation ditches to carry water from the melting polar caps to the planet's deserts. Lowell saw the canals grow darker, and sometimes grow double, as the Martian summer began, and he regarded this as additional evidence of an irrigation network (Figure 12-13).

On the other hand, not everyone saw the canals of Mars, even with the best telescopes. Edward Barnard of the Lick Observatory examined Mars repeatedly, but drew no canals or lines when he recorded his observations. As the advances of modern astrophysics revealed that Mars's surface temperature lies well below the freezing point of water, and a full 100 K below at night, the idea of 1000-kilometer canals full of water seemed far less believable. Furthermore, spectroscopic observations could detect no water vapor or oxygen in the Martian atmosphere and pointed toward an extremely thin

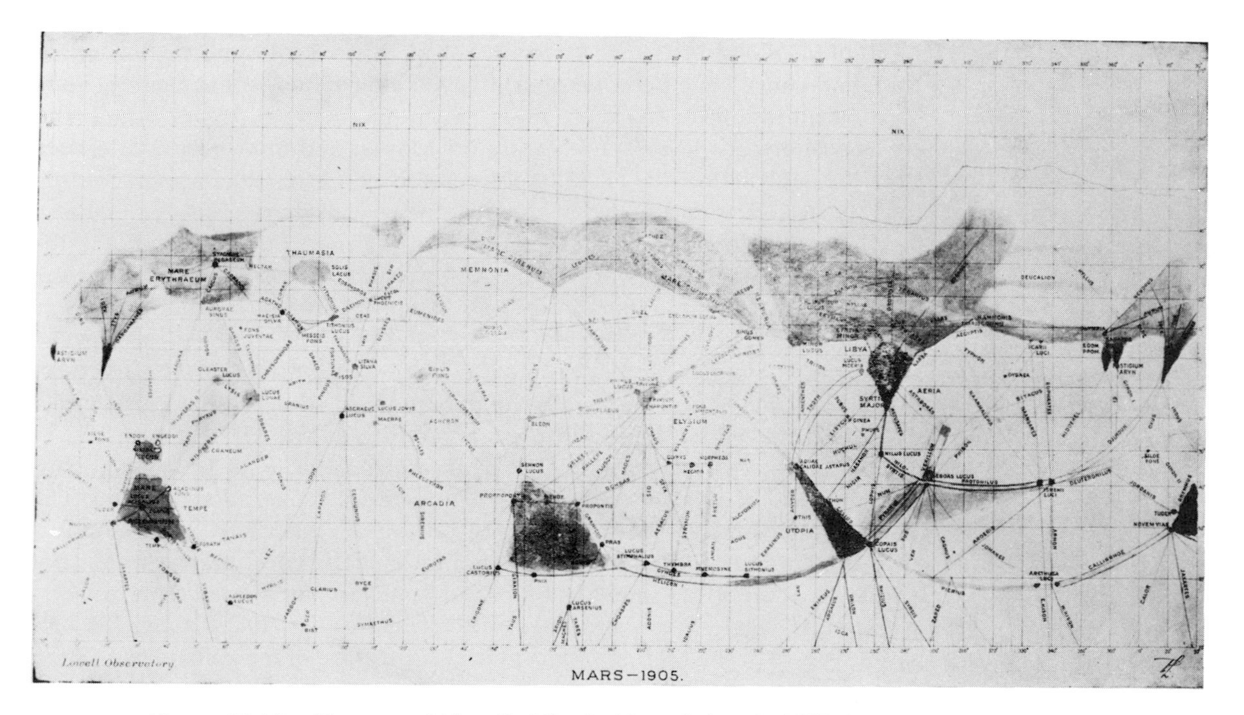

Figure 12-13 The map of Mars that Percival Lowell drew in 1905 shows the complex network of canals that Lowell observed on the Martian surface.

atmospheric blanket, incapable of keeping water in a liquid state (see page 428). Today we know that Lowell's canals must have been an optical illusion, lines produced in the observer's mind from small features on the planet's surface.

The fascination with Mars remained strong on Earth, and on Halloween evening in 1938, when Orson Welles broadcast a radio dramatization of H. G. Wells's *War of the Worlds*, he sent waves of panic along the Eastern seaboard. Thousands of listeners rushed from their homes to fight or flee the Martians who, Welles announced, were overrunning New Jersey. Reason soon prevailed, however, and within a few weeks it was difficult to find people who admitted to having been fooled by the broadcast. Twenty-five years later, the first terrestrial spacecraft to Mars took photographs from orbit, and a dozen years after that, the two Viking spacecraft landed on Mars.

Surface Conditions on Mars

The improved telescopic and spectroscopic observations of Mars, the photographic reconnaissance by Mariners 4, 6, 7, and 9, and the Viking orbital and landing investigations have given us a picture of the red planet far more detailed and far more tantalizing than the fragmentary data available to Lowell and Barnard (Figure 12-14 and Color Plate C). Mars is cold and its thin atmosphere, mostly carbon dioxide, is nothing like ours. No liquid water exists on the planet's surface, though some water is frozen in the polar caps and perhaps under its soil. Conditions on the Martian surface seem extremely hostile to life as we know it. Yet we have also found evidence that Mars may once have been warmer; that its atmosphere could have been at least 50 times thicker than it is now; that liquid water did once flow on its surface; and that the presence of life on Mars *now* cannot be ruled out entirely, though it appears an unlikely possibility. Let us take a more careful look at these facts and see what conclusions we can draw.

First and foremost, Mars is cold. The warmest spot on Mars, at the warmest moment of the warmest day in the Martian year, has a temperature of 300 K, about equal to that of Boston in summer. But on the same day, some sixteen hours later, the temperature at that spot falls to 185 K, as cold as winter in Antarctica! Moreover, at an average location on Mars, halfway between the equator and the poles, the maximum daytime temperature barely exceeds 275 K, while the nighttime temperature falls to 155 K.

Second, Mars has an extremely thin atmosphere, with an average surface pressure of about 6 millibars, 0.6% of the Earth's surface pressure. The Martian atmosphere consists mostly of carbon dioxide, with small amounts of nitrogen and argon, still less oxygen, and only trace amounts of water vapor. Thus the atmospheres of Mars and Venus resemble each other in composition, though Venus's atmosphere exerts more than 10,000 times the surface pressure of the Martian atmosphere. A pressure of 6 millibars equals the critical pressure at which water can no longer exist as a liquid. Instead, like frozen carbon dioxide ("dry ice"), frozen water will pass directly into the gaseous state as it vaporizes (*sublimes* is the technical word), because the at-

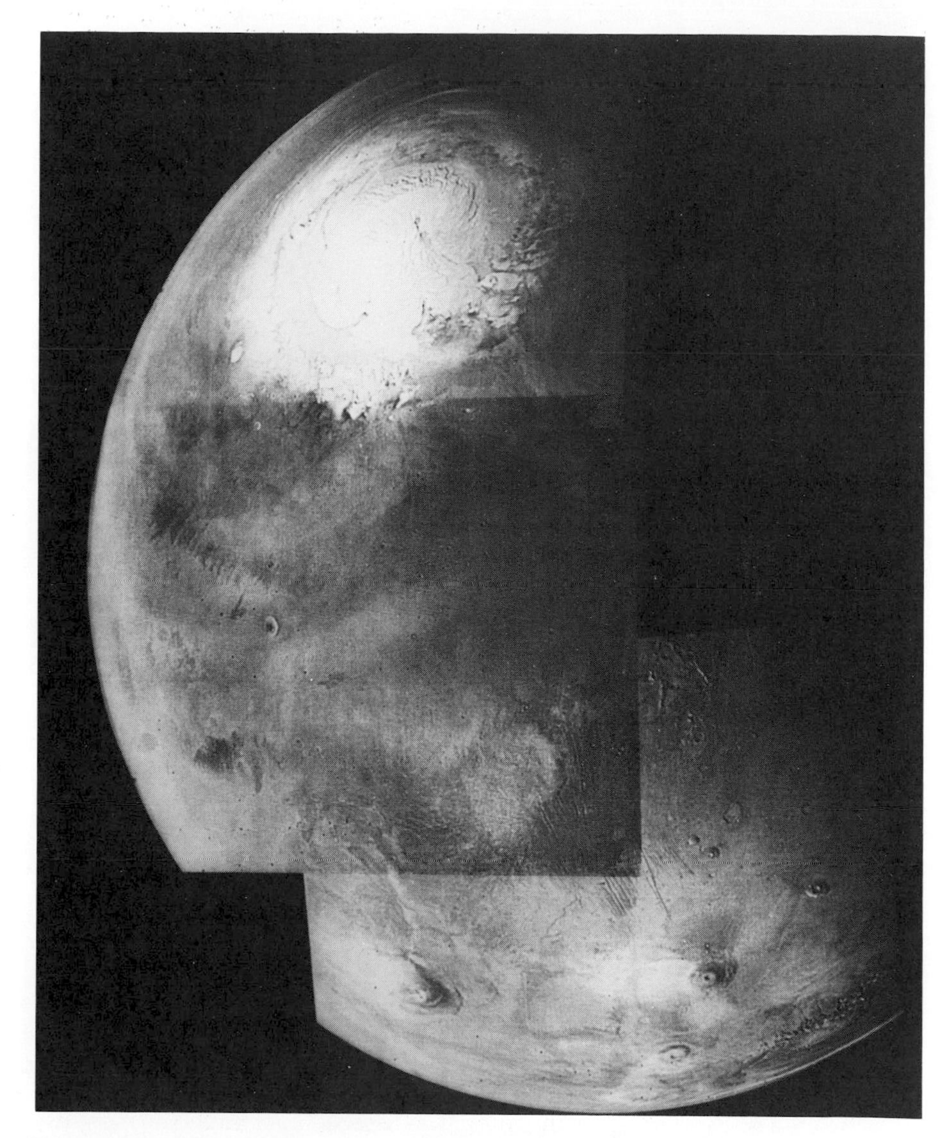

Figure 12-14 This mosaic of photographs of Mars taken by the Mariner 9 spacecraft shows the north polar cap (top), made of frozen ice and carbon dioxide, and three great volcanoes (bottom), including Olympus Mons, the largest known mountain in the solar system (bottom left).

mosphere can no longer put a "lid" of pressure on the liquid to keep it from vaporizing (Figure 12-15).

Because great differences in elevation exist on Mars, with as much as 12 kilometers between the high ridges and the low-lying plains, some low-altitude regions may have just enough atmospheric pressure for liquid water to exist. But certainly no such ponds or pools can be seen on Mars today (Figure 12-16), and the total amount of water in the Martian atmosphere falls below

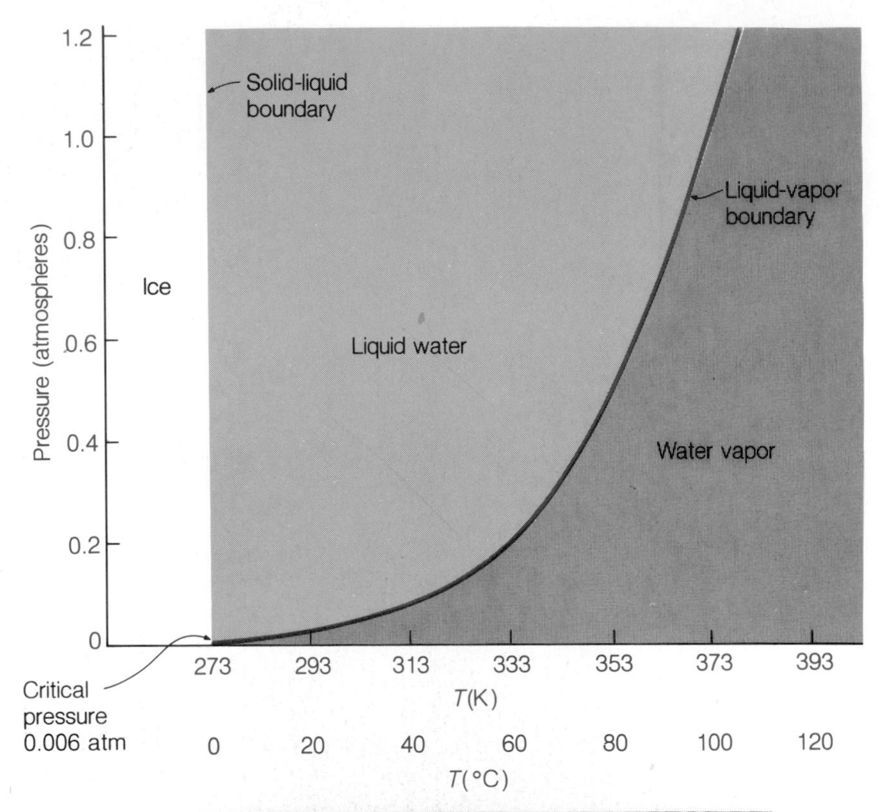

Figure 12-15
If the pressure falls below 6 millibars (0.6% of the atmospheric pressure at the Earth's surface), water cannot exist as a liquid. Instead, melting ice will sublime directly into vapor without passing through the liquid phase.

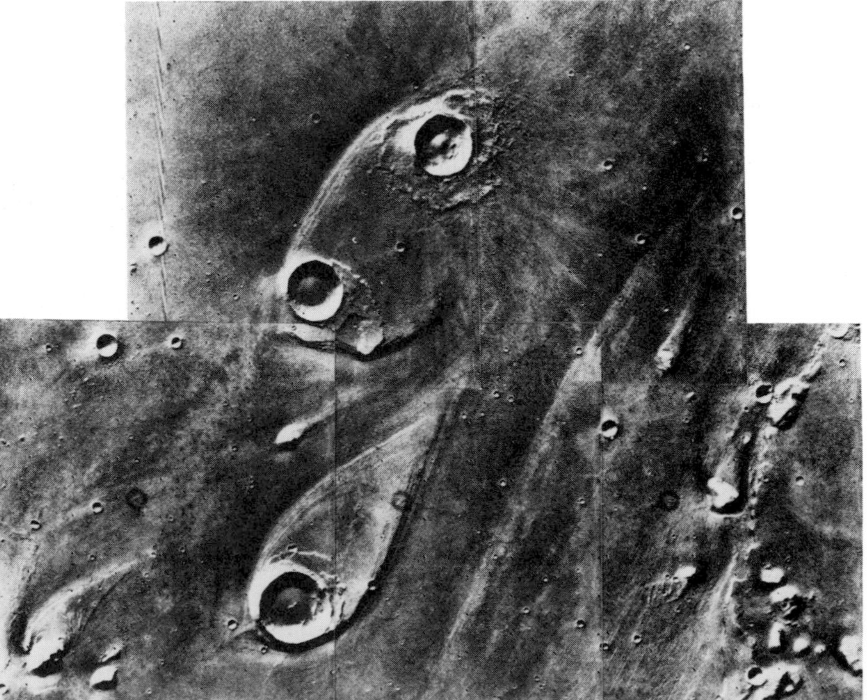

Figure 12-16
This region of Mars shows channels on the surface like those on Earth carved by running water. The presence of impact craters in some of the channels suggests that the channels are quite old, a billion years or more.

the amount in a single terrestrial lake such as Lake Tahoe! The Martian polar caps, which contain much more water than the atmosphere, nonetheless are mostly frozen carbon dioxide, testimony to the low temperatures near the poles, since carbon dioxide gas will freeze only at 150 K at the low atmospheric pressure on Mars. The thin Martian atmosphere provides only a tiny greenhouse effect to warm the planet's surface and does not have enough ozone to shield the planet's surface against ultraviolet light from the sun. The thin atmosphere produces almost no erosion (Figure 12-17).

Did Mars Once Have Running Water?

Thus our recent observations of Mars show an almost airless, freezing planet with no liquid water. But careful study of the photographs from the Mariner and Viking spacecraft have revealed evidence of old channels, seemingly carved by running water (Figure 12-16). Did liquid water once flow on Mars? The existence of immense Martian volcanoes, such as the mighty Olympus Mons (Figure 12-18), proves that the Martian crust has had tectonic activity within relatively recent eras. On Earth volcanoes have ejected most of the Earth's present atmosphere (Figure 12-19). Could it be that Mars often has a thicker atmosphere, and that our cosmic luck has taken us to Mars at the "wrong" time, when the atmosphere is thinner than usual and we find the red planet in an extreme ice age?

We do know that in past eras the angle by which Mars's rotation axis is tilted with respect to the planet's orbit around the sun has changed its value (Figure 12-20). During the course of many millions of years, Mars's rotation axis alternates between greater and lesser angles of tilt. The Earth, which now has an angle of inclination of its rotation axis quite similar to Mars's, undergoes far smaller changes in this angle, because our large moon suppresses what would otherwise be Mars-sized variations. With only two tiny satellites (see page 439), Mars experiences the full variation, by as much as 15° in both directions from its present angle of inclination, 24°.

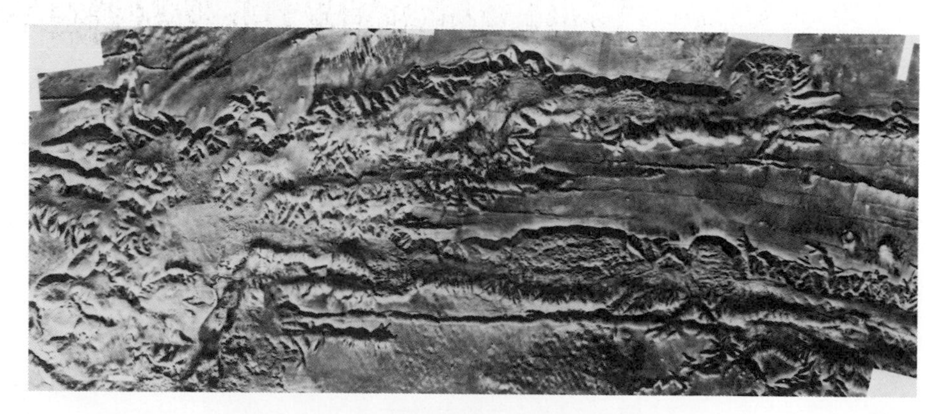

Figure 12-17 This mosaic photograph of the Valles Marineris, 4000 kilometers long, shows a huge, complex fault system with no evidence of erosion by water.

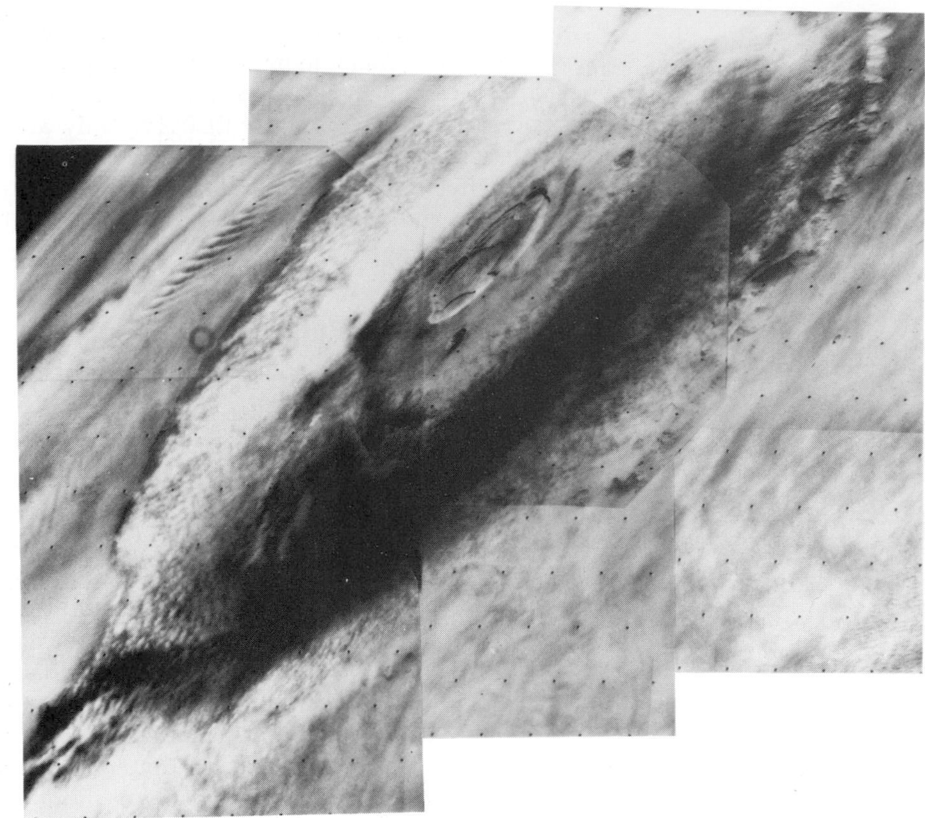

Figure 12-18
A mosaic of photographs shows a side view of Olympus Mons (top), nearly 600 kilometers wide at its base. The drawing below compares this giant volcano with Mount Everest and with Mauna Kea, the largest volcano on Earth.

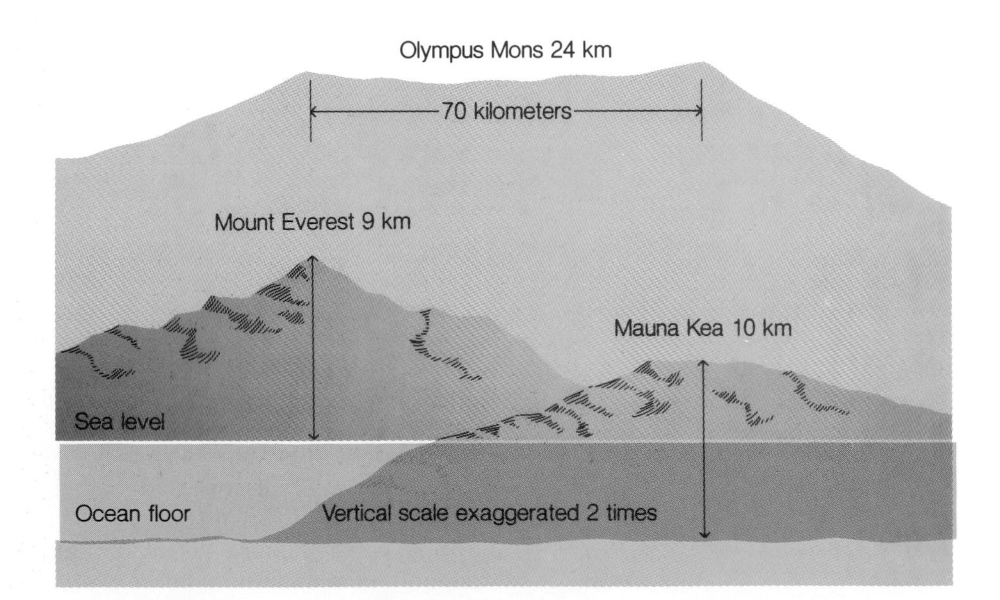

Figure 12-19
Volcanoes like Mount Saint Helens, seen here in eruption in 1980, produced most of the Earth's present atmosphere.

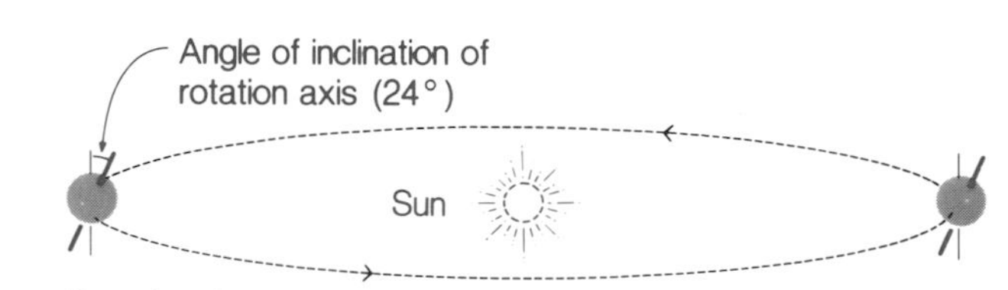

Figure 12-20
Mars's rotation axis now has an inclination of 24° from being perpendicular to the plane of Mars's orbit. If this angle of inclination were larger, the difference between the Martian summer and winter would be greater.

Consider what must have happened when the angle of inclination of Mars's rotation axis exceeded its current value. In those years, the change between summer and winter on Mars must have been more pronounced, and the polar caps might have disappeared completely in summer for a given hemisphere. This sublimation might release enough carbon dioxide gas to raise the atmospheric pressure to the point that liquid water could exist on the surface of Mars.

Table 12-2 Elemental Composition of Soil at the Two Viking Landing Sites on Mars

Element	Percentage of Total Composition of Soil	
	Site 1	*Site 2*
Silicon	20.9 ± 2.5	20.0 ± 2.5
Iron	12.7 ± 2.0	14.2 ± 2.0
Magnesium	5.0 ± 2.5	
Calcium	4.0 ± 0.8	3.6 ± 0.8
Sulfur	3.1 ± 0.5	2.6 ± 0.5
Aluminum	3.0 ± 0.9	
Chlorine	0.7 ± 0.3	0.6 ± 0.3
Titanium	0.5 ± 0.2	0.6 ± 0.2
All others (thought to be mostly oxygen)	50.1 ± 4.3	

Although this hypothesis seems appealing, the evidence speaks against the existence of liquid water on Mars during the past billion years or so. Careful photography from the Viking 1 orbiter, which was then searching for a safe landing site on Mars, showed that what seem to be dried-up stream beds have impact craters in the beds themselves (Figure 12-17). Because we know the rate at which meteoroids have made craters within the inner solar system, we can estimate with some certainty that at least a billion years have passed since water carved these channels. Thus we are forced to conclude that water has not existed on the Martian surface for a billion years or more.

Viking Lands on Mars

The high point of human exploration of the red planet came in 1976, when two automated spacecraft landed at widely separated points on Mars, sending to Earth detailed analysis of the surface and leaving in orbit cameras that mapped the planet as never before. Since the Viking landers would have been destroyed by landing on any rock larger than a basketball, careful radar studies and a good deal of luck were needed for both of them to reach the surface safely. But land they did, and the data from the mission continued to reach us more than five years after the Vikings left the Earth.

Color-filtered photographs from the Martian surface confirm that Mars *is* red, the result of iron oxides in the soil, and show that the Martian sky also has a reddish cast, because some of the soil particles are small enough to remain suspended in the atmosphere by wind action. A similar effect sometimes occurs on Earth when violent sandstorms put somewhat larger particles into the air and turn the sky yellow. Windblown dust also seems to be the cause of the wave of darkening that was once thought to suggest the presence of vegetation on Mars.

The photographic panoramas taken by Vikings 1 and 2 certainly show no signs of life: not a leaf, not a stalk; not a burrow, trail, or footprint; nothing but rocks, soil, sand, and the effect of the Martian winds (Figure 12-21). The two landing sites, separated by 7400 kilometers, show about the same sort of scenery. This general impression was confirmed by the chemical analysis made at the two sites: The composition of the soils 7400 kilometers apart is virtually identical (see Table 12-2). This analysis, which could recognize only chemical elements and not the compounds they might form, suggests that Martian soil closely resembles nontronite, an iron-rich clay found on Earth. If the elemental analysis of the two Martian landing sites proves typical of the entire planet, we may conclude that the abundance of elements on Mars strongly resembles that on Earth, where oxygen, silicon, and iron also dominate the planet's crust.

Why is Mars so Different from Earth?

Before we examine the Viking quest for life on Mars, we might pause to ask ourselves why Mars turned out to be so different from Earth. Unlike Venus, which almost equals the Earth in size and mass, Mars has only 53% of the Earth's diameter and 11% of the Earth's mass. The smallness of Mars, with an assist from its greater distance from the sun, explains the differences between the sun's third and fourth planets.

If Mars were larger, it might have gathered a larger supply of the volatile elements, such as carbon and nitrogen, that are essential for the existence of a dense, stable atmosphere within the inner solar system. A larger

Figure 12-21 Panoramic views taken by the cameras on the Viking 1 (top) and Viking 2 (bottom) landers show a desolate Martian landscape of rocks and sand. Over a period of months, astronomers could observe the effects of the Martian winds on the sand dunes seen behind the meteorology boom in the center of the Viking 1 panorama.

Figure 12-22 River systems on Earth develop lazy meanders and ox-bows within a few hundred thousand years, provided that water flows within them relatively slowly. This photograph shows the upper Colorado River, a rather young river by terrestrial standards.

planet would have released more volatile elements through volcanic outgassing, caused by a greater amount of tectonic activity, driven by the planet's greater internal heat (see page 381). This hypothetical Mars would no longer show so many craters on its surface, for the primitive crust would have disappeared, as on Earth, through plate-tectonic activity.

The real, small Mars once did have a denser atmosphere that seems to have permitted liquid water to exist. Made primarily of carbon dioxide, this early atmosphere began to disappear as carbonate rocks formed through the action of running water, while the nitrogen molecules broke apart into nitrogen atoms and escaped from the planet. Thus water on Mars eventually created the conditions under which the existence of water became impossible.

In the absence of tectonic activity, the carbonate rocks can never be broken apart to feed carbon dioxide back into the atmosphere. The greater distance of Mars from the sun keeps more of the small amount of atmospheric carbon dioxide frozen into the polar caps than would be the case if Mars were closer to the sun. Hence during the last billion years or more, Mars lost most of its atmospheric carbon dioxide and thus its ability to maintain liquid water. In fact, we have no evidence that large bodies of liquid water have *ever* existed on Mars. The branching channel systems show nothing like the meanders or "ox-bows" that typify mature, slow-moving river systems on Earth (Figure 12-22). Instead, we see what appear to be the scars of water moving in haste, as in the flash floods of Earth's desert regions, but on a vastly larger scale. Mars may in fact have large amounts of water frozen into its subsurface layers like terrestrial permafrost. Perhaps some of the scars on Mars are the result of geothermal activity, or meteoritic impact, that suddenly melted the permafrost and collapsed part of the Martian surface.

12.6 The Search for Life on Mars

Despite the vast amount of information about the chemical composition of Mars and its atmosphere provided by the Viking spacecraft, popular attention, and to a large degree scientific attention, has focused on the question that originally provided much of the impetus for the Viking project: Is there life on Mars? The canals of Mars had long been dismissed as an optical effect, the joining together of dark spots by the human eye and brain, and the low temperature and thin air on Mars spoke against an easy development of life. But the possibility that microscopic creatures, or even larger ones, could exist on the surface of Mars seemed reasonable. *Most* life on Earth is microscopic, and terrestrial microbes are surprisingly hardy, able to live in conditions we humans would call extremely hostile.

How to Try to Find Life on Mars

Scientists built into the Viking spacecraft the capacity to sample the Martian soil and to test for the presence of life. Previous tests for life had produced negative results: No radio signals had been received from Mars; nothing resembling signs of life (whatever they may be) had been seen; and no methane gas, the short-lived product of living organisms on Earth, had been found in the Martian atmosphere down to a level of 25 parts per billion.[2] But even if we eliminate the possibility of *larger* organisms, such as intelligent beings who build cities and highways, or cattle that release methane gas, we can still test for the presence of *microscopic* organisms in the Martian soil.

Two principles guided the life-detecting experiments: First, that any form of life must have an intimate involvement with its surroundings, and second, that life on Mars should be based on a carbon chemistry with water as a solvent (see page 491). With these principles in mind, scientists designed and built an amazingly compact laboratory that could perform three separate experiments to test for life at each of the landing sites. Figure 12-23 shows a representation of these three tests, called the gas-exchange (GEX), labeled release (LR), and pyrolitic release (PR) experiments.

In the GEX experiment, a soil sample was brought into contact with several dozen nutrients that were judged likely to be useful for life. These nutrients have such broad appeal for Earth-based life that the GEX came to be called the "chicken soup experiment." The goal of the GEX was the detection of changes in the composition of the gas above the Martian soil once the nutrient broth (the "chicken soup") had come into contact with it. On Earth, this approach would reveal changes in the amount of oxygen, carbon dioxide, or hydrogen in the air above the soil caused by the metabolic activity of organisms in the soil.

[2]In the presence of ultraviolet light, methane combines with oxygen to form carbon dioxide and water within a few years, even with the low abundance of oxygen in Mars's atmosphere. Thus if we found methane in the Martian atmosphere, we would need a source to replace the methane as it oxidizes.

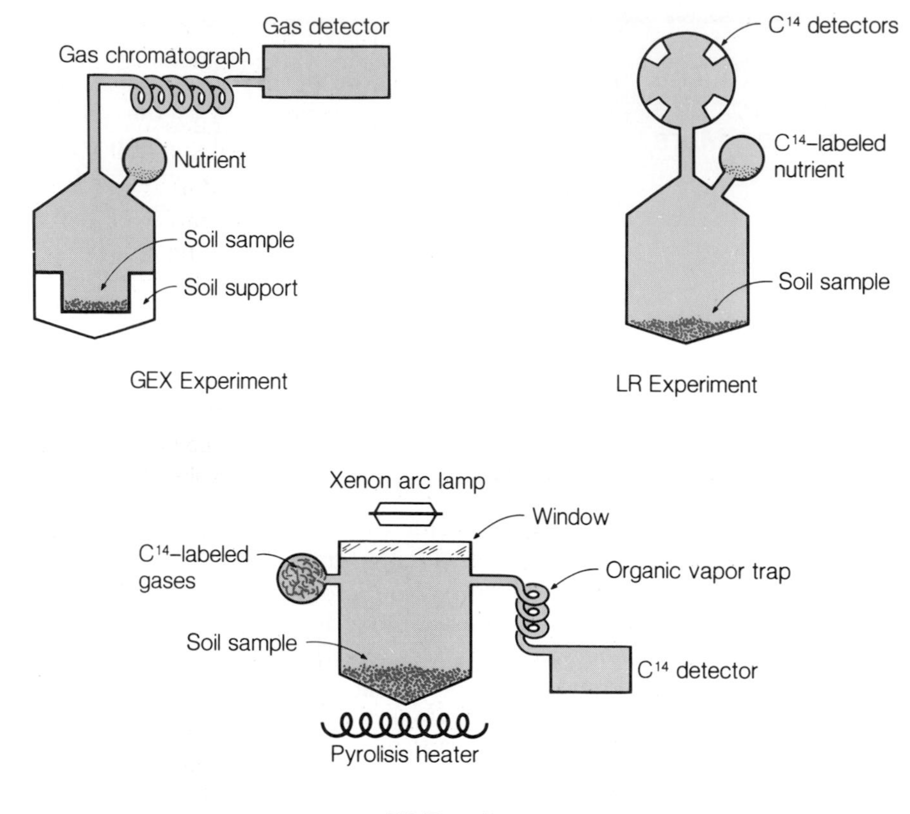

Figure 12-23 This schematic representation of the GEX, the LR, and the PR experiments shows that each of them analyzed the Martian soil in a different way to test for the presence of life. The GEX experiment dropped a broth of nutrients onto a few grams of soil and then looked for changes in the gas above the soil-and-nutrient mixture. The LR experiment tagged carbon-rich compounds with radioactive carbon-14 atoms in place of some of the usual carbon-12 atoms. These labeled compounds then dripped over the soil sample. Biological processes should have caused some tagged compounds to appear in the gas above the sample. The PR experiment replaced the normal Martian atmosphere with a similar set of gases labeled with radioactive carbon atoms. Any organism that ingested some of these labeled molecules would produce a radioactive signal after the soil in which they lived had been roasted.

The LR experiment attempted to find biological activity more directly, by using a set of compounds that had been "labeled" with radioactive carbon atoms substituted for some of the ordinary carbon atoms in the soil. This labeled mixture dripped onto the soil sample, and the gases above the sample were tested to see whether any radioactive gases, such as carbon dioxide or methane, had been released by the life processes of Martian microbes. On

Earth, we would call the LR experiment a respiration experiment that tests whether organisms are releasing gases into the atmosphere.

The PR experiment, unlike the GEX and LR, neither used a watery medium in its tests nor attempted to guess what Martian microbes might like to eat. Instead, Martian soil occupied an environment virtually identical to that on Mars's surface, except that the Martian carbon dioxide and carbon monoxide molecules were replaced with molecules that had been labeled with radioactive carbon atoms. After allowing any organisms to live for a while, the soil was heated to 1000 K. The volatile gases released by this heating passed into a vapor trap and then into a counting device (Figure 12-23). Once the carbon dioxide and carbon monoxide had passed all the way through this trap, the remaining gases were driven out of the trap by further heating and tested to see whether they contained any radioactivity. In other words, the PR experiment aimed to roast the corpses of any Martian microbes, thus releasing carbon atoms that the microbes had incorporated through biological activity.

Is There Life on Mars?

All three of the Viking biology experiments gave positive results, yet scientists still do not believe that there is life on Mars! The GEX, or chicken soup, experiment showed that the Martian soil released a large quantity of oxygen gas after it entered the nutrient. But this almost certainly occurred because of the humidity in the test chamber, more humidity than the Martian soil had known for billions of years. The LR experiment showed a sudden rise in the radioactivity above the Martian soil after just two drops of the radioactive nutrient had dripped onto it. But scientists realized that the radioactive gas, almost certainly carbon dioxide, could arise from simple chemical reactions involving peroxides, such as hydrogen peroxide (H_2O_2), if peroxides exist in the Martian soil.

The third experiment, hardest to fool, also came in positive. Radioactive carbon did indeed become part of the soil after an incubation time of five days. The carbon became only a small part, to be sure, about equal to what would be produced in Antarctic soil on Earth. But did this not show that a few tough microbes exist in the Martian soil too? This interpretation of the PR experiment lost favor when scientists arranged to heat the soil to 450 and 365 K for several hours before the radioactive gases were injected, and still found positive results. The higher temperature reduced the reaction by 90%, but the lower temperature had no effect. Since Mars's surface temperature never reaches even 310 K, scientists could not believe that life on Mars had adapted to survive three hours at 450 K, which few terrestrial organisms can do.

The consensus among the Viking researchers was that chemical, nonbiological reactions had mimicked the biological reactions that the experiments could detect. This conclusion gained strength from the results of the analysis of the Martian soil, which showed that the soil lacks organic material down to an abundance level of a few parts per billion. If Martian organisms exist but produce organic molecules in an abundance of less than a few parts per billion, then they must be immensely efficient scavengers, since the soil analysis showed no traces of their wastes, their food, or their corpses at the impressive level of sensitivity quoted above. Even the least biologically active soil on Earth would have enough organic material within it to have produced detectable amounts for the Viking laboratories.

We must not, of course, rule out life on Mars simply because two Viking landings apparently failed to detect it. We might have landed in the wrong place to find life, though the similarity of the soil at the two landing sites suggests that all of the surface has a general resemblance. This similarity arises because dust storms carry surface material from one place to another, eventually "gardening" the soil down to a depth of several meters.

The polar regions might prove a better bet in the search for life. Near the poles, relatively large amounts of water pass through a cycle of freezing, sublimation, and freezing again during each Martian year (Figure 12-24). The highest temperature near the poles, 240 K, hardly speaks in favor of an easy time for life. But precisely because the climate *remains* cold, organic material might concentrate there, since it could never heat up and blow away.

Even the polar regions of Mars show the characteristic red tint of the soil as the ice sublimates. Thus the soil there may also be "gardened" by the Martian winds and be much the same as the soil at lower latitudes. Perhaps we should look *under* the ice for life, or for evidence that life once existed when Mars had more of an atmosphere to warm the planet. While we make this search, we might also send automated "rovers" through the Martian canyons to beam pictures back to Earth, take samples of the soil, or even bring the samples back to our best laboratories here on Earth. Such missions to Mars, now being considered for possible development during the late 1980s, might finally resolve the question of whether life exists, or has ever existed, on the cratered, windblown surface of the red planet.

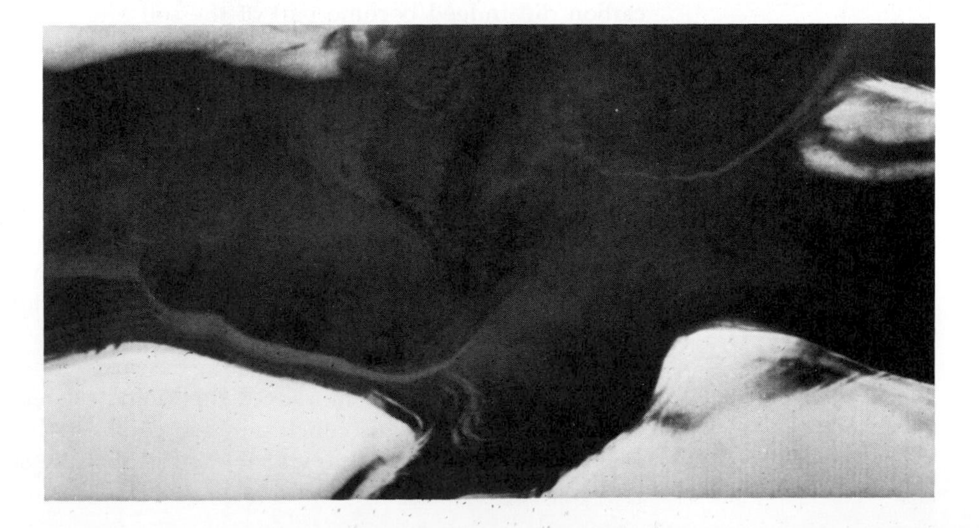

Figure 12-24
Water near the Martian poles may freeze, sublime into vapor, and freeze again year after year. This photograph shows the north polar cap of Mars during midsummer in the northern hemisphere. The white areas are ice, probably water and carbon dioxide frozen together. The dark areas are regions with no ice on the ground.

12.7 Phobos and Deimos

In contrast to the one large satellite of Earth, Mars has two tiny moons, Phobos and Deimos, named after the chariot horses of the god of war. The Martian moons were discovered by Asaph Hall in 1874, although Kepler, Voltaire, and Jonathan Swift had all speculated that Mars ought to have two moons, since the Earth had one and Jupiter (as far as was known until 1892) had four. Although the supposition that Mars had two moons simply fit a geometrical progression, some people have made fantastic speculations, crediting Kepler, Voltaire, and Swift with knowledge provided by extraterrestrials, or assigning to Mars an approach close enough to Earth to reveal tiny Phobos and Deimos.

With diameters of 15 and 10 kilometers, respectively, Phobos and Deimos are small enough to fit easily into greater Los Angeles. They are so small, in fact, that other speculative fancies once suggested they might be the artificial constructs of another civilization. Photographs from the Mariner and Viking missions, however, laid this idea to rest (Figure 12-25). As these photographs show, Phobos and Deimos do not have enough mass to form themselves into spheroidal shapes, as the planets and their larger satel-

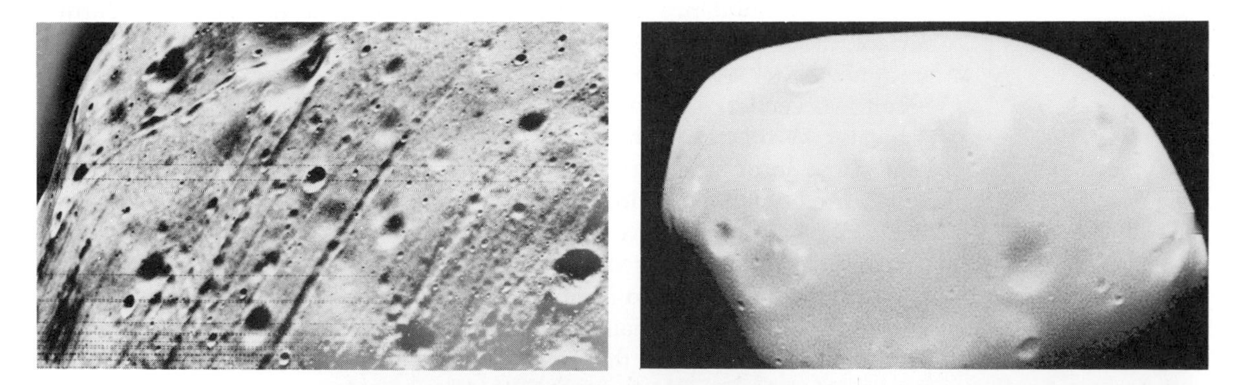

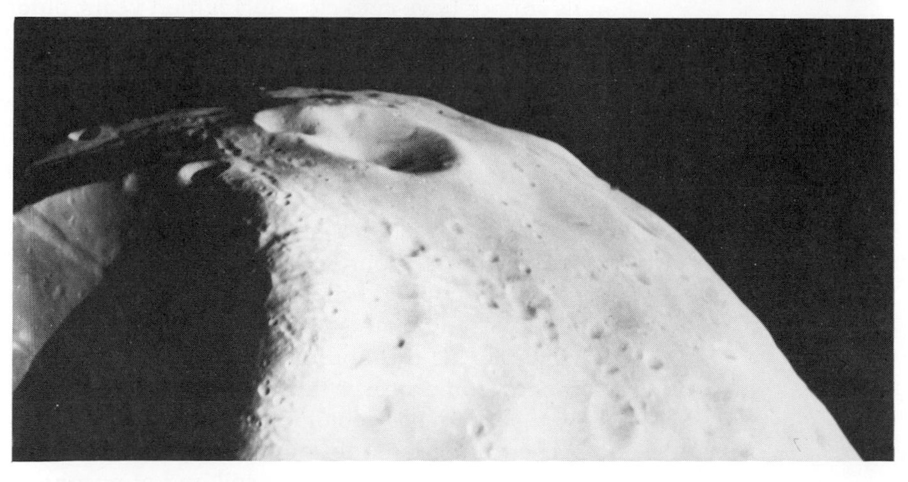

Figure 12-25
The Viking orbiters photographed Phobos (top left and bottom) and Deimos (top right). The presence of grooves and craters on Phobos, as well as its resemblance to other objects in the solar system, tends to argue against the hypothesis that Phobos is an artificial satellite of Mars.

lites have done. Battered by impacts from meteoroids, these moons probably look like most of the asteroids, and in fact may *be* asteroids that were captured by Mars from the asteroid belt beyond it.

Summary

The inner solar system contains four small, rocky planets, one large satellite (our moon), and two tiny satellites of Mars. All of these objects, save our Earth, have heavily cratered surfaces, relics of the intense meteoroid bombardment more than 4 billion years ago. Only the Earth has sufficient weathering and plate-tectonic activity to have erased the scars left from this phase soon after the solar system formed.

Mercury, smallest of the inner planets, has almost no atmosphere, consists preferentially of the heavier elements such as iron and nickel (though it has plenty of oxygen, silicon, and magnesium as well), and alternately bakes and freezes its surface during the long rotation period that exposes each half of the planet to the sun for 88 days. Venus, the planet closest to the Earth in size and mass, has a thick blanket of carbon dioxide that traps infrared photons. These photons, radiated from the surface after visible-light photons heat it, warm the lower atmosphere and surface to temperatures above 750 K, impressive evidence of the greenhouse effect that also heats the Earth's surface, though only by 25 K above what the temperature would be without an atmosphere. On Earth, most of the carbon dioxide near the surface has become locked in carbonate rocks through the action of living organisms in sea water. Venus lacked this chance to remove carbon dioxide from its atmosphere and shows the high-temperature results of its failure to do so.

Our moon shows a close resemblance to Mercury in its cratered surface and lack of an atmosphere. Small differences between the compositions of terrestrial and lunar soils show that the moon was never part of the Earth. But just how the moon did form—close by the Earth or in another part of the inner solar system—remains unknown. The absence of weathering and plate-tectonic activity on the moon explains why many of the rocks found on its surface have ages close to 4 billion years, older than any rock found on Earth.

Mars, like Venus, has a mostly carbon dioxide atmosphere, but one that is 100 times thinner than Earth's instead of 100 times thicker. At this low atmospheric pressure, liquid water simply cannot exist on Mars's surface, though photographs give evidence that channels carved by running water must have been made during the planet's history. A larger angle of inclination of Mars's rotation axis might have led to the melting of the polar caps (mostly frozen carbon dioxide, with some ice), thus raising the atmospheric pressure to the point where liquid water could have existed.

The heavily cratered surface of Mars suggests that little weathering has occurred during the 4 billion years since most of the craters formed. Gigantic volcanoes, some nearly three times the height of any mountain on Earth, show that Mars has geological activity that may extend to the present.

The Viking landings on Mars allowed us to conduct thorough chemical and biological analyses of the soil and atmosphere at two widely separated landing sites. The results of these studies suggest the presence of life on Mars, but a closer examination of this evidence reveals that nonbiological processes are more likely to have given the observed responses.

Key Terms

greenhouse effect
igneous
maria
regolith
runaway greenhouse effect
uplands
wave of darkening

Questions

1. In what ways do the surfaces of Mercury, Mars, and the moon resemble one another? How can we explain these resemblances?
2. Why does Mercury contain a larger proportion of iron-rich rocks than Venus and the Earth?
3. Why do we think that no plate-tectonic activity should occur on Mercury? Why does such activity occur on the Earth?
4. What is the greenhouse effect? Why does it have such a tremendous effect on Venus, compared to the Earth? Why is the Earth's much smaller greenhouse effect so important to us?
5. Why does the Earth have so little carbon dioxide in its atmosphere, compared to Venus? How does this affect the importance of the greenhouse effect on Earth?
6. Why are most of the lunar rock samples about 3 or 4 billion years old, while most terrestrial rock samples have much smaller ages?
7. How can we show that the moon never formed part of the Earth? What are the chief theories of where the moon formed?
8. What is the Martian wave of darkening? Why was it previously thought to signal the presence of life on Mars? How do we explain it now?
9. What are the Martian canals?
10. Why can water never exist as a liquid on Mars now? How could liquid water have existed in the past? What evidence do we have that Mars did indeed once have liquid water on its surface?
11. How did Mars's small size (compared to the Earth and Venus) affect the development of the planet's atmosphere and surface?
12. Why did the experiments to test for life on the Viking spacecraft concentrate on the search for microscopic life?
13. Why was the GEX (gas-exchange) experiment on the Viking landers called the "chicken soup" experiment? What did this experiment find out about the Martian soil?

14. Why might the regions of Mars close to the polar caps be more favorable to life than the regions close to the Martian equator? Why might they be less favorable to life?

Projects

1. *Observation of Venus.* Find out from an astronomy magazine, such as *Astronomy* or *Sky and Telescope*, whether Venus is a "morning" or an "evening" star, and find it in the sky at dawn or dusk. Were you surprised to find Venus so bright? With binoculars or a small telescope, you should be able to observe the crescent phase of Venus when it occurs. Work out the relative positions of Venus, the Earth, and the sun, using tennis balls as models, to explain the crescent appearance of Venus.
2. *Rotation of Mercury.* With a friend representing Mercury, and yourself as the sun, show that if Mercury rotates three times in every two orbits around the sun, the same face of Mercury will point toward the sun at *every other* point of closest approach (Figure 12-4).
3. *Rotation of the Moon.* If the preceding exercise seems too difficult, warm up by showing that if the moon rotates once in every orbit around the Earth, then the same face of the moon always points toward the Earth.
4. *The Canals of Mars.* Perform an experiment to show how the human mind tends to assemble lines from isolated points and splotches. Prepare a disk, meant to resemble Mars, and put some points on its surface in vaguely linear patterns. Then ask a friend to draw this disk from a distance of, say, 5 or 8 meters. Do this several times with several different "observers" and compare the results obtained from a single disk. Relate your experiment to Lowell's observation of canals on Mars.
5. *Observation of Mars.* Find Mars in the night sky by preparing a finding chart with the help of *Astronomy* or *Sky and Telescope* magazine. How bright does Mars seem, compared with the brightest stars? Why does Mars's apparent brightness change so much during the course of a year or two?

Further Reading

Arvidson, R. E.; Binder, A. B.; and Jones, K. L. 1978. "The Surface of Mars." *Scientific American* 238:3, 76.

Carr, Michael, 1976. "The Volcanoes of Mars." *Scientific American* 234:1, 32.

Horowitz, Norman. 1977. "The Search for Life on Mars." *Scientific American* 237:5, 52.

Leovy, Conway, 1977. "The Atmosphere of Mars." *Scientific American* 237:1, 34.

Moore, Patrick. 1977. *A Field Guide to Mars.* Boston: Houghton Mifflin.

Murray, Bruce, ed. 1973. *Mars and the Mind of Man.* New York: Harper & Row.

Scientific American. 1975. *The Solar System.* San Francisco: W. H. Freeman.

Veverka, Joseph. 1977. "Phobos and Deimos." *Scientific American* 236:2, 30.

Weaver, Kenneth. 1975. "Mariner Unveils Mercury and Venus." *National Geographic* 147: 858.

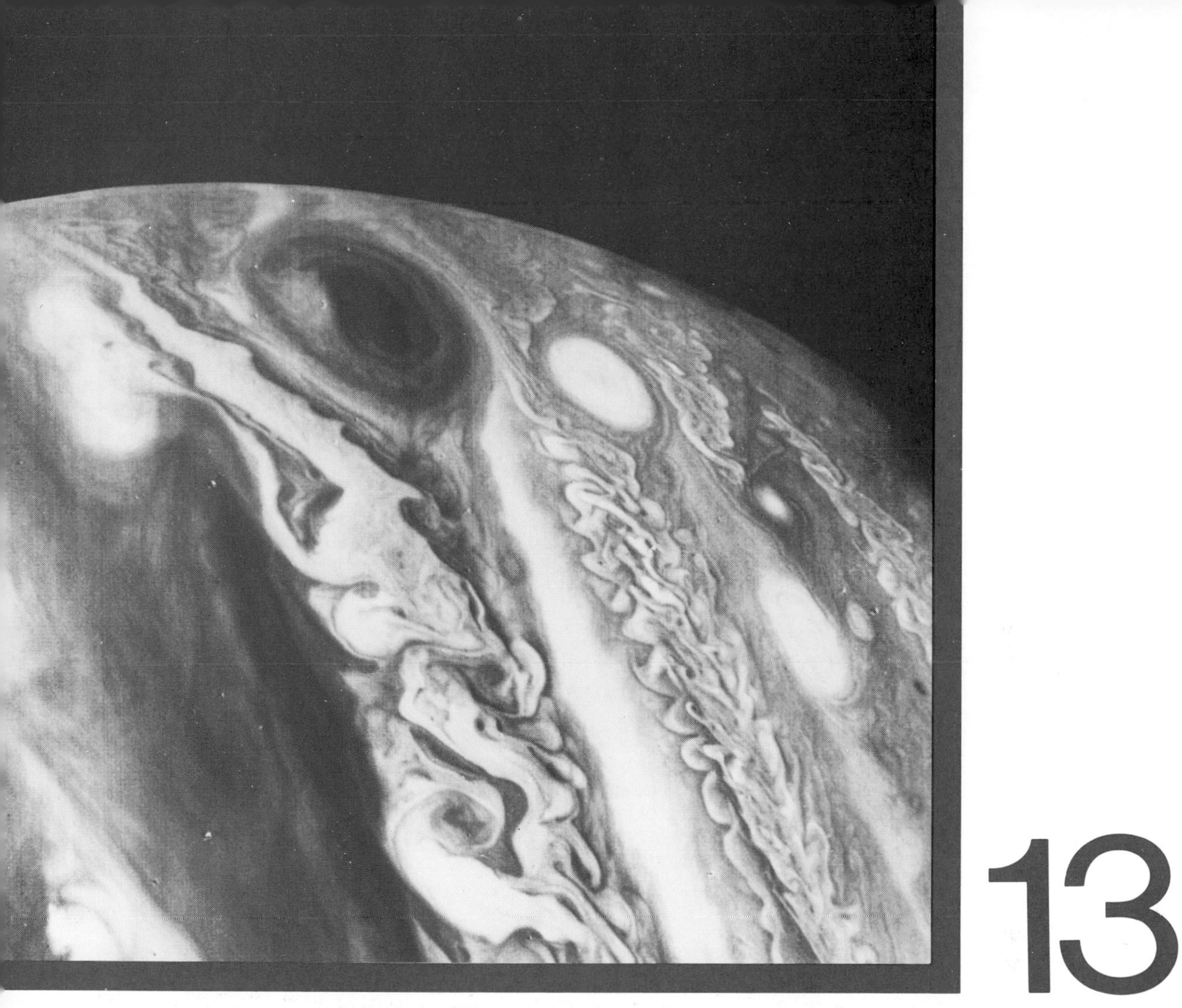

13

Jupiter, largest of the sun's planets, shows an ever-changing pattern of atmospheric circulation.

The Age of Exploration: The Outer Solar System

Far beyond the asteroid belt, 5 to 40 times the Earth's distance from the sun, orbit the sun's outer planets, Jupiter, Saturn, Uranus, Neptune, and Pluto. The first four of these are the giant planets, made mostly of hydrogen and helium. In contrast, the outermost planet, Pluto, has less mass than any other planet in the solar system and seems to be an icy object, possibly something like a giant cometary nucleus.

Human exploration of the outer solar system began in 1972, when the Pioneer 10 spacecraft started its billion-kilometer journey to Jupiter. The Pioneer 11 spacecraft, launched the next spring, also photographed Saturn. But

Figure 13-1 Voyager 1 sent back this photograph of Jupiter over almost a billion kilometers of distance. The planet's Great Red Spot can be seen at the lower left, amid the complex bands of atmospheric patterns. Two of Jupiter's four large moons also appear in the photograph.

the Pioneer results from both Jupiter and Saturn, impressive as they were, counted for little compared to the success of the Voyager 1 and Voyager 2 spacecraft. These two instrumented explorers reached Jupiter in 1979, sending to Earth photographs and measurements of exquisite detail (Figure 13-1).

The scientists who directed the Voyager flights to Jupiter arranged the spacecraft trajectories in such a way that Jupiter's gravitational force would help propel the spacecraft toward Saturn, where Voyager 1 arrived in 1980 and Voyager 2 in 1981. Owing to a favorable line-up of the giant planets, Voyager 2 should be able to proceed outward to study Uranus in 1986 and Neptune in 1990 (Figure 13-2). With the results from these encounters and from the Project Galileo mission to Jupiter, our knowledge of the giant planets should take another great step forward. For now, however, we can appreciate what has been learned from the Pioneer and Voyager probes of the two largest planets.

13.1 Jupiter

With 318 times the Earth's mass and more than 1000 times its volume, Jupiter ranks first among the giant planets and apparently falls only a few times

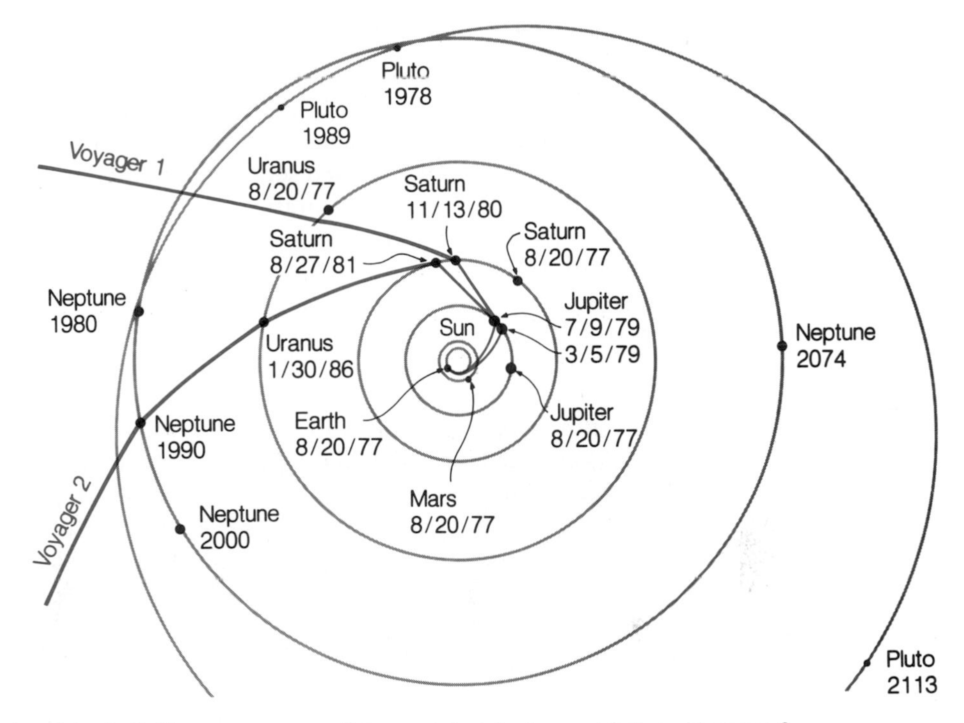

Figure 13-2 After both Voyager spacecraft have visited Jupiter and Saturn, Voyager 2 will continue along a trajectory that passes by Uranus in 1986 and by Neptune in 1990.

short of becoming a star. Because of its enormous mass (for a planet!), Jupiter held on to most of its primordial hydrogen and helium, as we can verify with observations and a computer experiment.

The Composition of Jupiter's Atmosphere

Consider a mixture of the chemical elements in the same proportions that exist in the sun and in other stars, and let these elements combine with one another to form molecules in every possible way. If we set the temperature and pressure, we can model the conditions in the atmospheres of the giant planets as they formed from the original cloud of gas and dust that made the solar system (see page 367). The chief compounds that we predict from this exercise match the ones that dominate the atmosphere of Jupiter: methane, ammonia, water vapor, and everywhere an excess of hydrogen. (Helium should form about 10% of Jupiter, and neon should be about as abundant as ammonia. But since these two inert gases do not form compounds readily, we should not expect to find them among the abundant molecules of Jupiter's atmosphere.)

Atmospheric Patterns on Jupiter

From what we have said about the molecules in the Jovian environment, we might expect that Jupiter would present a rather featureless broth of distasteful compounds. But the actual appearance of the largest planet presents us with a fantastically variegated set of swirls and colors (Figures 13-3, 13-1, and Color Plate D). The changes in the patterns of Jupiter's outer layers may arise from the planet's rapid rotation. Not only is Jupiter the largest planet, but it also rotates more rapidly than any other, spinning on its axis in less than 10 hours. Observations of the patterns at Jupiter's equator show that they take 9 hours 51 minutes to rotate while regions somewhat farther north and south rotate in 9 hours 56 minutes. The difference in rotation period testifies to the fact that we see a gaseous, rather than a solid, layer on Jupiter.

Since Jupiter's diameter exceeds 11 times the Earth's, while its period of rotation equals only 42% of the Earth's 23 hour 56 minute rotation, the speed with which the outer layers of Jupiter spin far exceeds the speed with which the Earth rotates. On our planet, the equator rotates at a speed of 1665 kilometers per hour, while New York's latitude moves at 1200 kilometers per hour. Even these relatively modest velocities produce a planet-wide circulation in which the prevailing jet-stream winds constantly circle the planet from west to east. On Jupiter, the equatorial rotation velocity reaches 44,000 kilometers per hour, equal to the speed of a rocket! Perhaps as the result of these enormous velocities, the clouds of Jupiter change on an hourly basis.

The Colors of Jupiter's Atmosphere

The colors of the belts in Jupiter's outer layers, which help to reveal the changes in the cloud patterns, have a mystery of their own. Instead of the white and gray we expect to see when sunlight reflects from ice crystals or ammonia, we find subtle tints of red and brown, especially in the famous Great Red Spot (Figure 13-4 and Color Plate D). What substances produce these colors? One school of thought attributes the coloration of Jupiter to **inorganic** compounds, such as those made from sulfur and phosphorus, and points to the known presence of phosphine (PH_3) in the Jovian atmosphere.

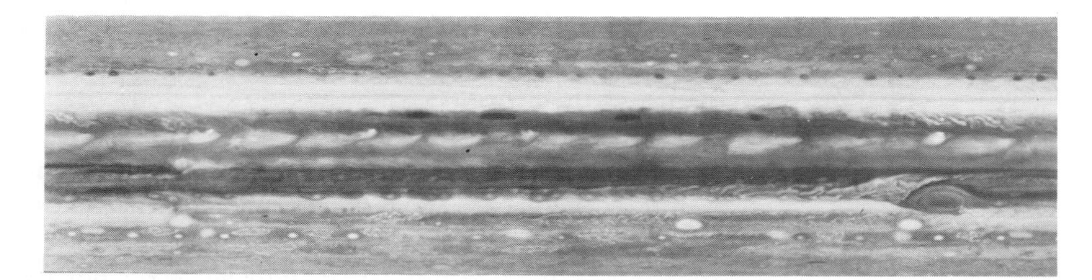

Figure 13-3 This map of Jupiter's appearance on February 1, 1979, shows the multicolored bands that encircle the largest planet.

Figure 13-4
The Great Red Spot, a semipermanent vortex in Jupiter's upper atmosphere, has a long axis of 25,000 kilometers and covers an area about as large as the Earth's surface. In this Voyager photograph, we can see turbulent eddies along the lower side of the spot, as well as a smaller oval of vortices that rotates in the opposite direction to the Great Red Spot itself.

Another school of scientists suggests that **organic** (carbon-containing) compounds are responsible. They cite laboratory experiments that duplicated the Jovian coloration by shining ultraviolet light on an appropriate mixture of methane and ammonia or by discharging lightning within such a mixture. These experiments invariably produce a wide range of organic compounds, some of which have the color of Jupiter's clouds.

The **Great Red Spot,** in existence for at least several centuries and perhaps a much longer time, appears to be the most nearly permanent feature of Jupiter's upper atmosphere, though its shape, size and location do vary somewhat from month to month and from year to year. This "spot," several times larger than the Earth, seems to be a sort of Jovian super-hurricane. Matter swirls around the Great Red Spot, sometimes joining it, and taking several days to do so.

The Interior of Jupiter

Unlike the Earth, Jupiter has no true surface. The hydrogen-helium mixture, together with other hydrogen compounds, simply thickens more and more as we descend (mentally!) into the planet, until finally, more than halfway to

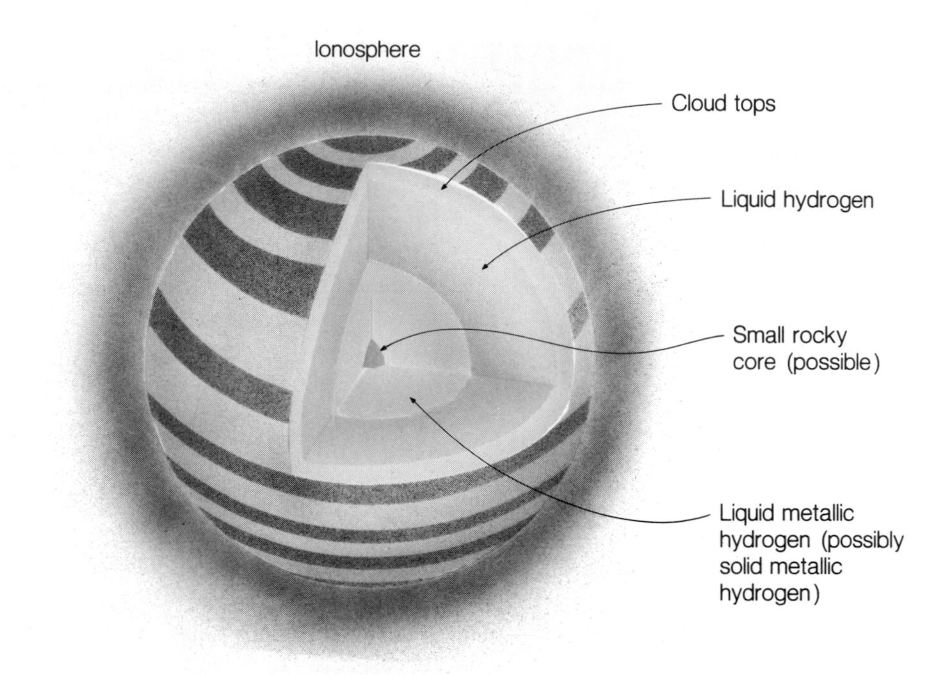

Figure 13-5 Models of Jupiter's interior and atmosphere, based on our observations and calculations, suggest that the bulk of the planet consists of liquid hydrogen, with a transition to the "metallic" phase of the liquid in the central third of the planet.

the center, we reach a transition to what is technically called "metallic" hydrogen (Figure 13-5). Metallic hydrogen has little resemblance to any metal we know on Earth, except in its ability to conduct electricity well. We should certainly not think that a metallic, Earthlike planet exists beneath the wrappings of the Jovian atmosphere. But long before the transition to metallic hydrogen would interrupt our descent, we would find an intriguing property of Jupiter's lower atmosphere: The temperature rises sharply as we descend.

Jupiter's Internal Source of Heat

The top of Jupiter's atmosphere, which we can observe from Earth and from spacecraft, has a temperature of about 125 K. Even this low temperature exceeds the value we expect in view of the fact that Jupiter receives only 4% as much sunlight energy per square centimeter of surface as the Earth does. At levels beneath the visible cloud tops, however, the temperature rises to Earthlike levels and then reaches far greater temperatures, 1000 K and more, as we pass farther inward. Let us first consider the effects of this increase in temperature and then examine its cause.

The existence of warm subsurface layers of the Jovian atmosphere produces a vertical convection pattern, with currents that carry material upward from the warmer layers to form the bright clouds (Figure 13-6). The descending currents allow us to see the darker clouds at somewhat lower altitudes. Thus all of the material in Jupiter's atmosphere eventually circulates between temperature levels of 150 K and 1000 K or more. The result of this circulation is the continual condensation and re-evaporation of water vapor and, at higher levels, of ammonia and ammonia-sulfur clouds. The latter are what we see, but underneath them, at a balmy temperature of about 300 K, we can expect to find clouds of familiar water vapor floating at a pressure just a few times the surface atmospheric pressure on Earth. Within the water vapor clouds, liquid water droplets provide the likelihood of thunderstorms, and indeed both Voyager spacecraft observed lightning discharges, presumably the result of charge separation within Jupiter's lower atmosphere.

What causes the temperature increase as we descend into the planet? The answer is that Jupiter possesses a significant *internal* source of heat, apparently the result of the slow contraction of the planet's interior. Jupiter has enough mass that its **self-gravitation** continues to compress the planet slowly. Part of the energy released by this compression goes into the heating of the particles within the planet, giving each particle a bit more energy of motion. The additional kinetic energy can be communicated to neighboring particles, and so diffused outward, through collisions. In an object with much more mass than Jupiter, the contraction through self-gravitation would eventually raise the temperature to the point that nuclear fusion reactions would begin (see page 394). Jupiter, with one-thousandth of the sun's mass, has too little mass and too little gravitational force ever to become a

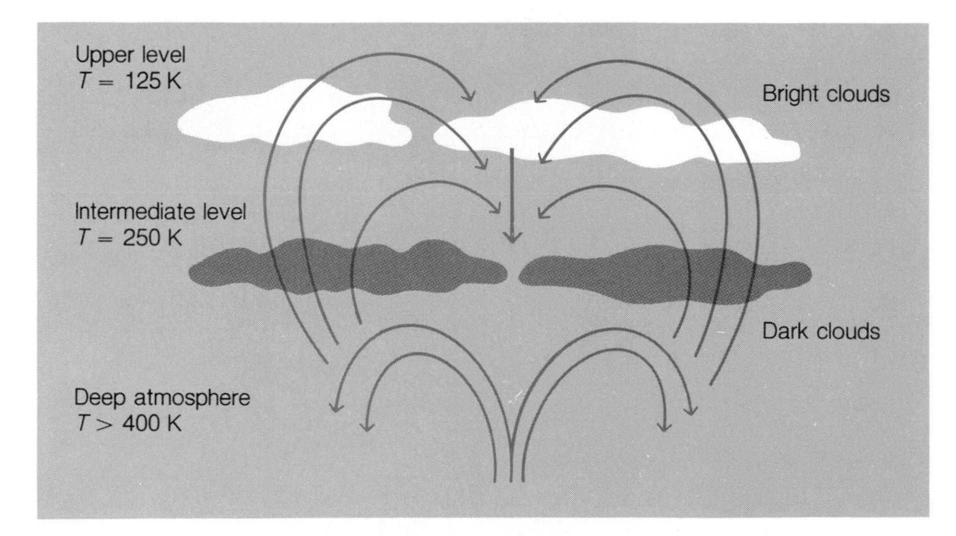

Figure 13-6 Vertical convection currents carry material upward from the warmer layers of Jupiter to form the bright clouds. Where the currents descend, we can see darker clouds at lower levels in the atmosphere.

star; the least massive stars known have 15 times Jupiter's mass. If the circumstances that formed the solar system had been a bit different, our sun might have a faint red-dwarf companion instead of a giant, still slowly contracting planet in orbit around it.

The Jovian Magnetic Field

Jupiter's interior produces still another interesting influence on its exterior: the planet's **magnetic field.** No one understands what produces planetary magnetic fields, but current theories tie them to the rotational energy of a planet and to the composition of the planetary core. Thus objects without liquid metallic cores, such as Mercury and the moon, should not have much of a magnetic field (as is indeed the case), and slowly rotating objects, such as Venus, should have weak magnetic fields even with metallic cores (and indeed Venus has quite a weak magnetic field). Jupiter, with the fastest rotation of any planet and a metallic hydrogen core, has the most intense magnetic field of any of the sun's planets. The strength of Jupiter's magnetic field exceeds that of the Earth by several times, and since it covers a much larger volume than the Earth's, the energy it embodies far outstrips that of the magnetic field around Earth.

The presence of a magnetic field causes remarkable, though hardly planet-shaking, effects. For both Jupiter and the Earth, charged particles in the solar wind become trapped by the magnetic field and spiral around the lines of magnetic force (Figure 13-7). Exceptionally large bursts of particles

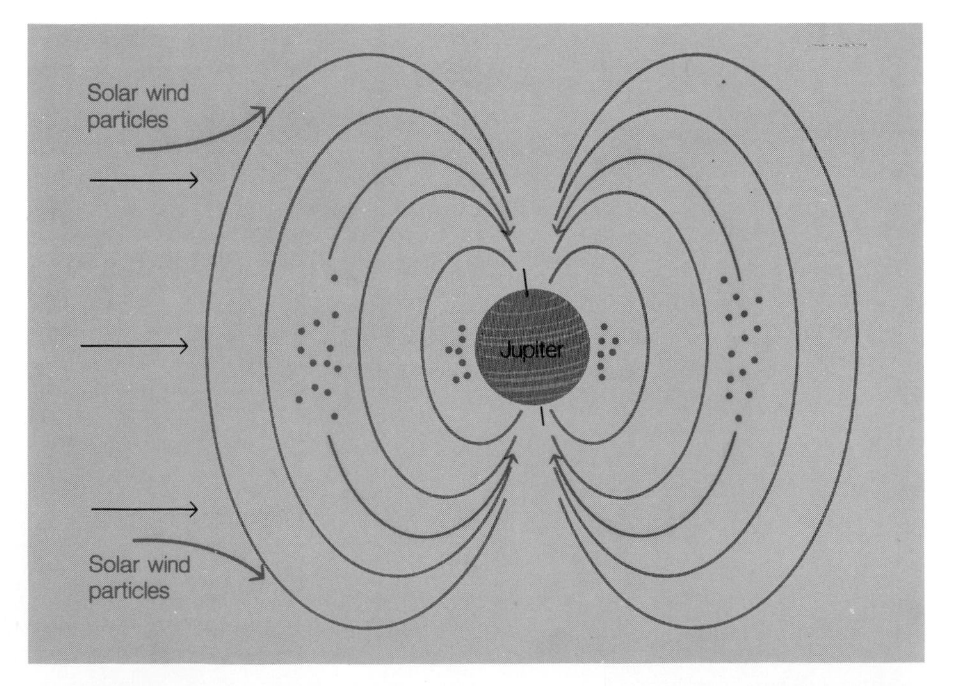

Figure 13-7
Jupiter's magnetic field, like the Earth's, traps charged particles, mostly electrons and protons from the solar wind. As the particles spiral around the magnetic lines of force, they emit radiation by the synchrotron process.

from the sun can produce magnetic storms of radio emission, since the spiraling particles can emit radio waves, and can also lead to the marvelous **auroras** seen on Earth as the particles interact with atoms and molecules in the Earth's upper atmosphere (Figure 8-46). Jupiter should also have impressive auroral displays, as indeed the Voyager observations have confirmed.

Radio Emission from Jupiter

Because Jupiter has an intense magnetic field that traps charged particles, the planet radiates large quantities of radio waves, most of them produced by the synchrotron emission process (see page 208). At certain times and at certain frequencies, Jupiter emits more radio power than the sun, so that a radio astronomer studying the solar system from another star might be more likely to detect Jupiter than the sun! An interesting sidelight to Jupiter's radio emission is that the planet's innermost large satellite, Io, has a significant perturbing effect on the magnetic field. As Io moves in its orbit around Jupiter, its perturbations of the magnetic field cause noticeable changes in the radio emission from the regions around the giant planet. The electrical currents that flow between the satellite and the outer atmosphere of Jupiter may also have an important effect on Io's surface (see page 455).

13.2 The Satellites of Jupiter

In addition to having the largest mass of any planet in the solar system, Jupiter has a tremendous number of known satellites—fourteen, with a fifteenth satellite possible but not confirmed. The four largest of these objects, much larger than the others, almost form worlds of their own, since three of them are larger than our moon and two of them are larger than Mercury (see Table 13-1). Of the remaining satellites, none has even one-thousandth the mass of the big four, and they must be much like the asteroids, made of rock. The four outer satellites orbit the planet in the direction opposite to the inner ten and in orbits that are highly inclined to the plane of Jupiter's equator, which also includes the orbits of the four large satellites (Figure 13-8). This implies that the four outer moons, and perhaps the next four inward (which also have highly inclined orbits, though these satellites move in the same direction as the large satellites), may well be objects captured from the belt of asteroids that orbit the sun between the orbits of Mars and Jupiter. Some of these asteroids have highly elongated orbits that carry them close to the giant planet, ripe targets for capture.

The Big Four: The Galilean Satellites

By far the most important of Jupiter's moons are the four **Galilean satellites,** Io, Europa, Ganymede, and Callisto. First discovered by Galileo in 1610 and named after four of the Greek god Jupiter's illicit loves, these satellites offer a fascinating diversity that deserves close examination, as well as the best

Table 13-1 Satellites in the Solar System

Planet	Satellite	Diameter (km)	Mean Distance From Planet (km)	Period of Revolution (Days)	Orbital Inclination (Degrees)[a]	Orbital Eccentricity	Mass (10^{24} gm)
Earth	Moon	3476	384,500	27.3217	23	0.055	73.5
Mars	Phobos	23	9,400	0.3189	1	0.021	
	Deimos	13	23,500	1.2624	2	0.003	
Jupiter[b]	Amalthea	240	180,000	0.4182	0.4	0.003	
	Io	3640	422,000	1.7691	0.0	0.000	73
	Europa	3130	671,000	3.5512	0.5	0.000	48
	Ganymede	5280	1,070,000	7.1546	0.2	0.001	154
	Callisto	4840	1,885,000	16.6890	0.2	0.007	95
	Leda	(10)[c]	11,110,000	240	26.7		
	Himalia	170	11,470,000	250.6	27.6	0.158	
	Lysithea	(20)	11,710,000	259.2	29.0	0.12	
	Elara	80	11,740,000	259.7	24.8	0.207	
	Ananke	(20)	20,700,000	617	147[e]	0.169	
	Carme	(30)	22,350,000	692	164[e]	0.207	
	Pasiphae	(40)	23,330,000	735	145[e]	0.40	
	Sinope	(30)	23,370,000	758	153[e]	0.275	
Saturn[d]	Mimas	(400)	187,000	0.9424	1.5	0.020	0.04
	Enceladus	(500)	238,000	1.3702	0.0	0.004	0.08
	Tethys	1000	295,000	1.8878	1.1	0.000	0.64
	Dione	1000	378,000	2.7369	0.0	0.002	1.1
	Rhea	1600	526,000	4.4175	0.4	0.001	2.3
	Titan	5100	1,221,000	15.945	0.3	0.029	137
	Hyperion	220	1,481,000	21.277	0.4	0.104	0.1
	Iapetus	1450	3,561,000	79.331	14.7	0.028	1.1
	Phoebe	(240)	12,960,000	550.33	150[e]	0.163	
Uranus	Miranda	(300)	130,000	1.414	3.4	0.00	0.1
	Ariel	(800)	192,000	2.5204	0.0	0.003	1.3
	Umbriel	(550)	267,000	4.1442	0.0	0.004	0.5
	Titania	(1000)	438,000	8.7059	0.0	0.002	4.3
	Oberon	(900)	587,000	13.463	0.0	0.001	2.6
Neptune	Triton	(4400)	354,000	5.8765	160[e]	0.00	140
	Nereid	(300)	5,600,000	359.88	27.6	0.75	
Pluto	Charon	(1600)	(20,000)	6.4	Large		

[a]The orbital inclination is measured with respect to the planet's axis of rotation. The Earth's moon has an orbit that is inclined by an angle of 5° with respect to the Earth's orbit around the sun. The inclination with respect to the Earth's rotation axis varies between 18½° and 28½°.

[b]Jupiter's fourteenth satellite has an orbit smaller than Amalthea's, but this orbit has not yet been well determined. The suspected fifteenth satellite has an orbit larger than that of the four large moons.

[c]Values enclosed by parentheses are approximate.

[d]In addition to nine satellites with diameters of at least 200 km, Saturn has at least six more small satellites, only recently discovered. Five of these satellites orbit inside the orbit of Mimas; the sixth has an orbit almost identical to that of Dione.

[e]Orbital inclinations greater than 90° indicate that the satellite orbits the planet backward with respect to the direction of the planet's orbit around the sun.

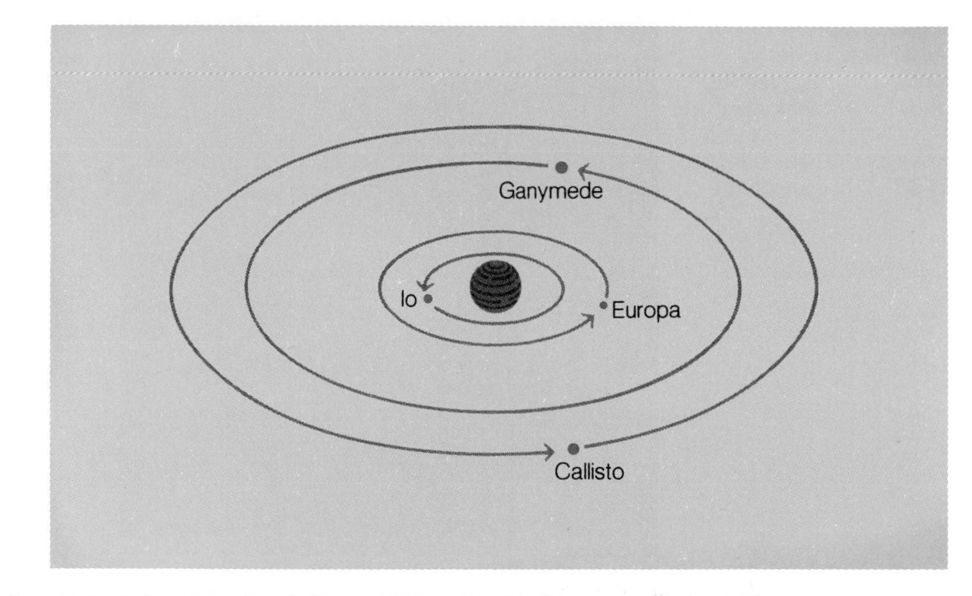

Figure 13-8 Jupiter's four largest satellites orbit the planet in the same plane and the same direction that Jupiter orbits the sun. The four outer satellites, however, move backward, relative to the large four, in much larger orbits that are highly inclined to the orbit of the four Galilean satellites. This suggests that the outer four satellites were captured from orbits typical of asteroids when they came too close to Jupiter to escape.

landing sites for any future expedition to the largest planet (see cover photograph).[1]

Io, Europa, Ganymede, and Callisto almost certainly formed along with Jupiter, for they orbit the planet in nearly circular orbits that coincide with Jupiter's equatorial plane. The densities of the four Galilean satellites roughly correspond to the densities in the solar system. The inner two have densities of 3.53 and 3.03 grams per cubic centimeter, something like the densities of the inner planets, and the outer two have densities of 1.93 and 1.79, much like the densities of the outer planets, suggesting that ice forms a large part of Ganymede and Callisto. Amalthea, the second-closest satellite to Jupiter, fits the pattern of densities, since it seems to be a rocky object, about the size of Long Island (Figure 13-9).

Io. Io, the innermost large satellite, has perhaps the most amazing surface of any moon in the solar system (Figure 13-10). A great range of colors and albedos (reflecting power) characterize Io's mottled exterior. The equatorial regions are mostly bright orange-red, with whitish patches, whereas the

[1]Galileo himself named the four large moons after the Medici family, from whom he may have hoped to obtain additional research funds. The names we use today were given by Galileo's contemporary Simon Marius.

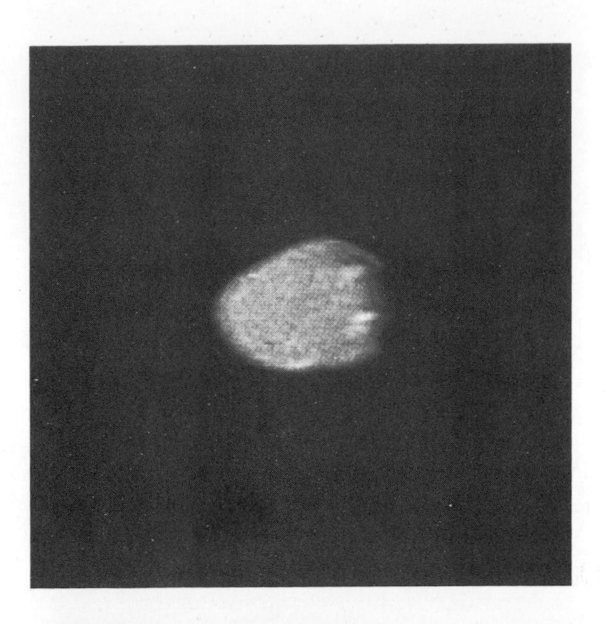

Figure 13-9
Amalthea, Jupiter's second closest satellite, shows few features in this photograph taken by Voyager 1 at a distance of 425,000 kilometers.

Figure 13-10
This mosaic of photographs of Io, the innermost of the four Galilean satellites, prompted the famous comment by Dr. Bradford Smith, the Voyager imaging team leader, "I've seen better looking pizzas." The absence of impact craters suggests that Io constantly produces thick flows of gas and liquid that cover and recover its surface with fresh deposits, rich in sodium and sulfur.

polar regions are darker (see cover photograph). The pits and other small markings on Io's surface do not resemble the impact craters found on Mercury, the moon, or Mars. The absence of impact craters, especially since Jupiter and its moons are close to the asteroid belt, implies that the surface must be continually renewed, on a time scale of a million years or less.

What could be covering and recovering Io with new layers of surface material? The answer emerged from the discovery by the Voyager spacecraft that *volcanoes* exist on Io (Figure 13-11). The scientists who examined the Voyager photographs have identified more than 100 volcanic depressions on Io, some as much as 200 kilometers in diameter. These volcano craters are much larger than those on Earth. From some of them, flows up to several hundred kilometers long have helped to cover Io with material from below the surface. The few mountains found on the satellite occur mainly in the polar regions. The lower latitudes of Io are relatively smooth, apparently covered with thick layers of sulfur-rich materials, which can produce numerous vivid colors, including red, yellow, black, and brown.

The discovery that active volcanism occurs on Io raises the question of the cause of such activity, since Io is too small to have anything like the

Figure 13-11 One of Io's active volcanoes can be seen at the edge of the satellite, spewing material to an altitude of several hundred kilometers.

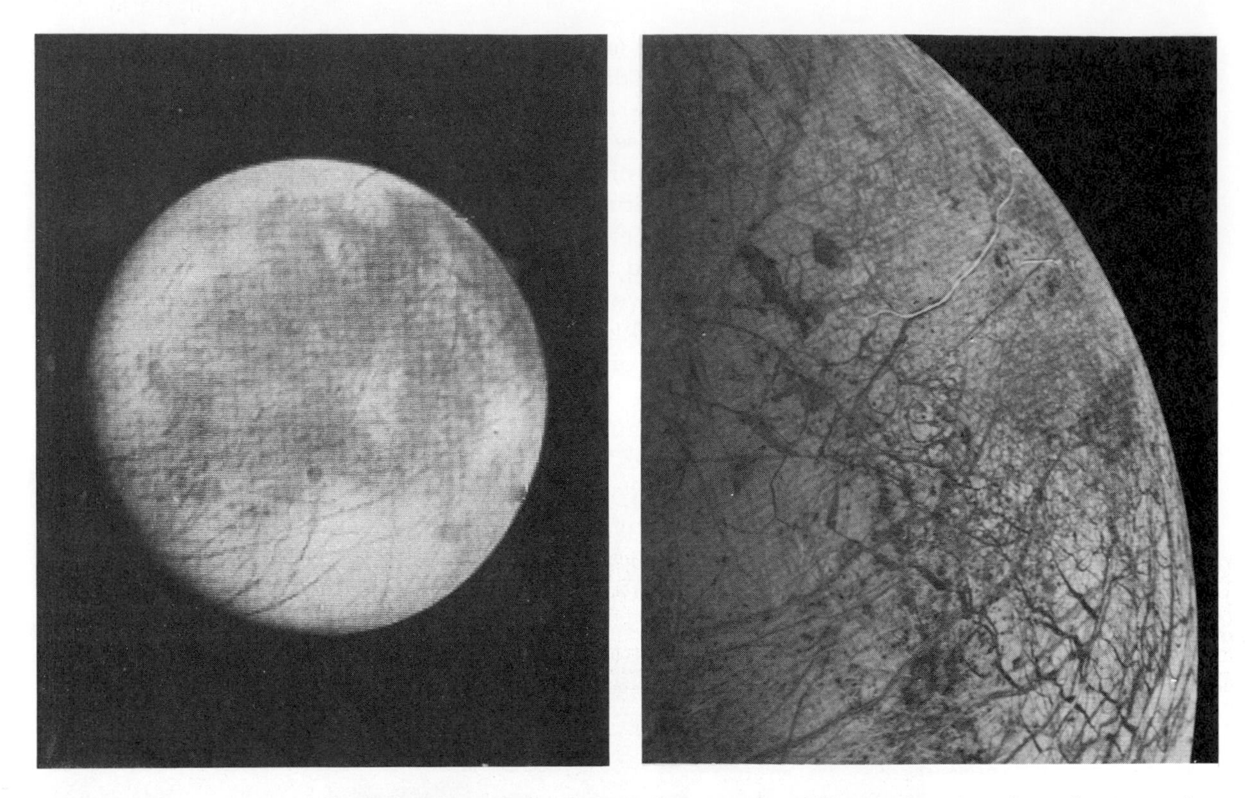

Figure 13-12
Europa has a diameter of 3130 kilometers, 90% of our moon's, and thus ranks as the smallest of the Galilean satellites. The close-up photograph (right) shows the network of lines that cross the satellite's surface.

Earth's radioactivity-driven heat engine. The answer involves the interplay of the gravitational forces of Jupiter and the other large satellites upon Io. As Io orbits Jupiter, it always presents the same side to the planet, whose gravity has locked on to Io in much the same way that the Earth has locked on to the moon. In addition, the next two large satellites, Europa and Ganymede, combine to produce an eccentricity in Io's orbit; that is, a deviation from a perfectly circular path around Jupiter. This eccentricity means that the tidal deformation of Io that Jupiter produces will vary as Io's distance from Jupiter varies. The resultant change in the amount of tidal stress produces enough heat within Io to maintain the volcanic action we observed with the Voyager spacecraft.

But Io has another source of heat as well. Jupiter's intense magnetic field extends past Io's orbit, and the satellite's motions influence the strength and direction of the field in its vicinity (see page 450). Electrical currents may arise between regions close to Jupiter and the surface of Io, and we may see the result of such currents when they produce "hot springs" in Io's subsurface layers. Thus the magnetic field of the giant planet, as well as the tidal stresses Jupiter produces, may help to heat Io, inducing the volcanic activity we have discovered with the Voyager flybys.

Europa. Europa, second of the large moons, is the smallest, most highly reflecting, and least pocked of the Galilean satellites (Figure 13-12). The surface shows no large impact basins, and the most striking features are intersecting lines several thousand kilometers long and about 100 kilometers wide. To scientists, these lines suggest tectonic processes; that is, motions of the satellite's crust. If this is true, then the global fractures of Europa's surface may be evidence of continuing activity caused by processes within Europa, processes we can only guess at for now. Europa may have an icy crust, in which case the lines should represent cracks along the crust.

Ganymede. Ganymede deserves a closer look, for it is the most massive satellite in the solar system, and has a diameter greater than the largest moons of Saturn and Neptune (Figure 13-13). The density of Ganymede, 1.93 grams per cubic centimeter, falls far below the densities of Io and Europa and may reflect the existence of a thick mantle of ice that surrounds a core made mostly of silicon and oxygen (Figure 13-14). Of the four large satellites,

Figure 13-13 Ganymede has 3½ times our own moon's volume and twice its mass.

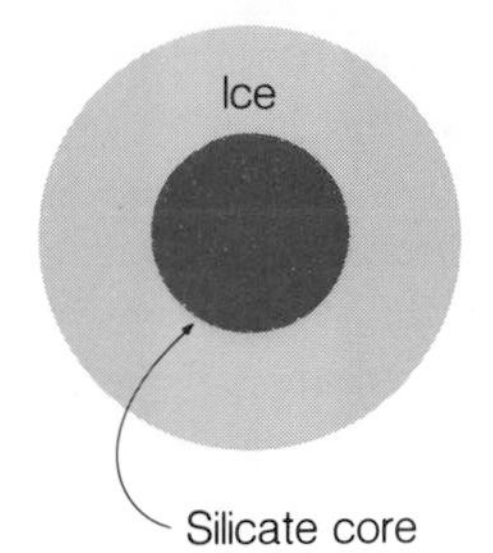

Figure 13-14
The internal structure of Ganymede and Callisto may include a silicon-oxygen core (that is, rocks) wrapped in a thick mantle of ice.

Ganymede most closely resembles our moon, since it appears to have many impact craters (Figure 13-13). Closer inspection reveals a crisscross pattern of stripes, and high-resolution photographs show that the dark background and lighter stripes represent two different kinds of surface, cratered and grooved. The cratered parts of Ganymede have many craters of several kilometers' diameter; the grooved areas show networks of shallow trenches, with several parallel grooves crossing one another (Figure 13-15). The fact that some of these grooved networks show sudden offsets suggests plate-tectonic activity on Ganymede.

The density of craters on the cratered areas of Ganymede resembles the density of craters on the ancient highland parts of the moon and on the other cratered objects in the inner solar system, except that Ganymede has

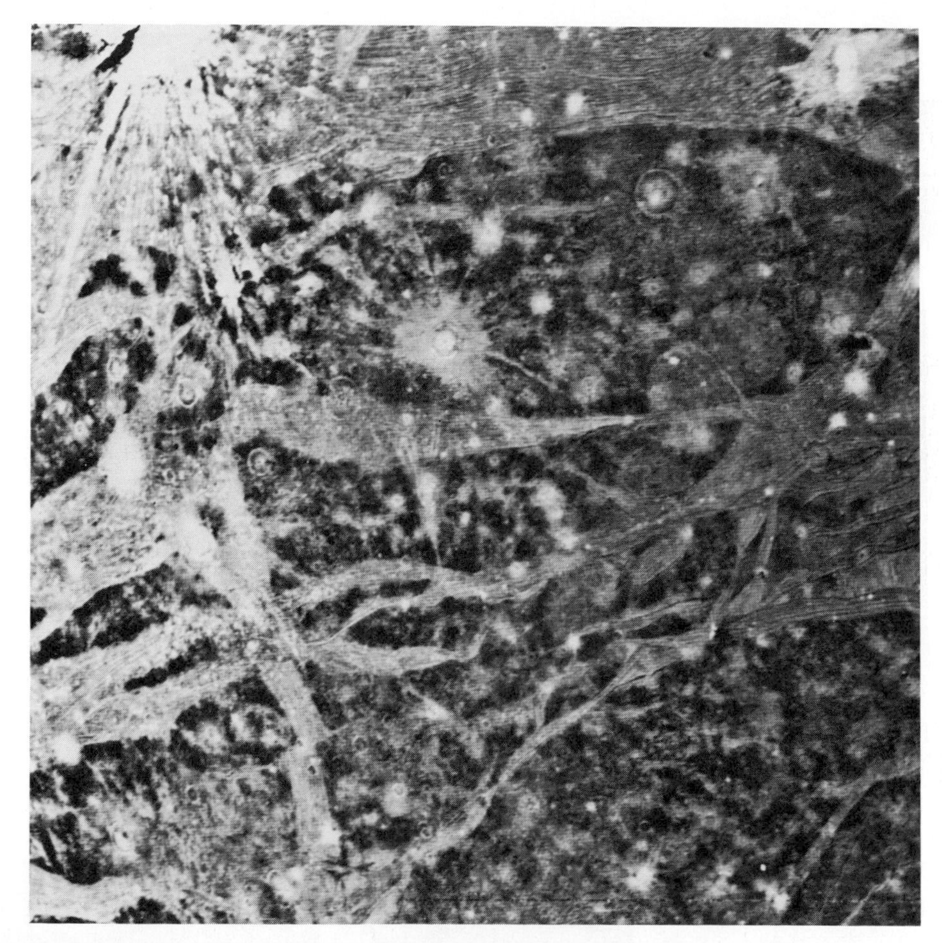

Figure 13-15 Detailed photographs of Ganymede show that its white spots appear to be bright craters with rays of material coming from them, presumably ejected by the impacts that made the craters. These rayed craters should be much younger than the dark craters with no rays. The surface also shows intricate patterns of intersecting lines.

few large craters. From this we may tentatively conclude that Ganymede's surface—at least in the cratered regions—has an age comparable to the 4 billion years, or a little more, that characterizes the familiar cratered surfaces closer to Earth. The grooved regions of Ganymede do have craters, but their density equals only 1/10 the density of craters in the cratered areas. This lower density of cratering approximates that of the lunar maria and of the oldest plains on Mars, and suggests that these areas formed after the initial burst of cratering had passed, but still several billion years ago.

Unlike the moon, Ganymede has no mountain ranges or large impact basins. The largest craters have diameters of about 50 kilometers, far below the 600-kilometer diameter of Mare Imbrium on the moon. The Voyager scientists suggest that Ganymede's crust remained active longer than the moon's did, and thus replaced some of the older, cratered regions with grooved areas, while washing out the original large- and medium-sized basins and mountain ranges.

Callisto. Callisto, fourth and least dense of the Galilean satellites, has the darkest surface, though it still reflects light twice as well as our moon. If you want to find craters, Callisto offers a fantastic opportunity (Figure 13-16). The satellite has not only a multitude of impact craters, but also a huge multiring basin that resembles those seen on Mercury and the moon. But in contrast to

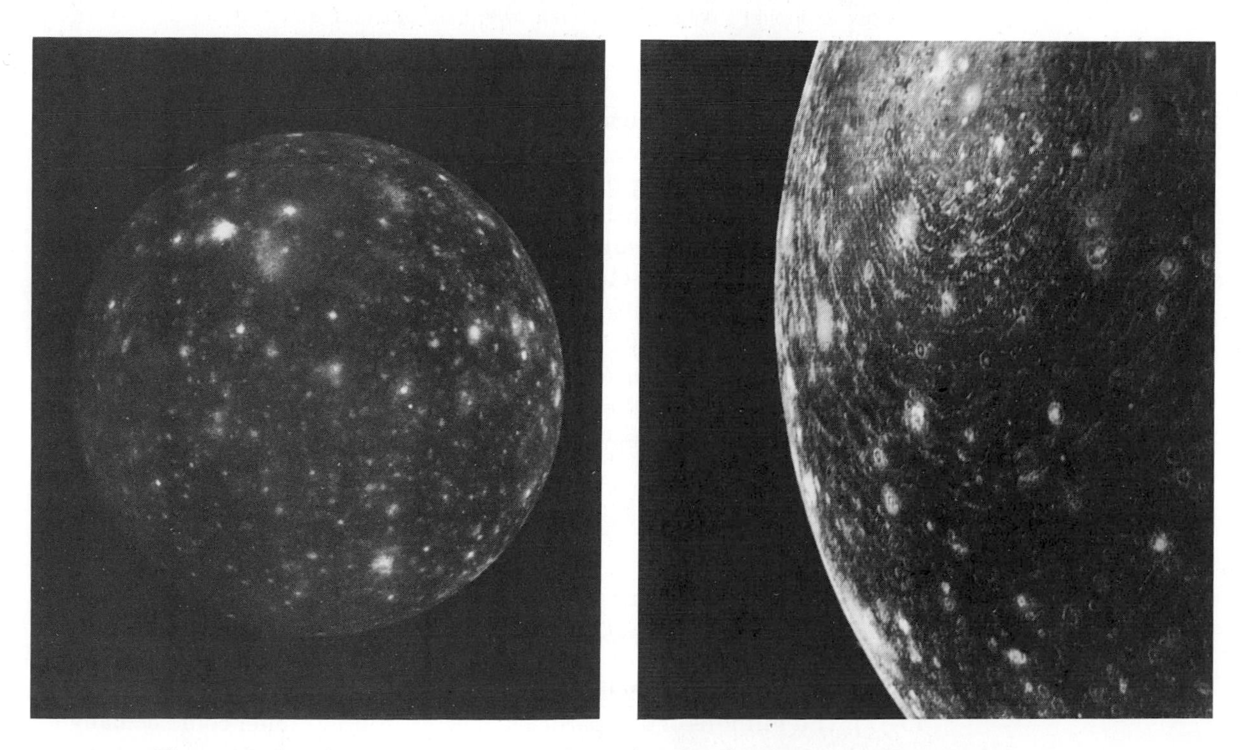

Figure 13-16 Callisto's surface has a fantastic array of craters, including one giant multiring system.

Figure 13-17 The rings of Jupiter, discovered by Voyager 1, were rephotographed by Voyager 2. Jupiter's rings are much narrower than Saturn's, and therefore were never discovered from Earth.

the Caloris basin on Mercury (Figure 12-3) and the great impact basins on the moon, Callisto's ringed pattern has no central basin. The relative flatness of the rings surrounding the central bright patch and the equal spacing of these rings contrast with the basins of the inner solar system. These differences in structure presumably arise from the differences in the materials that form the surfaces of Callisto (mostly ice) and of Mercury and the moon (mostly rocks). Since Callisto's density of matter, 1.79 grams per cubic centimeter, falls only slightly below Ganymede's, we cannot explain the marked differences in the satellites' surfaces on the basis of their bulk composition. Instead, we must await the results of the Galileo Project to make more detailed examination of these two moons and to search for new clues to their origin and structure.

The Rings of Jupiter

Before we leave the Jovian system, we ought to admire one more discovery made by Voyager: the **rings** of Jupiter. Much smaller than the famous rings of Saturn, these streams of rock and dust debris in orbit close to the planet were discovered by Voyager 1 and better photographed by Voyager 2 (Figure 13-17). The millions of tiny particles in orbit around Jupiter could actually be counted as moons, were we not limited by our need to keep a reasonable ceiling on the number of satellites in the solar system.

13.3 Saturn

Saturn, most beautiful among planets, owes its fame to the broad system of rings that circle this second-largest planet. In 1610, Galileo reported the first telescopic observations of Saturn. Since the rings appeared only indistinctly, Galileo concluded that Saturn had two large satellites, one on either side. "I have observed the farthest planet to be triple," Galileo wrote in an anagrammatic message. The riddle was resolved a generation later, when Christiaan Huygens used a better telescope to discover the true configuration of the rings. Three hundred years after Huygens, the Pioneer and Voyager spacecraft, inheritors of the exploring tradition that dominated the Holland of Huygens's era, reached Saturn, no longer the farthest planet known but still the farthest to be reached by human exploration.

Similarities Between Saturn and Jupiter

So far as we can tell, Saturn has a basic resemblance to Jupiter, though somewhat smaller and less dense. Like Jupiter, Saturn is a mixture of hydrogen and helium, with a sprinkling of the other less abundant and less dense elements in roughly cosmic proportions. Photographs of the planet show a system of banded patterns in the atmosphere, much like Jupiter's, though less intense in color and with far less of the red-brown coloration of the largest planet (Figure 13-18 and Color Plate E). Saturn further resembles Jupiter by its possession of an internal source of heat. Thus it should also have vertical convection currents that circulate material between the frigid upper layers and the warmer lower layers of its atmosphere.

The differences between Jupiter and Saturn arise primarily from Saturn's greater distance from the sun and its smaller size, which implies a less effective heat source. At 9.54 times the Earth's distance from the sun, Saturn

Figure 13-18
Saturn, the most flattened among planets, has a famous system of flat, wide rings. This spectacular close-up of the rings was taken by the Voyager 1 spacecraft.

Figure 13-19 Saturn's rings consist of millions of icy boulders and snowballs in orbit around the planet. These particles all move in almost the same orbital plane.

receives only 1/91 as much solar energy as the Earth on each square centimeter. Furthermore, Saturn's mass, though a full 95 times the Earth's, equals only 30% of Jupiter's mass, and the planet's average density, 0.7 gram per cubic centimeter, makes Saturn the least dense of planets. With its much lower mass and lesser self-gravitation, Saturn's slow contraction yields less heat than Jupiter's. On Saturn we would soon learn what cold really can be.

The surface layers of Saturn have a temperature of 95 K, a full 30° colder than Jupiter's surface. This may seem a small difference, especially when we deal with such low temperatures, but its significance lies in the fact that ammonia condenses much more effectively at 95 K than at 125 K. Hence we find Saturn with a thick layer of ammonia cirrus clouds that constantly hide from our view the warmer, lower atmosphere, which might well have characteristics like Jupiter's. Although we can often see part way through the upper atmosphere of Jupiter, with Saturn we cannot.

The Rings of Saturn

Saturn's rings are actually swarms of rock and ice particles that move around the planet in nearly circular orbits and in the plane of the planet's equator (Figure 13-19). The full extent of the ring system exceeds 200,000 kilometers, but all of the particles move within a belt no more than a few hundred meters thick! The reason for this remarkable thinness of the rings is that if one of the particles wanders above or below the plane of the rings (perhaps because of collisions with other particles), the collective gravitational pull

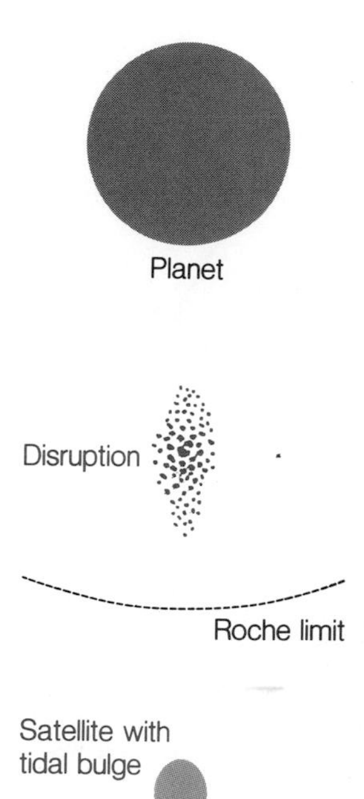

Figure 13-20
If an object is held together by its own gravitational force, it cannot survive in one piece within the Roche limit of a larger object. Satellites barely outside the Roche limit will develop large tidal bulges.

from the other particles, even though much weaker than Saturn's gravitational force, will gradually bring the errant particle back into the general swarm. In this process, the particles in the rings mimic the effects of stars in the plane of the Milky Way galaxy, on a scale just about a trillion times smaller (see page 153).

The ring particles apparently represent the remnants of a would-be satellite that could never form because the attempted point of formation lay too close to Saturn itself. If we calculate the *difference* in gravitational attraction exerted by the planet on two adjacent particles, we can compare this tidal disruptive force with the force of gravitational attraction between two particles. Within a certain distance of any planet, called the **Roche limit** (after its discoverer Edward Roche), the tidal disruptive force exceeds the gravitational attractive force between two adjacent particles (Figure 13-20). Inside the Roche limit, an object cannot hold itself together by self-gravitation. The millions of tiny satellites—rocks, boulders, snowballs, and dust particles—that form Saturn's rings are each held together not by self-gravitation, but rather by the electromagnetic forces involved in the chemical bonds of their silicate lattices. Similarly, electromagnetic forces can hold together an artificial Earth satellite well inside the Roche limit.

The total mass of the particles in the rings hardly equals that of the smallest known satellites of Saturn, but because their material spreads over such a wide area, they reflect light well. Thus they can be seen far more easily than a satellite, which keeps most of its matter *inside*, where it cannot reflect light. Even a small telescope, or a good pair of binoculars, can reveal Saturn's rings as well as the planet's largest satellite—second largest of all satellites—Titan.

Titan

Titan deserves attention, not so much for its impressive diameter—which exceeds that of the planet Mercury, though Titan's mass is just 41% of Mercury's—as for its atmosphere. The Galilean satellites of Jupiter may well have some extremely thin layers of gas around them. Io, we know, has a cloud of sodium, sulfur, and potassium ions around it, probably connected with the volcanic activity on that satellite, and a transient, local atmosphere of sulfur dioxide. But Titan's atmosphere, much thicker than Io's tenuous blanket, contains far more material, and exerts more pressure, than that of the Earth. This atmosphere consists mostly of nitrogen, and its temperature rises to a surprising (though hardly warm) 175 K in its upper reaches. Since radio measurements of Titan reveal a surface temperature of 92 K, something odd must be happening in Titan's atmosphere, which consists of nitrogen, methane, and traces of hydrogen cyanide, ethane, and acetylene. The presence of hydrogen cyanide is especially interesting because it implies that more complex organic molecules may exist (see Chapter 14). Titan's surface is so cold that lakes of liquid methane may exist, within which interesting chemical reactions could occur. Someday we may land automated spacecraft on Titan to investigate these possibilities.

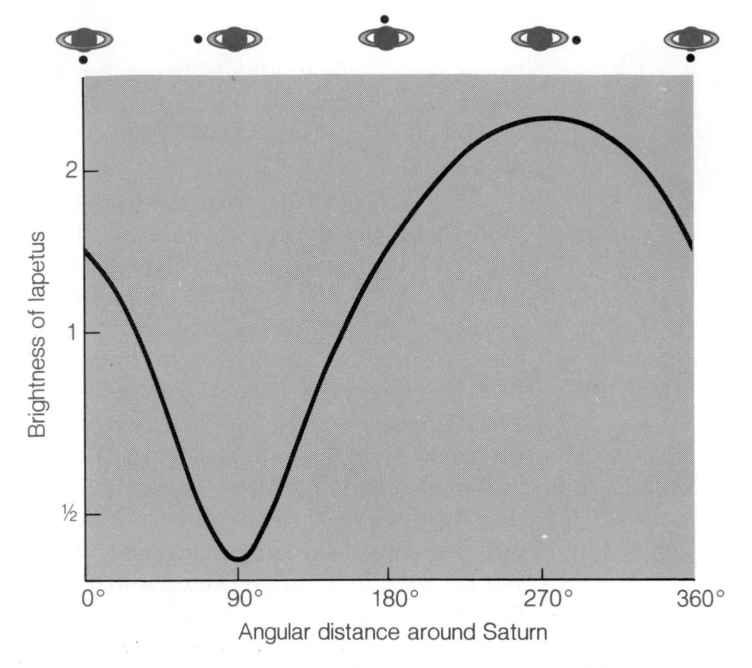

Figure 13-21 Iapetus, third-largest of Saturn's satellites, changes its apparent brightness by a factor of 7 during each orbit around the planet.

Saturn's Other Satellites

Saturn has at least fourteen other satellites besides Titan. Mimas, Dione, and Rhea are heavily cratered. The inner satellites consist largely of ice, which at extremely low temperatures becomes so rigid that sharply defined craters can persist. Farther out than Titan, the outer satellites appear to be somewhat denser than the inner five, and one of them seems quite odd indeed: Iapetus. This moon (whose name is pronounced Yă'-peh-tŭs) changes its brightness by a factor of 7 as it orbits Saturn, always keeping the same side toward the planet (Figure 13-21). One side of Iapetus, therefore, is seven times better at reflecting light than the other, which suggests that the surface is half ice and half rock.

13.4 Uranus

When we pass beyond Saturn in the solar system, we enter realms of distance from the sun far greater than those we have yet traversed. These distances are so great that the Voyager 2 spacecraft, traveling at 20 kilometers per second, will take more than four years to journey from Saturn to Uranus, more time (and a longer distance) than the Voyager needed to travel from the Earth out to Saturn (Figure 13-2). The three outermost planets of the solar system have such great distances from the sun that we see them only dimly, and the ancient astronomers who named the Earth's closer neighbors remained ignorant of Uranus, Neptune, and Pluto.

Uranus, barely bright enough to be seen without a telescope, had in fact been seen, but mistaken for a star, many times before William Herschel recognized the planet's disk in his telescope. Since Herschel made his discovery in 1781 in England, where he had immigrated from Germany, he suggested naming the new planet *Georgium Sidus* (George's Star) after King George III. Astronomers, however, soon replaced this with the mythological name Uranus, the Greek god of the heavens, father of Cronus (in Latin, Saturn) and husband of Gaea (Earth).

In a modern telescope, Uranus appears as a relatively featureless planet, though faint bands similar to those on Saturn have been reported (Figure 13-22). From spectroscopic studies, astronomers have found that Uranus's outer layers consist largely of methane and hydrogen. No doubt helium is present too. At the low temperature of 65 K, the result of Uranus's great distance from the sun, ammonia actually freezes out of the planet's atmosphere.

Uranus and its sister planet Neptune appear to be basically pale, small copies of Jupiter and Saturn, less than half as large as either and much colder, both because of their greater distances from the sun and because of their lesser sources of internal heat from contraction (see page 449). Without

Figure 13-22 Even the best photographs of Uranus taken from Earth show little detail on the planet. Labels mark Uranus's five satellites. The diffraction spikes arise in the telescope used to take the photographs.

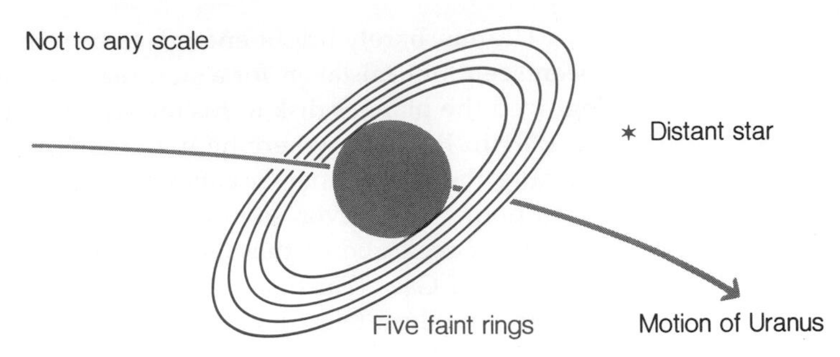

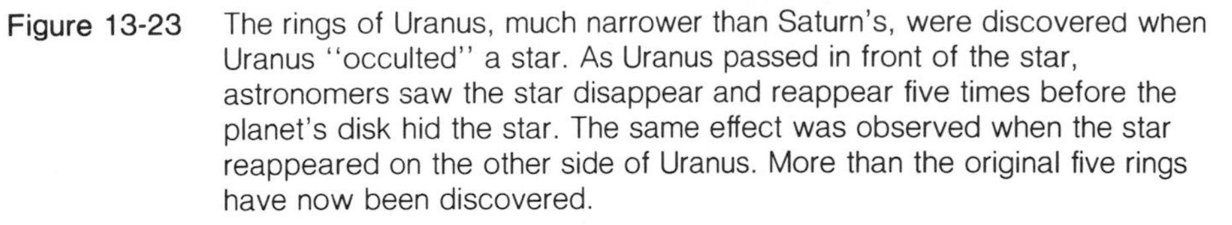

Figure 13-23 The rings of Uranus, much narrower than Saturn's, were discovered when Uranus "occulted" a star. As Uranus passed in front of the star, astronomers saw the star disappear and reappear five times before the planet's disk hid the star. The same effect was observed when the star reappeared on the other side of Uranus. More than the original five rings have now been discovered.

close-up photographs and spectroscopic observations, which will become available only during the late 1980s, we can say little more about these two giant planets. We can, however, note the strange rotation of Uranus. This planet's axis of rotation tips by a full 82° from being perpendicular to the plane of its orbit. In other words, Uranus rolls like a barrel as it orbits the sun. How the seventh planet acquired this rotation, which contrasts with the tendency of every other planet to keep its rotation axis roughly perpendicular to its orbital plane, remains a mystery.

Uranus has five moons, none particularly large, and all named after characters from Shakespeare's plays and from Alexander Pope's *Rape of the Lock*. In addition, the planet possesses a complex system of rings, millions of tiny satellites in orbit around the planet's equatorial plane (Figure 13-23). Astronomers first detected these rings in 1977, when Uranus passed directly in front of a star. During this **occultation**, the star seemed to disappear and reappear five times before the actual occultation by the planet's disk occurred, and the same phenomenon happened after the occultation, as the star reappeared on the opposite side of the planet. Since then, additional rings of Uranus have been found, though the total mass in all the rings probably does not equal more than the mass of a medium-sized mountain. Voyager 2 should help decide what the particles in the rings of Uranus are made of, and why they are concentrated into narrow intervals of distance from the planet, instead of being spread out over wide intervals like the rings of Saturn.

13.5 Neptune

The discovery of the eighth planet from the sun represented a remarkable triumph for the astronomers who had predicted Neptune's existence from studies of the motion of Uranus. These observations revealed that Uranus made small deviations from the orbit it would have if no planet beyond

Uranus perturbed it by gravitational force. Calculations made by John Couch Adams in England and by Urbain Leverrier in France showed where to look for the planet beyond Uranus that perturbed its orbit. Although Adams was first with his calculations, Leverrier had better connections with astronomical observatories, and the discovery of Neptune by Johann Galle in 1846 relied on Leverrier's work.

Neptune has a nearly circular orbit around the sun, at 30 times the Earth's distance from the sun, so the intensity of sunlight on each square centimeter of the planet equals 1/900 of that on Earth. At this distance, solar heating is so low that the planet's outer layers have a temperature just 50 K above absolute zero. So far as we can tell, Neptune closely resembles Uranus, though its mass is 19% larger and its density 35% higher (Figure 13-24). Neptune turns out to be by far the densest of the outer planets (though its density still equals only 31% of the Earth's), suggesting a relatively large core of rock and ice material. Neptune's outer layers contain methane, as well as hydrogen molecules and (presumably) helium atoms.

Since we know that Jupiter, Saturn, and Uranus possess ring systems, we can assume that Neptune also has swarms of small rock and ice particles in orbit close to the planet. We do know that Neptune has two moons, one large, one tiny. Triton, the larger moon, ranks fourth in size, behind only Ganymede, Titan, and Callisto among solar-system satellites. Triton has a peculiar orbit around Neptune which, although almost circular, carries this massive satellite in the reverse direction to Neptune's orbit. Every other satellite in the solar system, except for what appear to be captured asteroids that

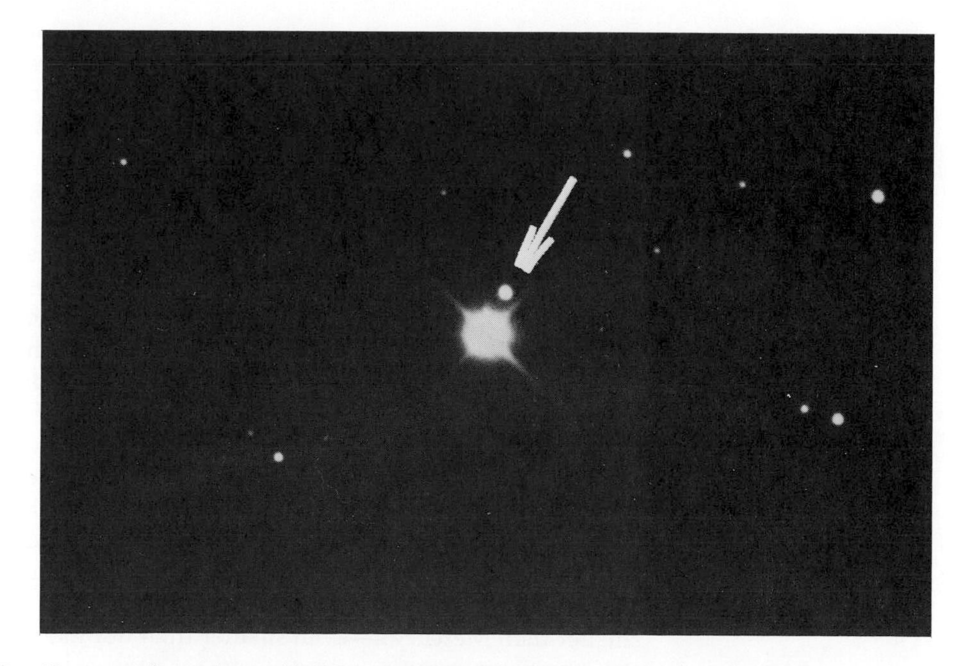

Figure 13-24 Neptune, shown here with Triton, its larger satellite (arrow), is almost twice as far from us as Uranus.

Figure 13-25
Two photographs of Pluto taken several days apart reveal the motion of the planet relative to the faraway stars within that time. However, such photographs cannot show us anything of the planet's surface.

orbit Jupiter and Saturn at great distances, moves around its planet in the same direction that the planet orbits the sun. Neptune's small moon, Nereid, also has an unusual orbit, with the largest eccentricity known for any satellite. The oddities of these two satellite orbits has prompted speculation that Pluto, the strange exception among the outer planets, may once have been a satellite of Neptune, and that the ejection of Pluto from the Neptune system may have left the unusual orbits we see for Triton and Nereid.

13.6 Pluto

Uranus was discovered by accident, Neptune by careful observation and computation. Quite fittingly for the modern era, Pluto was found by what hindsight shows to be a mixture of these activities. After Neptune had been observed to move through part of its orbit, some astronomers thought they had evidence that an outer planet was producing gravitational deviations in Neptune's path around the sun.[2] Their calculations suggested that such a planet should have a mass roughly equal to Neptune's. At the Lowell Observatory in Arizona, which had been originally built to search for life on Mars (see page 425), astronomers began a systematic search of the regions near the ecliptic, photographing each region on at least two different nights and then comparing the photographs to see if any points of light showed a motion relative to the fixed stars. In 1930, Clyde Tombaugh discovered Pluto (Figure 13-25), providing a ray of achievement in an otherwise dismal year. But Pluto appeared far dimmer than a Neptune-sized planet would, and for several decades astronomers wondered whether Pluto could include a Neptunelike mass in a much smaller volume, which would make Pluto by far the densest planet known. We now know that Pluto has a diameter much smaller than Mercury's—even smaller than our moon's—and that it has one-

[2]Notice that Neptune has yet to make its first complete orbit around the sun since its discovery in 1846!

seventh of the moon's mass. In short, Pluto seems to be an icy object, more like a satellite than a giant planet. But in that case, astronomers must have originally overestimated the planet's mass by a factor of 100 or more! In other words, the calculations that inspired the search for Pluto must have been just plain wrong—yet Pluto was there after all!

In 1977 James Christy discovered a satellite of Pluto, a tiny moon whose orbit allows us to compute Pluto's mass and, to some extent, its size. On the one hand, this satellite, named Charon after the boatman who ferried dead souls to Pluto's underworld, came as a welcome surprise. Astronomers could finally eliminate the notion of Pluto as a superdense planet. On the other hand, Pluto's satellite tends to demolish the attractive idea that Pluto was once a satellite of Neptune, since we would have to accept the idea that a satellite could have, or could acquire, its own satellite, or that both Pluto and its satellite could have escaped together and remained together.

When we consider that Pluto's orbit actually crosses *inside* Neptune's (Figure 13-2), we may be reluctant to abandon the former-satellite hypothesis for Pluto, especially since it would explain so well why Pluto differs markedly from the giant planets. But we may have to abandon this theory in the end and conclude that Pluto is a piece of icy debris that formed, as did its satellite, along with the rest of the solar system, and in a highly eccentric orbit that will keep Pluto closer to the sun than Neptune during the remainder of the twentieth century.

With Pluto we come to the end of the sun's planetary system. Suggestions have often been made of a trans-Plutonic planet, a tenth planet X. But the same kind of investigation that found Pluto has failed to reveal any such planet. We must therefore conclude that if such a planet exists, it must either be much farther from the sun than Pluto (say, 200 times the Earth's distance instead of an average of 39 for Pluto), or much smaller even than Pluto, or in an orbit around the sun that carries it far out of the ecliptic and thus out of the areas of the sky that have been carefully searched for a tenth planet.

Our excursions through the solar system have carried us past the realms explored by spacecraft, into the interstellar regions visited only by comets (see page 395). But little doubt exists that before long, we will venture out to Neptune, Pluto, and the region beyond, just as we have explored the sun's six inner planets. By that time, our understanding of where the sun and its planets came from may have grown enormously, and we may have detected the planets that orbit other stars, if not the life that may exist on these extrasolar planets.

Summary

The four giant planets, Jupiter, Saturn, Uranus, and Neptune, each far exceed the mass and size of the Earth, the largest of the inner planets. Unlike the rocky-iron composition of the four inner planets, the four giants are spheres of gas, mostly hydrogen and helium, resembling the stars in their composition far more than the inner planets do. The giant planets owe their gaseous, hydrogen-helium composition to the fact that they have retained much of their original matter, whereas the inner planets lost the lightest elements after the sun began to shine.

Jupiter, the largest and nearest of the giant planets, has received the closest scrutiny. In addition to hydrogen and helium, its atmosphere contains methane, ammonia, and water vapor, just the gases we would predict for an object that had a composition determined by the ingredients of the primitive solar system. But we also find colored material in the cloud layers, indicating the presence of more complex molecules, as yet undetermined. The most famous colored region, the Great Red Spot, may be a semipermanent cyclone driven by Jupiter's rapid rotation.

Jupiter (and Saturn as well) has an internal source of heat, the result of the slow contraction of its interior, similar to the prenuclear fusion stages of a protostar. The heat released inside Jupiter makes its lower atmosphere warmer than the upper atmosphere and drives vertical convection currents that help produce the changing patterns of the upper atmosphere we observe. Deep inside Jupiter, the hydrogen thickens into liquid, and eventually into a metallic phase, which may help to explain the planet's strong magnetic field. The other giant planets, less well studied than Jupiter, are thought to be smaller models of the basic structure of the giant among the sun's planets. Each of the giant planets has a ring system (except for Neptune, which may yet turn out to have rings), composed of debris in orbit close to the planet. But for unexplained reasons, Saturn's rings cover a wide range of distances from the planet, whereas those of Jupiter and Uranus do not.

The giant planets have extensive satellite systems, including, in the cases of Jupiter, Saturn, and Neptune, satellites larger than our own moon and almost as large (though far less dense) than Mercury. The large satellites, which almost certainly formed along with the planets, show a variety of surfaces that testify to different conditions among them. In particular, Io, Jupiter's innermost large satellite, has active volcanoes that cover and recover the surface with new flows of sulfur-rich material. Titan, Saturn's giant satellite, has a relatively thick atmosphere that gives it a reddish color, while Iapetus, the second-largest satellite of Saturn, shows an impressive change in brightness, for reasons still unknown.

Pluto, the outermost planet, has no resemblance to the giants. Smaller than Mercury, this planet might once have been a satellite of Neptune, though the fact that Pluto has a satellite of its own argues against this hypothesis. Pluto's density shows, as we would expect from Pluto's small size, that the planet consists of ice, not hydrogen and helium gases, but the nature of its surface has yet to be determined.

Key Terms

aurora
Galilean satellites
Great Red Spot
inorganic compounds
magnetic field
occultation
organic compounds
rings
Roche limit
self-gravitation

Questions

1. Why does Jupiter consist mostly of hydrogen and helium?
2. How can we explain the varied colors on Jupiter?
3. What is Jupiter's Great Red Spot? How long has it existed?
4. Why does Jupiter's atmosphere become warmer as we descend into it?
5. Why does Jupiter produce heat in its interior?
6. What effect does Jupiter's magnetic field have on the charged particles that come near the planet?
7. What causes the impressive mixture of colors on Io?
8. What evidence suggests that Ganymede's surface is as old as the surfaces of Mercury and the moon?
9. What do Saturn's rings consist of? Why are they so thin?
10. What is the Roche limit? How does it help explain the existence of Saturn's rings?
11. What is special about Saturn's largest satellite, Titan? What is special about its second-largest satellite, Iapetus?
12. How were Uranus's rings detected? How was Uranus itself found?
13. How was Neptune found? What does this tell us about our understanding of the gravitational forces among the planets?
14. How was Pluto found? What characteristic of Pluto does its moon allow us to determine? What impact does its moon have on theories of Pluto's origin?

Projects

1. *Observation of Jupiter and Saturn.* If you are observing in spring or summer, find Jupiter and Saturn in the constellations Virgo and Libra. About how long does it take each of these planets to make a complete circle around the ecliptic? Can you observe the motions of these planets, relative to the stars, after an interval of several weeks?
2. *Observation of the Galilean Satellites.* Observe Jupiter with a small telescope and plot the location of its four large satellites on several consecutive nights. With more observation, you can determine the orbital period of each of these satellites to a fair degree of precision and compare your results with those given in Table 12-1. *Sky and Telescope* shows the positions of the Galilean satellites for each night of a particular month.
3. *Telescopic Study of Saturn.* Observe Saturn with a small telescope and note the size of the ring system relative to the planet. You should be able to see the largest satellite, Titan, with little difficulty, and perhaps the next two, Iapetus and Rhea, if conditions are favorable.
4. *Observation of Uranus and Neptune.* Use the finding charts published each year in *Sky and Telescope* and a pair of binoculars to find Uranus and Neptune among the stars.

Further Reading

Cruikshank, Dale, and Morrison, David. 1976. "The Galilean Satellites of Jupiter." *Scientific American* 234:5, 108.

Fimmel, Robert; Swindell, William; and Burgess, Eric. 1974. *Pioneer Odyssey: Encounter with a Giant.* Washington, D.C.: NASA, U.S. Government Printing Office.

Ingersoll, Andrew. 1976. "The Meteorology of Jupiter." *Scientific American* 234:3, 46.

Sagan, Carl; Drake, Frank; Druyan, Ann; Ferris, Timothy; Lomberg, Jon; and Sagan, Linda. 1978. *Murmurs of Earth: The Voyager Interstellar Record.* New York: Random House.

Scientific American. 1975. *The Solar System.* San Francisco: W. H. Freeman.

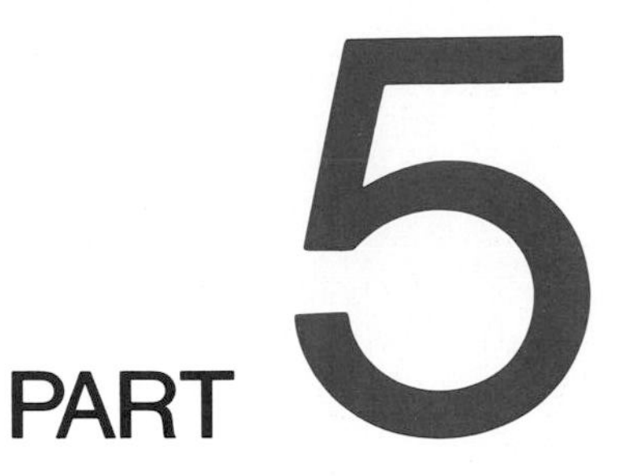

PART 5

The Search for Life in the Universe

For thousands of years, human beings have wondered about the origin of life on Earth and have considered the possibility that distant worlds may be teeming with other kinds of living creatures. When the Roman poet Lucretius wrote, "We are all sprung from heavenly seed," he was perfectly correct. The elemental composition of our bodies reflects rather faithfully the abundance of elements in the universe. Two thousand years later, we know that life on Earth began more than 3.5 billion years ago, apparently the result of molecular processes within "some warm little pond," to use Charles Darwin's phrase.

We do not know whether or not life has begun elsewhere in the universe, and how many times, and in how many places. Nor do we know the best way to attempt to find and to communicate with other intelligent forms of life, if they exist. We can, however, use our knowledge to guess at the answers to these problems and to debate whether we ought to make a serious effort to search for other civilizations in our galaxy.

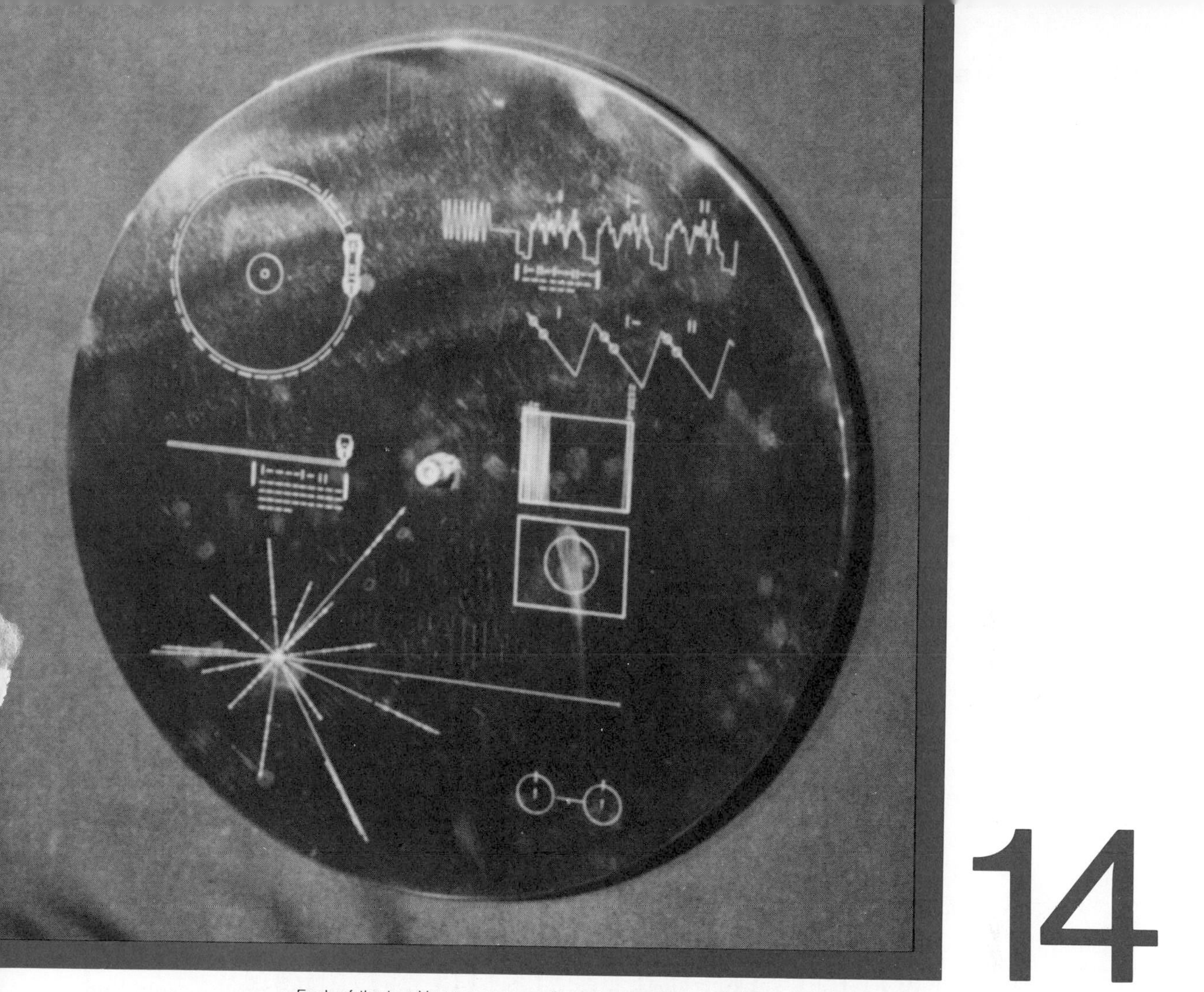

Each of the two Voyager spacecraft carries messages and pictures from Earth on gold-anodized records that will travel into interstellar space.

14

The Age of Exploration: The Search for Extraterrestrial Intelligence

Fifteen billion years of cosmic evolution, 4.6 billion years of solar system evolution, and more than 3 billion years of life on Earth have brought forth on this planet a species capable of exploring its surroundings. Since the human search for extraterrestrial life has begun only within the last hundred-millionth of the universe's history, we should not be surprised that we on Earth have still not found our closest neighbors with whom to communicate. In fact, we have yet to find another planet with life of any sort. We can now use our knowledge of the development of the universe to assess our chances of finding another civilization and the desirability of making such a search. The time may not be far away when our relationships with other galactic civilizations take on as much importance as the relationships among nations here on Earth.

Since our only knowledge of life comes from a single example, life on Earth, we cannot embark on a comparative exploration of different forms of life. Instead, we must examine our own planet and the results from our exploration of the solar system in an attempt to draw conclusions from a limited set of data.

14.1 What Is Life?

When we try to gauge the likelihood of life outside the Earth, using the forms of terrestrial life as a guide, we immediately face the questions, What *is* life, and how did life originate on Earth? The answer to the second question, if we knew it, might help us answer the first. But for the time being, we must impose a definition of life and try to understand how life began through informed guesswork and detective studies.

Since life seems to represent a quality of matter, we may try to define life through a description of the behavior and structure of living material. But does this pragmatic approach miss the heart of the matter, a mysterious "living force," not definable in terms of physics and chemistry, that is essential to make ordinary matter come alive? Scientists are willing to bypass the philosophical implications of such terminology. They take a descriptive approach and try to identify the characteristics that most clearly distinguish life from inanimate matter. The most striking feature of **life** is the ability to *reproduce* and *evolve.* A flame can produce other flames, and protons colliding with protons can, on occasion, produce more protons. But the defining property of life is that in addition to reproduction, life can transform itself continuously into new forms.

14.2 The Chemical Composition of Living Organisms

When we examine the elemental composition of life on Earth, a remarkable conclusion emerges that may have importance for our consideration of life elsewhere in the universe. The composition of living matter resembles the composition of the stars more closely than the composition of the planet on which we find ourselves (see Table 14-1). Just four elements—hydrogen, carbon, nitrogen, and oxygen—form 99% of what we call living matter. These are *not* the four most abundant elements on Earth, or in its crust, but they are the four most abundant elements in the universe (if we ignore helium and neon, which rarely form any compounds).

Table 14-1 The Most Abundant Elements (by Number) in the Sun, in the Earth, in the Earth's Crust, and in Living Organisms[a]

Sun		Earth		Earth's Crust		Bacteria		Human Beings	
Hydrogen	93.4%	Oxygen	50%	Oxygen	47%	Hydrogen	63%	Hydrogen	61%
Helium	6.5	Iron	17	Silicon	28	Oxygen	29	Oxygen	26
Oxygen	0.06	Silicon	14	Aluminum	8.1	Carbon	6.4	Carbon	10.5
Carbon	0.03	Magnesium	14	Iron	5.0	Nitrogen	1.4	Nitrogen	2.4
Nitrogen	0.011	Sulfur	1.6	Calcium	3.6	Phosphorus	0.12	Calcium	0.23
Neon	0.010	Nickel	1.1	Sodium	2.8	Sulfur	0.06	Phosphorus	0.13
Magnesium	0.003	Aluminum	1.1	Potassium	2.6			Sulfur	0.13
Silicon	0.003	Calcium	0.74	Magnesium	2.1				
Iron	0.002	Sodium	0.66	Titanium	0.44				
Sulfur	0.001	Chromium	0.13	Hydrogen	0.14				
Argon	0.0003	Phosphorus	0.08	Phosphorus	0.10				
Aluminum	0.0002			Manganese	0.10				
Calcium	0.0002			Fluorine	0.063				
Sodium	0.0002			Strontium	0.038				
Nickel	0.0001			Sulfur	0.026				
Chromium	0.00004								
Phosphorus	0.00003								

[a]For comparison, the abundances in the Earth's atmosphere are: nitrogen, 78%; oxygen, 21%; argon, 0.93%; carbon, 0.011%; neon, 0.0018%; and helium, 0.00052%.

We can explain the high abundance of hydrogen and oxygen in living creatures by the high percentage of water that they all contain. But carbon and nitrogen demand an additional explanation, which hinges on the chemical properties of carbon and nitrogen atoms. Carbon has the unique ability to form stable compounds in which as many as four other atoms link with each carbon atom. Nitrogen and oxygen atoms can each share more than one electron with a carbon atom, producing strong, though breakable, bonds that hold together complex molecules. Nitrogen and oxygen atoms have additional properties that make them useful in systems that change. For example, oxygen atoms easily combine with other atoms and with molecules in chemical compounds that release energy as they form. The major importance of finding hydrogen, carbon, nitrogen, and oxygen as the basic constituents of living matter lies in their cosmic abundance, which assures us that we are likely to find these atoms almost everywhere, and in the following conclusion, which we obtain from our look at how atoms interact: Certain types of atoms—those with useful chemical properties—are likely to be favored everywhere for the production of complex molecules, capable of interacting with one another in complicated ways.

In addition to the four basic elements, life on Earth typically contains calcium, phosphorus, and (still less abundantly) chlorine, sulfur, potassium, sodium, magnesium, iodine, and iron. Finally, various organisms have tiny amounts of *trace elements* such as manganese, molybdenum, silicon, fluorine, copper, and zinc. The concentrations of trace elements in bacteria, fungi, plants, and land animals show a marked correlation with the concentrations of these elements in sea water. This correlation suggests that life on Earth began in the seas of Earth, rather than arriving from another planet or from interstellar spores.

14.3 Biologically Important Molecules

We have listed the elements that comprise living creatures, but this does not tell us what life is, any more than a list of ingredients describes a cake. We must also know how the elements fit together into simple molecules, and how simple molecules join into more complex molecules.

Monomers and Polymers

At the molecular level, most forms of life consist of a small number of types of rather simple molecules called **monomers.** The best-known monomers are the twenty types of amino acids that form proteins. Other important monomers are sugars and fatty acids. The molecular units of living matter are mainly long, chainlike molecules called **polymers,** which are monomers strung together in a repetitive pattern, sometimes with small variations (Figure 14-1). In these polymers we often find ringlike structures and side chains, and the polymers themselves sometimes fold into elaborate, highly complex shapes. This ability to assume specific shapes, along with the ability to change shapes at appropriate times, allows some protein polymers to act as **catalysts,** sites where chemical reactions can occur much more rapidly than they could otherwise. Such catalysts are called **enzymes.**

In all these complex molecules, carbon is the element that allows the structure to exist. The complexity of molecular structure is essential for the storage and transmission of the information that allows one configuration of matter to reproduce itself, or to choose from many chemical compounds just the ones an organism needs to keep itself alive. In particular, all forms of living creatures on Earth rely on the same polymer, **DNA** (deoxyribonucleic acid) for the most basic of all processes, reproduction.

Part of a glycogen molecule

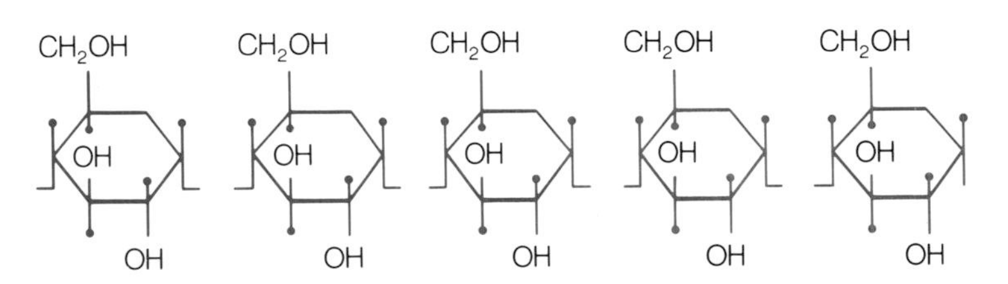

Figure 14-1 Large molecules called polymers, such as the glycogen molecule shown here in part, consist of small molecules called monomers strung together in a chain. The monomers in glycogen are glucose, each of which consists of twenty-two atoms. The chain in glycogen may repeat the glucose monomer thousands of times.

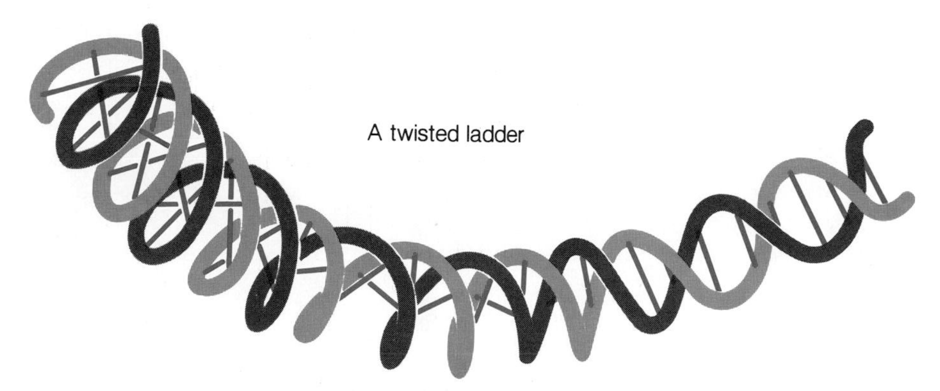

Figure 14-2 DNA molecules consist of two spirals, twisted around each other and connected by shorter molecules (adenine, cytosine, guanine, and thymine) joining in pairs.

DNA and the Genetic Code

Despite the diversity we see in the way animals and plants reproduce themselves, at a molecular level reproduction consists of nothing more nor less than the replication of DNA molecules (Figure 14-2). Furthermore, DNA molecules, together with their close relatives **RNA** (ribonucleic acid), tell the new organism, as well as the old one, how to function.

Each DNA molecule stores the **genetic code** that can tell the next generation of organisms how to carry out metabolism, grow, and reproduce. This code resides in the sequence of small molecules that runs along the inside of the double-stranded DNA (Figure 14-2). The two strands of each DNA molecule wind around each other like the handrails of a spiral staircase, the famed **double helix** of molecular biology. Joining one spiral with the other are pairs of four kinds of small molecules, adenine, cytosine, guanine, and thymine (Figure 14-3). These molecules cannot pair at random. Instead, adenine pairs only with thymine to form one kind of link, and guanine pairs only with cytosine to form the other kind. The *order* in which the pairs of molecules occur provides the genetic code.

During replication, each DNA molecule separates into two long strands as the adenine-thymine and guanine-cytosine pairs break apart. As each of the two strands then assembles the other half from available molecules, the new DNA molecules must have exactly the same pairs as the old one (Figure 14-3). The genetic code along each of the old strands completely determines the code along the newly formed strand, because only one kind of molecule will fit properly to make a pair. Thus each half of a DNA molecule can reproduce its partner exactly, creating two double strands of DNA identical with the "parent" molecule (Figure 14-3).

Each sequence of three pairing molecules—adenine, cytosine, guanine, or thymine—specifies a particular amino acid, in some sense a "word" of the genetic code. A sequence of 100 to 500 of these triplets along a strand of

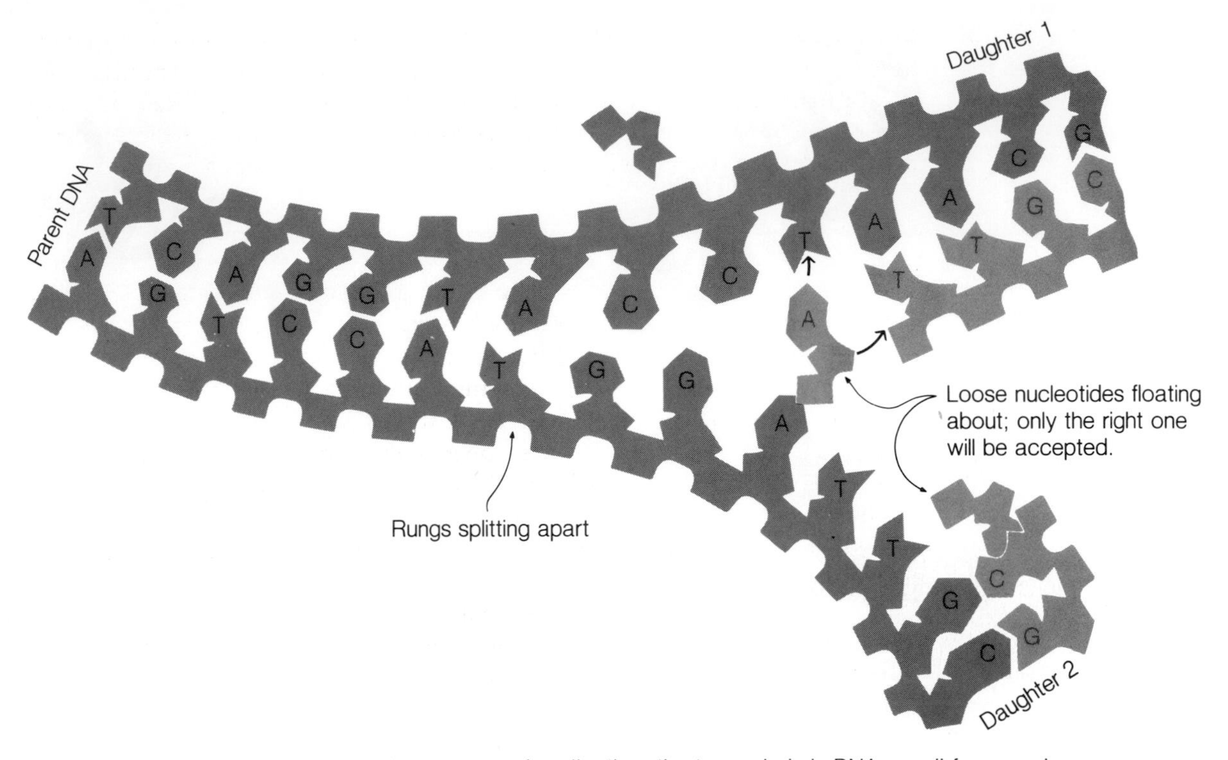

Figure 14-3 During the process of replication, the two spirals in DNA uncoil from each other and separate. When two daughter molecules form, one from each of the spirals, the restrictions on how the short molecules can link together make each daughter a copy of the parent molecule. Adenine can pair only with thymine, and cytosine can pair only with guanine. Thus the order of cross-linking molecules in each of the daughter DNAs duplicates that in the parent.

DNA can specify a particular protein or an RNA molecule. If such a sequence specifies a particularly important product for a living organism, it is called a **gene.** The steps by which DNA molecules govern the synthesis of protein molecules and other complex compounds are highly complicated and imperfectly understood, but they involve the use of DNA molecules as a sort of master blueprint. RNA molecules read this blueprint by assembling themselves along a strand of DNA. These **messenger RNAs** then react with another kind of RNA, small RNAs, to which amino acids attach themselves and link in the proper order to form a protein molecule. Information thus flows from the DNA molecule via the RNAs to the proteins.

This process of replication typifies the behavior of matter in living organisms. Complex, carbon-based polymers govern the formation and interaction of other complex polymers, using the smaller monomers as units. The energy that drives these processes comes directly or indirectly from the sun and is stored within the chemical bonds of particular polymers well suited to this purpose. All this activity occurs within the confines of living **cells,** membrane-bounded sacs of water in which these tiny yet complex molecules form, divide, and re-form (Figure 14-4). These cells allow the long polymers to float in one of the best **solvents** nature can provide—our familiar, life-giving reservoir of water. We shall consider the role of water, and its possible substitutes, at greater length as we examine the question of life on other planets. First we must see how life on our own planet began.

14.4 The Origin of Life on Earth

The basic unity of life on Earth—the fact that the same amino acids, the same proteins, and the same DNA and RNA occur in all living things—suggests a common origin of life. Since all living organisms have water-laden cells, and since the abundances of trace elements within these cells correspond with the abundances found in sea water, we might infer that life began in the seas. Our oldest evidence for life consists merely of microbial life, rather than large organisms (Figure 14-5). This suggests that life began as relatively small assemblages of matter, and only later evolved into large creatures.

But the most fundamental question of all—how atoms and molecules in the Earth's atmosphere and oceans ever combined into living matter—remains unanswered. No one has yet made a self-replicating, helical polymer from simple polymers without using products furnished by already living cells. The best we can do is to model what we think the Earth's early atmosphere and oceans were like and to find that relatively complex molecules *do* appear from natural causes within them. The steps from complex molecules to actual life must still be guessed.

Did Life Begin in Tide Pools?

The traditional scientific picture of the origin of life involves the spontaneous formation of complex molecules from atmospheric gases, and the molecules' later concentration in pools of water. These pools and ponds were

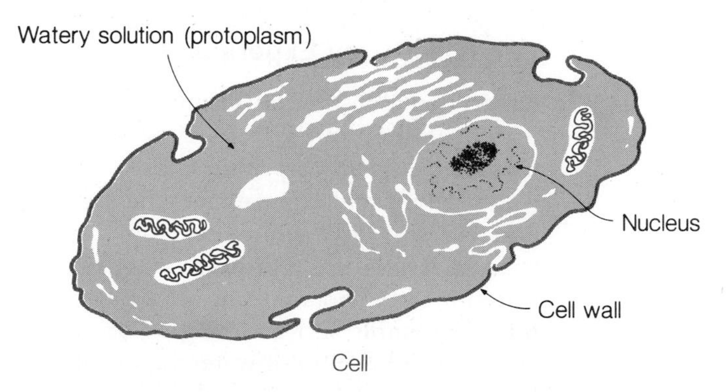

Figure 14-4 Cells provide localized environments in which large polymers float in a watery solution. Eukaryotic cells, such as the one shown here, have specialized subunits. In particular, these cells keep their DNA within a well-defined nucleus.

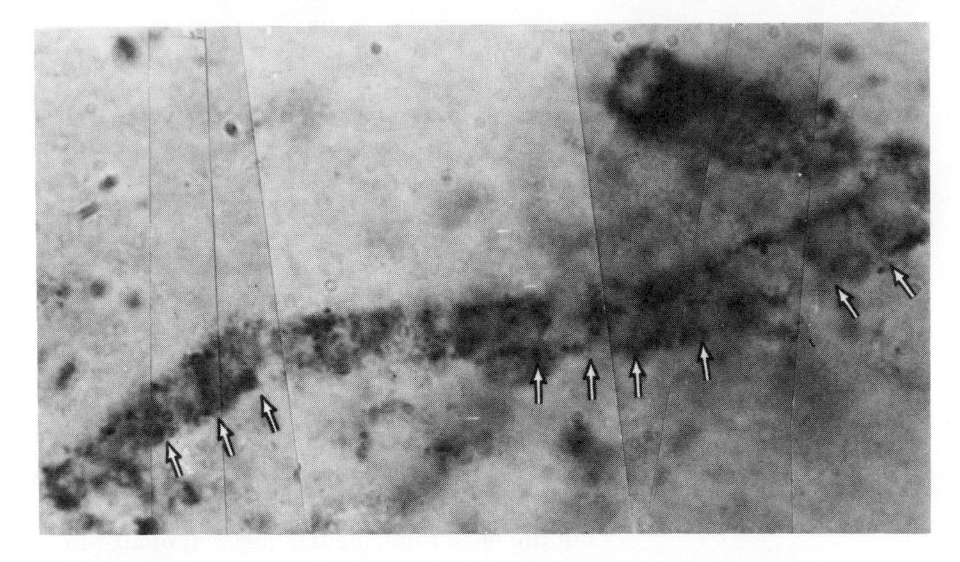

Figure 14-5 The oldest fossils yet found come from western Australia and are about 3.5 billion years old. Since this organism consists of numerous cells, we must conclude that the first (one-celled) forms of life appeared well over 3.5 billion years ago.

rich in organic (carbon-based) compounds and dissolved minerals. They probably served as the nurturing, protective environments in which molecules interacted at random, until some of the reactions produced compounds that could serve as guides in the formation of other molecules. Grains of clay at the edges of these ponds could have helped to organize small molecules into larger ones. This tide pool mixture of life-producing ingredients was dimly foreseen by Charles Darwin when he wrote of "some warm pond" in which life could have begun. Fifty years later, J.B.S. Haldane and A.I. Oparin independently developed this model in detail.

The Miller-Urey Experiment

Experimental testing of this model came in 1953, in a famous experiment by Stanley Miller and Harold Urey. As Figure 14-6 shows, Miller and Urey modeled the Earth's primitive oceans and atmosphere with a large flask of water, above which were methane, hydrogen, and ammonia to simulate the gases in the early atmosphere. Energy input came from an electrical discharge (to simulate lightning), a source of ultraviolet light (like sunlight before the Earth had an ozone atmospheric filter), or from shock waves (which might resemble surf breaking on an ocean beach). A heater and condenser provided circulation of water vapor like that of the Earth's evaporation-rainfall cycle.

The **Miller-Urey experiment** was a great success. Within hours after the experiment began, organic molecules appeared in significant amounts in

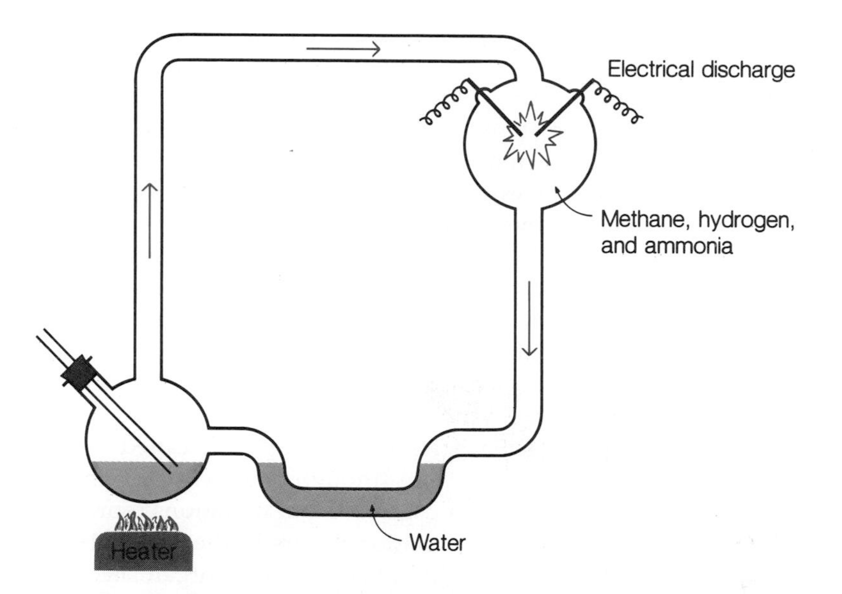

Figure 14-6 A schematic diagram of the experimental equipment used by Stanley Miller and Harold Urey shows the 5-liter flask that contained water, methane, hydrogen, and ammonia to simulate the Earth's atmosphere and oceans. Also shown is the electrical discharge apparatus that released energy into this mixture. After the first night of operation, a thin layer of hydrocarbon molecules formed on the water, and a few more weeks yielded several types of amino acids.

the flask containing the water. Chemical analysis showed that in addition to a large amount of unidentifiable organic "gunk," amino acids formed, including glycine, alanine, and numerous other compounds of interest. Not all of the twenty amino acids in living organisms could be produced by the Miller-Urey experiment, nor could the cross-linking molecules of DNA be made directly. But the results seem to suggest, all the same, that something like the Miller-Urey process, plus additional chemical reactions among its products, could explain the origin of life.

The Origins of Cells

The key events that led to the evolution of living cells are unknown. We do know, however, that organic polymers in high concentration are likely to join together and separate from the solution in the form of droplets. These droplets, called **coacervates,** could serve as the prototypes of cells. Laboratory studies of coacervates have shown that under appropriate conditions they can form systems within which simple chemical reactions occur. But no one has yet produced a self-maintaining system in this way that could form a model of the first cells.

14.5 Evolution

The history of life on Earth embraces the continual appearance of more complex creatures—larger, more specialized to particular environments, capable of new kinds of activities—from less advanced ones, many of which have nevertheless persisted. This **evolution** of life forms has brought human beings into existence where once only single-celled organisms could be found. Evolution has populated the Earth with millions of different species of plants, animals, and bacteria, and it has increased the rate at which different species appear with the passage of time.

For the purposes of assessing the prospects of life elsewhere, the most significant fact about the evolution of life on Earth lies in its apparent universality. All such evolution occurs because of random changes in DNA as it replicates. Such random changes are called **mutations.** No one knows the reason for mutations (though one theory assigns the cause to the cosmic rays we discussed on page 323), but mutations represent the driving force behind evolution. If a mutation occurs in the **germ plasm** of an organism, the part that governs the production of new organisms, it can change the characteristics of that individual's offspring. As Lewis Thomas wrote, "The capacity to blunder slightly is the real marvel of DNA. Without this special attribute, we would still be anaerobic bacteria and there would be no music."

Natural Selection

Some changes produced by mutations may aid an organism to survive and reproduce, while other changes may hinder these attempts and still others may have no effect at all. **Natural selection** is the term coined by Charles Darwin to describe the testing of these changes over the course of time. If a certain change does aid an organism in survival and reproduction, the organism will have proportionately more descendants than other members of that species without the mutationally induced change. Since the descendants can embody the change and pass it on to *their* descendants, natural selection makes new species of organisms appear as successive generations differ more and more from the ancestors before the mutation occurred.

The fossil record on Earth shows that although some species do not change, most do. Natural selection leads to the development of more complex species of life. In particular, natural selection has produced "intelligent" species—animals that can think about their environments and manipulate them in increasingly significant ways. These animals appeared because intelligence has great survival value: We can heal our sick, feed our hungry, move to any part of the planet, and pass from generation to generation the knowledge of how to achieve these goals. As we now speculate about other forms of life on other planets, we may reasonably conclude that natural selection should tend to produce intelligent species there too. But before we leap outward (in imagination!) to other planets around other stars, we ought to pause to summarize the more than 3 billion years on Earth that saw intelligent animals evolve from DNA in blue-green bacteria.

Prokaryotes: The Simplest Cells

The simplest living organisms, those that can survive and reproduce on their own, are the simplest cells, bacteria and blue-green algae.[1] These cells are **prokaryotes,** cells without special centers or nuclei. Prokaryotes contain long strands of DNA, each with several thousand genes, and thus already represent a fairly complex assemblage of organic molecules. As a result, prokaryotes are free-living entities. They can reproduce and manufacture proteins without a plant or animal host in which to live. The oldest evidence of prokaryotic life takes us back 3.5 billion years in the Earth's history (Figure 14-5), three-quarters of the way back to the formation of the Earth. Many scientists think it likely that prokaryotes appeared before this, perhaps only a few hundred million years after the Earth formed.

When prokaryotes first evolved, they had to draw energy from the organic compounds produced by the interaction of ultraviolet sunlight and other energy sources (lightning, surf, and so forth) with the chemical mixture in which they found themselves. Some prokaryotes, however, eventually evolved a far more direct way to use the sun's energy, one which immensely changed our planet's atmosphere: **photosynthesis.** This chemical process uses particular kinds of organic molecules, chlorophyll being the best known, to convert sunlight into stored chemical energy. The first forms of bacterial photosynthesis did not liberate oxygen as part of this process, but at some point in the Earth's history, somewhere between 2 and 2.5 billion years ago, certain prokaryotes evolved a more efficient form of photosynthesis that did release oxygen as a by-product (Figure 14-7).

The effects of oxygen-releasing photosynthesis can hardly be overemphasized in our analysis of life's history. It was *the* big change in the Earth's atmosphere, making possible all the larger creatures of Earth. The prokaryotes that released oxygen increased the abundance of that gas from a tiny fraction of a percent to more than 20% in the Earth's atmosphere today. Since this change occurred gradually, we may conclude that organisms that learned to use oxygen, as well as to tolerate it, became increasingly favored as life evolved new forms on Earth.

Eukaryotes: Complex Cells

About a billion years after the atmosphere became oxygen-rich, a new stage in the complexity of life appeared: **eukaryotes,** cells that contain true nuclei. These cells keep their **chromosomes,** the DNA-containing portion of the cell, within a membrane-bounded nucleus. In eukaryotic cells, the chromosomes typically contain hundreds of thousands of genes and 10 to 1000 times the percentage of DNA that occurs in bacteria. Eukaryotic cells thus store far

---

[1]Strictly speaking, these blue-green algae are not algae but a special kind of photosynthetic bacteria. We therefore must call them *blue-greens* or *blue-green bacteria* to be precise.

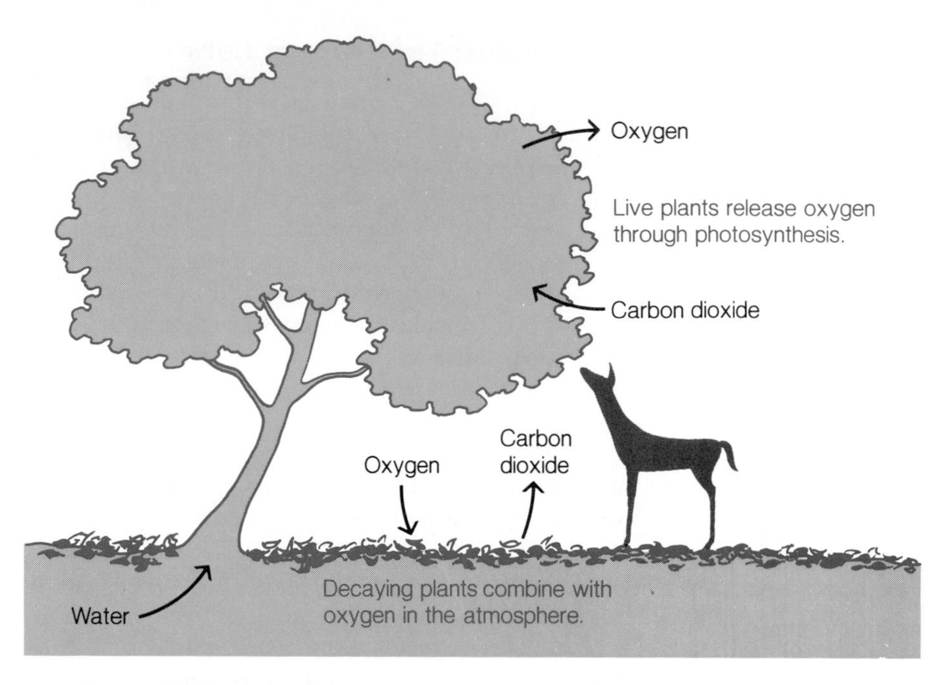

Figure 14-7 Plants use the process of photosynthesis to convert carbon dioxide and water into carbohydrate molecules, with the help of sunlight energy and the release of oxygen molecules. Animals that eat plants can use the energy stored by photosynthesis when they reverse the process, by breaking carbohydrate molecules into water and carbon dioxide with the release of stored energy. This process of oxidation also occurs when dead plants combine with oxygen in the Earth's atmosphere.

more genetic information than prokaryotic cells and show an increased complexity of cell structure and function. From the time when eukaryotes first appeared, somewhere between 700 million and 2 billion years ago, the processes of evolution have led to the increasing specialization of cells and to their incorporation into larger units, such as organs and massive individuals, which may each contain hundreds of trillions of cells.

As eukaryotic organisms continued to reproduce, they eventually developed the important technique of **sexual reproduction** (Figure 14-8). All eukaryotic cells can reproduce asexually, but sexual reproduction has the great advantage of mixing the genetic contribution from two parents in a single individual. This fact probably allows natural selection to proceed far more rapidly than it does for asexual reproduction.

The Cambrian Era

Just about 600 million years ago, some combination of circumstances, probably the result of the large amount of oxygen liberated by blue-green bacteria and plants, led to a sharp increase in the distribution, number, and variety of fossils that mark the beginning of the **Cambrian era.** During the 100 million years or so before this, soft-bodied organisms, much like today's jellyfish, should have evolved from the first, single-celled eukaryotes, but they left no fossils. Only at the start of the Cambrian era did organisms with "hard parts" (shells, carapaces, and exoskeletons) appear, able to leave well-preserved fossils in sedimentary rocks (Figure 14-9).

The trilobites of 600 million years ago had two eyes and a rather complex body structure, but were nowhere close to the complexity of whales and humans. Yet just one-sixth as much time was needed for mammals to evolve

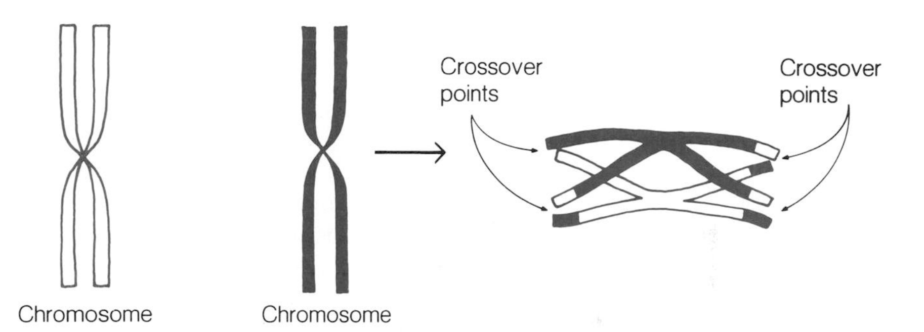

Figure 14-8 In sexual reproduction, the fertilized egg cell contains chromosomes with DNA molecules from each parent. When the egg cell divides and redivides, the chromosomes pair up to form connections at crossover points where the exchange of genetic material occurs. This shuffles the chromosomal material. After a number of these divisions, each new cell has a roughly equal mixture of maternal and paternal DNA.

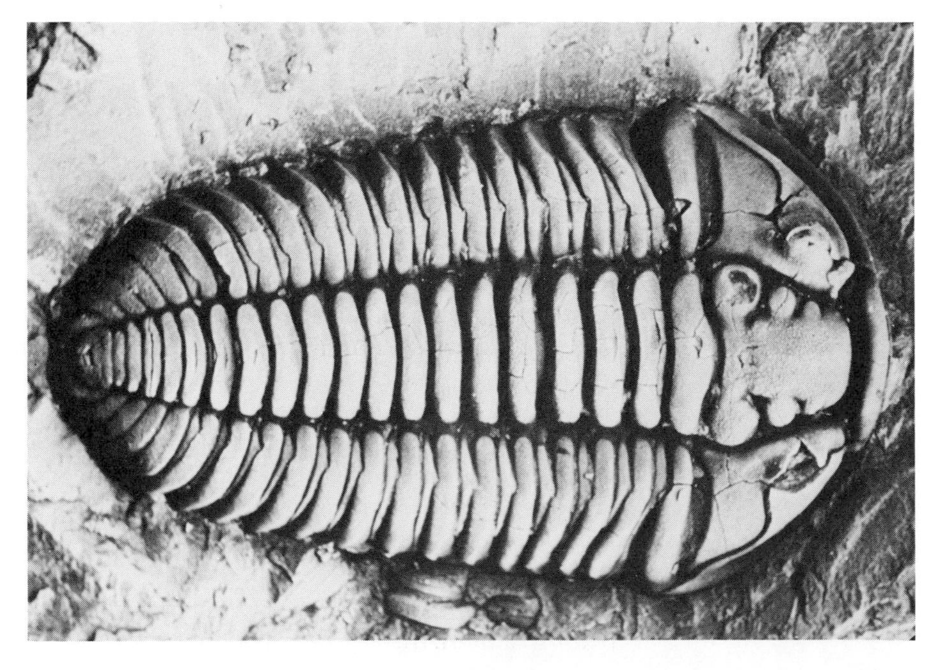

Figure 14-9
The trilobites that flourished in the seas at the start of the Cambrian era had two eyes and a rather complex body structure. This photograph shows the fossil of a trilobite from a somewhat later era.

from trilobites as was required for trilobites to evolve from the first prokaryotes. The evolution of *Homo Sapiens,* the modern human, from the first apelike hominids that moved from the jungle forests to the plains of Africa, took only 3 million years, less than a thousandth of the total history of life on Earth. This sort of ratio may have occurred, or be in the process of occurring, elsewhere in the universe. So when we do find another planet with life, the luck of the moment may well take us there at the trilobite stage (or even before) rather than at the stage of intelligent life.

14.6 Suitable Places for Life

Planets appear to be good places for life to begin and to flourish, because they provide favorable conditions for complex chemical reactions to occur.[2] If we expect to find life on other planets, then our discussion of life on Earth suggests an immediate, important prerequisite: Several billion years are needed for intelligent life to evolve from the first, simple forms. This means that planets around short-lived stars—those that last a billion years or less on the main sequence—do not provide good sites for life. Furthermore, stars that do last for at least several billion years may not *yet* have lived long enough, together with their planets, for intelligent life to have evolved around them.

Which Stars Last Long Enough?

A necessary, but not sufficient, condition for the appearance of intelligent life on a planet appears to be the existence of a star of spectral type F5 or cooler around which the planet orbits. Although the brightest stars in the sky are O, B, and A stars, by far the majority of stars are cooler than F5. Therefore we may not be limiting ourselves much in numbers by rejecting the possibility of finding intelligent life on planets (if they exist!) around Sirius, Vega, and Rigel.

The Habitable Zones Around Stars

If we look at the stars within 4 parsecs of the sun, we find that most of them meet our criterion of longevity. As Table 14-2 shows, most of these stars are K and M stars, capable of spending tens or hundreds of billions of years on the main sequence. The big problem with such stars comes not from their lifetimes but from their low energy output, because the planets must then nestle close to the parent star if their temperatures are to remain high enough for liquid water to exist. The zone in which this occurs has the name ecosphere (sphere of life) or **habitable zone** (Figure 14-10). If, for example, we hope to find liquid water (and life) near Barnard's star, we must hope to find planets within 3 to 5 million kilometers of the star, less than 1/10 Mercury's distance from the sun!

[2]See page 493 for possible exceptions to this rule.

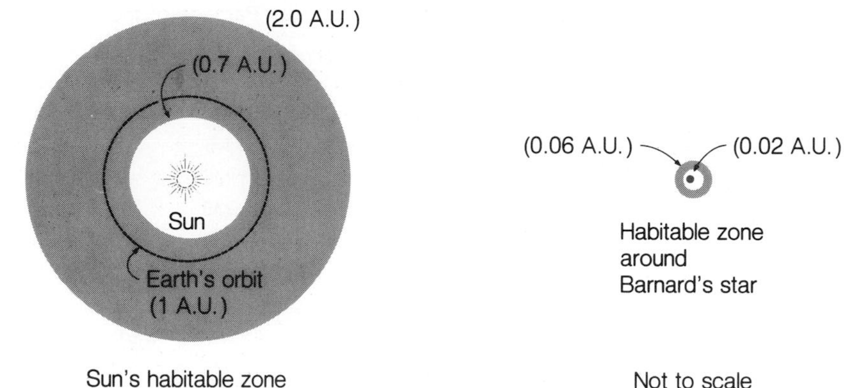

Figure 14-10 The habitable zone around a star is the spherical shell of space within which a planet should have the right temperature for life to exist. If this temperature range allows for water, ammonia, or methyl alcohol to be liquid (165 to 373 K), then the sun has a habitable zone that extends from 0.7 to 2.0 times the Earth's distance from the sun. The habitable zone around the much less luminous Barnard's star extends only from 0.02 to 0.06 times the Earth's distance from the sun.

The Importance of Temperature

Why do we insist on temperatures that will keep water liquid? What is so wonderful about water as a solvent, rather than ammonia, or methyl alcohol, or some other fluid? Water has noticeable advantages over any other solvent, especially over those that, like water, consist of cosmically abundant atoms. These advantages reside in water's ability to buffer temperature changes, thanks to its high heat capacity and high heat of vaporization. In addition, water's large surface tension, twice that of ammonia and three times the surface tension of methyl alcohol, means that water can better promote the concentration of solids at the interfaces of different media—for example, at the boundaries of cells. Water's advantages over ammonia and methyl alcohol hardly disqualify the latter two fluids as possible solvents. But it is clear why we would expect water to serve best in this capacity. Even if ammonia or methyl alcohol turned out to be the solvent for another form of life, the temperature zone we look for would expand only to slightly greater distances, since we would simply have a different set of temperature requirements to maintain the solvent liquid.

14.7 Are We Too Limited in Our Concept of Life?

The reader who has considered the question of life in other parts of the universe may feel that our discussion of life on other planets, based on a carbon chemistry with water as a solvent, is far too restrictive. Surely the universe offers far greater possibilities, perhaps beyond our imagination. Indeed a tremendous range of possibilities exists, but with scientific analysis we can see that some should be far likelier than others.

Table 14-2 Stars within Four Parsecs of the Sun

Star Name	Distance (parsecs)	Spectral Type	Luminosity (Sun = 1)	Mass Where Known (Sun = 1)
Alpha Centauri A	1.3	G2	1.53	1.1
B		K5	0.44	0.88
Proxima Centauri	1.3	M5	0.00006	0.1
Barnard's star[a]	1.8	M5	0.00043	
Wolf 359	2.3	M8	0.00002	
Lalande 211385[a]	2.5	M2	0.0054	0.35
Sirius A	2.6	A1	23.0	2.31
B		White dwarf	0.0020	0.98
Luyten 726-8 A	2.7	M5	0.00006	0.044
B		M5	0.00004	0.035
Ross 154	2.9	M5	0.0004	
Ross 248	3.2	M6	0.0001	
Epsilon Eridani	3.3	K2	0.30	0.8
Luyten 789-6	3.3	M7	0.00013	
Ross 128	3.3	M5	0.00033	
61 Cygni A[a]	3.4	K5	0.082	0.63
B		K7	0.039	0.6
Epsilon Indi	3.4	K8	0.138	
Procyon A	3.5	F5	7.6	1.77
B		White dwarf	0.0005	0.63
Sigma 2398 A	3.5	M4	0.003	0.4
B		M5	0.0014	0.4
Groombridge 34 A	3.5	M1	0.006	
B		M6	0.0004	
Lacaille 9352	3.6	M2	0.012	
Tau Ceti	3.7	G8	0.47	
BD +5° 1668[a]	3.8	M5	0.0015	
Luyten 725-32	3.8	M5	0.0003	
Lacaille 8760	3.8	M0	0.028	
Kapteyn's star	3.9	M3	0.0015	
Krüger 60 A	3.9	M3	0.0015	0.27
B		M4	0.0004	0.16

[a]These stars are thought to have unseen companions.

Nonchemical Life: Life on a Neutron Star

So far we have examined only the possibilities of chemical life; that is, of life based upon the interaction of atoms and molecules. Chemical interactions typically proceed at temperatures of hundreds, or thousands of degrees above absolute zero (roughly in the range 50 to 5000 K). At lower temperatures, atoms move more slowly and interact only rarely. At higher temperatures, particles move so rapidly that collisions destroy any complex molecules that might otherwise appear. So long as we restrict ourselves to chemical interactions, we may well conclude that carbon represents the superior atomic element for making complex structures, and that water provides the best solvent for such molecules to float in.

But must we restrict our thoughts to chemistry? No, we need not. Consider, for example, the possibility of life on the surface of a neutron star, where the temperatures are hundreds of thousands of degrees. No molecules will be found here, and chemical interactions, based upon electromagnetic forces among atoms, have little relevance. What counts at these temperatures and densities are the strong forces that hold nuclei together, and any forms of life we imagine will be assemblages of protons and neutrons temporarily bound together by these strong forces.

As the American astronomer Frank Drake first pointed out, we can imagine that in the course of the formation and destruction of superlarge nuclei, millions of times the size of the largest nuclei found on Earth, something we might call life could appear: nuclei capable of reproduction and evolution. The only hitch is that the typical size of such nuclei would be 10^{-10} centimeter, and the typical lifetime, before another kind of nucleus appears, would be of the order of 10^{-14} second. In other words, our hypothetical superlarge nuclei would have sizes and lifetimes less than one-trillionth of our own! If this were true, entire civilizations could come and go more rapidly than our brains could think about them. We would face grave difficulties in communicating with such a civilization, since both the time scales and the wavelengths used for communication would be much less than the values familiar to us.

Life in Interstellar Clouds

In addition to nonchemical life, such as that on neutron stars, life might originate within interstellar clouds, where the temperatures roughly equal those on the outer planets of the sun. In fact, astronomers have discovered many types of simple molecules in interstellar clouds, and they feel confident that more complex molecules might form there. We should realize, however, that chemical reactions—atoms and molecules colliding with one another and interacting through electromagnetic forces among them—occur far more slowly in interstellar clouds than on Earth or on another planet. The low density of matter in interstellar clouds makes it more difficult for particles to meet. If life on Earth took perhaps a billion years to appear, then it might take thousands of millions of times longer to emerge in interstellar clouds, where molecules interact at intervals millions of times greater than on Earth.

14.8 The Number of Civilizations in the Milky Way Galaxy

If we agree that planets offer the most likely sites for the origin and development of life, and that any other sites will simply be an additional bonus in our search, we may proceed to estimate the number of civilizations in our galaxy now. We limit ourselves, at least for the time being, to our own galaxy because we would like to communicate with other civilizations. A message from Earth would take 160,000 years to reach another galaxy—and we would have to wait *another* 160,000 years to receive a reply! This time factor, as well as the greater difficulty in communicating between larger distances (see page 500), suggests that we restrict ourselves to civilizations in the Milky Way.

The Number of Perfect Restaurants

To estimate the number of civilizations, we must perform a series of multiplications, each of which carries us one step further in defining the conditions that produce a civilization. Consider, as an analogy, the number of restaurants in the United States just like our favorite restaurant. To estimate this number, we must multiply the number of cities (where we think restaurants are likely to exist) by the average number of restaurants per city. Then we must multiply this product by the fraction of restaurants with the right location, to obtain the number of restaurants in the country with the right sort of location to meet our specifications. Finally, we must multiply this product by the fraction of restaurants that serve the right food, and multiply this by the fraction that have the right prices. We thus obtain the following result:

$$\text{Number of perfect restaurants} = \text{number of cities} \times \text{number of restaurants per city} \times \text{fraction of restaurants with right location} \times \text{fraction of restaurants with right food} \times \text{fraction of restaurants with right prices}$$

If there are 5000 cities in the United States with an average of 200 restaurants per city, and if the fraction with the right location, the fraction of those with the right food, and the fraction of those with the right prices all equal ¼, we find our estimate to be:

$$\begin{aligned}\text{Number of perfect restaurants} &= 5000 \times 200 \times \frac{1}{4} \times \frac{1}{4} \times \frac{1}{4} \\ &= 15{,}625\end{aligned}$$

Notice the cumulative effect of the restrictions imposed by demanding the right location, the right food, and the right prices. If each of these fractions were, say, ⅛ instead of ¼, our final product would be 8 times less, and we would estimate that 1953.125 perfect restaurants exist instead of 15,625. Of course, since this is only a rough estimate, we would round off the figure to 2000.

The Drake Equation

Similar considerations apply to our estimate of the number of civilizations in the Milky Way. Again we must multiply a series of numbers, many of them fractions that represent some restriction upon planets that we think must be essential for life. The equation that expresses this multiplication has the name **Drake equation** from the man who first wrote it down, Frank Drake. The Drake equation may be written as follows:

$$\begin{array}{c}\text{Number of}\\ \text{civilizations in}\\ \text{our galaxy now}\end{array} = \begin{array}{c}\text{number of}\\ \text{stars in}\\ \text{our galaxy}\end{array} \times \begin{array}{c}\text{average number}\\ \text{of planets}\\ \text{per star}\end{array} \times \begin{array}{c}\text{fraction of stars}\\ \text{that last long}\\ \text{enough for life}\\ \text{to emerge}\end{array}$$

$$\times \begin{array}{c}\text{fraction of}\\ \text{planets that are}\\ \text{suitable for life}\end{array} \times \begin{array}{c}\text{fraction of those}\\ \text{planets where life}\\ \text{actually develops}\end{array}$$

$$\times \begin{array}{c}\text{fraction of planets}\\ \text{with life where}\\ \text{civilizations arise}\end{array} \times \frac{\begin{array}{c}\text{average lifetime}\\ \text{per civilization}\end{array}}{\text{lifetime of galaxy}}$$

The first two terms in the Drake equation give us the number of planets in the Milky Way, and multiplication by the fraction of stars that last long enough (say, 4 or 5 billion years) for life to appear and evolve gives us the number of planets around suitable stars. The next three fractions reduce our number, first to the number of planets suitable for life, next to the number where life actually develops (perhaps only a tiny part of the suitable planets), and finally to the number not only with life but with civilizations, which we may define as life capable of interstellar communication. The final term, extremely important, provides the fraction of the galaxy's lifetime that a given civilization exists, since we seek the number of civilizations *now*, which we consider a representative moment in galactic history.

Let us attempt to estimate the numbers that we need to make our grand multiplication. The number of stars in the Milky Way equals about 400 billion. We don't know the average number of planets per star, but if our own solar system is typical, the number would be 9. Let us round off to 10 for purposes of estimation. The fraction of stars that last long enough for life to appear equals a healthy ½, probably even more. The fraction of planets suitable for life is unknown; again, using the solar system as a guide, a fraction close to 1/10 seems reasonable. The fraction of planets suitable to life upon which life actually develops cannot easily be decided, but if we believe that life appears naturally, given "world enough and time," we must set the fraction not far below unity, say ½ to be modest. Similarly, our own example suggests that where life appears, civilizations eventually appear, so the fraction of planets with life that acquire civilizations should likewise not be much less than unity, say again ½. Finally, we cannot estimate the average lifetime of a civilization. Our own has lasted about 70 years since it acquired the ability to communicate across interstellar distances with radio waves. We may last 70 more, or 7 million more years before we disappear. Let us, for the time being at least, simply denote the average lifetime of a civilization by L, measured in years. We do know that the Milky Way has an age of about 10 billion years.

When we insert these values into the Drake equation, we obtain an interesting result:

$$\text{Number of civilizations in our galaxy now} = 400 \text{ billion} \times 10 \times \frac{1}{2} \times \frac{1}{10} \times \frac{1}{2} \times \frac{1}{2} \times \frac{L}{10 \text{ billion}} = 5L$$

In other words, the number of civilizations in the Milky Way now should be equal to five times the average lifetime of a civilization, measured in years. If the lifetime turns out to be 1000 years, then 5000 civilizations should exist. If the lifetime equals 1 million years, then 5 million civilizations should sprinkle the galaxy at any given time, since "now" marks a fairly typical time in Milky Way history. Notice that this result depends on having all of the fractions fairly close to unity in our estimate. If any one of them were, say, one-millionth, our final product would be almost a million times less than our result of $N = 5L$, where N is the number of civilizations in our galaxy now. In view of the great uncertainties involved in our estimate, we might round off our answer to a simple $N = L$, and say that *the number of civilizations in our galaxy now should approximately equal the average lifetime of a civilization, measured in years.*

The Distance Between Neighboring Civilizations

The number of civilizations in our galaxy has tremendous importance in determining the average *distance* between civilizations, which in turn determines how difficult it will be to establish contact between neighboring civilizations. If $N = L$ and $L = 100{,}000$ years, civilizations in the Milky Way should be spaced about 100 parsecs (326 light years) apart, so that round-trip radio messages would need 650 years. This may seem a long time, and a long way to the nearest neighbor (past about 400,000 stars that lie within 100 parsecs), but the distance involved spans only 1/300 of our galaxy's diameter. We may, however, be too optimistic when we assume that $N = L$ and $L = 100{,}000$ years. If more accurate figures should give $N = L/100$ and $L = 100$ years, then at any given time, only *one* civilization should exist in the Milky Way galaxy! This would provide an unfavorable situation for interstellar communication.

14.9 Methods of Interstellar Communication

Suppose that we take an optimistic view in our estimate of the number of civilizations and assume that $N = L$, or even $N = 5L$, and that L, the average lifetime of a civilization, exceeds 100,000 years. Then we may expect to find civilizations spaced by no more than a few hundred parsecs, and we should be able to come into contact with our nearest neighbors if we put our minds to it. But how does such communication start? There seem to be basically three answers: Go and find them; let them come and find us; or send radio messages.

Interstellar Spacecraft

The first plan involves the construction of spacecraft capable of traveling interstellar distances. This feat may eventually become possible, but we would do well to remember the immense distances that separate the stars. Our fastest spacecraft travel at 0.01% of the speed of light and would require nearly 100,000 years to reach the closest star to the sun.

We may someday improve our spacecraft to the point that we can travel at a significant fraction of the speed of light, but the laws of physics imply that the energy costs of operating such a spacecraft would be enormous. Far from any star, as would inevitably be the case on interstellar journeys, we would need an immense fuel supply to accelerate to near-light velocities, decelerate at the far end, accelerate toward home, and decelerate when we return (Figure 14-11). Such a fuel supply has yet to be discovered. Even if we could produce and contain enough antimatter to use matter-antimatter annihilation to propel the spacecraft, we would require thousands of times more fuel than payload to make an interstellar journey at nearly the speed of light.

Why Not Let *Them* Visit *Us*?

If we cannot make interstellar trips, at least for the present, perhaps we should wait for more advanced civilizations to visit us. A large body of public opinion holds that such civilizations have in fact overcome the obstacles to interstellar spaceflight and often visit the Earth (see page 504). In view of the scientific objections to the feasibility of such journeys, and in the absence of convincing evidence that we are in fact being visited, we would do well to remain skeptical of the notion that we can sit back and let the others do the traveling. There is something a bit too simple about this approach to interstellar communication. Furthermore, we must consider that our galaxy contains 400 billion stars, and that the number of civilizations is hardly likely to exceed 1 billion (if $N = 10L$ and $L = 100$ million years). What, then, would be the special lure of the Earth to explain even one visit per year, let alone the many visits implied by UFO reports (page 505)? We may well

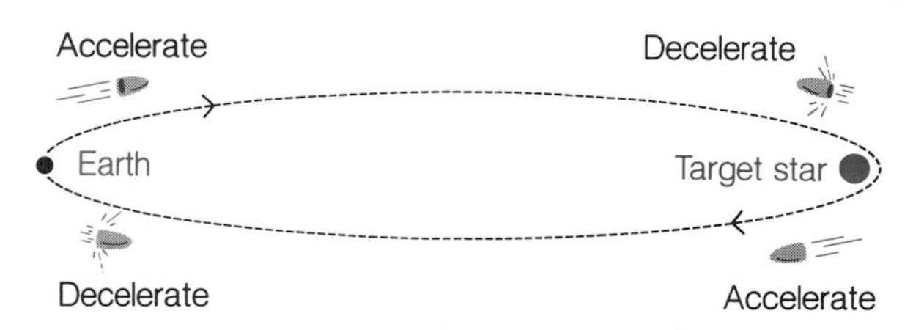

Figure 14-11 To travel to another star and back, we must make four separate acceleration or deceleration maneuvers. First we must reach a high velocity for travel and then slow down at our destination. To return we must accelerate and then decelerate at Earth. If each such maneuver requires a large amount of fuel, the total fuel requirements become enormous, especially if we must carry fuel for the entire journey along with us.

doubt that any civilization has become so energy-rich that it can send probe after probe to Earth, however attractive we think we may be to faraway civilizations.

Interstellar Communication by Photons

The third approach to interstellar communication has the support of most of the scientists who have considered this question. If interstellar spaceflight seems impossible, or nearly so, why not use photons—light waves, radio waves, and other forms of electromagnetic radiation—which all travel at the speed of light and cost nearly nothing? Radio waves now provide the fastest means of sending messages, and with relatively low cost per message. To send men to the moon cost billions of dollars; to send the radio signals that guided them cost a few cents per message.

Radio messages may not have the excitement of personal visits, but they can carry as much information as any person can, and at speeds of 300,000 kilometers per second. We have an understandable human prejudice in favor of exploration in person, and our movies and television programs reflect this bias. Similar prejudices may well have arisen in other civilizations, but logic seems to dictate that the civilizations that achieve contact do so by overcoming their biases and making contact by radio.

The Advantages of Radio

Why do we talk of radio messages, rather than communication by light waves, infrared waves, or x rays? All of these travel at the speed of light in empty space, but radio waves, which can also be used to send television pictures, have two key advantages.

Radio waves can penetrate the interstellar medium. Although light waves also have this ability to some extent, other forms of electromagnetic radiation, such as x rays, ultraviolet waves, infrared waves, and gamma rays, cannot travel for great distances without being absorbed by interstellar atoms and electrons. As Figure 14-12 shows, the best-favored radio waves are those with frequencies between 1000 and 10,000 megahertz (MHz). Radio waves with lower frequencies (longer wavelengths) must compete with the general background of radio emission from our galaxy. This background emission consists mostly of synchrotron emission from fast-moving charged particles that accelerate in the magnetic field of the galaxy (see page 208). Radio waves with higher frequencies (shorter wavelengths) than our favored range are absorbed by water vapor molecules in the Earth's atmosphere and cannot reach the Earth's surface.

The second advantage of radio waves is low cost. Although both radio waves and light waves avoid the problem of interstellar absorption, a message sent by radio costs far less than one sent with light waves, perhaps by laser flashes. Each photon in a message—each individual bundle of energy that travels at the speed of light—can carry, at least in theory, one "bit" of information. These "bits" correspond to a minimum amount of information,

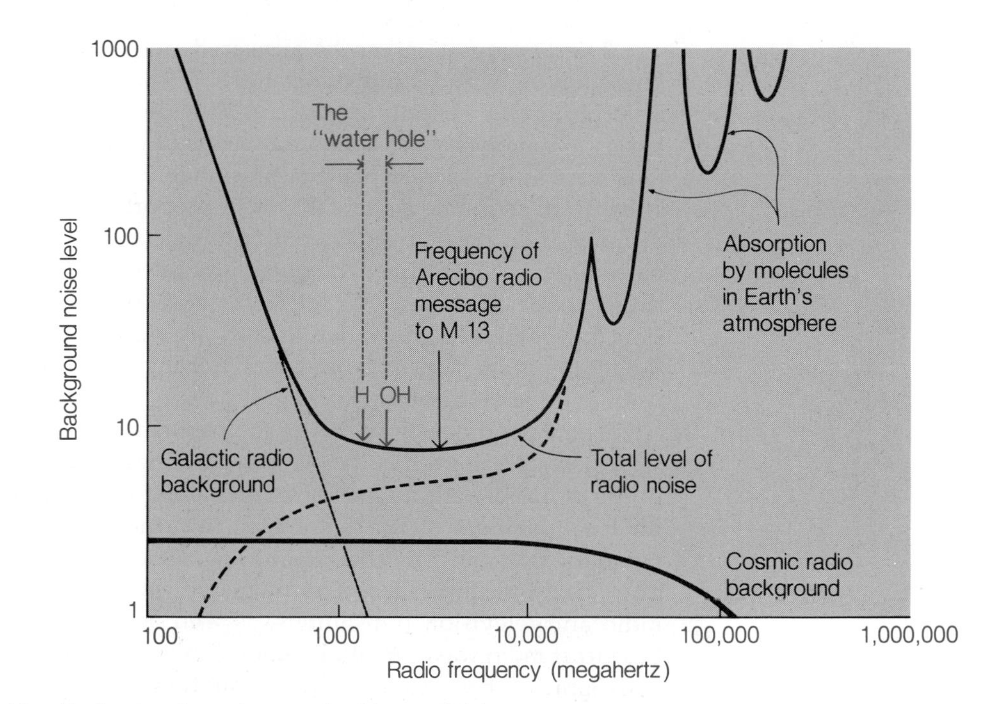

Figure 14-12 Radio signals sent or received in our vicinity must compete with low-frequency radio emission from natural processes in our galaxy, and with the absorption by molecules in our planet's atmosphere. The best frequencies to send messages from Earth, or to detect them on Earth, lie between 1000 and 10,000 megahertz. The part of this frequency domain between 1420 and 1620 megahertz is called the water hole. The Arecibo radio message to M13 is described on page 501.

such as whether a switch is on or off, whether a number is odd or even, and so forth. From the individual bits, an entire set of numbers, words, or books can be constructed, bit by bit. The cost of sending *any* set of information—any set of bits—varies almost in direct proportion to the number of bits in the message, if we are considering the cheapest possible ways to send information. Since the energy in a light-wave photon exceeds that in a radio-wave photon by perhaps a million times, the cost of sending information, in energy terms, will increase in about the same proportion. No civilization can be infinitely rich in energy, and we may expect that in choosing a means of intercommunication, others will also be led to low-energy methods.

14.10 How Can We Make a Radio Search for Other Civilizations?

If we agree that civilizations are most likely to communicate through radio messages (which can carry entire television programs), we face some simple-to-put, difficult-to-answer questions. In which directions should we look for extraterrestrial radio messages? Which frequencies carry the messages? And what sorts of messages should we expect?

The first question can be answered in general terms only. We do not know where to find other civilizations, but we expect to find them on planets orbiting stars resembling the sun. The same considerations of energy that led us to radio waves as a favored means of communication suggest that we look first at the closest planets, presumably orbiting the closest stars. The strength of any radio signal falls off in proportion to 1 over the square of the distance from the source. Therefore to detect a signal at a distance of 1000 parsecs requires 100 times the sensitivity needed to detect the same signal at a distance of 100 parsecs. We might begin by searching the two dozen closest stars (see Table 14-2). If we fail to find any signals, we can proceed to look at the more distant stars in progression, hoping eventually to find radio signals from another civilization.

The second question, that of the frequency at which radio messages are sent, poses greater difficulty. We must distinguish here between signals sent in the deliberate effort to achieve interstellar communication and those used for local communication that escape into interstellar space. The latter case raises the possibility of **eavesdropping,** detecting another civilization by the photons it leaks into space. Our own civilization, for example, uses powerful radio and television transmitters, as well as military and civil radars, that broadcast radio waves in all directions, though primarily parallel to the horizon. Some of these waves escape through the Earth's atmosphere and radiate into space, surrounding the Earth with an ever-expanding cocoon of mixed signals.

The Water Hole

If we set out to broadcast to another civilization or to search for deliberate signals, we might be guided in our frequency search by the fact that the universe does contain one primary radio frequency, that emitted by hydrogen atoms (see page 193). The frequency of 1420 megahertz that characterizes this radiation provides a cosmic guidepost in our frequency search. We must not expect to find the signals of extraterrestrial civilizations at precisely this frequency, however, since 1420 megahertz, and 1 or 2 megahertz on either side, are jammed with natural emission from hydrogen atoms, some of which has a Doppler shift when we detect it.

If we wonder how far from the 1420-megahertz frequency we should search, we should remember that most of the life in the universe may be based on water as a solvent (see page 491). If this proves true, we may expect the frequency domain between the natural hydrogen (H) emission and the natural hydroxyl (OH) emission to form a cosmic pool of frequency. Water molecules themselves have an extremely complex spectrum of radio emission, with no single primary frequency, but OH molecules, like hydrogen atoms, have a definite frequency (or a band of frequencies, near 1620 megahertz, in the case of OH) that could serve as a marker. The physicist Bernard Oliver has named the frequency region between 1420 and 1620 megahertz the galactic **water hole,** the place (in frequency) where galactic civilizations can meet in peace.

Of course, even if we accept the water-hole concept, we still have 200 megahertz of frequency to search. Considerations of how radio signals travel through interstellar space suggest that any message must be at least 0.1 hertz wide. We therefore face the problem of searching through 200 megahertz, 0.1 hertz at a time, in every direction we choose to examine, provided that the water-hole approach has merit. Luckily, however, modern technology seems capable of allowing us to build antennas and receiver systems that can analyze the signals from a given direction in several hundred million frequency channels at the same time.

Interstellar Radio Messages

What sorts of messages might one civilization broadcast to announce that "we are here"? It may surprise you to learn that our own astronomers have already sent a few such messages. The most important of these, sent from the Arecibo Observatory on November 16, 1974, contains 1679 bits of information. These bits each consist of a photon pulse at one of two frequencies close to 2380 megahertz. (Notice that this frequency lies outside the water hole but was used anyway because it was extremely convenient for sending.)

The use of 1679 bits, no more, no less, suggests (to mathematicians!) a factorization into two prime numbers, 23 and 73. We might therefore arrange the bits into a pattern of 23 columns and 73 rows, or 23 rows and 73 columns. The latter produces nothing interesting, but if we chose the former, and color each bit in black or white, depending on which of the two neighboring frequencies had the photon pulse, we find the pattern shown in Figure 14-13. This diagram conveys a large amount of information: the atomic numbers of the atoms most important to life (1, 6, 7, 8, 15); the chemical formulas of important simple molecules; a stab at the structure of DNA, the all-important polymer; a representation of a human being, together with its size (in units of the wavelength of the message); a map of the solar system, with the third planet marked as special; and a crude picture of the Arecibo radio telescope, with its size in message wavelengths.

Not a bad job if we can decipher it! The question remains, How many messages like this are percolating through the galaxy at this very moment? The Arecibo message, directed toward the famous globular cluster in Hercules (page 176), will take about 25,000 years to arrive, so we can expect a return message in 50,000 years or so from whichever of the million or so stars' planets choose to answer. But can we detect such messages? Or can we find another civilization simply by eavesdropping? The answer is, almost. We now have the technological capability to detect a message similar in strength and form to that sent from Arecibo in 1974 emitted from anywhere within a distance of a thousand parsecs, provided we know the frequency and direction at which to observe. This implies that we could find a civilization that sent such a message anywhere among the nearest 100 million stars if we looked hard and long enough, and if we looked in the right direction at just the right time. Since our own message was sent from Arecibo for only a few minutes out of a total of a million minutes or so since 1974, our chances are slim of finding another civilization in this manner.

Should We Try to Eavesdrop on Other Civilizations?

A far better, though more expensive, way to locate our galactic neighbors consists of eavesdropping. The same advantages of low cost that argue in favor of radio waves for interstellar communication have made radio and

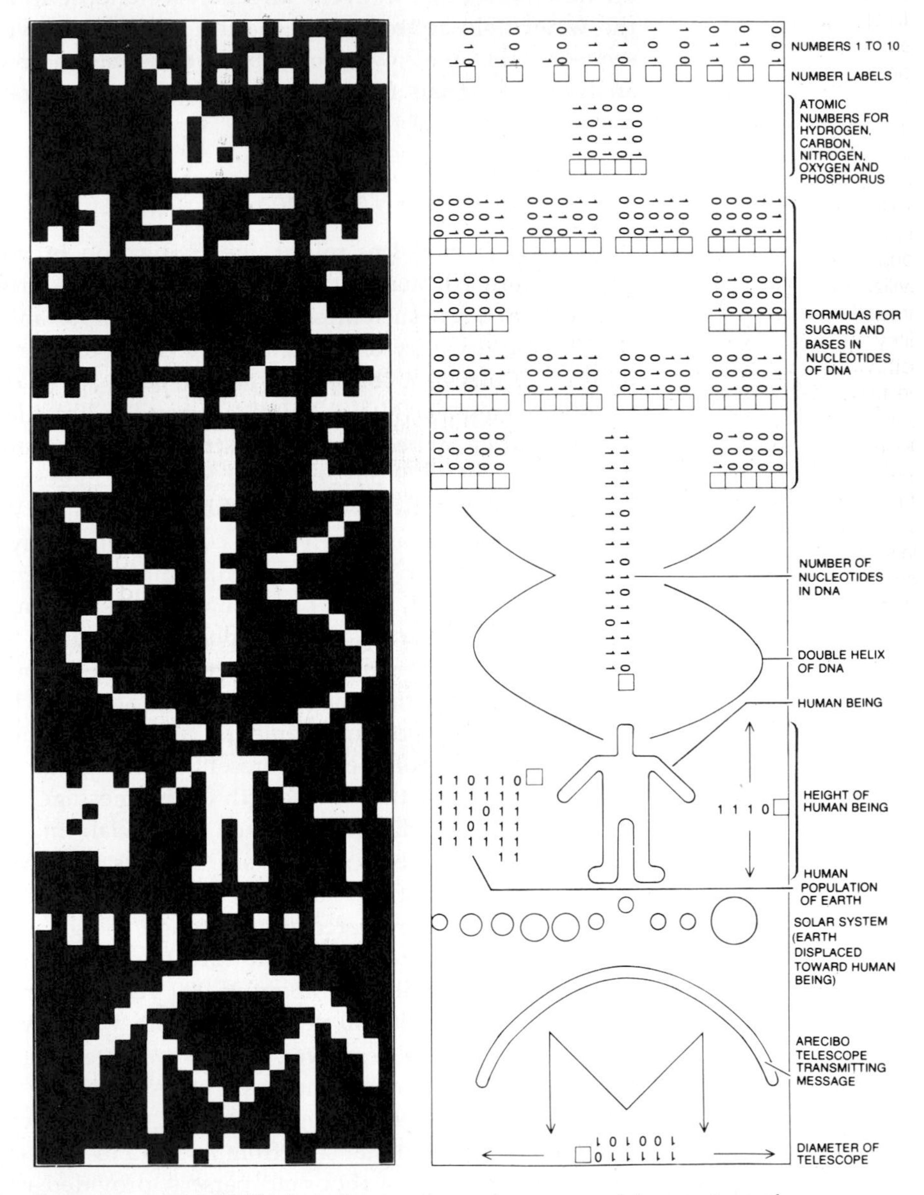

Figure 14-13 If we arrange the 1679 bits in the Arecibo radio message into a pattern of 23 columns and 73 rows, we find the pattern shown on the left. For those who have trouble understanding it (on Earth!), the explanation appears on the right. From "The Search for Extraterrestrial Intelligence," by Carl Sagan and Frank Drake. Copyright © 1975 by Scientific American, Inc. All rights reserved.

Figure 14-14
The Arecibo radio telescope in Puerto Rico, 300 meters across, can be used to send a well-directed radio beam out into space. Without such beaming, a radio signal from Earth cannot travel many light years without becoming too weak to be detected by a technological civilization similar to ours. With such beaming, we could communicate with a civilization like our own throughout much of the Milky Way, if we knew both where to look and the frequency at which to listen. The Arecibo Observatory is part of the National Astronomy and Ionosphere Center, operated by Cornell University under contract with the National Science Foundation.

television the favored means of communication on Earth, and we may expect on other planets as well. Moreover, instead of broadcasting for only a tiny fraction of the year, as has been the case for our own attempts at interstellar recognition signals, radio, television, and radar emission goes on continuously. The key problems, and serious ones, concern the lack of beaming of these messages (they are aimed nowhere in particular) and the fact that we don't know *how long* a civilization will continue to broadcast radio messages as it develops. This makes them much harder to distinguish from the background "noise" of natural emission than a beamed signal would be (Figure 14-14).

If we hope to detect another civilization by eavesdropping, we must build far more sensitive antennas and receiver systems than we now have. The technology to do so exists, but the complete system would cost several billion dollars. We would need several large radio antennas to collect radio signals, and an elaborate analysis of what we detect, in several hundred million frequency channels at the same time. If the initial antennas failed to detect anything that we would call a civilization, we could build additional antennas, if we decided to pay the price for greater sensitivity (Figure 14-15). We may yet find that a "galactic club" of civilizations already exists, and that we, as latecomers to the community, must pass some sort of test before being asked to join. The bubble of old radio and television programs, many light years in diameter around us, has already passed thousands of stars and may explain why we have not yet been asked! Or we may yet—and this is still more depressing—prove to be the wittiest and wisest in the galaxy. We shall never know until we find who is out there.

Figure 14-15
The complete Cyclops system to search for other civilizations would consist of 1000 radio antennas, each 100 meters in diameter, and all connected together to function as a single large receiver system. We could start by building only a few antennas, and then add more to increase the sensitivity of the system.

14.11 UFO Reports

The desire to establish contact with another civilization may affect our ability to analyze dispassionately the evidence that other civilizations have visited, and continue to visit, the Earth. Thousands of reports of unidentified flying objects (UFOs) have been made by serious observers who saw something they could not explain. Unfortunately, in many cases we can neither explain rationally what the observer saw nor confirm that something beyond the known laws of science was involved.

We *can* say, however, that most UFO reports turn out to involve natural objects, in particular meteor showers (see page 405) and the planets Venus, Mars, and Jupiter, which can appear unusually bright, especially if we are not used to looking at the sky. Venus alone has caused more UFO reports than all other sources combined. Even Jimmy Carter once filed such a report (before he became president). Also, we know that charlatans and tricksters find fertile ground in making UFO reports, since they can be sure of a receptive audience. Finally, enthusiasm leads many people, especially in crowds, to mis-see events and to misreport what they have seen, as documented by many conflicting eyewitness reports of crimes.

We simply have no hard evidence that UFOs represent spacecraft. The eyewitness testimony turns out to be either reliable accounts of fairly unusual (but hardly unexplainable) events or unreliable accounts of tremendously unusual events. Eyewitness testimony will never allow us to resolve the UFO question unless we are extremely lucky. Of photographic evidence

Figure 14-16 These photographs, taken near McMinnville, Oregon, in 1950, are often cited as the best evidence we have that UFOs are spacecraft. They were taken by the husband of a woman who had reported seeing UFOs on several previous occasions, "but no one would believe me," and were said to have been retrieved from under a sofa by a reporter who came to interview the husband and wife.

we have little. The "best" photographs offered in support of the theory that UFOs are something beyond Earthly explanation resemble those in Figure 14-16. However, even these will convince no one with finality, for we cannot rule out deception as the origin of the strange images they show.

Summary

Life, defined as a system capable of reproduction and evolution, appears to have arisen on Earth more than 3 billion years ago as a natural consequence of the interactions among atoms and molecules on the surface of the planet. In this process, simple molecules (monomers) linked up into the long-chain, carbon-based polymers that characterize living organisms. These linkages occurred as the molecules floated within the basic solvent, water, which still forms most of all living cells.

The extent to which the conditions on Earth that led to life have recurred throughout our galaxy, and indeed throughout the universe, cannot be determined now. We can say, however, that we expect most sunlike stars to have planets, and many of these planets should fall within the habitable zones of the stars, those in which water or another solvent (ammonia or methyl alcohol) could promote a carbon-based chemistry. The most likely sites for life should be the surfaces of planets like the Earth, though the thick atmospheres of Jupiterlike planets and the interiors of dense interstellar clouds also provide opportunity for the origin of life.

To estimate the number of civilizations now in our galaxy with whom we might communicate, astronomers use a summary called the Drake equation, in which each of the probability factors deals with a particular step in the evolution of an intelligent civilization. The key unknown in the Drake equation is the lifetime of a civilization once it has developed the ability to maintain communication over interstellar distances. If our own civilization provides a representative example, this lifetime equals at least 70 years, but it could be much longer, in which case the number of civilizations, and our chance for contacting them, increases dramatically.

Radio contact rather than interstellar voyages seems the most likely way to meet other civilizations because of the tremendous difficulty of interstellar spaceflight compared to radio messages. The scientific plans for interstellar contact hinge on the construction of large systems of antennas and receivers that could search among likely target areas and through the many radio frequencies that might carry a beamed message or a civilization's internal communications upon which we might "eavesdrop." In contrast to this scientific approach, a simpler method of making contact is to wait for another civilization to visit us. Reports of UFOs often involve an intense desire to believe that such visits now occur regularly on Earth, but scientific analysis of UFO sightings has not yet found any convincing proof that extraterrestrial beings have come to us.

Key Terms

Cambrian era
catalyst
cell
chromosome
coacervates
DNA
double helix
Drake equation
eavesdropping
enzymes
eukaryotes
evolution
gene
genetic code
germ plasm
habitable zone
life
messenger RNA
Miller-Urey experiment
monomers
mutation
natural selection
photosynthesis
polymers
prokaryotes
RNA
sexual reproduction
solvent
water hole

Questions

1. Are you satisfied with the definition of life given on page 478? Can you propose a better definition?
2. How does the fact that life on Earth consists primarily of hydrogen, carbon, nitrogen, and oxygen affect the possibility of finding life elsewhere in the universe?
3. Why does carbon seem to be the best element for holding complex molecules together?
4. What is the genetic code? How does the replication of DNA molecules preserve this code from one generation of molecules to the next?
5. Why are solvents important to life?
6. What does the Miller-Urey experiment suggest about the origin of life on Earth?
7. What is the difference between prokaryotes and eukaryotes?
8. What is the habitable zone around a star? To what extent does this definition reflect our Earth-based prejudices?
9. What would life on the surface of a neutron star be like, if such life existed?
10. What is the Drake equation? Why does it include the lifetime of a civilization with communications ability and desire?
11. Why do radio waves seem far superior to spacecraft for interstellar communication? Which radio frequencies seem best suited to this purpose? Why?
12. Why are radio waves more difficult to detect at greater distances from the radio transmitter?
13. Why is "eavesdropping" on another civilization's internal radio communication more difficult than detecting a signal beamed in a particular direction?
14. For spacecraft velocities v that are nearly equal to the speed of light c, the energy needed to reach a given velocity varies in proportion to the ratio $c/(c - v)$. How much more energy is needed to accelerate a spacecraft to a velocity of 99.99% of the speed of light than is needed to reach a velocity of 99% of the speed of light?

Projects

1. *The Age of Fossils.* Visit a museum of natural history and determine the age of the oldest fossils on exhibit. Why does this fall far short of the 3.4-billion-year estimate for the beginning of life on Earth?
2. *Natural Selection.* Try to imagine, and to construct on paper, a history of natural selection for an Earth with no land—only oceans. Would intelligent sea creatures have developed? Would they resemble octopi, or would they be more likely to resemble whales and dolphins?
3. *Lifetime of a Civilization.* Take a poll among people whose opinion you respect to determine their best estimate for the lifetime, L, of our own

civilization after achieving interstellar communications ability. Ask these people whether this value of L should provide an accurate estimate of the galactic average, or whether our civilization should be notably better or worse than the average. What does this tell you about human attitudes toward the surrounding cosmos?

4. *Frequencies.* Consult a book on radio engineering or an article on radio and television in an encyclopedia to find out why AM, FM, and television broadcasts employ the frequencies they do. Does this have any bearing on the likelihood of another civilization to use similar frequencies?

5. *UFOs.* Interview your friends to find out if any of them has seen something he or she would classify as a UFO. Find out as much as possible about the circumstances of the observation, and make a list of conclusions about the nature of the UFO that would fit these circumstances. Among the possible explanations are planets (which ones?), weather balloons, meteors, search lights, hoaxes, and extraterrestrial spacecraft.

Further Reading

Asimov, Isaac. 1979. *Extraterrestrial Civilizations.* New York: Crown.

Bracewell, Ronald. 1975. *The Galactic Club.* San Francisco: W. H. Freeman.

Goldsmith, Donald. 1980. *The Quest for Extraterrestrial Life: A Book of Readings.* Mill Valley, Calif.: University Science Books.

Goldsmith, Donald, and Owen, Tobias. 1980. *The Search for Life in the Universe.* Menlo Park, Calif.: Benjamin/Cummings.

Hoyle, Fred. 1957. *The Black Cloud.* New York: Signet.

Klass, Philip. 1974. *UFOs Explained.* New York: Vintage.

Lovelock, J. E. 1980 *Gaia: A New Look at Life on Earth.* Oxford: Oxford University Press.

Morrison, Philip; Billingham, John; and Wolfe, John, eds. 1978. *SETI: The Search for Extraterrestrial Intelligence.* New York: Harper & Row.

Sagan, Carl, ed. 1973. *Communication with Extraterrestrial Intelligence.* Cambridge, Mass.: MIT Press.

Story, Ronald D., ed. 1980. *The Encyclopedia of UFOs.* New York: Doubleday.

Appendix 1 Stargazing

Naked-Eye Astronomy

Because the Earth orbits the sun, our view of the sky at night changes throughout the year (see page 43). The circumpolar constellations never set, but they present a different orientation at, for example, 9 P.M. in June and 9 P.M. in December. During the summer, the Big Dipper passes nearly overhead in the early evening; in winter, Cassiopeia takes the position the Big Dipper had six months earlier (see maps on endpapers).

South of the circumpolar constellations are the parts of the celestial sphere that rise and set. Maps 1 to 4 show the sky as it appears at 8:30 P.M. (standard time) on March 22, June 22, September 22, and December 22. The Earth's motion around the sun makes the stars rise two hours earlier with each successive month (see page 44). We can use this fact to extrapolate between the dates for which the maps are drawn, or for times other than 8:30 P.M. The maps have been drawn for an observer at 34° north latitude, but will serve well anywhere between 25° and 50° north latitude.

The Circumpolar Stars

The circumpolar constellations include the Big Dipper (part of Ursa Major, the Great Bear), the Little Dipper (part of Ursa Minor, the Little Bear), Draco, Cassiopeia, and Cepheus. Of these, the Big Dipper has the brightest stars, which can serve as a useful guide to other constellations. The two stars in the bowl of the Big Dipper that are opposite the dipper's handle are the Pointers. They point the way to Polaris, the north star. If you draw an imaginary line on the sky from the lower of the two stars (bottom of the Big Dipper's bowl) through the upper star, and extend this line for five times the distance between the two Pointers, you will come close to Polaris. Polaris lies within 1° of the north celestial pole and is the end of the handle of the Little Dipper. Notice that the two dippers are almost "pouring into each other." Do not be discouraged if you cannot see all of the Little Dipper. Its faint stars do not appear in full glory until you leave city lights behind.

Draco, the Dragon, winds its way between the two Dippers, then wraps itself around the Little Dipper on the side away from the Big Dipper.

Cepheus, with five rather dim stars, can be found by extending the line from the Pointers of the Big Dipper through Polaris. Cassiopeia, a sprawling W made of six fairly bright stars, lies on the line from the Big Dipper's handle through Polaris, continued almost as far as the distance from the handle to Polaris.

The Skies of March (Map 1)

The most visible constellations in March are Orion, Gemini, Leo, and the Big and Little Dogs (Canis Major and Canis Minor). Find Orion near the western horizon by noting the three bright stars of the belt, close to one another and almost parallel to the horizon. Still brighter stars lie on either side of the belt: Betelgeuse above and northward, Rigel below. Betelgeuse forms an almost equilateral triangle, the Winter Triangle, with Sirius and Procyon. Sirius, the brightest star in the sky, is the heart of Canis Major, and Procyon, not quite as bright, is the center of Canis Minor. If you extend an imaginary line from Rigel through Betelgeuse, you will come to Gemini, whose two bright stars, Castor and Pollux, mark the heads of the Twins. A line from Rigel through Procyon brings you to Leo, the Lion, a large constellation with a particularly bright star (Regulus) in the lion's forepaws. Leo's head stretches northward from Regulus, while his hindquarters lie eastward of Regulus, terminating in a small triangle of stars.

The Skies of June (Map 2)

In summer, look for the Big Dipper and follow the arc of its handle to reach Arcturus, the brightest star in Bootes, the Herdsman. After you "arc over to Arcturus," continue to follow the curving trajectory you have established to "speed on to Spica." Spica marks the center of Virgo, the Virgin, a loose group of stars spread along the ecliptic. The second brightest star in Virgo, Beta Virginis, lies close to the autumn equinox on the sky.

In the east, the three bright stars of the Summer Triangle are rising (Figure 1-1). Vega, marginally the brightest of the three, rises first. Between Vega and Arcturus are Hercules, not so easy to see, and Corona Borealis, the Northern Crown. Corona Borealis consists of a tight half-circle of stars, opening toward the northeast.

To the south, the red star Antares marks the heart of Scorpius, the Scorpion, whose claws extend northwest toward Virgo, and whose fishhook tail, ending in a close pair of stars almost parallel to the horizon, may be lost in the murk near the horizon. Between Scorpius and Virgo is the ecliptic constellation Libra, the Scales. At the best of times you can see four stars in Libra, which make a sort of kite-shaped rectangle whose long axis points to the south. Sagittarius, the Archer, just to the east of Scorpius, represents a centaur shooting arrows into the Scorpion. The center of the Milky Way lies in the direction of the Sagittarius-Scorpius boundary.

The Skies of September (Map 3)

As summer passes into fall, the Summer Triangle appears more toward the west, passing overhead in the early evening. The three bright stars of the Summer Triangle dominate the sky and can be used to locate other constellations. Behind the Summer Triangle is Pegasus, the Winged Horse whom Perseus rode. The "great square" of Pegasus, made of second-magnitude stars, can be used to find Andromeda (the maiden whom Perseus rescued) and the Andromeda galaxy (see page 12). To the south lie a number of dim stars in Sagittarius, Capricornus, and Aquarius, and toward the southeast is the bright star Fomalhaut, in Piscis Austrinis (Southern Fish). The ecliptic constellation of Pisces, not easy to see, lies southeast of Pegasus.

The Skies of December (Map 4)

The winter skies bring the rising of Orion, the Hunter, followed by his two dogs, marked by their brightest stars, Sirius and Procyon. The three stars in Orion's belt, now almost perpendicular to the eastern horizon, point downward (eastward) to Sirius and upward (westward) to Aldebaran, the orange star in Taurus, the Bull. Aldebaran marks one tip of a V of stars—the Hyades star cluster (see page 181). Still farther from Orion's belt along the path through Aldebaran are the Pleiades, a compact open cluster, in which seven stars can be seen without a telescope.

Northward from Orion we find Gemini, two parallel lines of stars whose brightest, Castor and Pollux, are the farthest in the Twins from Orion. Halfway between Orion and Polaris, the north star, is the constellation Auriga, the Charioteer, whose brightest star, Capella, is as bright as Rigel and Betelgeuse in Orion. Auriga and Perseus (between Auriga and Cassiopeia) are almost circumpolar constellations for the northern half of the United States, and are indeed circumpolar in Great Britain.

Planetary Motions

Because the orbits of the planets lie close to the ecliptic plane, the planets always appear on the sky near the ecliptic (page 46). The position of the planets along the ecliptic changes noticeably from week to week, and even—at least for Mercury, Venus, and Mars—from day to day. If you once locate Venus, Mars, Jupiter, or Saturn, keep track of the planet's motion and note whether it is "normal" or "retrograde" in direction. You might want to subscribe to the monthly *Sky Calendar* of the Abrams Planetarium, Michigan State University, East Lansing, Michigan 48824 ($2 per year), which gives useful information on locating the planets. Figure 2-17 shows the positions of Jupiter among the stars for the years 1981–1983. During all of this period, Saturn will be in the constellation Virgo.

The ecliptic itself can be fairly easily located among the stars. In addition to the planets, the sun, and the moon, which are always close to the ecliptic, look for the bright stars Regulus, Spica, and Aldebaran, which are all within 5° of the ecliptic. In Scorpius, the ecliptic passes close to the northern claw of the Scorpion, and about 4° north of Antares.

Buying a Telescope

Many telescopes are purchased each year only to be rendered inoperative through lack of care or ignored through lack of interest. Be sure you consider whether your investment in a telescope really makes sense, and take the time to choose carefully. With reasonable care, your telescope can last indefinitely.

The best place to find out about telescopes is a store that specializes in selling them. Look under "Telescopes" in the Yellow Pages to find the stores in your area. Another good way to obtain information is to join an astronomy club. For a list of the astronomy clubs in your area, write to the Astronomical Society of the Pacific, 1290 24th Avenue, San Francisco, CA 94122.

There are two types of telescopes, refracting and reflecting. Reflecting telescopes have the virtue of sturdiness; their lenses seal a tube that requires a minimum of care. Refractors typically have somewhat longer tubes than reflectors of the same aperture (lens or mirror diameter). A refractor with a 75-millimeter (3-inch) aperture will probably be 1.2 meters long, and a refractor with a 100-millimeter (4-inch) aperture will measure about 1.6 meters. Length may be an important consideration if you plan to transport your telescope in an automobile, an increasingly necessary step in finding clear skies.

Reflecting telescopes offer greater apertures—and thus greater light-gathering power—than refracting telescopes of the same price. Reflectors, especially those with apertures of 100 to 160 millimeters, are the preferred instrument of most amateur astronomers. They do, however, require somewhat more care than refractors, since their tubes are open at one end (in the simpler models).

Large reflectors are quite unwieldy, especially if designed in the Newtonian mode (see page 70). More expensive reflectors use Cassegrain or other "folded" optical paths to minimize the length of the telescope. As an extreme case, for several thousand dollars you can purchase a reflector with a 175-millimeter aperture that fits inside a case only 80 centimeters long! If you can live with a telescope tube more than 2 meters long, you can obtain the same aperture at one-fifth the price.

No telescope is better than its mounting, which determines the steadiness of the image you see. Although it is just as difficult to manufacture a good mounting system as to make a good optical system, most first-time telescope buyers concentrate on the latter. Before you buy a telescope, set it up as you would in observation—preferably at night, but in daylight if necessary. Look through the telescope with various eyepieces and check the ease of changing eyepieces. See how readily the telescope reacts to slight jars. If it vibrates for more than a fraction of a second, the mounting is poor. Check the "slow motion" controls that allow you to make fine adjustments in the direction the telescope points. Do they work smoothly? Are they easy to find in the dark? Check the focusing system. Does it, too, work without vibration? Does it hold its position well, even when the eyepiece is held in a vertical position? If the telescope has a motor drive to follow the stars, does it work smoothly and accurately? (Here there is no substitute for actually tracking a star with the telescope, and preferably several stars at different distances from the celestial pole.)

Finally, do some comparative shopping, either by telephone or in person, before you buy. The popular astronomy journals *Sky and Telescope, Star and Sky,* and *Astronomy* each list used telescopes for sale through classified advertisements. Some of these are amateur-made telescopes, which range from very poor to extremely good in performance and value. These journals also contain advertisements by the major telescope manufacturers and some of the large telescope stores.

Appendix 2 The Metric System

Scientists use the metric system of units because it is much simpler than the English system. In the metric system, we measure length in meters (m), centimeters (cm), and kilometers (km). These basic units are related to each other and to the English units in the following way.

$$1\text{m} = 100 \text{ cm}$$
$$1 \text{ km} = 1000 \text{ m} = 10^5 \text{ cm}$$

$$1 \text{ cm} = 0.3937 \text{ inch}$$
$$1 \text{ m} = 39.37 \text{ inches}$$
$$1 \text{ km} = 0.6214 \text{ mile}$$

$$1 \text{ inch} = 2.54 \text{ cm}$$
$$1 \text{ mile} = 1.6093 \text{ km}$$

Other important units in the metric system are the millimeter (1 mm = 10^{-1} cm = 10^{-3} m), the micron (10^{-6} m), and the angstrom (1 Å = 10^{-8} cm = 10^{-10} m).

Mass is measured in grams (gm) and kilograms (kg). One kilogram equals 1000 grams. The relationship with English units is

$$1 \text{ gm} = 0.0353 \text{ ounce}$$
$$1 \text{ kg} = 2.2046 \text{ pounds}$$

$$1 \text{ ounce} = 28.3495 \text{ gm}$$
$$1 \text{ pound} = 0.4536 \text{ kg} = 453.6 \text{ gm}$$

One cubic centimeter of water at 0° C weighs 1 gram. A volume of 1000 cubic centimeters is called 1 liter. One liter of water at 0° C weighs 1 kilogram. One liter equals 1.0567 quarts.

The metric system takes the second as the basic unit of time, and uses seconds, minutes, and hours, as in the English system.

The metric unit of force is the dyne. One dyne is the amount of force that will give a 1-gram mass an acceleration of 1 cm/sec^2.

In the metric system of units, the basic units of energy are the erg and the joule. One erg is the amount of energy expended by a 1-dyne force acting through a distance of 1 centimeter. It is also the amount of kinetic energy possessed by a 2-gram mass moving at a speed of 1 cm/sec. One joule equals 10^7 ergs.

Appendix 3 Important Physical and Astronomical Constants

Velocity of light	c	=	$2.99792458 \times 10^{10}$ cm/sec
Gravitational constant	G	=	6.672×10^{-8} dyne cm^2/gm^2
Proton mass	m_p	=	1.672×10^{-24} gm
Electron mass	m_e	=	9.1096×10^{-28} gm
Stefan-Boltzmann constant	σ	=	5.670×10^{-5} erg/cm^{2}K^4 sec
Astronomical unit	A.U.	=	$1.495978707 \times 10^{13}$ cm
Parsec	pc	=	206,265 A.U. = 3.086×10^{18} cm
Light year	ly	=	63,240 A.U. = 9.4605×10^{17} cm
Mass of sun	$M_\odot$	=	1.989×10^{33} gm
Radius of sun	$R_\odot$	=	6.960×10^{10} cm
Luminosity of sun	$L_\odot$	=	3.83×10^{33} erg/sec
Mass of Earth	$M_\oplus$	=	5.974×10^{27} gm
Equatorial radius of Earth	$R_\oplus$	=	6378 km = 6.378×10^8 cm
Sidereal year		=	365.256366 days = 3.155815×10^7 sec
Tropical year (equinox to equinox)		=	365.242199 days = 3.155693×10^7 sec

Appendix 4 The Messier Catalog

Messier Number	Constellation	Type of Object	Special Name
1	Taurus	Supernova remnant	Crab Nebula
2	Aquarius	Globular cluster	
3	Canes Venatici	Globular cluster	
4	Scorpius	Globular cluster	
5	Serpens Caput	Globular cluster	
6	Scorpius	Open cluster	
7	Scorpius	Open cluster	
8	Sagittarius	Diffuse nebula	Lagoon Nebula
9	Ophiuchus	Globular cluster	
10	Ophiuchus	Globular cluster	
11	Scutum	Open cluster	
12	Ophiuchus	Globular cluster	
13	Hercules	Globular cluster	
14	Ophiuchus	Globular cluster	
15	Pegasus	Globular cluster	
16	Serpens Cauda	Open cluster	
17	Sagittarius	Diffuse nebula	Omega Nebula
18	Sagittarius	Open cluster	
19	Ophiuchus	Globular cluster	
20	Sagittarius	Diffuse nebula	Trifid Nebula
21	Sagittarius	Open cluster	
22	Sagittarius	Globular cluster	
23	Sagittarius	Open cluster	
24	Sagittarius	Open cluster	
25	Sagittarius	Open cluster	

Messier Number	Constellation	Type of Object	Special Name
26	Scutum	Open cluster	
27	Vulpecula	Planetary nebula	Dumbbell Nebula
28	Sagittarius	Globular cluster	
29	Cygnus	Open cluster	
30	Capricorn	Globular cluster	
31	Andromeda	Spiral galaxy	Andromeda galaxy
32	Andromeda	Elliptical galaxy	
33	Triangulum	Spiral galaxy	
34	Perseus	Open cluster	
35	Gemini	Open cluster	
36	Auriga	Open cluster	
37	Auriga	Open cluster	
38	Auriga	Open cluster	
39	Cygnus	Open cluster	
40	Ursa Major	(Probably two stars)	
41	Canis Major	Open cluster	
42	Orion	Diffuse nebula	Orion Nebula
43	Orion	Diffuse nebula	
44	Cancer	Open cluster	Praesepe (Beehive cluster)
45	Taurus	Open cluster	Pleiades
46	Puppis	Open cluster	
47	Puppis	Open cluster	
48	Hydra	Open cluster	
49	Virgo	Elliptical galaxy	
50	Monoceros	Open cluster	
51	Canes Venatici	Spiral galaxy	Whirlpool galaxy
52	Cassiopeia	Open cluster	
53	Coma Berenices	Globular cluster	
54	Sagittarius	Globular cluster	
55	Sagittarius	Globular cluster	
56	Lyra	Globular cluster	
57	Lyra	Planetary nebula	Ring Nebula
58	Virgo	Spiral galaxy	
59	Virgo	Elliptical galaxy	
60	Virgo	Elliptical galaxy	
61	Virgo	Spiral galaxy	
62	Ophiuchus	Globular cluster	
63	Canes Venatici	Spiral galaxy	

Messier Number	Constellation	Type of Object	Special Name
64	Coma Berenices	Spiral galaxy	
65	Leo	Spiral galaxy	
66	Leo	Spiral galaxy	
67	Cancer	Open cluster	
68	Hydra	Globular cluster	
69	Sagittarius	Globular cluster	
70	Sagittarius	Globular cluster	
71	Sagitta	Globular cluster	
72	Aquarius	Globular cluster	
73	Aquarius	Open cluster	
74	Pisces	Spiral galaxy	
75	Sagittarius	Globular cluster	
76	Perseus	Planetary nebula	
77	Cetus	Spiral galaxy	
78	Orion	Diffuse nebula	
79	Lepus	Globular cluster	
80	Scorpius	Globular cluster	
81	Ursa Major	Spiral galaxy	
82	Ursa Major	Irregular galaxy	
83	Hydra	Spiral galaxy	
84	Virgo	Elliptical galaxy	
85	Coma Berenices	Spiral galaxy	
86	Virgo	Elliptical galaxy	
87	Virgo	Elliptical galaxy	
88	Coma Berenices	Spiral galaxy	
89	Virgo	Elliptical galaxy	
90	Virgo	Spiral galaxy	
91	Coma Berenices	(Probably a comet)	
92	Hercules	Globular cluster	
93	Puppis	Open cluster	
94	Canes Venatici	Spiral galaxy	
95	Leo	Spiral galaxy	
96	Leo	Spiral galaxy	
97	Ursa Major	Planetary nebula	Owl Nebula
98	Coma Berenices	Spiral galaxy	
99	Coma Berenices	Spiral galaxy	
100	Coma Berenices	Spiral galaxy	
101	Ursa Major	Spiral galaxy	

Messier Number	Constellation	Type of Object	Special Name
102	Draco	Spiral galaxy	
103	Cassiopeia	Open cluster	
104[a]	Virgo	Spiral galaxy	Sombrero galaxy
105[a]	Leo	Spiral galaxy	
106[a]	Canes Venatici	Spiral galaxy	
107[a]	Ophiuchus	Globular cluster	
108[a]	Ursa Major	Spiral galaxy	
109[a]	Ursa Major	Spiral galaxy	

[a]These objects were added to Messier's original catalog by other astronomers.

Glossary

absolute brightness The intrinsic brightness, or luminosity, of an object.

absolute magnitude The apparent magnitude an object would have if its distance from us were 10 parsecs.

absolute scale of temperature Temperature measured on a scale that begins with zero at absolute zero and increases by the same units as the centigrade scale, so that water freezes at 273.16 K and boils at 373.16 K.

absolute zero The lowest point on the temperature scale, the temperature at which all motion ceases (except for certain quantum-mechanical effects). Absolute zero occurs at -273.16 on the centigrade scale and at 0 K.

absorption The removal of photons of a particular energy, frequency, and wavelength from a beam of photons, usually the result of the photons' interaction with atoms or molecules.

absorption line A limited region of the spectrum of photons within which the intensity of radiation falls below that of neighboring spectral regions.

accretion disk The immediate surroundings of a powerful source of gravitation, such as a compact star or black hole, within which matter is spiraling toward the object in ever-tightening orbits.

angstrom A unit of length equal to 10^{-8} centimeter and abbreviated Å, often used in spectroscopy.

angular momentum A measure of the amount of spin possessed by an object, determined by the mass of the object times its rate of spin times the square of its size in the direction perpendicular to the axis of spin.

angular size The fraction of a circle (360°) over which an object appears to extend as we see it, measured in degrees, minutes of arc, or seconds of arc.

annular eclipse An eclipse of the sun in which the moon does not appear to cover the sun's disk completely, leaving a thin ring, or annulus, of the photosphere visible around the moon.

antimatter The complementary form of ordinary matter, made of antiparticles possessing the same amount of mass but the opposite electric charge as the particles they complement.

antineutrino The antimatter complement of the neutrino, with no mass and no electric charge. An antineutrino and a neutrino will interact with other particles in slightly different ways.

antiparticle The antimatter complement of a particle, with the same mass but opposite electric charge as the particle.

apparent brightness The brightness an object appears to have as we observe it.

apparent magnitude An object's apparent brightness as measured in the magnitude system, in which 5 magnitudes correspond to a brightness ratio of 100. Larger magnitudes correspond to lesser apparent brightnesses.

asteroid One of the small objects, ranging in diameter from 1000 kilometers to less than 1 kilometer, that orbit the sun, mainly between the orbits of Mars and Jupiter.

asteroid belt The region of the solar system between the orbits of Mars and Jupiter, within which orbit the bulk of the asteroids.

astrology Study of the positions of celestial objects that purports to predict their influence on the course of human events.

astronomical unit The average distance from the Earth to the sun, equal to 149,600,000 kilometers.

astrophysics The physics of astronomical phenomena.

atom The smallest unit of an element, consisting of a nucleus of protons and neutrons surrounded by one or more electrons in orbit around the nucleus. The number of electrons always equals the number of protons in the nucleus.

aurora Luminous displays in the night skies, more visible in the Earth's polar regions, thought to be the result of solar-wind particles interacting with upper-atmosphere molecules.

Balmer series The absorption lines produced by hydrogen atoms when the electron in an atom jumps from the second-smallest orbit into a larger orbit. Also, the emission lines produced at the same frequencies when a hydrogen atom's electron jumps into the second-smallest orbit from a larger orbit.

barred spiral galaxy A spiral galaxy in which the spiral arms unwind from a spindle-shaped "bar" of stars that forms the central regions of the galaxy.

big bang The primeval explosion, 15 or 20 billion years ago, which began the universe in its continuing state of expansion.

binary pulsar A pulsar in orbit with a star; that is, a pulsar that belongs to a binary-star system.

binary star Two stars in orbit around their common center of mass as the result of their mutual gravitational attraction.

black dwarf A burnt-out star that has exhausted all ways to radiate photon energy.

black hole An object with such enormous gravitational force at its surface that neither light nor matter can escape from it.

black-hole radius The critical radius for any object, equal to 3 kilometers times the object's mass in units of the sun's mass. Any object that contracts within its black-hole radius must become a black hole.

blue shift A Doppler shift to higher frequencies and shorter wavelengths, caused by a relative velocity of approach between the source and the observer.

calendar A system for reckoning time in which the beginning and length of the year are specified together with a means of subdividing the year.

Cambrian era A geological epoch about 600 million years ago, during which vast numbers of new species of plants and animals appeared within a few tens of millions of years.

carbonaceous chondrite A meteorite with carbon-rich inclusions or chondrules.

carbon cycle A series of nuclear reactions that converts hydrogen nuclei to helium nuclei by using carbon nuclei as a catalyst, resulting in the release of new kinetic energy.

catalyst A substance that modifies (usually increases) the rate at which a process occurs without being consumed itself.

cD galaxy A galaxy with an extremely large mass located within a rich cluster of galaxies. Such a galaxy has presumably engulfed other galaxies in the cluster.

celestial equator The circle on the celestial sphere that lies directly above the Earth's equator and equidistant between the celestial poles.

celestial poles The two points on the celestial sphere that lie directly above the Earth's north and south poles.

celestial sphere The imaginary sphere of the sky, centered at the Earth's center, on which the sun, moon, planets, and stars may be visualized all at the same distance from the Earth.

cell The smallest structural unit of a living organism, consisting of a membrane that surrounds a watery fluid within which organic processes occur.

center of mass The point within an object or group of objects that makes the quantity of mass times distance the same on any side of that point in any direction.

centigrade (Celsius) scale of temperature The scale of temperature that registers the freezing point of water at 0° C and the boiling point of water at 100° C.

Cepheid variable star A star whose luminosity varies periodically in a manner similar to that of Delta Cephei. Such stars show a relationship between their luminosities and periods of light variation.

chondrite A meteorite with small round inclusions, called chondrules.

chromatic aberration Distortion of an image produced by a lens, the result of the lens's focusing different colors of light at different points.

chromosome The DNA-containing part of a cell, responsible for the determination and transmission of the hereditary characteristics of the organism.

chromosphere The part of the sun, or of another star, between the photosphere and corona, about 10,000 kilometers in thickness.

circumpolar stars Those stars close enough to the celestial pole so that for a given latitude, the stars are always above the horizon as the Earth rotates.

coacervate A cluster of droplets, rich in polymeric molecules, formed within a liquid medium.

color index The ratio of brightness of an object as observed in two different colors, often expressed as the difference in the object's magnitude in blue and yellow light.

coma The region surrounding the nucleus of a comet.

comet A fragment of primitive solar system material, made of ice and dust, typically in a huge, elongated orbit around the sun.

conservation of angular momentum A statement of the fact that an object acted upon by no net outside force will keep the same value of its angular momentum.

continental drift The old, informal name for the theory of plate tectonics, which explains the shifting of continental and oceanic plates of the Earth's crust.

continuum emission The emission of photons in a smooth spectrum, without absorption or emission lines.

convergent point The point on the sky toward which the proper motions of stars in a cluster appear to be converging.

corona The outermost parts of the sun, or of other stars, millions of kilometers in extent, consisting of highly rarefied gas at temperatures of millions of degrees.

coronal hole Region of the corona of relatively low density.

cosmic microwave background The sea of photons produced everywhere in the universe during its early moments, now characterized by an ideal-radiator spectrum and a temperature of 2.7 K.

cosmic rays Particles moving at almost the speed of light, perhaps the result of supernova explosions. Most cosmic rays are electrons, protons, and helium or other nuclei.

cosmology The study of the universe as a whole, and of its structure and evolution.

critical density The value of the average density of matter in the universe that determines the future of the universal expansion. If the actual value of the density exceeds the critical value, the universe will eventually contract; if the density falls below the critical value, the universe will expand forever.

decoupling The loss by photons in the early universe of enough energy to interact significantly with atoms, either by ionizing them or by exciting them from their ground states.

degenerate matter Matter in which the exclusion principle plays a significant role in determining how the particles in the matter can move.

degree of arc 1/360 of a full circle.

density-wave pattern A pattern of alternatingly greater and lesser densities that is thought to rotate around the centers of spiral galaxies and that is responsible for the formation of new stars that continue to mark out the spiral arms.

deuterium An isotope of hydrogen, of which each nucleus contains one proton and one neutron.

deuteron A nucleus that consists of one proton and one neutron, held together by strong forces.

differentiation The separation of different elements, or different compounds of elements, at different depths within an object such as a planet.

DNA (deoxyribonucleic acid) The part of the chromosomal material that determines inherited characteristics. Each DNA molecule consists of two long helical chains joined by cross-linking pairs of molecules.

Doppler effect The change in energy, frequency, and wavelength observed for photons arriving from a source that has a relative velocity of approach or recession along the observer's line of sight.

Doppler shift The change in energy, frequency, and wavelength of photons as caused by the Doppler effect. Also, the amount of change caused by the Doppler effect.

double helix The twin-spiral form of the DNA molecule.

double quasar A quasar that appears as two images with almost the same Doppler shift, presumably because a gravitational lens has bent light from the quasar along two different paths to us.

Drake equation The equation, first devised by Frank Drake, that summarizes our estimate of the number of civilizations capable of interstellar communication that now exist in the Milky Way.

dwarf elliptical galaxy A loose collection of about a million, or a few tens of millions, of stars, typically a few thousand parsecs in diameter.

dynamics The study of objects in motion and of the relationship between forces on objects and the motions these forces produce.

dyne A unit of force in the metric system; the amount of force needed to give a 1-gram mass an acceleration of 1 centimeter per second per second.

eavesdropping The technique of detecting, or attempting to detect, other civilizations by capturing some of the photons these civilizations may use for internal communications.

eclipse The partial or total obscuration of one celestial object by another, usually of the sun by the moon or of the moon by the Earth.

eclipse of the moon Passage of the moon into the Earth's shadow, resulting in the cutting off of sunlight from the lunar surface.

eclipse of the sun Passage of the moon between the sun and the Earth, resulting in the cutting off of sunlight from part of the Earth's surface.

eclipsing binary star A binary-star system in which one member periodically passes between us and the other member.

ecliptic The path that the sun appears to follow around the celestial sphere during the course of a year.

electromagnetic forces Forces that act between electrically charged particles, either of repulsion (between particles with the same sign of electric charge) or of attraction (between particles with opposite signs of electric charge).

electromagnetic radiation Streams of photons carrying energy from a source of radiation.

electron An elementary particle and one of the basic particle types in an atom with one unit of negative electric charge and a mass of 9.1×10^{-28} gram.

element The set of all atoms that have the same number of protons in the atomic nucleus.

ellipse A curve defined by the property that each point on it has a constant sum of its distances from two fixed points called the foci of the ellipse.

elliptical galaxy A galaxy with an ellipsoidal distribution of stars; hence, a galaxy whose shape as seen on a photograph is that of an ellipse.

emission line A narrow region of the spectrum at which especially large numbers of photons are emitted within a small range of frequency and wavelength, typically as the result of an atom's electron jumping into a smaller orbit.

energy The capacity to do work. In physics, work is specified by a given amount of force acting through a given distance.

energy of mass The energy equivalent to a given amount of mass, equal to the mass times the square of the speed of light.

energy of motion Energy associated with motion; kinetic energy. For objects moving at speeds much less than the speed of light, an object's energy of motion equals one-half its mass times the square of its velocity.

enzyme A protein that catalyzes and regulates processes within a living organism.

equinox One of the two points on the celestial sphere where the ecliptic intersects the celestial equator. On the two days of the year when the sun reaches one of the equinoxes, day and night have equal length all over the Earth.

erg A unit of energy in the metric system, equal to the work done by 1 dyne of force acting through a distance of 1 centimeter.

eukaryote A cell with a well-defined nucleus, within which the DNA-containing material resides.

evolution The process by which groups of organisms, called species, change with the passage of time so that their descendants differ in structure and appearance from their ancestors; the result of the process of natural selection.

excited state A state of an atom in which at least one electron is in an orbit larger than the smallest allowed orbit. An atom in an excited state can pass back into the ground state with the emission of a photon as the electron jumps into the smallest allowed orbit.

exclusion principle The quantum-mechanical rule that for certain kinds of elementary particles (in particular, for protons, neutrons, and electrons), no two identical particles can be at almost the same location and have almost the same velocity.

exploding galaxy A galaxy within which a violent outburst has occurred within the last few million years.

exponent A number or symbol placed to the right of and above another number or symbol. The exponent denotes the power to which the other number of symbol is to be raised.

filament A solar prominence seen projected against the disk of the sun, appearing as a thin, dark line.

flux A flow of energy from a source.

frequency The number of times a photon oscillates or vibrates each second, measured in units of Hertz (Hz) or cycles per second.

galactic cannibalism The engorgement of a smaller galaxy by a larger one as the result of gravitational forces.

galaxy A large group of stars, usually together with some gas and dust, held together by the mutual gravitational attraction of the stars. Most galaxies contain from a few million up to one trillion times the mass of the sun.

Galilean satellites The four largest satellites of the planet Jupiter.

gamma rays Photons with the largest energies, usually defined as photons with energies in excess of 10^{-6} erg and frequencies above 1.6×10^{20} hertz.

gene The functional hereditary unit of a chromosome, occupying a fixed position, with a specific influence on specific characteristics of an organism's offspring and capable of mutation.

geocentric Centered on the center of the Earth.

germ plasm The totality of hereditary material, or genes, within a cell.

giant planets The four planets Jupiter, Saturn, Uranus, and Neptune, each of which has several times the Earth's diameter and contains much more mass than the Earth.

globular star cluster A spherical or ellipsoidal collection of stars, usually a few parsecs in radius, representing some of the first agglomerations of matter to condense as a galaxy formed.

gravitational forces Forces of attraction that act between all particles. For any two particles with mass, the amount of gravitational force is proportional to the product of the two particle's masses divided by the square of the distance between their centers.

gravitational lens A compact object that exerts enough gravitational force to have a significant effect on light rays passing close by the object, thereby focusing the light from a distant source.

gravitational radiation Streams of gravitons emitted when two objects move around their common center of mass. More massive objects moving in smaller orbits will emit greater amounts of gravitational radiation.

gravitational red shift A decrease in the energy and frequency of photons leaving the surface of an object. The shift to lower frequency and longer wavelength arises from the loss of energy in opposing the force of gravity and is proportional to the mass of the object divided by the radius of its surface.

graviton A particle with no mass and no electric charge, which always travels at the speed of light and which plays the same role in gravitational radiation that photons play in electromagnetic radiation.

gravity waves See *gravitational radiation*.

Great Red Spot A semipermanent feature on the visible surface of Jupiter, apparently a sort of cyclone several times the size of the Earth.

greenhouse effect The trapping of infrared radiation by a planet's atmosphere, which raises the temperature of the surface above the value it would have if infrared radiation could escape.

ground state A state of an atom in which all of the electrons are in the smallest allowed orbits.

H I region A region of intersellar space in which hydrogen atoms are not ionized.

H II region A region of ionized hydrogen and other gases around a bright hot star or stars. The stars emit ultraviolet photons that ionize the hydrogen atoms, which produce visible-light photons as they recombine.

halo The part of a galaxy that extends far above, below, and beyond the nucleus and disk of the galaxy.

heat The collective kinetic energy of random motion of a group of particles contained within a particular volume.

heliocentric Centered on the sun.

helium flash The sudden increase in luminosity that occurs within an aging star as the star's central regions become hot enough to make helium nuclei fuse together.

hertz A unit of frequency, corresponding to one cycle per second, abbreviated Hz.

Hertzsprung-Russell (H-R) diagram A graph of the luminosity or absolute magnitude of stars against the stars' surface temperatures or spectral types.

high-velocity star A star in the Milky Way with a large velocity relative to the sun. Such stars are typically moving in highly elongated orbits around the galactic center.

Hubble's constant The constant of proportionality in Hubble's law that relates the velocity of recession to the distance. This constant equals approximately 75 kilometers per second for each megaparsec of distance.

Hubble's law The summary of Hubble's discovery of the expanding universe, which states that the recession velocities of galaxies equal a constant (Hubble's constant) times their distances from us.

ideal radiator An object in complete equilibrium with its surroundings, emitting and absorbing the same number of photons each second at all frequencies.

igneous Formed from a molten or partially molten state.

infrared Radiation made of photons with slightly longer wavelengths and slightly lower frequencies than visible-light photons, usually defined as photons with wavelengths between about 7000 Å and 1 millimeter.

initial singularity The moment of the big bang, when the universe began its present expansion.

inner planets The four planets Mercury, Venus, Earth, and Mars, all of which are small, dense, and rocky compared to the four giant planets.

inorganic Not involving organic life or the chemistry on which organic life is based; in particular, non-carbon-based.

instability strip The region of the Hertzsprung-Russell diagram that contains stars that are far along in their evolution and that are likely to pulsate in size and luminosity.

interstellar clouds Collections of interstellar gas and dust, typically a few parsecs or a few dozen parsecs in diameter, with densities that vary from a few to a few million atoms or molecules per cubic centimeter.

interstellar dust Particles made of a few hundred thousand or a few million atoms and molecules, perhaps in the atmospheres of red-giant stars, found in interstellar space.

interstellar gas The fraction of a galaxy not found in stars, mostly made of hydrogen and helium atoms and hydrogen molecules, but also containing some atoms of other elements, especially carbon, oxygen, nitrogen, neon, and iron.

interstellar reddening Preferential absorption of blue light over red light produced by interstellar dust particles, which causes light to be reddened by its passage through regions containing interstellar dust.

inverse-square law A relationship in which a given quantity, such as a force, varies in proportion to 1 over the square of the distance. Gravitational and electromagnetic forces follow inverse-square laws.

ion An atom that has lost one or more electrons, either through collisions or photon impact.

ionize To make an atom into an ion.

irregular galaxy A galaxy whose shape appears irregular to astronomers, and which cannot be classified as an elliptical or a spiral galaxy.

isotope One of a group of atoms that all have the same number of protons in the nucleus, but that differ in the number of neutrons.

Kelvin scale The absolute scale of temperature, on which the lowest temperature is 0 K (absolute zero) and water freezes at 273.16 K and boils at 373.16 K.

Kepler's three laws of planetary motion Kepler's description of the orbits of planets, which states that these orbits are ellipses; that the planet-sun line sweeps over equal areas in equal amounts of time; and that the squares of the planets' orbital periods are proportional to the cubes of the orbital semi-major axes.

kiloparsec One thousand parsecs.

kinetic energy Energy associated with motion. For objects moving at speeds much less than the speed of light, the kinetic energy equals one-half the object's mass times the square of its velocity.

Large Magellanic Cloud The Milky Way's largest satellite galaxy, an irregular galaxy about 55 kiloparsecs from the sun.

latitude The angular distance on Earth north or south of the equator, reaching 90° at the poles.

law of universal gravitation Newton's expression of the gravitational attraction between two objects with mass, equal to a universal constant *(G)* times the product of the objects' masses, divided by the square of the distance between the objects' centers.

life A property of matter evidenced by the ability to reproduce and to evolve.

light year The distance light travels in one year, equal to 9.46×10^{12} kilometers.

line emission The emission of photons in a narrow range of frequency and wavelength.

Local Group The small cluster of about twenty galaxies to which the Milky Way belongs.

luminosity The total energy emitted in photons per second by a given object.

luminosity class One of the classes of stars organized by the stars' positions on the Hertzsprung-Russell diagram, such as red supergiant stars and main-sequence stars.

lunar eclipse An eclipse of the moon.

Magellanic Clouds The two relatively large irregular galaxies that are satellites of the Milky Way, each containing a few billion stars.

magnetic field A field of force in space, created by a magnet or by an electric current, that changes the trajectories of electrically charged particles.

magnitude A measure of the relative intensity or brightness of objects, on which scale an increase by one unit of magnitude signifies a decrease in brightness by a factor of 2.512. See *absolute magnitude* and *apparent magnitude*.

main sequence The region of the Hertzsprung-Russell diagram that stars occupy during the main part of their lifetimes, while they are fusing protons into helium nuclei at a constant rate.

maria Large, smooth areas on the moon, the result of ancient lava flows.

mass A measure of the amount of matter in an object, which can be determined by the object's resistance to being accelerated by a given force.

matter That which occupies space and can exert gravitational force on other objects.

megaparsec One million parsecs.

meridian The circle on the sky as seen by any observer that passes from the due north point of the horizon through the zenith and on to the due south point of the horizon.

messenger RNAs A kind of RNA that carries genetic information needed in the synthesis of protein molecules in cells.

meteor A luminous streak of light produced by frictional heating when meteoroids enter the Earth's atmosphere.

meteorite The fragment of a meteoroid that survives passage through the atmosphere and lands on Earth.

meteoroid An object made of rock or metal, or a rock and metal mixture, in orbit around the sun, smaller than an asteroid.

meteor shower An especially large number of meteors, caused by the Earth's intersection in its orbit with the orbits of an especially large number of meteoroids.

microwaves Electromagnetic radiation consisting of photons whose frequencies and wavelengths place them in the high-frequency, or short-wavelength, part of the radio domain of the spectrum; photons with wavelengths between 1 millimeter and a few centimeters.

Milky Way The galaxy to which the sun belongs, whose central regions appear as a band of light on the sky as seen from Earth.

mini-black hole A black hole with a mass millions of times less than the Earth's mass, presumably formed during the early moments of the universe, and hence sometimes called a primordial black hole.

minute of arc One-sixtieth of a degree of arc.

model A mental image of a physical entity, used to understand and to predict properties of the entity under consideration.

molecule A stable grouping of two or more atoms, bound together by the electromagnetic forces among the electrons and nuclei in the atoms.

momentum An object's tendency to keep moving, usually defined (for so-called linear momentum) as the product of the object's mass and velocity.

monomer A relatively small molecule that can form a single unit of a long-chain molecule or polymer.

mutation Any inheritable change in the genes or chromosomes of a living organism.

natural selection The selective increase among the individuals in a population of those types whose characteristics are better adapted for their own survival and for the survival of their offspring.

neap tides Tides of the lowest range, which occur when the sun and moon pull on the Earth in perpendicular directions, near the times of first-quarter moon and last-quarter moon.

nebulae Diffuse masses of interstellar gas and dust, usually lit from within by stars.

neutrino A particle with no mass and no electric charge, which always travels at the speed of light, characteristically emitted or absorbed in particle interactions governed by weak forces.

neutron An elementary particle with a mass of 1.6747×10^{-24} gram and no electric charge, stable when forming part of an atomic nucleus but subject to rapid decay into a proton, electron, and antineutrino when not part of an atomic nucleus.

neutron star A tremendously compact object, formed from a collapsed star, in which almost all of the protons and electrons have combined to form neutrons, and which is supported against further collapse by the exclusion principle.

nodes The two points on the celestial sphere at which the orbit of an object (most commonly of the moon) intersects the ecliptic.

nonthermal radiation Photon radiation that does not have the spectrum of an ideal radiator.

nova A star that shows a sudden increase in brightness; presumably, one star in a binary system that suddenly receives an infall of nuclear fuel from the other star.

nucleic acid One of a class of complex molecules, of which DNA is the best example, involved in the functioning and reproduction of living cells.

nucleus The central region of an atom, composed of one or more protons and of none or more neutrons. Also, the central region of a cell or galaxy, and the basic part of a comet.

occultation The passage of one celestial object in front of another, as seen from the Earth; typically, the passage of the moon in front of a star or planet.

1.4-solar-mass limit The upper limit on the mass that any white-dwarf star can have. Stars with more than 1.4 times the sun's mass simply cannot become white dwarfs and are likely to collapse at the end of their life cycles.

open star cluster A loose grouping of stars containing several dozen to several hundred members, a few parsecs in diameter, typically younger than a globular star cluster.

optics The study of the propagation of light and of the means of focusing and collecting light.

organic Referring to chemical compounds containing carbon as an important structural element. Also, having properties associated with life.

ozone Molecules made of three oxygen atoms, which shield the Earth's surface against ultraviolet radiation.

parallax effect The apparent displacement in position caused by the motion of an observer; for stars, the apparent displacement in position on the celestial sphere caused by the Earth's motion around the sun.

parallax shift The amount of displacement in a star's position, back and forth from the average position during the course of a year, caused by the parallax effect.

parsec A unit of length, equal to 3.262 light years or about 30 trillion kilometers, equal to the distance an object would have if its parallax shift were equal to 1 second of arc.

periodic variable star A star whose luminosity varies in a cyclical manner, with a well-defined, repetitive period of variation.

phase One of the cyclically reappearing forms or apparent shapes of the moon or of one of the planets.

photoionization Ionization produced by photon impact.

photon an elementary particle with no mass and no electric charge, which forms electromagnetic radiation and which always travels at the speed of light, 299,792 kilometers per second.

photosphere The visible surface of a star, from which most of the star's energy output escapes directly into space.

photosynthesis The process by which plants convert incident sunlight into chemically stored energy and make organic compounds, especially carbohydrates, from carbon dioxide and water, with the simultaneous release of oxygen.

planet One of the nine largest objects in orbit around the sun. Also, similar objects that may be in orbit around other stars.

planetary nebula A shell of gas surrounding an

aging star, heated and partially ionized by the photons emitted from the hot surface of the star, and which has been ejected from the star itself.

plate tectonics Slow motions of the plates of the Earth's crust, which cause the plates to collide with one another, leading to the formation of new mountain ranges.

polymer A long-chain molecule composed of up to millions of repeated units called monomers.

Population I stars Stars with ages ranging from about 10 billion down to only a few million years, characterized by a relatively large fractional abundance (about 1 percent) of elements heavier than hydrogen and helium.

Population II stars Stars with ages of 8 to 15 billion years, and a relatively low abundance of elements heavier than hydrogen and helium. These stars are believed to have formed before large numbers of supernova explosions had enriched the Milky Way with heavy elements.

positron The antiparticle of an electron, with the same mass as an electron but with one unit of positive electric charge.

power The rate at which work is done or energy is expended, often measured in the metric system in ergs per second or in joules (10^7 ergs per second).

powers of 10 A method of notation by which extremely large or extremely small numbers can be expressed by writing 10 raised to a given power. See *exponent*.

precession The motion of the Earth's rotation axis, caused by the gravitational forces from the sun and the moon, which slowly changes the location of the equinoxes on the ecliptic.

primitive atmosphere The atmosphere of a planet soon after the time of its formation.

prokaryote A single-celled organism that does not possess a well-defined nucleus containing the genetic material.

prominence A region of glowing gas visible at the edge of the solar disk, rising above the solar photosphere.

proper motion A star's apparent motion through space as seen from Earth after the effects of the parallax shift have been subtracted.

protein A molecule made of amino acids, the basic structural units of living organisms.

protogalaxy A galaxy in the process of formation.

proton An elementary particle with a mass of 1.6724×10^{-24} gram and one unit of positive electric charge; one of the basic constituents of atomic nuclei.

proton-proton cycle The series of three nuclear fusion reactions by which most stars fuse protons into helium nuclei and release new kinetic energy from energy of mass.

protoplanet A planet in the process of formation.

protostar A star in the process of formation.

pulsar An object that emits regularly spaced pulses of electromagnetic radiation, typically of radio waves, but sometimes of visible light, x rays, and gamma rays as well; thought to be produced by a rotating neutron star.

pulsating variable star A star in the later stages of its evolution, during which it alternately expands and contracts, changing its luminosity in a periodic manner as it does so.

quantum mechanics The description of the structure of atoms and of their interaction with photons, and of the ways in which elementary particles move and interact.

quasar (quasi-stellar radio source) An object almost starlike in appearance, but whose spectrum shows large red shifts, thought to be the result of tremendous recession velocities arising from the object's enormous distance from us.

radiation pressure Pressure exerted by photons.

radio Electromagnetic radiation composed of photons with energies less than about 10^{-20} erg and wavelengths greater than about 1 millimeter; the lowest-energy form of electromagnetic radiation.

radio astronomy That part of astronomy concerned with the detection and analysis of sources of radio waves.

radio galaxy A galaxy that emits at least as much power in radio waves as in visible-light radiation.

recombination The addition of an electron to an ion to remake an atom.

red giant A star that has evolved through the main-sequence stage and has expanded to a much greater size than it had as a main-sequence star.

red shift A shift to lower photon energies, and thus to lower frequencies and longer wavelengths. See *Doppler effect*.

red supergiant A particularly large and luminous red-giant star.

reflecting telescope A telescope that employs one or more mirrors to produce an image.

refracting telescope A telescope in which the light to be focused passes first through a lens.

regolith A powdery layer of soil on the lunar surface, thought to be the result of meteoritic bombardment.

retrograde Backward in terms of the ordinary motion of the planets as projected on the celestial sphere. The usual (prograde) motion is from west to east among the stars; retrograde motion carries the planet from east to west relative to the stars.

ring A swarm of particles in almost the same orbit around a planet, known to exist for the planets Jupiter, Saturn, and Uranus.

RNA (ribonucleic acid) A single-stranded polymer whose ordered structure determines protein synthesis within living cells.

Roche limit The minimum distance from an object that a satellite held together by self-gravitation must have to avoid being disrupted by tidal forces from the primary object.

Roche lobe In a binary-star system, the teardrop-shaped region around each of the two stars, within which matter will remain gravitationally bound to that star.

RR Lyrae variable star A star that varies periodically in luminosity, having a slightly smaller absolute brightness and a slightly faster cycle of light variation than a Cepheid variable star.

runaway greenhouse effect A greenhouse effect that builds on itself as the heating of a planet's surface produces greater evaporation; the evaporation products increase the heating, which in turn promotes more evaporation.

S0 galaxy A galaxy with a disklike shape, reminiscent of spiral galaxies but with no spiral arms, thought to be a type intermediate between spiral and elliptical galaxies.

second of arc One-sixtieth of a minute of arc, hence 1/3600 of a degree.

self-gravitation The gravitational force that an object exerts on itself; the gravitational attraction of one part of an object for all the other parts.

semidetached binary star A binary star system in which one of the components has not swelled to fill its Roche lobe.

semi-major axis Half of the major axis (longest diameter) of an ellipse.

sexual reproduction Reproduction that involves two sexes and the exchange of chromosomal material between two parents.

Seyfert galaxy A spiral galaxy with an unusually bright, compact nucleus, often the source of intense visible-light emission lines and of infrared and radio emission.

Small Magellanic Cloud The smaller of the two irregular satellite galaxies of the Milky Way, about 48 kiloparsecs from Earth.

solar eclipse An eclipse of the sun.

solar system The sun and the objects in orbit around it, including planets, asteroids, meteoroids, and comets.

solar wind The outermost parts of the solar corona, which expand past the Earth and the other planets as a stream of electrons and ions, mostly hydrogen and helium nuclei.

solstice One of the two points on the celestial sphere that mark the sun's greatest excursions on the ecliptic to the north and to the south of the celestial equator.

solvent A liquid capable of dissolving another substance; a liquid in which molecules can float and interact within living cells.

spectroscopic binary star A binary-star system in which the existence of one or more of the component stars has been deduced from the Doppler shift it produces in the spectrum of the observed star as both stars orbit their common center of mass.

spectroscopic parallax The distance to a star as determined by using the star's spectrum to locate the star's position on the Hertzsprung-Russell diagram, and thus to deduce the star's luminosity. The luminosity and apparent brightness allow the determination of the star's distance.

spectroscopy The observation and analysis of the spectra of celestial objects.

spectrum The distribution of photons emitted by a particular source, often shown as a graph of the number of photons with each particular frequency or wavelength.

sphere of influence See *Roche lobe.*

spicule A relatively small jet of gas rising from the solar photosphere, several thousand kilometers in altitude, which disappears after a few minutes.

spin flip The change in orientation of the spin of the electron in a hydrogen atom from being parallel to the proton's spin to antiparallel, or from antiparallel into the parallel position.

spiral arms The spiral features seen within the disks of spiral galaxies, outlined by the youngest, hottest stars and the H II regions they produce.

spiral galaxy A galaxy characterized by a flattened disk of stars within a much larger halo, and by spiral arms within the disk of the galaxy.

spring tides The highest and lowest tides in a month, the result of the sun and moon combining their tide-raising forces at the times of full moon and new moon.

standard candle A light source of standard luminosity, used in the determination of distances by comparing the apparent brightnesses of such sources.

star A self-luminous mass of gas held together by gravitation, in which the energy released by nuclear fusion balances the inward pull of gravitational force.

steady-state theory A theory of the universe that suggests that the overall appearance of the universe does not change with time, because new matter appears as the universe expands.

Stefan-Boltzmann constant The constant of proportionality in the Stefan-Boltzmann law.

Stefan-Boltzmann law The radiation law stating that the total emitted power from a hot object varies in proportion to a constant times the fourth power of the object's temperature.

stellar association A loose grouping of stars that were born at the same place and time, which does not remain a single unit for more than a few million years.

strong forces Attractive forces that act among protons and neutrons, but which have an effect only for distances approximately equal to the diameter of the nucleus of an atom, falling rapidly to zero for larger distances.

sunspot A relatively dark and cool area on the sun's surface, which owes its darkness to the smaller amount of light it emits as a result of being cooler than the rest of the photosphere.

supermassive black hole A black hole with more than a million times the sun's mass.

supernova An exploding star, visible for a few weeks or months over enormous distances because of its tremendous luminosity, then slowly fading into obscurity.

synchrotron emission Photons emitted when electrically charged particles moving at almost the speed of light accelerate or decelerate in the presence of a magnetic field.

synchrotron radiation See *synchrotron emission.*

temperature The measure of the average kinetic energy within a group of particles. On the absolute or Kelvin temperature scale, the temperature is directly proportional to the average kinetic energy per particle.

temperature-luminosity diagram See *Hertzsprung-Russell diagram.*

thermal emission Electromagnetic radiation whose spectrum corresponds to that of an ideal radiator.

tides A bulge produced in a deformable object by the differences in gravitational force from a nearby object on different parts of the first object. Specifically, the bulges of the Earth's oceans produced by the sun and the moon.

total eclipse An eclipse of the sun in which the moon completely covers the solar disk; or an eclipse of the moon in which the moon completely enters the Earth's shadow.

true brightness See *luminosity.*

T Tauri star A member of a class of young stars that are still in the process of expelling some of the layers of gas from which they formed.

turnoff point The point on the main sequence of the Hertzsprung-Russell diagram of a star cluster at which the stars have the highest surface temperature and greatest luminosity. The main-sequence stars originally more luminous have already become red giants.

21-centimeter emission Radio emission produced by hydrogen atoms at a wavelength of 21.1 centimeters when the atoms' electrons flip from the parallel position into the antiparallel position. See *spin flip.*

ultraviolet Photons with frequencies somewhat greater than the frequencies of visible light, usually defined as photons with frequencies between 10^{15} and 10^{17} hertz.

universe Everything that exists.

uplands The parts of the moon at relatively high elevation, as opposed to the low-elevation maria.

visible light Photons with wavelengths between 3000 and 7000 Å, which our eyes have evolved to detect.

visual binary star A binary-star system in which both stars can be seen in a telescope.

water hole The domain of the radio spectrum between the characteristic emission frequency of hydrogen (1420 MHz) and the frequencies emitted by OH (hydroxyl) molecules (near 1620 MHz).

wavelength The distance between two successive wave crests or two successive wave troughs in a series of oscillations; for photons, the distance a photon travels during a single oscillation.

wave of darkening A phenomenon observed on Mars: Regions progressively closer to the Martian equator grow darker in succession as spring progresses in a given hemisphere.

wave theory of light The theory that light consists of oscillations or waves, rather than of compact particles.

weak forces Forces that act among certain types of elementary particles over distances less than the size of an atomic nucleus.

weak reaction An interaction among elementary particles in which weak forces are important.

white-dwarf star A star that has fused helium nuclei into carbon nuclei before becoming degenerate in its interior, so that the exclusion principle supports the star against further collapse as the star radiates its remaining store of energy.

Wien displacement law In an ideal radiator, the relationship between the peak photon emission frequency and the temperature. Also, a similar relationship between the wavelength at which the maximum photon emission occurs and the temperature of the ideal radiator.

work In physics, the measure of energy expended by a force, defined as the amount of force times the distance through which the force acts.

x rays Photons with energies greater than those of ultraviolet but less than those of gamma rays, with frequencies in the range of 10^{17} hertz to 1.6×10^{20} hertz.

zenith The point on the celestial sphere that is directly overhead, as seen by a given observer.

zodiac The band of twelve constellations that include the ecliptic on the celestial sphere, and through which the sun, moon, and planets appear to move.

zone of avoidance The part of the celestial sphere on which galaxies cannot be seen because of the effect of interstellar absorption within our own Milky Way galaxy.

Index

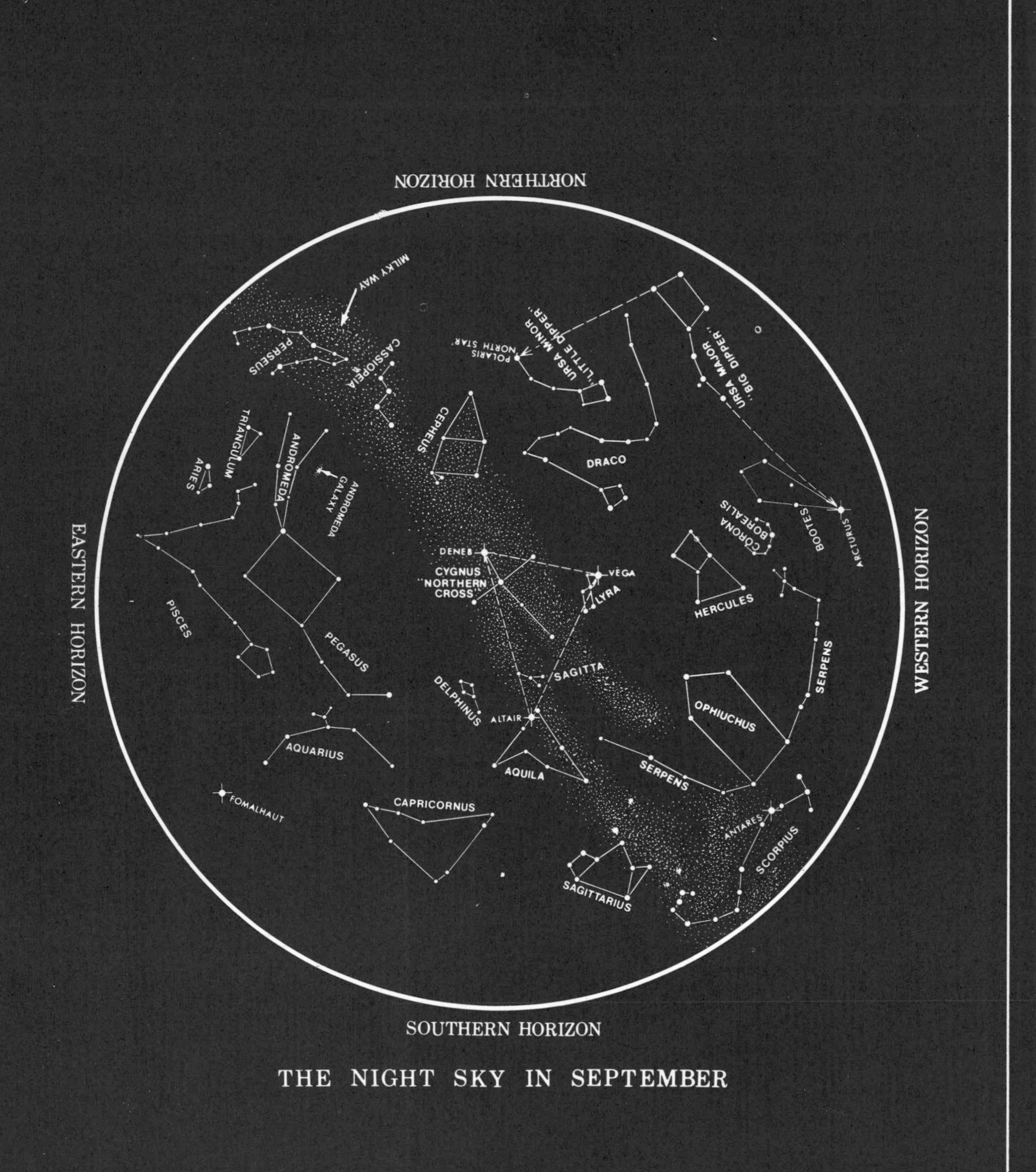

THE NIGHT SKY IN SEPTEMBER